Semigroups, Formal Languages and Groups

NATO ASI Series

Advanced Science Institutes Series

A Series presenting the results of activities sponsored by the NATO Science Committee, which aims at the dissemination of advanced scientific and technological knowledge, with a view to strengthening links between scientific communities.

The Series is published by an international board of publishers in conjunction with the NATO Scientific Affairs Division

A Life Sciences **B Physics**	Plenum Publishing Corporation London and New York
C Mathematical and Physical Sciences **D Behavioural and Social Sciences** **E Applied Sciences**	Kluwer Academic Publishers Dordrecht, Boston and London
F Computer and Systems Sciences **G Ecological Sciences** **H Cell Biology** **I Global Environmental Change**	Springer-Verlag Berlin, Heidelberg, New York, London, Paris and Tokyo

PARTNERSHIP SUB-SERIES

1. Disarmament Technologies	Kluwer Academic Publishers
2. Environment	Springer-Verlag / Kluwer Academic Publishers
3. High Technology	Kluwer Academic Publishers
4. Science and Technology Policy	Kluwer Academic Publishers
5. Computer Networking	Kluwer Academic Publishers

The Partnership Sub-Series incorporates activities undertaken in collaboration with NATO's Cooperation Partners, the countries of the CIS and Central and Eastern Europe, in Priority Areas of concern to those countries.

NATO-PCO-DATA BASE

The electronic index to the NATO ASI Series provides full bibliographical references (with keywords and/or abstracts) to more than 50000 contributions from international scientists published in all sections of the NATO ASI Series.
Access to the NATO-PCO-DATA BASE is possible in two ways:

– via online FILE 128 (NATO-PCO-DATA BASE) hosted by ESRIN, Via Galileo Galilei, I-00044 Frascati, Italy.

– via CD-ROM "NATO-PCO-DATA BASE" with user-friendly retrieval software in English, French and German (© WTV GmbH and DATAWARE Technologies Inc. 1989).

The CD-ROM can be ordered through any member of the Board of Publishers or through NATO-PCO, Overijse, Belgium.

Series C: Mathematical and Physical Sciences – Vol. 466

Semigroups, Formal Languages and Groups

edited by

John Fountain
Department of Mathematics,
University of York,
Heslington, York
U.K.

Kluwer Academic Publishers

Dordrecht / Boston / London

Published in cooperation with NATO Scientific Affairs Division

Proceedings of the NATO Advanced Study Institute on
Semigroups, Formal Languages and Groups
University of York, U.K.
7–21 August 1993

A C.I.P. Catalogue record for this book is available from the Library of Congress.

ISBN 0-7923-3540-6

Published by Kluwer Academic Publishers,
P.O. Box 17, 3300 AA Dordrecht, The Netherlands.

Kluwer Academic Publishers incorporates the publishing programmes of
D. Reidel, Martinus Nijhoff, Dr W. Junk and MTP Press.

Sold and distributed in the U.S.A. and Canada
by Kluwer Academic Publishers,
101 Philip Drive, Norwell, MA 02061, U.S.A.

In all other countries, sold and distributed
by Kluwer Academic Publishers Group,
P.O. Box 322, 3300 AH Dordrecht, The Netherlands.

Printed on acid-free paper

Contents

Foreword

This volume presents the core of invited expository lectures given at the 1993 NATO ASI held at the University of York. The subject matter of the ASI was the interplay between automata, semigroups, formal languages and groups. The invited talks were of an introductory nature but at a high level and many reached the cutting edge of research in the area. The lectures were given to a mixed group of students and specialists and were designed to be accessible to a broad audience. The papers were written in a similar spirit in the hope that their readership will be as wide as possible. With one exception they are all based on the talks which the lecturers gave at the meeting. The exception is caused by the fact that due to unanticipated progress the topic of John Rhodes' talk is now in such a state of flux that it has not been possible to produce a paper giving a clear picture of the situation. However, we do include an article by a member of the "Rhodes school", namely Christopher Nehaniv, expanding on a contributed talk he gave. It generalizes the celebrated Krohn-Rhodes theorem for finite semigroups to all semigroups.

For many years there has been a strong link between formal language theory and the theory of semigroups. Each subject continues to influence the other. Jean-Eric Pin's first article gives an elegant introduction to the basic ideas and fundamental results such as the theorems of Kleene and Schützenberger. It continues with an account of Eilenberg's theory of varieties and a brief glance at some recent developments. His second article is an exposition of some of the deepest work to emerge from the interplay between semigroups and languages. This involves the various characterizations of the variety of block groups. In a third article written together with Dominique Perrin we are given an introduction to the theory of infinite words emphasizing the combinatorial and algebraic aspects of this theory. As well as the classical results, this paper contains a review of the new work of Wilke.

Relatively free profinite monoids are the subject of a paper by Jorge Almeida and Pascal Weil. The connections between profinite monoids and implicit operations are discussed first; the last two sections provide an account of very recent work inspired by geometrical ideas associated with the Cayley graphs of free profinite groups.

In the last ten years an exciting offshoot of the connections between semigroups and groups has arisen in the complexity theory of Boolean circuits. Various logics are also involved, building on the classical results which describe regular and star-free languages in terms of monadic second-order logic and first-order logic over word models. All these strands are brought together in an introductory paper by Howard Straubing, Denis Thérien and Wolfgang Thomas.

Stuart Margolis, John Meakin and Mark Sapir give an interesting and extensive review of algorithmic problems in algebra concentrating on semigroups, groups and inverse semigroups. They survey combinatorial, geometric and topological methods and illustrate their use with a wealth of results. Related ideas are discussed and applied in the paper by Steve Pride where the geometric approach of "pictures" which has proved to be very powerful in group theory is adapted for semigroups.

An important development of the past few years is the interplay between geometry, group theory and the theory of formal languages. This has included the introduction of automatic groups of various kinds and hyperbolic groups. The basic ideas of this approach are outlined in the first part of a paper by Hamish Short. In the second part he initiates the study of a new class of groups called transducer groups. The definition is similar to that of an automatic group but uses transducers instead of acceptors.

Another very active area of group theory is the theory of groups acting on Λ-trees. Ian Chiswell gives a detailed account of some aspects of the basic theory and in the final section, a survey of recent results on free actions of groups on Λ-trees.

The theory of reductive monoids is due to Mohan Putcha and Lex Renner. It brings together semigroup theory and the theory of groups of Lie type. Louis Solomon has written a coherent and intelligible introduction to the subject that will make it available to a wider audience. This is followed by articles by Putcha and Renner on representations of reductive monoids.

Thanks must go first of all to NATO; the meeting would not have taken place without its generous support. The organisers are also grateful to the London Mathematical Society for financial support which allowed us to invite participants who would not otherwise have been able to attend. Both the NSF and ISF gave grants to other participants.

On a more personal level I would like to extend my heartfelt thanks to my fellow organisers, Simon Goberstein, Victoria Gould, Peter Higgins, John Howie and Norman Reilly for their cheerful willingness to help whenever asked. Victoria, as the other local organiser, had a heavy burden of paperwork before the conference and at the meeting gave great support to me and all the participants. Simon handled much of the North American work and was a constant source of support before, during and after the meeting and in the preparation of these proceedings.

The success of the proceedings depended, of course, on the investment of a great deal of work by the authors to produce the excellent articles which appear in it. I am very grateful to them and to all the other speakers who helped to make the meeting a success. Another major factor in this success was the staff of Langwith College and the Conference Office at the University of York. In particular, Margaret Silcock was always available for help and guidance. Several of the participants arrived early and were generous in the help they gave in the final stages of preparation.

I would like to give special mention to Diane Watson, the Mathematics Department secretary, for her endless toil which she undertook cheerfully and enthusiastically and without which the meeting would have foundered. She was ably assisted by Gill Hopwood and Wendy Williamson.

Without the unfailing support of my wife, Margaret, and family, I would not have survived the work of preparing for the conference or the running of the actual meeting.

Finally, I express my sincere thanks to Michael Beaty for his careful work and patience in making final corrections to all the papers in this proceedings and for his painstaking efforts to ensure the uniform high quality of the appearance of the articles.

John Fountain

FINITE SEMIGROUPS AND RECOGNIZABLE LANGUAGES: AN INTRODUCTION

JEAN-ERIC PIN
LITP/IBP, Université Paris VI et CNRS, Tour 55-65, 4 Place Jussieu, 75252 Paris Cedex 05, France

1. Foreword

This paper is an attempt to share with a larger audience some modern developments in the theory of finite automata. It is written for the mathematician who has a background in semigroup theory but knows next to nothing about automata and languages. No proofs are given, but the main results are illustrated by several examples and counterexamples.

What is the topic of this theory? It deals with languages, automata and semigroups, although recent developments have shown interesting connections with model theory in logic, symbolic dynamics and topology. Historically, in their attempt to formalize natural languages, linguists such as Chomsky gave a mathematical definition of natural concepts such as words, languages or grammars: given a finite set A, a word on A is simply an element of the free monoid on A, and a language is a set of words. But since scientists are fond of classifications of all sorts, language theory did not escape this mania. Chomsky established a first hierarchy, based on his formal grammars. In this paper, we are interested in the *recognizable* languages, which form the lower level of the Chomsky hierarchy. A recognizable language can be described in terms of finite automata while, for the higher levels, more powerful machines, ranging from pushdown automata to Turing machines, are required. For this reason, problems on finite automata are often under-estimated, according to the vague — but totally erroneous — feeling that "if a problem has been reduced to a question about finite automata, then it should be easy to solve".

Kleene's theorem [22] is usually considered as the foundation of the theory. It shows that the class of recognizable languages (i.e., those recognized by finite automata), coincides with the class of *rational* languages, which are given by rational expressions. Rational expressions can be thought of as a generalization of polynomials involving three operations: union (which plays the role of addition), product and star. An important corollary of Kleene's theorem is that rational languages are closed under complement. In the sixties, several classification schemes for the rational languages were proposed, based on the number of nested uses of a particular operator (star or product, for instance). This led to the natural notions of star height, extended star height, dot-depth and concatenation level. However, the first natural questions attached to these notions — "do they define strict hierarchies ?", "given a rational language, is there an algorithm for computing its star height, ex-

J. Fountain (ed.), Semigroups, Formal Languages and Groups, 1–32.

tended star height?", etc. — appeared to be extremely difficult. Actually, several of them, like the hierarchy problem for the extended star height, are still open.

A break-through was made by Schützenberger in the mid sixties [52]. Schützenberger established the equivalence between finite automata and finite semigroups and showed that a finite monoid, called the *syntactic monoid*, is canonically attached to each recognizable language. Then he made a non trivial use of this invariant to characterize the languages of extended star height 0, also called *star-free* languages. Schützenberger's theorem states that a language is star-free if and only if its syntactic monoid is aperiodic. Two other "syntactic" characterizations were obtained in the early seventies: Simon [56] proved that a language is of concatenation level 1 if and only if its syntactic monoid is $\mathcal{J}$-trivial and Brzozowski-Simon [9] and independently, McNaughton [28] characterized an important subfamily of the languages of dot-depth one, the locally testable languages. These successes settled the power of the semigroup approach, but it was Eilenberg who discovered the appropriate framework to formulate this type of result [17].

Recall that a variety of finite monoids is a class of monoids closed under the taking of submonoids, quotients and finite direct products. Eilenberg's theorem states that varieties of finite monoids are in one to one correspondence with certain classes of recognizable languages, the varieties of languages. For instance, the rational languages correspond to the variety of all finite monoids, the star-free languages correspond to the variety of aperiodic monoids, and the piecewise testable languages correspond to the variety of $\mathcal{J}$-trivial monoids. Numerous similar results have been established during the past fifteen years and the theory of finite automata is now intimately related to the theory of finite semigroups. This had a considerable influence on both theories: for instance algebraic definitions such as the graph of a semigroup or the Schützenberger product were motivated by considerations of language theory. The same thing can be said for the systematic study of power semigroups. In the other direction, Straubing's wreath product principle has allowed us to obtain important new results on recognizable languages. The open question of the decidability of the dot-depth is a good example of a problem that is of interest to both theories (and also formal logic!).

The paper is organized as follows. Sections 2 and 3 present the necessary material to understand Kleene's theorem. The equivalence between finite automata and finite semigroups is detailed in section 4. The various hierarchies of rational languages, based on star height, extended star height, dot-depth and concatenation level are introduced in section 5. The syntactic characterization of star-free, piecewise testable and locally testable languages are formulated in sections 6, 7 and 8, respectively. The variety theorem is stated in section 9 and some examples of its application are given in section 10. Other results about the hierarchies are analyzed in section 11 and recent developments are reported in section 12. The last section 13 contains the conclusion of this article.

2. Rational and recognizable sets

The terminology used in the theory of automata originates from various sources. Part of it came from linguistics, some other parts were introduced by physicists or by logicians. This gives sometimes a curious mixture but it is rather convenient in practice.

An *alphabet* is a finite set whose elements are *letters*. Alphabets are usually denoted by capital letters: A, B, ... and letters by lower case letters from the beginning of the latin alphabet: a, b, c,... A *word* (over the alphabet A) is a finite sequence $(a_1, a_2, \ldots, a_n)$ of letters of A; the integer n is the *length* of the word. In practice, the notation $(a_1, a_2, \ldots, a_n)$ is shortened to $a_1a_2 \cdots a_n$. The empty word, which is the unique word of length 0, is denoted by 1. Given a letter a, the number of occurrences of a in a word u is denoted by $|u|_a$. For instance, $|abbab|_a = 2$ and $|abbab|_b = 3$.

The (concatenation) *product* of two words $u = a_1a_2 \cdots a_p$ and $v = b_1b_2 \cdots b_q$ is the word $uv = a_1a_2 \cdots a_pb_1b_2 \cdots b_q$. This product is an associative operation on words. The set of all words over the alphabet A is denoted by A^*. Equipped with the product of words, it is a monoid, with the empty word as an identity. It is in fact the free monoid on the set A. The set of non-empty words is denoted by A^+; it is the free semigroup on the set A.

A *language* of A^* is a set of words over A, that is, a subset of A^*. The *rational operations* on languages are the three operations union, product and star, defined as follows

(1) Union: $L_1 + L_2 = \{u \mid u \in L_1 \text{ or } u \in L_2\}$

(2) Product: $L_1L_2 = \{u_1u_2 \mid u_1 \in L_1 \text{and } u_2 \in L_2\}$

(3) Star: $L^* = \{u_1 \cdots u_n \mid n \geq 0 \text{ and } u_1, \ldots, u_n \in L\}$.

It is also convenient to introduce the operator

$$L^+ = LL^* = \{u_1 \cdots u_n \mid n > 0 \text{ and } u_1, \ldots, u_n \in L\}.$$

Note that L^+ is exactly the subsemigroup of A^* generated by L, while L^* is the submonoid of A^* generated by L. The set of rational languages of A^* is the smallest set of languages of A^* containing the finite languages and closed under finite union, finite product and star. For instance, $(a+ab)^*ab+(ba^*b)^*$ denotes a rational language over the alphabet $\{a, b\}$.

The set of *rational languages* of A^+ is the smallest set of languages of A^+ containing the finite languages and closed under finite union, product and plus. It is easy to verify that the rational languages of A^+ are exactly the rational languages of A^* that do not contain the empty word.

It may seem a little awkward to have two separate definitions for the rational languages: one for the free monoid A^* and another one for the free semigroup A^+. There are actually two parallel theories and although the difference between them may appear of no great significance at first sight, it turns out to be crucial. The reason is that the algebraic classification of rational languages, as given in the forthcoming sections, rests on the notion of varieties of finite monoids (for languages of the free monoid) or varieties of finite semigroups (for languages of the free semigroup). And varieties of finite semigroups cannot be considered as varieties of finite monoids. The simplest example is the variety of finite nilpotent semigroups, which, as we shall see, characterizes the finite or cofinite languages of the free semigroup. If one tries, in a naive attempt, to add an identity to convert each nilpotent semigroup into a monoid, the variety of finite monoids obtained in this way is the variety of all finite monoids whose idempotents commute with every element. But this variety of monoids does *not* characterize the finite-cofinite languages of the free monoid.

Rational languages are often called *regular* sets in the literature. However, in the author's opinion, this last term should be avoided for two reasons. First, it interferes with the standard use of this word in semigroup theory. Second, the term *rational* has a sound mathematical foundation. Indeed one can extend the theory of languages to series with non commutative variables over a commutative ring or semiring[1] k. Such series can be written as $s = \sum_{u \in A^*}(s,u)u$, where (s,u) is an element of k. In this context, languages appear naturally as series over the boolean semiring. Now the rational series form the smallest set of series $\mathcal{R}$ satisfying the following conditions:

(1) every polynomial is in $\mathcal{R}$,
(2) $\mathcal{R}$ is a semiring under the usual sum and product of series,
(3) if s is a series in $\mathcal{R}$ such that $(s,1) = 0$, then $s^* = \sum_{n \geq 0} s^n$ belongs to $\mathcal{R}$.

Note that if k is a ring, then $s^* = (1-s)^{-1}$. In particular, in the one variable case, this definition coincides with the usual definition of rational series, which explains the terminology. We shall not detail any further this nice extension of the theory of languages, but we refer the interested reader to [4] for more details.

3. Finite automata and recognizable sets

A finite (non-deterministic) automaton is a quintuple $\mathcal{A} = (Q, A, E, I, F)$ where Q is a finite set (the set of *states*), A is an alphabet, E is a subset of $Q \times A \times Q$, called the set of *transitions* and I and F are subsets of Q, called the set of *initial* and *final* states, respectively. Two transitions (p,a,q) and (p',a',q') are *consecutive* if $q = p'$. A *path* in $\mathcal{A}$ is a finite sequence of consecutive transitions

$$e_0 = (q_0, a_0, q_1),\ \ e_1 = (q_1, a_1, q_2),\ \ \ldots\ ,\ \ e_{n-1} = (q_{n-1}, a_{n-1}, q_n)$$

also denoted

$$q_0 \xrightarrow{a_0} q_1 \xrightarrow{a_1} q_2 \ \cdots\ q_{n-1} \xrightarrow{a_{n-1}} q_n$$

The state q_0 is the *origin* of the path, the state q_n is its *end*, and the word $x = a_0 a_1 \cdots a_{n-1}$ is its *label*. It is convenient to have also, for each state q, an empty path of label 1 from q to q. A path in $\mathcal{A}$ is *successful* if its origin is in I and its end is in F.

The language *recognized* by $\mathcal{A}$ is the set, denoted $L^*(\mathcal{A})$, of the labels of all successful paths of $\mathcal{A}$. A language X is *recognizable* if there exists a finite automaton $\mathcal{A}$ such that $X = L^*(\mathcal{A})$. Two automata are said to be *equivalent* if they recognize the same language. Automata are conveniently represented by labelled graphs, as in the example below. Incoming arrows indicate initial states and outgoing arrows indicate final states.

[1] A semiring is a set k equipped with an addition and a multiplication. It is a commutative monoid with identity 0 for the addition and a monoid with identity 1 for the multiplication. Multiplication is distributive over addition and 0 satisfies $0x = x0 = 0$ for every $x \in k$. The simplest example of a semiring which is not a ring is the boolean semiring $\mathbf{B} = \{0,1\}$ defined by $0+0=0$, $0+1=1+1=1+0=1$, $1.1=1$ and $1.0=0.0=0.1=0$.

Example 3.1 Let $\mathcal{A} = (\{1,2\},\{a,b\},E,\{1\},\{2\})$ be an automaton, with $E = \{(1,a,1),(1,b,1),(1,a,2)\}$. The path $(1,a,1)(1,b,1)(1,a,2)$ is a successful path of label aba. The path $(1,a,1)(1,b,1)(1,a,1)$ has the same label but is unsuccessful since its end is 1.

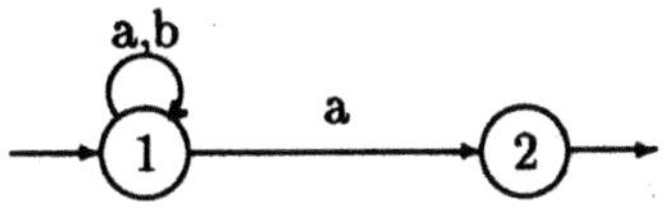

Figure 3.1. An automaton.

The set of words accepted by $\mathcal{A}$ is $L^*(\mathcal{A}) = A^*a$, the set of all words ending with an a.

In the case of the free semigroup, the definitions are the same, except that we omit the empty paths of label 1. In this case, the language recognized by $\mathcal{A}$ is denoted $L^+(\mathcal{A})$. Kleene's theorem states the equivalence between automata and rational expressions.

Theorem 3.1 *A language is rational if and only if it is recognizable.*

In fact, there is one version of Kleene's theorem for the free semigroup and one version for the free monoid. The proof of Kleene's theorem can be found in most books on automata theory [20].

An automaton is *deterministic* if it has exactly one initial state, usually denoted q_0, and if E contains no pair of transitions of the form $(q,a,q_1),(q,a,q_2)$ with $q_1 \neq q_2$.

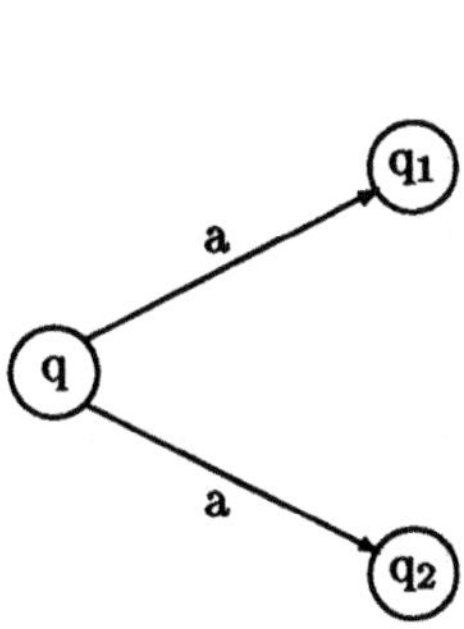

Figure 3.2. The forbidden pattern in a deterministic automaton.

In this case, each letter a defines a partial function from Q to Q, which associates with every state q the unique state $q.a$, if it exists, such that $(q, a, q.a) \in E$. This can be extended into a right action of A^* on Q by setting, for every $q \in Q$, $a \in A$ and $u \in A^*$:

$$q.1 = q$$

$$q.(ua) = \begin{cases} (q.u).a & \text{if } q.u \text{ and } (q.u).a \text{ are defined} \\ \text{undefined} & \text{otherwise.} \end{cases}$$

One can show that every finite automaton is equivalent to a deterministic one, in the sense that they recognize the same language.

States which cannot be reached from the initial state or from which one cannot access any final state are clearly useless. This leads to the following definition. A deterministic automaton $\mathcal{A} = (Q, A, E, q_0, F)$ is *trim* if for every state $q \in Q$ there exist two words u and v such that $q_0.u = q$ and $q.v \in F$. It is not difficult to see that every deterministic automaton is equivalent to a trim one.

Deterministic automata are partially ordered as follows. Let $\mathcal{A} = (Q, A, E, q_0, F)$ and $\mathcal{A}' = (Q', A, E', q_0', F')$ be two deterministic automata. Then $\mathcal{A} \leq \mathcal{A}'$ if there is a surjective function $\varphi : Q \to Q'$ such that $\varphi(q_0) = q_0'$, $\varphi^{-1}(F') = F$ and, for every $u \in A^*$ and $q \in Q$, $\varphi(q.u) = \varphi(q).u$. One can show that, amongst the trim deterministic automata recognizing a given recognizable language L, there is a minimal one under this partial order. This automaton is called the *minimal automaton* of L. Again, there are standard algorithms for minimizing a given finite automaton [20].

4. Automata and semigroups

In this section, we turn to a more algebraic definition of the recognizable sets, using semigroups in place of automata. Although this definition is more abstract than the definition using automata, it is more suitable to handle the fine structure of recognizable sets. Indeed, as illustrated in the next sections, semigroups provide a powerful and systematic tool to classify recognizable sets. We treat the case of the free semigroup. For free monoids, just replace every occurrence of "A^+" by "A^*" and "semigroup" by "monoid" in the definitions below.

The abstract definition of recognizable sets is based on the following observation. Let $\mathcal{A} = (Q, A, E, I, F)$ be a finite automaton. To each word $u \in A^+$, there corresponds a boolean square matrix of size $Card(Q)$, denoted by $\mu(u)$, and defined by

$$\mu(u)_{p,q} = \begin{cases} 1 & \text{if there exists a path from } p \text{ to } q \text{ with label } u \\ 0 & \text{otherwise.} \end{cases}$$

It is not difficult to see that μ is a semigroup morphism from A^+ into the multiplicative semigroup of square boolean matrices of size $Card(Q)$. Furthermore, a word u is recognized by $\mathcal{A}$ if and only if $\mu(u)_{p,q} = 1$ for some initial state p and some final state q. Therefore, a word is recognized by $\mathcal{A}$ if and only if

$\mu(u) \in \{m \in \mu(A^+) \mid m_{p,q} = 1 \text{ for some } p \in I \text{ and } q \in F\}$. The semigroup $\mu(A^+)$ is called the ***transition semigroup*** of $\mathcal{A}$, denoted $S(\mathcal{A})$.

Example 4.1 Let $\mathcal{A} = (Q, A, E, I, F)$ be the automaton represented below.

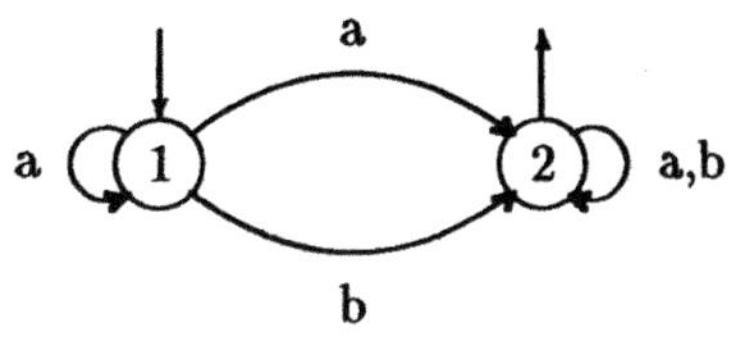

Figure 4.1. A non-deterministic automaton.

Here $Q = \{1, 2\}$, $A = \{a, b\}$ and $E = \{(1, a, 1), (1, a, 2), (2, a, 2), (2, b, 1), (2, b, 2)\}$, $I = \{1\}$, $F = \{2\}$, whence

$$\mu(a) = \begin{pmatrix} 1 & 1 \\ 0 & 1 \end{pmatrix} \qquad \mu(b) = \begin{pmatrix} 0 & 0 \\ 1 & 1 \end{pmatrix} \qquad \mu(aa) = \mu(a)$$

$$\mu(ab) = \begin{pmatrix} 1 & 1 \\ 1 & 1 \end{pmatrix} \qquad \mu(ba) = \mu(bb) = \mu(b)$$

Thus $\mu(A^+) = \{\left(\begin{smallmatrix} 0 & 0 \\ 1 & 1 \end{smallmatrix}\right), \left(\begin{smallmatrix} 1 & 1 \\ 0 & 1 \end{smallmatrix}\right), \left(\begin{smallmatrix} 1 & 1 \\ 1 & 1 \end{smallmatrix}\right)\}$.

This leads to the following definition. A semigroup morphism $\varphi : A^+ \to S$ ***recognizes*** a language $L \subset A^+$ if $L = \varphi^{-1}\varphi(L)$, that is, if $u \in L$ and $\varphi(u) = \varphi(v)$ implies $v \in L$. This is also equivalent to saying that there is a subset P of S such that $L = \varphi^{-1}(P)$. By extension, a semigroup S recognizes a language L if there exists a semigroup morphism $\varphi : A^+ \to S$ that recognizes L. As shown by the previous example, a set recognized by a finite automaton is recognized by the transition semigroup of this automaton.

Proposition 4.1 *If a finite automaton recognizes a language L, then $S(\mathcal{A})$ recognizes L.*

The previous computation can be simplified if $\mathcal{A}$ is deterministic. Indeed, in this case, the transition semigroup of $\mathcal{A}$ is naturally embedded into the semigroup of partial functions on Q under composition.

Example 4.2 Let $\mathcal{A}$ be the deterministic automaton represented below in Figure 4.2. The transition semigroup $S(\mathcal{A})$ of $\mathcal{A}$ contains five elements which correspond to the words a, b, ab, ba and aa. If one identifies the elements of $S(\mathcal{A})$ with these words, one has the relations $aba = a$, $bab = b$ and $bb = aa$. Thus $S(\mathcal{A})$ is the aperiodic Brandt semigroup BA_2.

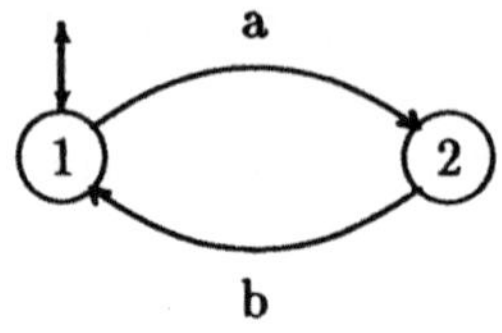

Figure 4.2. A deterministic automaton

Here is the transition table of $\mathcal{A}$:

	1	2
a	2	-
b	-	1
aa	-	-
ab	1	-
ba	-	2

Conversely, given a semigroup morphism $\varphi : A^+ \to S$ recognizing a subset X of A^+, one can build a finite automaton recognizing X as follows. Denote by S^1 the monoid equal to S if S has an identity and to $S \cup \{1\}$ otherwise. Take the right representation of A on S^1 defined by $s.a = s\varphi(a)$. This defines a deterministic automaton $\mathcal{A} = (S^1, A, E, \{1\}, P)$, where $E = \{(s, a, s.a) \mid s \in S^1, a \in A\}$.

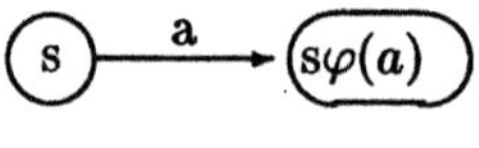

Figure 4.3. The transitions of $\mathcal{A}$.

This automaton recognizes L and thus, the two notions of recognizable sets (by finite automata and by finite semigroups) are equivalent.

Example 4.3 Let $A = \{a, b, c\}$ and let $S = \{1, a, b\}$ be the three element monoid defined by $a^2 = a$, $b^2 = b$, $ab = b$ and $ba = a$. Let $\varphi : A^+ \to S$ be the semigroup morphism defined by $\varphi(a) = a$, $\varphi(b) = b$ and $\varphi(c) = 1$ and let $P = \{a\}$. Then $\varphi^{-1}(P) = A^*ac^*$ and the construction above yields the automaton represented in Figure 4.4.

Now, Kleene's theorem can be reformulated as follows.

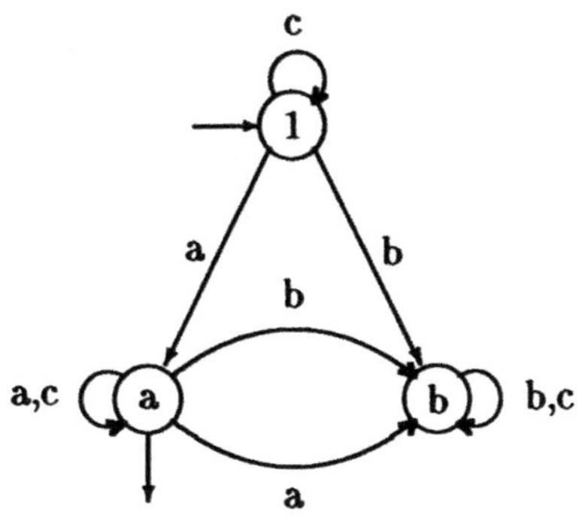

Figure 4.4. The automaton associated with S.

Theorem 4.2 *Let L be a language of A^+. The following conditions are equivalent.*
(1) L is recognized by a finite automaton,
(2) L is recognized by a finite deterministic automaton,
(3) L is recognized by a finite semigroup,
(4) L is rational.

Kleene's theorem has important consequences.

Corollary 4.3 *Recognizable languages are closed under finite boolean operations[2], inverse morphisms and morphisms.*

The trick is that it is easy to prove the last property (closure under morphisms) for rational sets and the other ones for recognizable sets. Here are two examples to illustrate these techniques:

Example 4.4 (Closure of recognizable sets under morphism). Let $\varphi : \{a,b\}^+ \to \{a,b,c\}^+$ be the semigroup morphism defined by $\varphi(a) = aba$ and $\varphi(b) = ca$ and let $L = a^*b + bab$ be a rational set. Then $\varphi(L) = (aba)^*ca + caabaca$ is a rational set.

Example 4.5 (Closure of recognizable sets under complement). Let L be a recognizable set. Then there exists a finite semigroup S, a semigroup morphism $\varphi : A^+ \to S$ and a subset P of S such that $L = \varphi^{-1}(P)$. Now $A^+ \setminus L = \varphi^{-1}(S \setminus P)$ and thus the complement of L is recognizable.

The patient reader can, as an exercise, prove the remaining properties by using either semigroups or automata. The impatient reader may consult [16, 36].

Let L be a recognizable language of A^+. Amongst the finite semigroups that recognize X, there is a minimal one (with respect to division). This finite semigroup

[2] Boolean operations comprise union, intersection, complementation and set difference.

is called the *syntactic semigroup* of L. It can be defined directly as the quotient of A^+ under the congruence $\sim_L$ defined by $u \sim_L v$ if and only if, for every $x, y \in A^*$, $xuy \in L \iff xvy \in L$. It is also equal to the transition semigroup of the minimal automaton of L. This last property is especially useful for practical computations. It is a good exercise to take a rational expression at random, to compute the minimal automaton of the language represented by this rational expression and then to compute the syntactic semigroup of the language. See examples 6.1 and 7.2 below for outlines of such computations.

5. Early attempts to classify recognizable languages

Kleene's theorem shows that recognizable languages are closed under complementation. Therefore, every recognizable language can be represented by a *extended rational expression*, that is, a formal expression constructed from the letters by means of the operations union, product, star and complement. In order to keep concise algebraic notation, we shall denote by L^c the complement of the language L[3], by 0 the empty language and by u the language $\{u\}$, for every word u. In particular, the language $\{1\}$, containing the empty word, is denoted 1. These notations are coherent with the intuitive formulæ $1L = L1 = L$ and $0L = L0 = 0$ which hold for every language L. For instance, if $A = \{a, b\}$, the expression $\left(\left(0^c(ab+ba)0^c\right)^c + (aba)^*\right)^c$ represents the language $(A^*abA^* \cup A^*baA^*) \setminus (aba)^*$ of all words containing the factors ab and ba which are not powers of aba.

Thus we have an algebra on A with four operations: $+$, $.$, * and ^{c}. Now a natural attempt to classify recognizable languages is to find a notion analogous with the degree of a polynomial for these extended rational expressions. It is a remarkable fact that all the hierarchies based on these "extended degrees" suggested so far lead to some extremely difficult problems, most of which are still open.

The first proposal concerned the star operation. The *star height* of an extended rational expression is defined inductively as follows:

(1) The star height of the basic languages is 0. Formally

$$sh(0) = 0 \qquad sh(1) = 0 \quad \text{and} \quad sh(a) = 0 \text{ for every letter } a.$$

(2) Union, product and complement do not affect star height. If e and f are two extended rational expressions, then

$$sh(e+f) = sh(ef) = \max\{sh(e), sh(f)\} \qquad sh(e^c) = sh(e).$$

(3) Star increases star height. For each extended rational expression e,

$$sh(e^*) = sh(e) + 1.$$

Thus the star height counts the number of nested uses of the star operation. For instance

$$\left((a^* + b^c a^*)^* + (b^* ab^*)^*\right)^* (b^* a^* + b)^*$$

---

[3] If L is a language of A^*, the complement of L is $A^* \setminus L$; if L is a language of A^+, the complement is $A^+ \setminus L$.

is an extended rational expression of star height 3. Now, the *extended star height*[4] of a recognizable language L is the minimum of the star heights of the extended rational expressions representing L

$$esh(L) = \min\{sh(e) \mid e \text{ is an extended rational expression for } L\}.$$

The difficulty in computing the extended star height is that a given language can be represented in many different ways by an extended rational expression!

The languages of extended star height 0 (or *star-free languages*) are characterized by a beautiful theorem of Schützenberger that will be presented in section 6.1. Schützenberger's theorem implies the existence of languages of extended star height 1, such as $(aa)^*$ on the alphabet $\{a\}$, but, as surprising as it may seem, nobody has been able so far to prove the existence of a language of extended star height greater than 1, although the general feeling is that such languages do exist. In the opposite direction, our knowledge of the languages proven to be of extended star height ≤ 1 is rather poor (see [45, 50, 51] for recent advances on this topic).

The *star height* of a recognizable language is obtained by considering rational expressions instead of extended rational expressions [15],

$$sh(L) = \min\{\text{star height}(e) \mid e \text{ is a rational expression for } L\}.$$

That is, one simply removes complement from the list of the basic operations. This time, the corresponding hierarchy was proved to be infinite by Dejean and Schützenberger [14].

Theorem 5.1 *For each $n \geq 0$, there exists a language of star height n.*

It is easy to see that the languages of star height 0 are the finite languages, but the effective characterization of the other levels was left open for several years until Hashiguchi first settled the problem for star height 1 [18] and a few years later for the general case [19].

Theorem 5.2 *There is an algorithm to determine the star height of a given recognizable language.*

Hashiguchi's first paper is now well understood, although it is still a very difficult result, but volunteers are called for to simplify the very long induction proof of the second paper.

The second proposal for constructing hierarchies was to ignore the star operation (which amounts to working with star-free languages) and to consider the concatenation product or, more precisely, a variation of it, called the *marked concatenation product*. Given languages $L_0, L_1, \dots, L_n$ and letters $a_1, a_2, \dots, a_n$, the product of $L_0, \dots L_n$ marked by $a_1, \dots, a_n$ is the language $L_0a_1L_1a_2 \cdots a_nL_n$. As product is often denoted by a dot, Brzozowski defined the "dot-depth" of languages of the free semigroup [5]. Later on, Thérien (implicitly) and Straubing (explicitly) introduced a similar notion (often called the *concatenation level* in the literature) for the languages of the free monoid. The languages of dot-depth 0 are the finite

[4] also called *generalized* star height.

or cofinite languages, while the languages of concatenation level 0 are A^* and the empty language 0. Otherwise, the two hierarchies are constructed in the same way and count the number of alternations in the use of the two different types of operations: boolean operations and marked product. More precisely, the languages of dot-depth (resp. concatenation level) $n+1$ are the finite boolean combinations of marked products of the form

$$L_0 a_1 L_1 a_2 \cdots a_k L_k$$

where $L_0, L_1, \ldots, L_k$ are languages of dot-depth (resp. concatenation level) n and $a_1, \ldots, a_k$ are letters.

Note that a language of dot-depth (resp. concatenation level) m is also a language of dot-depth (resp. concatenation level) n for every $n \geq m$. Brzozowski and Knast [8] have shown that the hierarchy is strict: if A contains at least two letters, then for every n, there exist some languages of dot-depth (resp. concatenation level) $n+1$ that are not of dot-depth (resp. concatenation level) n.

It is still an outstanding open problem to know whether there is an algorithm to compute the dot-depth (resp. concatenation level) of a given star-free language. The problem has been solved positively, however, for the dot-depth (resp. concatenation level) 1: there is an algorithm to decide whether a language is of dot-depth (resp. concatenation level) 1. These results are detailed in sections 7 and 11. The other partial results concerning these hierarchies are briefly reviewed in section 11. Another remarkable fact about these hierarchies is their connections with some hierarchies of formal logic. See the article of W. Thomas in this volume or the survey article [40].

But it is time for us to hark back to Schützenberger's theorem on star-free sets.

6. Star-free languages

The set of *star-free* subsets of A^* is the smallest set of subsets of A^* containing the finite sets and closed under finite boolean operations and product. For instance, A^* is star-free, since A^* is the complement of the empty set. More generally, if B is a subset of the alphabet A, the set B^* is also star-free since B^* is the complement of the set of words that contain at least one letter of $B' = A \setminus B$. This leads to the following star-free expression

$$B^* = A^* \setminus A^*(A \setminus B)A^* = (0^c(A \setminus B)0^c)^c = (0^c(A^c \cup B)^c 0^c)^c$$

If $A = \{a, b\}$, the set $(ab)^*$ is star-free, since $(ab)^*$ is the set of words not beginning with b, not finishing by a and containing neither the factor aa, nor the factor bb. This gives the star-free expression

$$(ab)^* = A^* \setminus \left(bA^* \cup A^*a \cup A^*aaA^* \cup A^*bbA^*\right) = \left(b0^c + 0^c a + 0^c aa0^c + 0^c bb0^c\right)^c$$

Readers may convince themselves that the sets $\{ab, ba\}^*$ and $\left(a(ab)^*b\right)^*$ also are star-free but may also wonder whether there exist any non star-free rational sets. In fact, there are some, for instance the sets $(aa)^*$ and $\{b, aba\}^*$, or similar examples that can be derived from the algebraic approach presented below.

Let S be a finite semigroup and let s be an element of S. Then the subsemigroup of S generated by s contains a unique idempotent, denoted s^ω. Recall that a finite semigroup M is *aperiodic* if and only if, for every $x \in M$, $x^\omega = x^{\omega+1}$. This notion is in some sense "orthogonal" to the notion of group. Indeed, one can show that a semigroup is aperiodic if and only if it is $\mathcal{H}$-trivial, or, equivalently, if it contains no non-trivial subgroup. The connection between aperiodic semigroups and star-free sets was established by Schützenberger [52].

Theorem 6.1 *A recognizable subset of A^* is star-free if and only if its syntactic monoid is aperiodic.*

There are essentially two techniques to prove this result. The original proof of Schützenberger [52, 36, 21], slightly simplified in [31], is by induction on the $\mathcal{J}$-depth of the syntactic semigroup. The second proof [11, 30] makes use of a weak form of the Krohn-Rhodes theorem: every aperiodic finite semigroup divides a wreath product of copies of the monoid $U_2 = \{1, a, b\}$, given by the multiplication table $aa = a$, $ab = b$, $ba = b$ and $bb = b$.

Corollary 6.2 *There is an algorithm to decide whether a given*[5] *recognizable language is star-free.*

Given the minimal automaton $\mathcal{A}$ of the language, the algorithm consists of checking whether the transition monoid of M is aperiodic. The complexity of this algorithm is analyzed in [10, 57].

Example 6.1 Let $A = \{a, b\}$ and consider the set $L = (ab)^*$. Its minimal automaton is represented below:

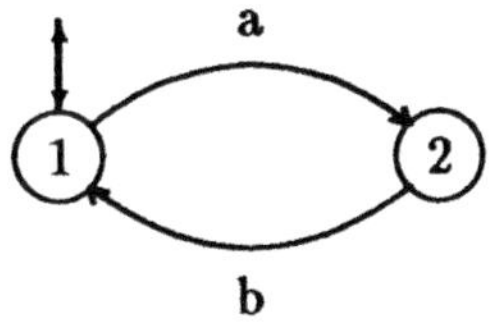

Figure 6.1. The minimal automaton of $(ab)^*$.

The transitions and the relations defining the syntactic monoid M of L are given in the following tables

[5] A recognizable set can be given either by a finite automaton, by a finite semigroup or by a rational expression since there are standard algorithms to pass from one representation to the other.

1	1	2
a	2	-
b	-	1
aa	-	-
ab	1	-
ba	-	2

$$a^2 = b^2 = 0$$
$$aba = a$$
$$bab = b$$

Since $a^2 = a^3$, $b^2 = b^3$, $(ab)^2 = ab$ and $(ba)^2 = ba$, M is aperiodic and thus L is star-free. Consider now the set $L' = (aa)^*$. Its minimal automaton is represented below:

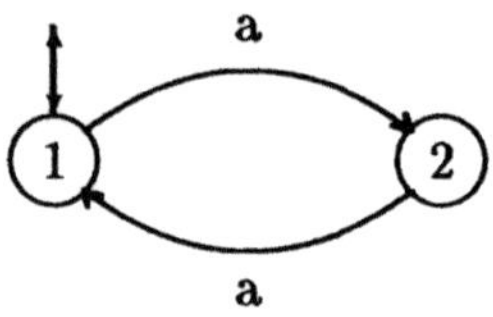

Figure 6.2. The minimal automaton of $(aa)^*$.

The transitions and the relations defining the syntactic monoid M' of L' are given in the following tables

1	1	2
a	2	1
b	-	-
aa	1	2

$$a^3 = a$$
$$b = 0$$

Thus M' is not aperiodic and hence L' is not star-free.

7. Piecewise testable languages

Recall that the languages of concatenation level 0 of A^* are A^* and 0. According to the general definition, the languages of concatenation level 1 are the finite boolean combinations of the languages of the form $A^*a_1A^*a_2A^* \cdots A^*a_kA^*$, where $k \geq 0$ and $a_i \in A$. The languages of this form are called ***piecewise testable***. Intuitively, such a language can be recognized by an automaton that one could call a *Hydra automaton*. Such an automaton has a finite number h of heads, each of which can read a letter of the input word. The heads are ordered, so that together they may read a subword (in the sense of a subsequence of non necessarily consecutive letters) of the input word. The automaton computes in this way the set of all subwords of length $\leq h$ of the input word. This set is then compared to the finite collection of sets of words contained in the memory. If it occurs in the memory, the word is

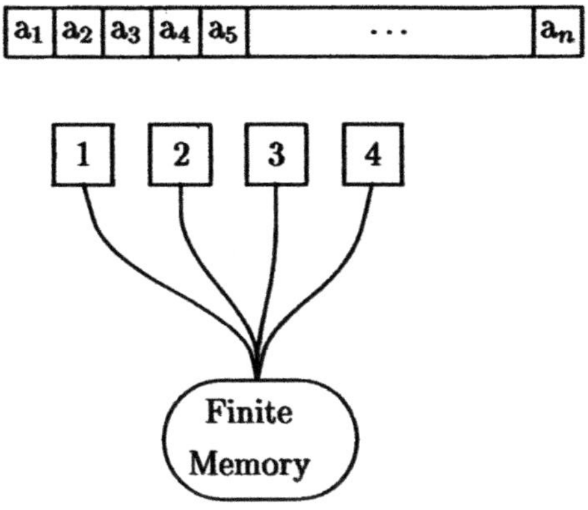

Figure 7.1. A Hydra automaton with four heads.

accepted, otherwise it is rejected. For instance, for the language $(A^*aA^*bA^*aA^* \cap A^*bA^*bA^*aA^*) \setminus (A^*aA^*bA^*bA^* \cup A^*bA^*bA^*bA^*)$, the memory would contain the collection of all sets whose members are words of length 3 and which contain *aba* and *bba* but do not contain *abb* or *bbb*. Piecewise testable languages are characterized by a deep result of I. Simon [56].

Theorem 7.1 *A language of A^* is piecewise testable if and only if its syntactic monoid is $\mathcal{J}$-trivial, or, equivalently, if it satisfies the equations $x^\omega = x^{\omega+1}$ and $(xy)^\omega = (yx)^\omega$.*

Corollary 7.2 *There is an algorithm to decide whether a given star-free language is of concatenation level* 1.

Given the minimal automaton $\mathcal{A}$ of the language, the algorithm consists in checking whether the transition monoid of M is $\mathcal{J}$-trivial. Actually, this condition can be directly checked on $\mathcal{A}$ in polynomial time [10, 57].

There exist several proofs of Simon's theorem [2, 56, 68, 57]. The central argument of Simon's original proof [56] is a careful study of the combinatorics of the subword relation. Stern's proof [57] borrows some ideas from model theory. The proof of Straubing and Thérien [68] is the only one that avoids totally combinatorics on words. In the spirit of the proof of Schützenberger, it works by induction on the cardinality of the syntactic monoid. The proof of Almeida [2] is based on implicit operations (see the papers of J. Almeida and P. Weil in this volume for more details).

Example 7.1 Let $A = \{a, b, c\}$ and let $L = A^*abA^*$. The minimal automaton of L is represented below

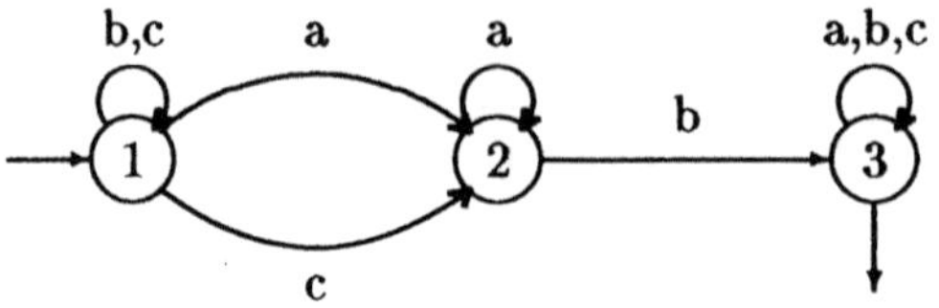

Figure 7.2. The minimal automaton of L.

The transitions and the relations defining the syntactic monoid M of L are given in the following tables

1	1	2	3
a	2	2	3
b	1	3	3
c	1	1	3
ab	3	3	3
ba	2	3	3

$$a^2 = a$$
$$ab = 0$$
$$ac = c$$
$$b^2 = b$$
$$bc = b$$
$$ca = a$$
$$cb = c$$
$$c^2 = c$$

The $\mathcal{J}$-class structure of M is represented in the following diagram.

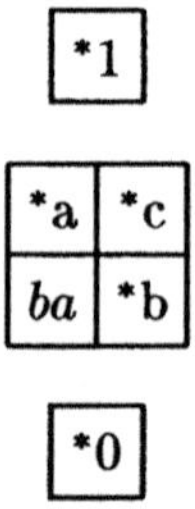

Figure 7.3. The $\mathcal{J}$-classes of M.

In particular, $a \; \mathcal{J} \; c$ and thus M is not $\mathcal{J}$-trivial. Therefore L is not piecewise testable.

Example 7.2 Consider now the language $L' = A^*abA^*$ on the alphabet $A = \{a, b\}$. Then the minimal automaton of L' is obtained from that of L by erasing the transitions with label c (see Figure 7.4). The transitions and the relations defining the

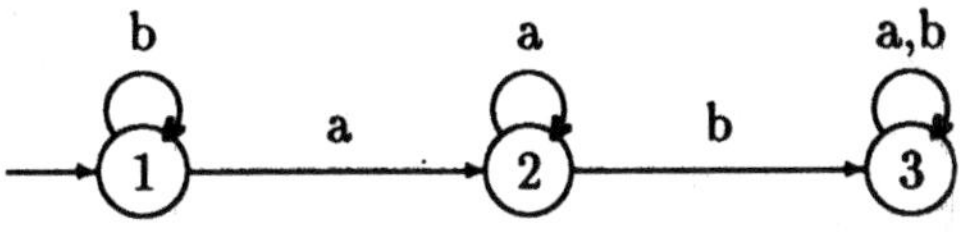

Figure 7.4. The minimal automaton of L'.

syntactic monoid M' of L' are given in the following tables

1	1	2	3
a	2	2	3
b	1	3	3
ab	3	3	3
ba	2	3	3

$$a^2 = a \qquad ab = 0 \qquad b^2 = b$$

The $\mathcal{J}$-class structure of M' is represented in the following diagram. Thus M'

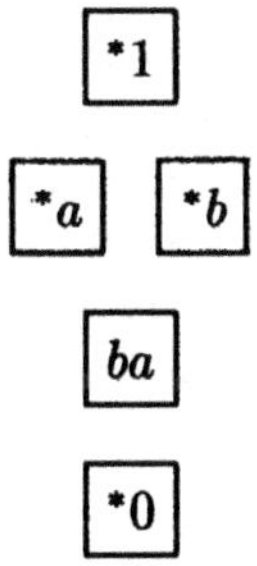

Figure 7.5. The $\mathcal{J}$-classes of M'.

is $\mathcal{J}$-trivial and L' is piecewise testable. In fact $L' = A^*aA^*bA^*$.

Simon's theorem also has some nice consequences for pure semigroup theory. An *ordered monoid* is a monoid equipped with a stable order relation. An ordered monoid $(M, \leq)$ is called 1-*ordered* if, for every $x \in M$, $x \leq 1$. A finite 1-ordered monoid is always $\mathcal{J}$-trivial. Indeed, if $u \mathrel{\mathcal{J}} v$, there exist $x, y, z, t \in M$ such that $u = xvy$ and $v = zyt$. Now $x \leq 1$, $y \leq 1$ and thus $u = xvy \leq v$ and similarly, $v \leq u$ whence $u = v$. The converse is not true: there exist finite $\mathcal{J}$-trivial monoids which cannot be 1-ordered.

Example 7.3 Let M be the monoid with zero presented on $\{a, b, c\}$ by the relations $aa = ac = ba = bb = ca = cb = cc = 0$. Thus $M = \{1, a, b, c, ab, bc, abc, 0\}$ and M is $\mathcal{J}$-trivial. However, M is not a 1-ordered monoid. Otherwise, one would have on the one hand, $b \leq 1$, whence $abc \leq ac = 0$ and on the other hand, $0 \leq 1$, whence $0 = 0.abc \leq 1.abc = abc$, a contradiction since $abc \neq 0$.

However, Straubing and Thérien [68] proved that 1-ordered monoids generate all the finite $\mathcal{J}$-trivial monoids in the following sense.

Theorem 7.3 *A monoid is $\mathcal{J}$-trivial if and only if it is a quotient of a* 1*-ordered monoid.*

Actually, it is not difficult to establish that this result is equivalent to Simon's theorem. But Straubing and Thérien also gave an ingenious direct proof of their result by induction on the cardinality of the monoid. This gives in turn a proof of Simon's theorem. Straubing [62] also observed the following connection with semigroups of relations.

Theorem 7.4 *A monoid is $\mathcal{J}$-trivial if and only if it divides a monoid of reflexive relations on a finite set.*

8. Locally testable languages

A language of A^+ is *locally testable* if it is a boolean combination of languages of the form uA^*, A^*v or A^*wA^* where $u, v, w \in A^+$. For instance, if $A = \{a, b\}$, the language $(ab)^+$ is locally testable since $(ab)^+ = (aA^* \cap A^*b) \setminus (A^*aaA^* \cup A^*bbA^*)$. These languages occur naturally in the study of the languages of dot-depth one. Actually they form the first level of a natural subhierarchy of the languages of dot-depth one (see [35] for more details). Locally testable languages also have a natural interpretation in terms of automata. They are recognized by *scanners*. A scanner is a machine equipped with a finite memory and a window of size n to scan the input word.

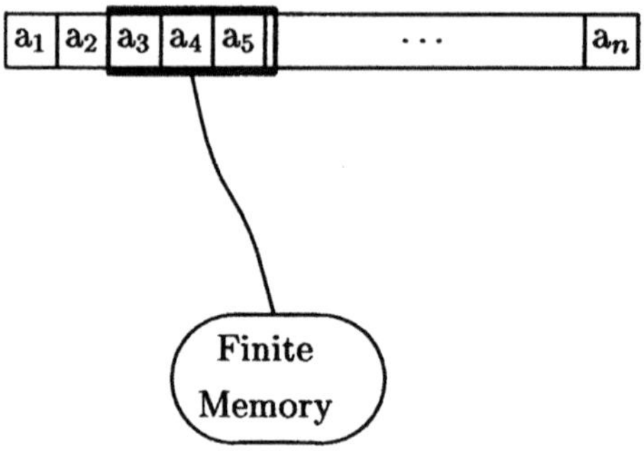

Figure 8.1. A scanner.

The window can also be moved beyond the first and last letter of the word, so that the prefixes and suffixes of length $< n$ can be read. For instance, if $n = 3$, and $u = abbaaabab$, the different positions of the window are represented on the following diagrams:

[*a*]*bbaaabab* [*ab*]*baaabab* [*abb*]*aaabab* *a*[*bba*]*aabab* $\cdots$ *abbaaaba*[*b*]

At the end of the scan, the scanner memorizes the prefixes and the suffixes of length $< n$ and the set of factors of length n of the input word, but does not count the multiplicities. That is, if a factor occurs several times, it is memorized just once. This information is then compared to a collection of permitted sets of prefixes, suffixes and factors contained in the memory. The word is accepted or rejected, according to the result of this test.

The algebraic characterization of locally testable languages is slightly more involved than for star-free or piecewise testable languages. Recall that a finite semigroup S is said to have a property *locally*, if, for every idempotent e of S, the subsemigroup $eSe = \{ese \mid s \in S\}$ has the property. In particular, a semigroup is *locally trivial* if, for every idempotent e of S, $eSe = e$ and is *locally idempotent and commutative* if, for every idempotent e of S, eSe is idempotent and commutative. Equivalently, S is locally idempotent and commutative if, for every $e, s, t \in S$ such that $e = e^2$, $(ese)^2 = (ese)$ and $(ese)(ete) = (ete)(ese)$. The following result was proved independently by Brzozowski and Simon [9] and by McNaughton [28].

Theorem 8.1 *A recognizable language of A^+ is locally testable if and only if its syntactic semigroup is locally idempotent and commutative.*

This result, or more precisely the proof of this result, had a strong influence on pure semigroup theory. The reason is that Theorem 8.1 can be divided into two separate statements.

Proposition 8.2 *A recognizable language of A^+ is locally testable if and only if its syntactic semigroup divides a semidirect product of a semilattice by a locally trivial semigroup.*

Proposition 8.3 *A semigroup divides a semidirect product of a semilattice by a locally trivial semigroup if and only if it is locally idempotent and commutative.*

The proof of Proposition 8.2 is relatively easy, but Proposition 8.3 is much more difficult and relies on an interesting property. Given a semigroup S, form a graph $G(S)$ as follows: the vertices are the idempotents of S and the edges from e to f are the elements of the form esf. Then one can show that a semigroup divides a semidirect product of a semilattice by a locally trivial semigroup if and only if its graph is locally idempotent and commutative in the following sense: if p and q are loops around the same vertex, then $p = p^2$ and $pq = qp$. We shall encounter another condition on graphs in Theorem 11.1. This type of graph condition is now well understood, although numerous problems are still pending. The graph of a semigroup is a special instance of a derived category and is deeply connected with the study of the semidirect product (see Straubing [67] and Tilson [70]).

9. Varieties, another approach to recognizable languages.

In 1974, the syntactic characterizations of the star-free, piecewise testable and locally testable languages had already established the power of the semigroup approach. However, these theorems were still isolated. In 1976, Eilenberg presented in his book a unified framework for these three results. The cornerstone of this approach is the concept of variety.

Recall that a *variety of finite semigroups* (or *pseudovariety*) is a class of semigroups $\mathbf{V}$ such that:
(1) if $S \in \mathbf{V}$ and if T is a subsemigroup of S, then $T \in \mathbf{V}$,
(2) if $S \in \mathbf{V}$ and if T is a quotient of S, then $T \in \mathbf{V}$,
(3) if $(S_i)_{i \in I}$ is a finite family of semigroups of $\mathbf{V}$, then $\prod_{i \in I} S_i$ is also in $\mathbf{V}$.
Varieties of finite monoids are defined in the same way.

Condition (3) can be replaced by the conjunction of conditions (3.a) and (3.b):

(3.a) the trivial semigroup 1 belongs to $\mathbf{V}$,

(3.b) if S_1 and S_2 are semigroups of $\mathbf{V}$, then $S_1 \times S_2$ is also in $\mathbf{V}$.

Indeed, condition (3.a) is obtained by taking $I = \emptyset$ in (3).

Recall that a semigroup T *divides* a semigroup S if T is a quotient of a subsemigroup of S. Division is a transitive relation on semigroups and thus conditions (1) and (2) can be replaced by condition (1′)

(1′) if $S \in \mathbf{V}$ and if T divides S, then $T \in \mathbf{V}$.

Given a class $\mathcal{C}$ of semigroups, the intersection of all varieties containing $\mathcal{C}$ is still a variety, called the variety generated by $\mathcal{C}$, and denoted by $\langle \mathcal{C} \rangle$. In a more constructive way, $\langle \mathcal{C} \rangle$ is the class of all semigroups that divide a finite product of semigroups of $\mathcal{C}$.

Example 9.1

(1) The class $\mathbf{M}$ of all finite monoids forms a variety of finite monoids.
(2) The smallest variety of finite monoids is the trivial variety, denoted by I, consisting only of the monoid 1.
(3) The class **Com** of all finite commutative monoids form a variety of finite monoids.
(4) The class $\mathbf{J_1}$ of all finite idempotent and commutative monoids (or semilattices) forms a variety of finite monoids.
(5) The class $\mathbf{A}$ of all finite aperiodic monoids forms a variety of finite monoids.
(6) The class $\mathbf{J}$ of all finite $\mathcal{J}$-trivial monoids forms a variety of finite monoids.
(7) The class of $\mathbf{LI}$ of all finite locally trivial semigroups forms a variety of finite semigroups.
(8) The class $\mathbf{LJ_1}$ of all finite locally idempotent and commutative semigroups forms a variety of finite semigroups.

Equations are a convenient way to define varieties. For instance, the variety of finite commutative semigroups is defined by the equation $xy = yx$, the variety of aperiodic semigroups is defined by the equation $x^\omega = x^{\omega+1}$. Of course, $x^\omega = x^{\omega+1}$

is not an equation in the usual sense, since ω is not a fixed integer ... However, one can give a rigorous meaning to this "pseudoequation". Since J. Almeida and P. Weil present this topic in great detail in this volume, we refer the reader to their article for more information. For our purpose, it suffices to remember that equations (or pseudoequations) give an elegant description of the varieties of finite semigroups, but are sometimes very difficult to determine. We shall now extend this purely algebraic approach to recognizable languages.

If $\mathbf{V}$ is a variety of semigroups, we denote by $\mathcal{V}(A^+)$ the set of recognizable languages of A^+ whose syntactic semigroup belongs to $\mathbf{V}$. This is also the set of languages of A^+ recognized by a semigroup of $\mathbf{V}$.

A $+$-*class* of recognizable languages is a correspondence which associates with every finite alphabet A, a set $\mathcal{C}(A^+)$ of recognizable languages of A^+. Similarly, a $*$-*class* of recognizable languages is a correspondence which associates with every finite alphabet A, a set $\mathcal{C}(A^*)$ of recognizable languages of A^*. In particular, the correspondence $\mathbf{V} \to \mathcal{V}$ associates with every variety of semigroups a $+$-class of recognizable languages. Eilenberg gave a combinatorial description of the classes of languages that occur in this way.

If X is a language of A^+ and if $u \in A^*$, the *left quotient* (resp. *right quotient*) of X by u is the language

$$u^{-1}X = \{v \in A^+ \mid uv \in X\} \qquad (\text{resp. } Xu^{-1} = \{v \in A^+ \mid vu \in X\}).$$

Left and right quotients are defined similarly for languages of A^* by replacing A^+ by A^* in the definition.

A $+$-*variety* is a class of recognizable languages such that

(1) for every alphabet A, $\mathcal{V}(A^+)$ is closed under finite boolean operations (finite union and complement),

(2) for every semigroup morphism $\varphi : A^+ \to B^+$, $X \in \mathcal{V}(B^+)$ implies $\varphi^{-1}(X) \in \mathcal{V}(A^+)$,

(3) If $X \in \mathcal{V}(A^+)$ and $u \in A^+$, then $u^{-1}X \in \mathcal{V}(A^+)$ and $Xu^{-1} \in \mathcal{V}(A^+)$.

Similarly, a $*$-*variety* is a class of recognizable languages such that

(1) for every alphabet A, $\mathcal{V}(A^*)$ is closed under finite boolean operations,

(2) for every monoid morphism $\varphi : A^* \to B^*$, $X \in \mathcal{V}(B^*)$ implies $\varphi^{-1}(X) \in \mathcal{V}(A^*)$,

(3) If $X \in \mathcal{V}(A^*)$ and $u \in A^*$, then $u^{-1}X \in \mathcal{V}(A^*)$ and $Xu^{-1} \in \mathcal{V}(A^*)$.

We are ready to state Eilenberg's theorem.

Theorem 9.1 *The correspondence $\mathbf{V} \to \mathcal{V}$ defines a bijection between the varieties of semigroups and the $+$-varieties.*

The variety of finite semigroups corresponding to a given $+$-variety is the variety of semigroups generated by the syntactic semigroups of all the languages $L \in \mathcal{V}(A^+)$, for every finite alphabet A. There is, of course, a similar statement for the $*$-varieties.

Theorem 9.2 *The correspondence $\mathbf{V} \to \mathcal{V}$ defines a bijection between the varieties of monoids and the $*$-varieties.*

Varieties of finite semigroups or monoids are usually denoted by boldface letters and the corresponding varieties of languages are denoted by the corresponding cursive letters.

We already know four instances of Eilenberg's variety theorem.

(1) By Kleene's theorem, the $*$-variety corresponding to $\mathbf{M}$ is the $*$-variety of rational languages.
(2) By Schützenberger's theorem, the $*$-variety corresponding to $\mathbf{A}$ is the $*$-variety of star-free languages.
(3) By Simon's theorem, the $*$-variety corresponding to $\mathbf{J}$ is the $*$-variety of piecewise testable languages.
(4) By Theorem 8.1, the $+$-variety corresponding to $\mathbf{LJ_1}$ is the $+$-variety of locally testable languages.

To clear up any possible misunderstanding, note that the four theorems mentioned above (Kleene, Schützenberger, etc.) are *not* corollaries of the variety theorem. For instance, the variety theorem indicates that the languages corresponding to the finite aperiodic monoids form a $*$-variety; it doesn't say that this $*$-variety is the variety of star-free languages ... Actually, it is often a difficult problem to find an explicit description of the $*$-variety of languages corresponding to a given variety of finite monoids, or, conversely, to find the variety of finite monoids corresponding to a given $*$-variety.

However, the variety theorem provided a new direction to classify recognizable languages. Systematic searches for the variety of monoids (resp. languages) corresponding to a given variety of languages (resp. monoids) were soon undertaken. A partial account of these results is given into the next section.

10. Bestiary

We review in this section a few examples of correspondence between varieties of finite monoids (or semigroups) and varieties of languages. A *boolean algebra* is a set of languages containing the empty language and closed under finite union, finite intersection and complement.

Let us start our visit of the zoo with the subvarieties of the variety **Com** of all finite commutative monoids: the variety **Acom** of commutative aperiodic monoids, the variety **Gcom** of commutative groups, the variety $\mathbf{J_1}$ of idempotent and commutative monoids (or semilattices) and the trivial variety **I**.

Proposition 10.1 *For every alphabet A, $\mathcal{I}(A^*) = \{0, A^*\}$.*

Proposition 10.2 *For every alphabet A, $\mathcal{J}_1(A^*)$ is the boolean algebra generated by the languages of the form A^*aA^* where a is a letter. Equivalently, $\mathcal{J}_1(A^*)$ is the boolean algebra generated by the languages of the form B^* where B is a subset of A.*

Proposition 10.3 *For every alphabet A, $\mathcal{G}com(A^*)$ is the boolean algebra generated by the languages of the form*

$$L(a, k, n) = \{u \in A^* \mid |u|_a \equiv k \bmod n\}$$

where $a \in A$ and $0 \leq k < n$.

Proposition 10.4 *For every alphabet A, $\mathcal{A}com(A^*)$ is the boolean algebra generated by the languages of the form*

$$L(a,k) = \{u \in A^+ \mid |u|_a = k\}$$

where $a \in A$ and $k \geq 0$.

Proposition 10.5 *For every alphabet A, $\mathcal{C}om(A^*)$ is the boolean algebra generated by the languages of the form*

$$L(a,k) = \{u \in A^+ \mid |u|_a = k\} \quad \text{or} \quad L(a,k,n) = \{u \in A^+ \mid |u|_a \equiv k \bmod n\}$$

where $a \in A$ and $0 \leq k < n$.

Consider now the variety **LI** of all locally trivial semigroups and its subvarieties $\mathbf{L^rI}$, $\mathbf{L^\ell I}$ and **Nil**. A finite semigroup S belongs to **LI** if and only if, for every $e \in E(S)$ and every $s \in S$, $ese = e$. The asymmetrical versions of this condition define the varieties $\mathbf{L^rI}$ and $\mathbf{L^\ell I}$. Thus $\mathbf{L^rI}$ (resp. $\mathbf{L^\ell I}$) is the variety of all finite semigroups S such that $se = e$ (resp. $es = e$). Equivalently, a semigroup belongs to **LI** (resp. $\mathbf{L^rI}$, $\mathbf{L^\ell I}$) if it is a nilpotent extension of a rectangular band (resp. a right rectangular band, a left rectangular band). Finally **Nil** is the variety of nilpotent semigroups, defined by the condition $es = se = e$ for every $e \in E(S)$ and every $s \in S$. Recall that a subset F of a set E is cofinite if its complement in E is finite.

Proposition 10.6 *For every alphabet A, $\mathcal{N}il(A^+)$ is the set of finite or cofinite languages of A^+.*

Proposition 10.7 *For every alphabet A, $\mathcal{L}^r\mathcal{I}(A^+)$ (resp. $\mathcal{L}^\ell\mathcal{I}(A^+)$) is the set of languages of the form $A^*X \cup Y$ (resp. $XA^* \cup Y$), where X and Y are finite subsets of A^+.*

Proposition 10.8 *For every alphabet A, $\mathcal{LI}(A^+)$ is the set of languages of the form $XA^*Y \cup Z$, where X, Y and Z are finite subsets of A^+.*

Note that the previous characterizations do not make use of the complement, although the sets $\mathcal{N}\mathrm{il}(A^+)$, $\mathcal{L}^r\mathcal{I}(A^+)$, $\mathcal{L}^\ell\mathcal{I}(A^+)$ and $\mathcal{LI}(A^+)$ are closed under complement. Actually, the following characterizations hold.

Proposition 10.9 *For every alphabet A,*
(1) $\mathcal{L}^r\mathcal{I}(A^+)$ is the boolean algebra generated by the languages of the form A^*u, where $u \in A^+$,
(2) $\mathcal{L}^\ell\mathcal{I}(A^+)$ is the boolean algebra generated by the languages of the form uA^*, where $u \in A^+$,
(3) $\mathcal{LI}(A^+)$ is the boolean algebra generated by the languages of the form uA^* or A^*u, where $u \in A^+$.

It would be too long to state in full detail all known results on varieties of languages. Let us just mention that the languages corresponding to the following varieties of finite semigroups or monoids are known: all varieties of bands ([44] for the lower levels and [55] for the general case), the varieties of $\mathcal{R}$-trivial (resp. $\mathcal{L}$-trivial) monoids [17, 7, 36], the varieties of p-groups (resp. nilpotent groups) [17], the varieties of solvable groups [59], the varieties of monoids whose groups are commutative [53, 25], nilpotent [17], solvable [59], the variety of monoids with commuting idempotents [26], the variety of $\mathcal{J}$-trivial monoids with commuting idempotents [3], the variety of monoids whose regular $\mathcal{J}$-classes are rectangular bands [54], the variety of block groups (see the author's article "**BG** = **PG**, a success story" in this volume) and many others which follow in particular from the general results given in section 12.

11. Back to the early attempts

As the variety approach proved to be successful in many different situations, it was expected to shed some new light on the difficult problems mentioned in section 5. The reality is more complicated. In brief, varieties do not seem to be helpful for investigating star height, it is so far the most successful approach for studying dot-depth and the concatenation levels and, with regard to the extended star height, it seems to be a useful tool, but probably nothing more. Let us comment on this judgment in more detail.

Varieties do not seem to be helpful when considering star height, simply because the languages of a given star height are not closed under inverse morphisms between free monoids and thus do not form a variety of languages. However, the notion of syntactic semigroup arises in the proof of Hashiguchi's theorems.

Schützenberger's theorem shows that the languages of extended star height 0 form a variety. However, it seems unlikely that a similar result holds for the languages of extended star height 1. Indeed, one can show [32] that every finite monoid divides the syntactic monoid of a language of the form L^*, where L is finite. It follows that if the languages of extended star height 1 form a variety of languages, then this variety is the variety of all rational languages. In particular, this would imply that every recognizable language is of extended star height 0 or 1.

Varieties are much more useful in the study of the concatenation product. We have already seen the syntactic characterization of the languages of concatenation level 1. There is a similar result for the languages of dot-depth one. It is easy to see from the general definition that a language of A^+ is of "dot-depth one" if it is a boolean combination of languages of the form

$$u_0A^*u_1A^*u_2\cdots A^*u_{k-1}A^*u_k$$

where $k \geq 0$ and $u_i \in A^*$. The syntactic characterization of these languages was settled by Knast [23, 24]. The premises of the condition in the following theorem are illustrated in Figure 11.1.

Theorem 11.1 *A language of A^+ is of dot-depth one if and only if the graph of its syntactic semigroup satisfies the following condition : if e and f are two vertices, p and r edges from e to f, and q and s edges from f to e, then $(pq)^\omega ps(rs)^\omega = (pq)^\omega(rs)^\omega$.*

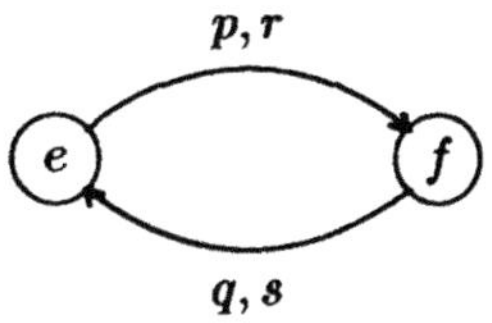

Figure 11.1.

More generally, one can show that the languages of dot-depth n form a $+$-variety of languages. The corresponding variety of finite semigroups is usually denoted by $\mathbf{B}_n$. Similarly, the languages of concatenation level n form a $*$-variety of languages and the corresponding variety of finite monoids is denoted $\mathbf{V}_n$. The two hierarchies are strict [8].

Theorem 11.2 *For every $n \geq 0$, there exists a language of dot-depth $n+1$ which is not of dot-depth n and a language of concatenation level $n+1$ which is not of concatenation level n.*

An important connection between the two hierarchies was found by Straubing [66]. Given a variety of finite monoids $\mathbf{V}$ and a variety of finite semigroups $\mathbf{W}$, denote by $\mathbf{V} * \mathbf{W}$ the variety of finite semigroups generated by the semidirect products $S * T$ with $S \in \mathbf{V}$ and $T \in \mathbf{W}$ such that the action of T on S is right unitary.

Theorem 11.3 *For every $n > 0$, one has $\mathbf{B}_n = \mathbf{V}_n * \mathbf{LI}$ and $\mathbf{V}_n = \mathbf{B}_n \cap \mathbf{M}$.*

In particular $\mathbf{B}_1 = \mathbf{J} * \mathbf{LI}$. It follows also, thanks to a deep result of Straubing [66] that $\mathbf{B}_n$ is decidable if and only if $\mathbf{V}_n$ is decidable. However, it is still an open problem to know whether the varieties $\mathbf{V}_n$ are decidable for $n \geq 2$. The case $n = 2$ is especially frustrating, but although several partial results have been obtained [43, 67, 71, 69, 73, 13], the general case remains open.

12. Recent developments

We shall not discuss in detail the numerous developments of the theory since Eilenberg's variety theorem, but we shall indicate the main trends. A quick glance at the known examples shows that the combinatorial description of a variety of languages follows most often the following pattern: the variety is described as the smallest variety closed under a given class of operations, such as boolean operations, product, etc. Varieties of semigroups are also often defined with the help of operators: join, semidirect products, Malcev products, etc. In view of Eilenberg's theorem, one may expect some relationship between the operators on languages (of combinatorial nature) and the operators on semigroups (of algebraic nature).

$$\begin{array}{ccc} \mathbf{V} & \xrightarrow{\text{Operations on semigroups}} & \mathbf{W} \\ \downarrow & & \downarrow \\ \mathcal{V} & \xrightarrow{\text{Operations on languages}} & \mathcal{W} \end{array}$$

In the late seventies, several results of this type were established, in particular by H. Straubing. We first consider the marked product. One of the most useful tools for studying this product is the *Schützenberger product* of n monoids, which was originally defined by Schützenberger for two monoids [52], and extended by Straubing [63] for any number of monoids.

Given a monoid M, the set of subsets of M, denoted $\mathcal{P}(M)$, is a semiring under union as addition and the product of subsets as multiplication, defined, for all $X, Y \subset M$ by $XY = \{xy \mid x \in X \text{ and } y \in Y\}$.

Let $M_1, \ldots, M_n$ be monoids. We denote by M the product monoid $M_1 \times \cdots \times M_n$, k the semiring $\mathcal{P}(M)$ and by $M_n(k)$ the semiring of square matrices of size n with entries in k. The Schützenberger product of $M_1, \ldots, M_n$, denoted $\lozenge_n(M_1, \ldots, M_n)$ is the submonoid of the multiplicative monoid $M_n(k)$ composed of all the matrices P satisfying the three following conditions:

(1) if $i > j$, $P_{i,j} = 0$,
(2) if $1 \leq i \leq n$, $P_{i,i} = \{(1, \ldots, 1, s_i, 1, \ldots, 1)\}$ for some $s_i \in M_i$,
(3) if $1 \leq i \leq j \leq n$, $P_{i,j} \subset 1 \times \cdots \times 1 \times M_i \times \cdots \times M_j \times 1 \cdots \times 1$.

Condition (1) indicates that the matrices of the Schützenberger product are upper triangular, condition (2) enables us to identify the diagonal coefficient $P_{i,i}$ with an element s_i of M_i and condition (3) shows that if $i < j$, $P_{i,j}$ can be identified with a subset of $M_i \times \cdots \times M_j$. With this convention, a matrix of $\lozenge_3(M_1, M_2, M_3)$ will have the form

$$\begin{pmatrix} s_1 & P_{1,2} & P_{1,3} \\ 0 & s_2 & P_{2,3} \\ 0 & 0 & s_3 \end{pmatrix}$$

with $s_i \in M_i$, $P_{1,2} \subset M_1 \times M_2$, $P_{1,3} \subset M_1 \times M_2 \times M_3$ and $P_{2,3} \subset M_2 \times M_3$.

Notice that the Schützenberger product is not associative, in the sense that in general the monoids $\lozenge_2(M_1, \lozenge_2(M_2, M_3))$, $\lozenge_3(M_1, M_2, M_3)$ and $\lozenge_2(\lozenge_2(M_1, M_2), M_3)$ are pairwise distinct.

The following result shows that the Schützenberger product is the algebraic operation on monoids that corresponds to the marked product.

Proposition 12.1 *Let $L_0, L_1, \ldots, L_n$ be languages of A^* recognized by monoids $M_0, M_1, \ldots, M_n$ and let $a_1, \ldots, a_n$ be letters of A. Then the marked product $L_0 a_1 L_1 \cdots a_n L_n$ is recognized by the monoid $\lozenge_{n+1}(M_0, M_1, \ldots, M_n)$.*

This result was extended to varieties by Reutenauer [49] for $n = 1$ and by the author [35] in the general case (see also [72] for a simpler proof). Let $\mathbf{V}_0, \ldots, \mathbf{V}_n$ be varieties of finite monoids and let $\lozenge_{n+1}(\mathbf{V}_0, \mathbf{V}_1, \ldots, \mathbf{V}_n)$ be the variety of finite monoids generated by the Schützenberger products of the form $\Diamond_{n+1}(M_0, M_1, \dots, M_n)$ with $M_0 \in \mathbf{V}_0$, $M_1 \in \mathbf{V}_1$, ..., $M_n \in \mathbf{V}_n$.

Theorem 12.2 *Let $\mathcal{V}$ be the $*$-variety corresponding to the variety of finite monoids $\Diamond_{n+1}(\mathbf{V}_0, \mathbf{V}_1, \dots, \mathbf{V}_n)$. Then, for all alphabets A, $\mathcal{V}(A^*)$ is the boolean algebra generated by all the marked products of the form $L_0a_1L_1 \cdots a_nL_n$ where $L_0 \in \mathcal{V}_0(A^*)$, ..., $L_n \in \mathcal{V}_n(A^*)$ and $a_1, \dots, a_n \in A$.*

If $\mathbf{V}_0 = \mathbf{V}_1 = \dots = \mathbf{V}_n = \mathbf{V}$, the variety $\Diamond_{n+1}(\mathbf{V}, \mathbf{V}, \dots, \mathbf{V})$ is denoted $\Diamond_{n+1}\mathbf{V}$ and $\Diamond\mathbf{V} = \bigcup_{n>0} \Diamond_n\mathbf{V}$ denotes the union of all $\Diamond_n\mathbf{V}$. Given a $*$-variety of languages $\mathcal{V}$, the *extension of $\mathcal{V}$ under marked product* is the $*$-variety $\mathcal{V}'$ such that, for every alphabet A, $\mathcal{V}'(A^*)$ is the boolean algebra generated by the marked products of the form $L_0a_1L_1 \cdots a_nL_n$ where $L_0, L_1, \dots, L_n \in \mathcal{V}(A^*)$ and $a_1, \dots, a_n \in A$. The *closure of $\mathcal{V}$ under marked product* is the smallest $*$-variety $\bar{\mathcal{V}}$ such that, for every alphabet A, $\bar{\mathcal{V}}(A^*)$ contains $\mathcal{V}(A^*)$ and all the marked products of the form $L_0a_1L_1 \cdots a_nL_n$ where $L_0, L_1, \dots, L_n \in \bar{\mathcal{V}}(A^*)$ and $a_1, \dots, a_n \in A$. The $*$-variety corresponding to $\Diamond\mathbf{V}$ is described in the following theorem.

Theorem 12.3 *Let $\mathbf{V}$ be a monoid variety and let $\mathcal{V}$ be the corresponding $*$-variety. Then the $*$-variety corresponding to $\Diamond\, \mathbf{V}$ is the extension of $\mathcal{V}$ under marked product.*

Corollary 12.4 *A $*$-variety is closed under marked product if and only if the corresponding variety of monoids $\mathbf{V}$ satisfies $\mathbf{V} = \Diamond\, \mathbf{V}$.*

The Schützenberger product has a remarkable algebraic property [63, 38]. Let $M_1, \dots, M_n$ be monoids and let π be the monoid morphism from $\Diamond_n(M_1, \dots, M_n)$ into $M_1 \times \cdots \times M_n$ that maps a matrix onto its diagonal.

Theorem 12.5 *For every idempotent e of $M_1 \times \cdots \times M_n$, the semigroup $\pi^{-1}(e)$ is in the variety $\mathbf{B}_1$.*

Given a variety of finite semigroups $\mathbf{V}$, a finite monoid M is called a *$\mathbf{V}$-extension of a finite monoid N* if there exists a surjective morphism $\varphi : M \to N$ such that, for every idempotent e of N, $\varphi^{-1}(e) \in \mathbf{V}$. Theorem 12.5 shows that the Schützenberger product of n finite monoids is a $\mathbf{B}_1$-extension of their product. Given a variety of finite monoids $\mathbf{W}$, the Malcev product $\mathbf{V} \circledM \mathbf{W}$ is the variety of finite monoids generated by all the $\mathbf{V}$-extensions of monoids of $\mathbf{W}$. This gives the following relation between the $\mathbf{V}_n$.

Theorem 12.6 *For every $n \geq 0$, $\mathbf{V}_{n+1}$ is contained in $\mathbf{B}_1 \circledM \mathbf{V}_n$.*

It is conjectured that $\mathbf{V}_{n+1} = \mathbf{B}_1 \circledM \mathbf{V}_n$ for every n. If this conjecture were true, it would reduce the decidability of the dot-depth to a problem on the Malcev products of the form $\mathbf{B}_1 \circledM \mathbf{V}$.

Malcev products actually play an important role in the study of the marked product. For instance, Straubing has established the important following result, which gives support to the previous conjecture.

Theorem 12.7 *Let* $\mathbf{V}$ *be a monoid variety and let* $\mathcal{V}$ *be the corresponding* $*$*-variety. Then the* $*$*-variety corresponding to* $\mathbf{A} \circledM \mathbf{V}$ *is the closure of* $\mathcal{V}$ *under marked product.*

Example 12.1 Let **H** be a variety of finite groups (for instance, the variety of all finite commutative groups, nilpotent groups, solvable groups, etc.). Denote by $\bar{H}$ the variety of all monoids whose subgroups (that is, $\mathcal{H}$-classes containing an idempotent) belong to **H**. One can show that $\mathbf{A} \circledM \bar{H} = \bar{H}$. Therefore, the corresponding $*$-variety is closed under marked product.

The marked product $L = L_0 a_1 L_1 \cdots a_n L_n$ of n languages $L_0, L_1, \ldots, L_n$ is ***unambiguous*** if every word u of L admits a unique factorization of the form $u_0 a_1 u_1 \cdots a_n u_n$ with $u_0 \in L_0$, $u_1 \in L_1$, ..., $u_n \in L_n$. The following result was established in [34, 45] as a generalization of a former result of Schützenberger [54].

Theorem 12.8 *Let* $\mathbf{V}$ *be a monoid variety and let* $\mathcal{V}$ *be the corresponding* $*$*-variety. Then the* $*$*-variety corresponding to* $\mathbf{LI} \circledM \mathbf{V}$ *is the closure of* $\mathcal{V}$ *under unambiguous marked product.*

The extension of a given $*$-variety is also characterized in [45]. Other variations of the marked product have been considered [54, 34, 48]. They lead to some interesting algebraic constructions.

Another operation on semigroups has a natural counterpart in terms of languages. Given a variety of finite monoids $\mathbf{V}$, denote by $\mathbf{PV}$ the variety of finite monoids generated by all the monoids of the form $\mathcal{P}(M)$, for $M \in \mathbf{V}$. A monoid morphism $\varphi : B^* \to A^*$ is *length preserving* if it maps a letter of B onto a letter of A. Given a $*$-variety of languages $\mathcal{V}$, the *extension of* $\mathcal{V}$ *under length preserving morphisms* is the smallest $*$-variety $\mathcal{V}'$ such that, for every alphabet A, $\mathcal{V}'(A^*)$ contains the languages of the form $\varphi(L)$ where $L \in \mathcal{V}(B^*)$ and $\varphi : B^* \to A^*$ is a length preserving morphism. The *closure of* $\mathcal{V}$ *under length preserving morphisms* is the smallest $*$-variety $\bar{\mathcal{V}}$ containing $\mathcal{V}$ such that, for every length preserving morphism $\varphi : B^* \to A^*$, $L \in \bar{\mathcal{V}}(A^*)$ implies $\varphi^{-1}(L) \in \bar{\mathcal{V}}(B^*)$. We can now state the result found independently by Reutenauer [49] and Straubing [61].

Theorem 12.9 *Let* $\mathbf{V}$ *be a monoid variety and let* $\mathcal{V}$ *be the corresponding* $*$*-variety. Then the* $*$*-variety corresponding to* $\mathbf{PV}$ *is the extension of* $\mathcal{V}$ *under length preserving morphisms.*

Corollary 12.10 *A* $*$*-variety is closed under length preserving morphisms if and only if the corresponding variety of monoids* $\mathbf{V}$ *satisfies* $\mathbf{V} = \mathbf{PV}$.

These results motivated the systematic study of the varieties of the form $\mathbf{PV}$, which is not yet achieved. See the survey article [37] for the known results prior to 1986 and the book of J. Almeida [1] for the more recent results.

The Schützenberger product and the power monoid are actually particular cases of a general construction which gives the monoid counterpart of a given operation on languages [41, 42, 39]. This general construction works for most operations on languages, with the notable exception of the star operation, but its presentation would take us too far afield. We conclude this section with a few results on the semidirect product of two varieties.

We have already defined the semidirect product $\mathbf{V} * \mathbf{W}$ of a variety of finite monoids $\mathbf{V}$ and a variety of finite semigroups $\mathbf{W}$. One can define similarly the semidirect product of two varieties of finite monoids or of two varieties of finite semigroups. For instance, if $\mathbf{V}$ and $\mathbf{W}$ are two varieties of finite monoids, $\mathbf{V} * \mathbf{W}$ is the variety of finite monoids generated by the semidirect products $M * N$ with $M \in \mathbf{V}$ and $N \in \mathbf{W}$ such that the action of N on M is unitary. This variety is also generated by the wreath products $M \circ N$ with $M \in \mathbf{V}$ and $N \in \mathbf{W}$.

Straubing has given a general construction to describe the languages recognized by the wreath product of two finite monoids. Let $M \in \mathbf{V}$ and $N \in \mathbf{W}$ be two finite monoids and let $\eta : A^* \to M \circ N$ be a monoid morphism. We denote by $\pi : M \circ N \to N$ the monoid morphism defined by $\pi(f, n) = n$ and we put $\varphi = \pi \circ \eta$. Thus φ is a monoid morphism from A^* into N. Let $B = N \times A$ and $\sigma : A^* \to B^*$ be the map (which is not a morphism!) defined by

$$\sigma(a_1a_2 \cdots a_n) = (1, a_1)(\varphi(a_1), a_2) \cdots (\varphi(a_1a_2 \cdots a_{n-1}), a_n)$$

Then Straubing's "wreath product principle" can be stated as follows.

Theorem 12.11 *If a language L is recognized by $\eta : A^* \to M \circ N$, then L is a finite boolean combination of languages of the form $X \cap \sigma^{-1}(Y)$, where $Y \subset B^*$ is recognized by M and where $X \subset A^*$ is recognized by N.*

Conversely, the finite boolean combinations of languages of the form $X \cap \sigma^{-1}(Y)$ are not necessarily recognized by $M \circ N$, but are certainly recognized by a monoid of the variety $\mathbf{V} * \mathbf{W}$. Therefore, a careful study of the languages of the form $\sigma^{-1}(Y)$ usually suffices to give a combinatorial description of the languages corresponding to $\mathbf{V} * \mathbf{W}$. A similar wreath product principle holds when $\mathbf{V}$ or $\mathbf{W}$ are varieties of finite semigroups. Examples of the application of this technique include Proposition 8.2 and the proof of Schützenberger's theorem based on the fact that every finite aperiodic monoid divides a wreath product of copies of U_2. Straubing also has successfully used this principle to describe the variety of languages corresponding to solvable groups (solvable groups are wreath products of commutative groups) and in his proof of the equality $\mathbf{B}_n = \mathbf{V}_n * \mathbf{LI}$.

13. Conclusion

We have centred our presentation around the notion of variety and voluntarily left out several aspects of the theory which are developed extensively in other articles of this volume: H. Straubing, D. Thérien and W. Thomas survey the connections with formal logic and boolean circuits, J. Almeida and P. Weil present the implicit operations, D. Perrin and the author treat the theory of automata on infinite words, the topological aspects are mentioned in the author's account of the success story $\mathbf{BG} = \mathbf{PG}$, S. W. Margolis J. Meakin and M. Sapir cover the extensions of automata theory to inverse monoids and demarcate the border between decidable and undecidable and H. Short shows that automata are also useful in group theory. Some other extensions are not covered at all in this volume, in particular the connections with variable length codes, rational and recognizable sets on arbitrary monoids and the extension of the theory to power series and algebras.

We hope that the reading of the articles of this volume will convince the reader that the algebraic theory of automata is a recent but flourishing subject. It is intimately related to the theory of finite semigroups and certainly one of the most convincing applications of this theory.

Acknowledgements

I would like to thank Pascal Weil, Marc Zeitoun and Monica Mangold for many useful suggestions.

References

1. J. Almeida, *Semigrupos Finitos e Álgebra Universal*, Publicaçoes do Instituto de Matematica e Estatística da Universidade de Saõ Paulo, (1992).
2. J. Almeida, *Implicit operations on finite $\mathcal{J}$-trivial semigroups and a conjecture of I. Simon*, J. Pure Appl. Algebra **69**, (1990) 205–218.
3. C. J. Ash, T.E. Hall and J.E. Pin, *On the varieties of languages associated to some varieties of finite monoids with commuting idempotents*, Information and Computation **86**, (1990) 32–42.
4. J. Berstel and Ch. Reutenauer, *Rational Series and Their Langages*, Springer, Berlin (1988).
5. J. A. Brzozowski, *Hierarchies of aperiodic languages*, RAIRO Inf. Theor. **10** (1976) 33–49.
6. J. A. Brzozowski, *Open problems about regular languages*, Formal language theory, perspectives and open problems (R.V. Book editor), Academic Press, New York, (1980) 23–47.
7. J. A. Brzozowski and F. E. Fich, *Languages of $\mathcal{R}$-trivial monoids*, J. Comput. System Sci. **20** (1980) 32–49.
8. J. A. Brzozowski and R. Knast, *The dot-depth hierarchy of star-free languages is infinite*, J. Comput. System Sci. **16**, (1978) 37–55.
9. J. A. Brzozowski and I. Simon, *Characterizations of locally testable languages*, Discrete Math. **4**, (1973) 243–271.
10. S. Cho and D. T. Huynh, *Finite automaton aperiodicity is PSPACE-complete*, Theoret. Comput. Sci. **88**, (1991) 99–116.
11. R. S. Cohen and J. A. Brzozowski, *On star-free events*, Proc. Hawaii Int. Conf. on System Science, Honolulu, (1968) 1–4.
12. R. S. Cohen and J. A. Brzozowski, *Dot-depth of star-free events*, J. Comput. System Sci. **5**, (1971) 1–15.
13. D. Cowan, 1993, *Inverse monoids of dot-depth* 2, Int. Jour. Alg. and Comp. to appear.
14. F. Dejean and M. P. Schützenberger, *On a question of Eggan*, Information and Control **9**, (1966), 23–25.
15. L. C. Eggan, *Transition graphs and the star height of regular events*, Michigan Math. J. **10**, (1963) 385–397.
16. S. Eilenberg, *Automata, languages and machines*, Vol. A, Academic Press, New York (1974).
17. S. Eilenberg, *Automata, languages and machines*, Vol. B, Academic Press, New York (1976).
18. K. Hashiguchi, *Regular languages of star height one*, Information and Control **53** (1982) 199–210.
19. K. Hashiguchi, *Representation theorems on regular languages*, J. Comput. System Sci. **27**, (1983) 101–115.
20. J. E. Hopcroft and J. D. Ullman, *Introduction to Automata Theory, Languages and Computation*, Addison Wesley, (1979).
21. J. Howie, *Automata and Languages*, Oxford Science Publications, Clarendon Press, Oxford, 1991.

22. S. C. Kleene, *Representation of events in nerve nets and finite automata*, in "Automata Studies" C. E. Shannon and J. McCarthy eds.), Princeton University Press, Princeton, New Jersey, (1956) 3–42.
23. R. Knast, *A semigroup characterization of dot-depth one languages*, RAIRO Inform. Théor. **17** (1983) 321–330.
24. R. Knast, *Some theorems on graph congruences*, RAIRO Inform. Théor. **17** (1983) 331–342.
25. G. Lallement, *Semigroups and combinatorial applications*, Wiley, New York (1979).
26. S. W. Margolis and J.E. Pin, *Languages and inverse semigroups*, 11th ICALP, Lecture Notes in Computer Science **199**, Springer, Berlin (1985) 285–299.
27. S. W. Margolis and J.E. Pin, *Inverse semigroups and varieties of finite semigroups*, J. Algebra **110** (1987) 306–323.
28. R. McNaughton, *Algebraic decision procedures for local testability*, Math. Syst. Theor. **8**, (1974) 60–76.
29. R. McNaughton and S. Papert, *Counter-free Automata*, MIT Press, (1971).
30. A. R. Meyer, *A note on star-free events*, J. Assoc. Comput. Mach. **16**, (1969) 220–225.
31. D. Perrin, *Automata*, Chapter 1 in Handbook of Theoretical Computer Science (J. Van Leeuwen, ed.), Vol B: Formal Models and Semantics, Elsevier (1990).
32. J.-E. Pin, *Sur le monoïde de L^* lorsque L est un langage fini*, Theor. Comput. Sci. **7** (1978) 211–215.
33. J.-E. Pin, *Variétés de langages et monoïde des parties*, Semigroup Forum **20** (1980) 11–47.
34. J.-E. Pin, *Propriétés syntactiques du produit non ambigu*, 7th ICALP, Lecture Notes in Computer Science **85** (1980) 483–499.
35. J.-E. Pin, *Hiérarchies de concaténation*, RAIRO Informatique Théorique **18**, (1984) 23–46.
36. J.-E. Pin, *Variétés de langages formels*, Masson, Paris (1984); English translation: *Varieties of formal languages*, Plenum, New-York (1986).
37. J.-E. Pin, *Power semigroups and related varieties of finite semigroups*, Semigroups and Their Applications (S. M. Goberstein and P. M. Higgins, D. Reidel eds.), (1986) 139–152.
38. J.-E. Pin, *A property of the Schützenberger product*, Semigroup Forum **35**, (1987) 53–62.
39. J.-E. Pin, *Relational morphisms, transductions and operations on languages*, in J.E. Pin (ed.), *Formal properties of finite automata and applications*, Lect. Notes in Comp. Sci. **386**, Springer, (1989) 120–137.
40. J.-E. Pin, *Logic, Semigroups and Automata on Words*, Proceedings of the Spring School on Logic and Computer Science, ed. S. Grigorieff, to appear.
41. J.-E. Pin and J. Sakarovitch, *Operations and transductions that preserve rationality*, 6th GI Conf., Lecture Notes in Computer Science **145**, Springer, Berlin (1983) 277–288.
42. J.-E. Pin and J. Sakarovitch, *Une application de la représentation matricielle des transductions*, Theor. Comput. Sci. **35**, (1985) 271–293.
43. J.-E. Pin and H. Straubing, *Monoids of upper triangular matrices*, Colloquia Mathematica Societatis Janos Bolyai 39, Semigroups, Szeged, (1981) 259–272.
44. J.-E. Pin, H. Straubing and D. Thérien, *Small varieties of finite semigroups and extensions*, J. Austral. Math. Soc. **37** (1984) 269–281.
45. J.-E. Pin, H. Straubing and D. Thérien, *Locally trivial categories and unambiguous concatenation*, J. Pure and Appl. Algebra **52** (1988) 297–311.
46. J.-E. Pin, H. Straubing and D. Thérien, *New results on the generalized star height problem*, STACS 89, Lecture Notes in Computer Science **349**, (1989), 458–467.
47. J.-E. Pin, H. Straubing and D. Thérien, *Some results on the generalized star height problem*, Information and Computation, **101** (1992), 219–250.
48. J.-E. Pin and D. Thérien, *The bideterministic concatenation product*, to appear in Int. Jour. Alg. and Comp.
49. Ch. Reutenauer, *Sur les variétés de langages et de monoïdes*, Lecture Notes in Computer Science **67**, Springer, Berlin, (1979) 260–265.
50. J. M. Robson, *Some Languages of Generalised Star Height One*, LITP Technical Report **89-62**, 1989, 9 pages.
51. J. M. Robson, *More Languages of Generalised Star Height* 1, Theor. Comput. Sci. **106**, (1992) 327–335.
52. M. P. Schützenberger, *On finite monoids having only trivial subgroups*, Information and Control **8**, (1965) 190–194.
53. M. P. Schützenberger, *Sur les monoïdes finis dont les groupes sont commutatifs*, RAIRO Inf.

Theor. **1** (1974) 55–61.

54. M. P. Schützenberger, *Sur le produit de concaténation non ambigu*, Semigroup Forum **13** (1976) 47–75.
55. H. Sezinando, *The varieties of languages corresponding to the varieties of finite band monoids*, Semigroup Forum **44** (1992) 283–305.
56. I. Simon, *Piecewise testable events*, Proc. 2nd GI Conf., Lect. Notes in Comp. Sci. **33**, Springer, Berlin, (1975) 214–222.
57. J. Stern, *Characterization of some classes of regular events*, Theoret. Comp. Sci. **35**, (1985) 17–42.
58. J. Stern, *Complexity of some problems from the theory of automata*, Information and Control **66**, (1985) 63–176.
59. H. Straubing, *Families of recognizable sets corresponding to certain varieties of finite monoids*, J. Pure Appl. Algebra, **15** (1979) 305–318.
60. H. Straubing, *Aperiodic homomorphisms and the concatenation product of recognizable sets*, J. Pure Appl. Algebra, **15** (1979) 319–327.
61. H. Straubing, *Recognizable sets and power sets of finite semigroups*, Semigroup Forum, **18** (1979) 331–340.
62. H. Straubing, *On finite $\mathcal{J}$-trivial monoids*, Semigroup Forum **19**, (1980) 107–110.
63. H. Straubing, *A generalization of the Schützenberger product of finite monoids*, Theoret. Comp. Sci., **13** (1981) 137–150.
64. H. Straubing, *Relational morphisms and operations on recognizable sets*, RAIRO Inf. Theor. **15** (1981) 149–159.
65. H. Straubing, *The variety generated by finite nilpotent monoids*, Semigroup Forum **24**, (1982) 25–38.
66. H. Straubing, *Finite semigroup varieties of the form $V * D$*, J. Pure Appl. Algebra **36**,(1985) 53–94.
67. H. Straubing, *Semigroups and languages of dot-depth two*, Theoret. Comp. Sci. **58**, (1988) 361–378.
68. H. Straubing and D. Thérien, *Partially ordered finite monoids and a theorem of I. Simon*, J. Algebra **119**, (1985) 393–399.
69. H. Straubing and P. Weil, *On a conjecture concerning dot-depth two languages*, Theoret. Comp. Sci. **104**, (1992) 161–183.
70. B. Tilson, *Categories as algebras*, J. Pure and Appl. Algebra **48**, (1987) 83–198.
71. P. Weil, *Inverse monoids of dot-depth two*, Theoret. Comp. Sci. **66**, (1989) 233–245.
72. P. Weil, *Closure of varieties of languages under products with counter*, Journal of Computer and System Sciences **45** (1992) 316–339.
73. P. Weil, *Some results on the dot-depth hierarchy*, Semigroup Forum **46** (1993) 352–370.

BG = PG: A SUCCESS STORY

JEAN-ERIC PIN
LITP/IBP, Université Paris VI et CNRS, Tour 55-65, 4 Place Jussieu, 75252 Paris Cedex 05, France

1. Introduction

All semigroups and monoids except free monoids and free groups are assumed finite in this paper. A monoid M *divides* a monoid N if M is a quotient of a submonoid of N. The set of idempotents of a monoid M is denoted $E(M)$. The word "variety" will be used for pseudo-variety — that is, a collection of finite monoids closed under division and finite direct product.

A block group is a monoid such that every $\mathcal{R}$-class (resp. $\mathcal{L}$-class) contains at most one idempotent. Block groups form a variety of monoids, denoted **BG**, which is the topic of this paper. More precisely, we want to make some comments on the following cryptic line, which states that six apparently distinct varieties of monoids are in fact all equal:

$$\Diamond\mathbf{G} = \mathbf{PG} = \mathbf{J} * \mathbf{G} = \mathbf{J} \textcircled{m} \mathbf{G} = \mathbf{BG} = \mathbf{EJ}$$

As **BG** has already been introduced, let us now briefly present the other five actors. More detailed definitions will be given in the forthcoming sections. The variety $\Diamond\mathbf{G}$ is the variety generated by Schützenberger products of groups, **PG** is the variety generated by power groups, $\mathbf{J} * \mathbf{G}$ is the variety generated by semidirect products of a $\mathcal{J}$-trivial monoid by a group, $\mathbf{J} \textcircled{m} \mathbf{G}$ is the variety generated by Malcev products of a $\mathcal{J}$-trivial monoid by a group and finally **EJ** is the variety of monoids M such that $E(M)$ generates a $\mathcal{J}$-trivial monoid.

All but one of these equalities were obtained by Margolis and the author around 1984 [16, 17], but the last missing inclusion $\mathbf{J} \textcircled{m} \mathbf{G} \subset \mathbf{J} * \mathbf{G}$ resisted proof for seven years. During these seven years, several fundamental results were obtained on the Malcev product and the semidirect product and in particular on the varieties of the form $\mathbf{V} * \mathbf{G}$ or $\mathbf{V} \textcircled{m} \mathbf{G}$. On the positive side, it was shown that the equality $\mathbf{V} * \mathbf{G} = \mathbf{V} \textcircled{m} \mathbf{G}$ holds when the variety **V** is *local* in the sense of categories [39, 15]. On the negative side, it was observed that in general, $\mathbf{V} * \mathbf{G}$ is a proper subvariety of $\mathbf{V} \textcircled{m} \mathbf{G}$. The case of the variety $\mathbf{V} = \mathbf{J}$ seemed to be a really difficult problem. On the one hand, a deep result of Knast [13] showed that the variety **J** was not local and this led Margolis and the author to believe that the inclusion $\mathbf{J} \textcircled{m} \mathbf{G} \subset \mathbf{J} * \mathbf{G}$ might be proper. On the other hand, if the inclusion had been proper, power groups would have satisfied other (pseudo)identities than the ones of **BG**.

As it happens often with difficult problems, the solution came from a more general result, namely Ash's proof of the Rhodes "cover conjecture" [3, 4]. An account of

J. Fountain (ed.), Semigroups, Formal Languages and Groups, 33–47.

the discovery and the consequences of this result is given in [11]. The final solution given by Henckell and Rhodes [10, 11] combines Ash's theorem, derived categories and Knast's result with a subtle computation on weak inverses.

The article is organized as follows. After a very short section of general definitions, block groups are defined in section 3 and power groups in section 4. The Schützenberger product is introduced in section 5 and the part of the proof that makes use of language theory is presented in section 6. Section 7 is devoted to Malcev products and contains a short outline of the proof of Henckell and Rhodes. A consequence of the equality $\mathbf{BG} = \mathbf{PG}$ for ordered monoids is discussed in section 8 and connections with topology are briefly mentioned in section 9.

2. General definitions

We refer the reader to [5, 6, 12, 14, 22] for additional background material. Given a subset P of a monoid M, $\langle P \rangle$ denotes the submonoid of M generated by P. The exponent of a semigroup S, that is, the smallest integer n such that s^n is idempotent for all $s \in S$, is usually denoted ω. The wreath product of two monoids M and N is denoted $M \circ N$. Finally, $|X|$ denotes the number of elements of a finite set X.

3. Block groups

The term "block" refers to a little known result of R. L. Graham [7] on 0-simple semigroups.

Theorem 3.1 *Let S^0 be a 0-simple semigroup and let T be the subsemigroup of S^0 generated by the union of all maximal groups of S. Then the regular $\mathcal{D}$-classes B_1, ..., B_n of T have the following properties:*
(1) each B_i is generated by the union of its maximal subgroups,
(2) for all $i \neq j$, $B_iB_j = B_jB_i = 0$,
(3) for all $s \in S \setminus \bigcup_{1\leq i\leq n} B_i$, $s^2 = 0$.

The B_i's are called the *blocks* of S^0. It is easy to compute the blocks given the egg-box picture of S. You are given two different kind of tokens, stars and circles. First put a star and a circle in each $\mathcal{H}$-class which is a group (that is, containing an idempotent). Next, play the following game, as long as you can: each time there is a star in $H_{i,j}$ and circles in $H_{i,j'}$ and $H_{i',j}$, add a circle in $H_{i',j'}$.

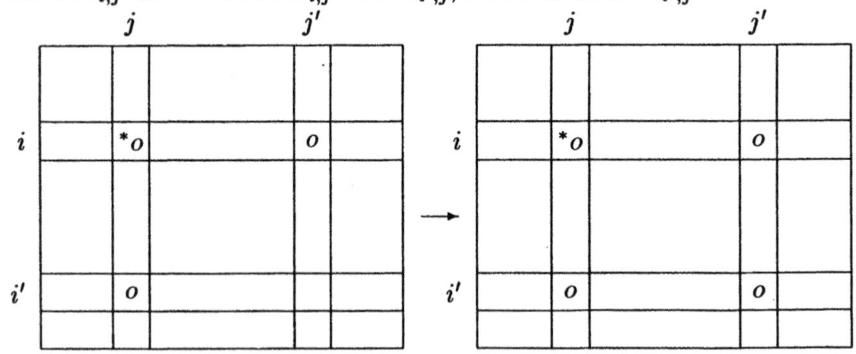

When the game is over, one gets the block structure of S.

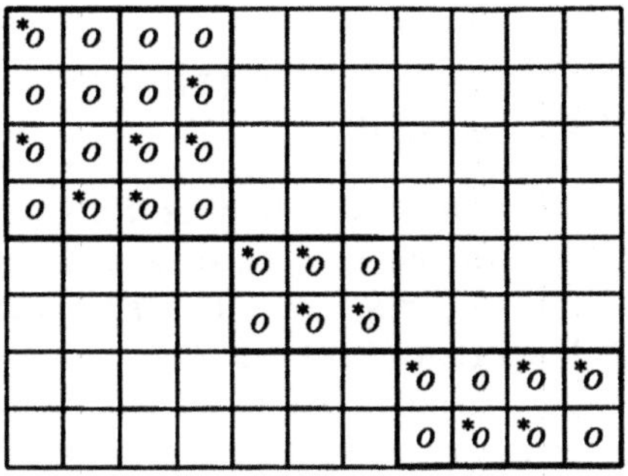

Figure 3.1. The block structure of a 0-simple semigroup.

Let D be a regular $\mathcal{D}$-class of a semigroup S. Then it is well-known that $D^0 = D \cup \{0\}$ is a 0-simple semigroup. Thus, one may define the ***blocks of the*** $\mathcal{D}$***-class*** D as the blocks of D^0. A ***block group*** is a monoid in which the blocks of every regular $\mathcal{D}$-class are groups. In this case, every element has at most one semigroup inverse and every regular $\mathcal{D}$-class can be represented as follows.

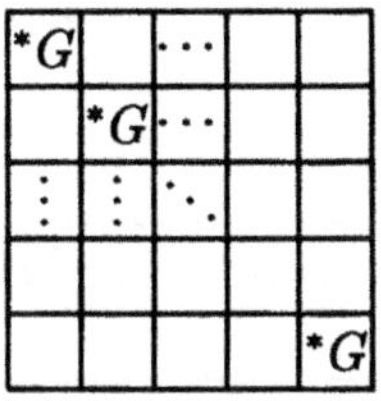

Figure 3.2. A regular $\mathcal{D}$-class of a block group.

Note that a regular monoid is a block group if and only if it is inverse. It follows that M is a block group if and only if every $\mathcal{R}$-class and every $\mathcal{L}$-class has at most one idempotent. The following theorem of [16] summarizes some equivalent formulations.

Theorem 3.2 *Let M be a monoid. The following conditions are equivalent:*
(1) M is a block group,
(2) for every regular $\mathcal{D}$-class D of M, D^0 is a Brandt semigroup,
(3) for all $e, f \in E(M)$, $e \,\mathcal{R}\, f$ implies $e = f$ and $e \,\mathcal{L}\, f$ implies $e = f$,
(4) for all $e, f \in E(M)$, $efe = e$ implies $ef = e = fe$.

One can also convert these conditions into equations as follows [16]. Recall that x^ω can be interpreted as the unique idempotent of the subsemigroup of M generated by x.

Theorem 3.3 *Let M be a monoid. The following conditions are equivalent:*
(1) M is a block group,
(2) for all $x, y \in M$, $(x^\omega y)^\omega = (yx^\omega)^\omega$,
(3) for all $x, y \in M$, $(x^\omega y^\omega)^\omega = (y^\omega x^\omega)^\omega$,
(4) for all $x, y \in M$, $(x^\omega y^\omega)^\omega x^\omega = (x^\omega y^\omega)^\omega = y^\omega (y^\omega x^\omega)^\omega$,
(5) the submonoid $\langle E(M) \rangle$ is $\mathcal{J}$-trivial.

The most difficult part of this theorem is (1) implies (5). One first shows by induction on n that if M is a block group and $e_1, \ldots, e_n$ are idempotents of M, then, for $0 \leq i \leq n$, $(e_1 \cdots e_n)^\omega = (e_1 \cdots e_n)^\omega e_1 \cdots e_i$. This shows that $\langle E(M) \rangle$ is $\mathcal{R}$-trivial and a symmetrical proof shows that $\langle E(M) \rangle$ is $\mathcal{L}$-trivial.

It is easy to see that the collection **BG** of all block groups is a variety. If one denotes by **EJ** the variety of monoids M such that $E(M)$ generates a $\mathcal{J}$-trivial monoid, we obtain our first equality:

Proposition 3.4 ***BG = EJ.***

4. Power groups

Given a monoid M, denote by $\mathcal{P}(M)$ (resp. $\mathcal{P}'(M)$) the monoid of subsets (resp. non empty subsets) of M under the multiplication of subsets, defined, for all $X, Y \subset M$ by $XY = \{xy \mid x \in X \text{ and } y \in Y\}$. For instance, if G is the trivial monoid, then $\mathcal{P}(M)$ is equal to the monoid $U_1 = \{1, 0\}$ with multiplication given by the rules $1.1 = 1$ and $1.0 = 0.1 = 0$. If M is the cyclic group of order 3, $\mathbb{Z}/3\mathbb{Z} = \{0, 1, -1\}$, then $\mathcal{P}'(M)$ has the following structure:

$^*\{1\}, \{0\}, \{-1\}$

$\{1,0\}, \{0,-1\}, \{1,-1\}$

$^*\{1,0,-1\}$

Figure 4.1. The $\mathcal{D}$-classes of $\mathcal{P}'(\mathbb{Z}/3\mathbb{Z})$.

The monoids $\mathcal{P}(M)$ and $\mathcal{P}'(M)$ are related as follows

Proposition 4.1 *For every monoid M, $\mathcal{P}'(M)$ is a submonoid of $\mathcal{P}(M)$ and $\mathcal{P}(M)$ divides $\mathcal{P}'(M) \times U_1$.*

The following results summarize the local structure of $\mathcal{P}'(M)$ when M is a group [20, 16].

Proposition 4.2 *Let G be a group and let $X, Y \in \mathcal{P}'(G)$. Then*
(1) the idempotents of $\mathcal{P}'(G)$ are the subgroups of G,

(2) $X \mathcal{R} Y$ *(resp.* $X \mathcal{L} Y$*) if and only if there exists* $g \in G$ *such that* $Xg = Y$ *(resp.* $gX = Y$*),*
(3) $X \mathcal{D} Y$ *if and only if there exist* $g_1, g_2 \in G$ *such that* $g_1 X g_2 = Y$*. In particular, if* $X \mathcal{D} Y$*, then* $|X| = |Y|$*.*

Corollary 4.3 *Let* G *be a group and let* H *and* K *be two subgroups of* G*.*
(1) H *and* K *are* $\mathcal{D}$*-equivalent in* $\mathcal{P}'(G)$ *if and only if they are conjugate.*
(2) Let D *be the* $\mathcal{D}$*-class of* $\mathcal{P}'(G)$ *containing* H*. Then* D^0 *is a Brandt semigroup of size* $|G : N(H)|$ *with structure group* $N(H)/H$*, where* $N(H)$ *is the normalizer of* H*.*

This last property shows that $\mathcal{P}'(G)$ is a block group [20].

Corollary 4.4 *If* G *is a group, then* $\mathcal{P}'(G)$ *and* $\mathcal{P}(G)$ *are block groups.*

Given a monoid M, denote by $\mathcal{P}_1(M)$ the submonoid of $\mathcal{P}(M)$ consisting of all subsets of M containing the identity.

Theorem 4.5 *For every monoid* M*,* $\mathcal{P}_1(M)$ *is a* $\mathcal{J}$*-trivial monoid.*

Proof. Let $X, Y \in \mathcal{P}_1(M)$. If $X \mathcal{J} Y$, then $AXB = Y$ and $CYD = X$ for some $A, B, C, D \in \mathcal{P}_1(M)$. In particular, A, B, C and D contain 1 and thus $X = 1X1 \subset AXB = Y$ and $Y = 1Y1 \subset CYD = X$. Thus $X = Y$. □

For a group G, there is a nice connection between $\mathcal{P}_1(G)$ and $\mathcal{P}'(G)$.

Proposition 4.6 *If* G *is a group, then* $\mathcal{P}'(G)$ *is a quotient of a semidirect product of the form* $\mathcal{P}_1(G) * G$*.*

Proof. Let G act on the left on $\mathcal{P}_1(G)$ by conjugacy. That is, set $g \cdot X = gXg^{-1}$ for all $g \in G$ and $X \in \mathcal{P}_1(G)$. This defines a semidirect product of $\mathcal{P}_1(G)$ by G. Now, one verifies that the map $\pi : \mathcal{P}_1(G) * G \rightarrow \mathcal{P}'(G)$ defined by $\pi(X, g) = Xg$ is a surjective morphism. □

Given a variety of monoids **V** denote by **PV** the variety generated by all monoids of the form $\mathcal{P}(M)$, where $M \in$ **V**. Thus in particular **PG** is the variety generated by all power monoids of groups. By Corollary 4.4, this variety is contained in **BG**. Denote by **J** $*$ **G** the variety generated by all semidirect products of a $\mathcal{J}$-trivial monoid by a group. Then **J** $*$ **G** contains U_1, since U_1 is $\mathcal{J}$-trivial, and by Propositions 4.5 and 4.6, it contains the monoids of the form $\mathcal{P}'(G)$, where G is a group. These results can be summarized as follows.

Proposition 4.7 *The following inclusions hold:* ***PG*** $\subset$ ***J*** $*$ ***G*** *and* ***PG*** $\subset$ ***BG****.*

The author is not aware of any direct proof of the opposite inclusions. The inclusion **J** $*$ **G** $\subset$ **PG** is proved in section 6 by arguments of language theory and a purely semigroup theoretic proof is still wanted.

5. Schützenberger product

We introduce in this section the variety $\Diamond\mathbf{G}$. Given a monoid M, $\mathcal{P}(M)$ is not only a monoid but also a semiring under union as addition and the product of subsets as multiplication.

Let $M_1, \ldots, M_n$ be monoids. Denote by M the product monoid $M_1 \times \cdots \times M_n$, by k the semiring $\mathcal{P}(M)$ and by $M_n(k)$ the semiring of square matrices of size n with entries in k. The Schützenberger product of $M_1, \ldots, M_n$, denoted $\Diamond_n(M_1, \ldots, M_n)$ is the submonoid of the multiplicative monoid $M_n(k)$ composed of all the matrices P satisfying the three following conditions:
(1) If $i > j$, $P_{i,j} = 0$.
(2) If $1 \le i \le n$, $P_{i,i} = \{(1, \ldots, 1, s_i, 1, \ldots, 1)\}$ for some $s_i \in S_i$.
(3) If $1 \le i \le j \le n$, $P_{i,j} \subset 1 \times \cdots \times 1 \times M_i \times \cdots \times M_j \times 1 \cdots \times 1$.
Condition (1) shows that the matrices of the Schützenberger product are upper triangular, condition (2) enables us to identify the diagonal coefficient $P_{i,i}$ with an element s_i of M_i and condition (3) shows that if $i < j$, $P_{i,j}$ can be identified with a subset of $M_i \times \cdots \times M_j$. With this convention, a matrix of $\Diamond_3(M_1, M_2, M_3)$ will have the form

$$\begin{pmatrix} s_1 & P_{1,2} & P_{1,3} \\ 0 & s_2 & P_{2,3} \\ 0 & 0 & s_3 \end{pmatrix}$$

with $s_i \in M_i$, $P_{1,2} \subset M_1 \times M_2$, $P_{1,3} \subset M_1 \times M_2 \times M_3$ and $P_{2,3} \subset M_2 \times M_3$.

Given a variety of monoids $\mathbf{V}$, $\Diamond\mathbf{V}$ denotes the variety of monoids generated by all Schützenberger products of the form $\Diamond_n(M_1, \ldots, M_n)$ with $n > 0$ and $M_1, \ldots, M_n \in \mathbf{V}$.

6. A tour of language theory

We present in this section a complete proof of the inclusions $\mathbf{J} * \mathbf{G} \subset \Diamond\mathbf{G}$ and $\Diamond\mathbf{G} \subset \mathbf{PG}$, first given in [16, 17]. The reader is referred to our article [28] in this volume for an introduction to language theory. The idea is the following: by Eilenberg's theorem, varieties of monoids are in one-to-one correspondence with varieties of languages, and thus it suffices to establish the inclusions $\mathcal{V}_1 \subset \mathcal{V}_2 \subset \mathcal{V}_3$, where $\mathcal{V}_1$, $\mathcal{V}_2$ and $\mathcal{V}_3$ denote respectively the varieties of languages corresponding to $\mathbf{J} * \mathbf{G}$, $\Diamond\mathbf{G}$ and $\mathbf{PG}$. We first describe the three varieties of languages $\mathcal{V}_1$, $\mathcal{V}_2$ and $\mathcal{V}_3$. Recall that a ***group language*** is a language recognized by a group.

Let us start with the variety $\mathcal{V}_2$. Given languages $L_0, L_1, \ldots, L_n$ of A^* and letters $a_1, a_2, \ldots, a_n$, the product of $L_0, \ldots L_n$ marked by $a_1, \ldots, a_n$ is the language $L_0a_1L_1a_2 \cdots a_nL_n$. It is known (see [28], Theorem 12.3) that the Schützenberger product is the algebraic operation on monoids that corresponds to the marked product. The following theorem is a simple application of this general result.

Theorem 6.1 *For all alphabets A, $\mathcal{V}_2(A^*)$ is the boolean algebra generated by all the marked products of the form $L_0a_1L_1 \cdots a_nL_n$ where $L_0, \ldots, L_n$ are group languages of A^* and $a_1, \ldots, a_n \in A$.*

Next, there exists a description of the variety of languages corresponding to **PV**, obtained independently by Reutenauer [32] and Straubing [36]. This result can be applied in particular to the variety **PG** as follows. A monoid morphism $\varphi : B^* \to A^*$ is *length preserving* if it maps each letter of B onto a letter of A.

Theorem 6.2 *The variety $\mathcal{V}_3$ is the smallest variety of languages $\mathcal{V}$ such that, for each alphabet A, $\mathcal{V}(A^*)$ contains the languages of the form $\varphi(L)$ where L is a group language of B^* and $\varphi : B^* \to A^*$ is a length preserving morphism.*

The variety $\mathcal{V}_1$ can be described by Straubing's "wreath product principle", which is a general construction to describe the languages recognized by the wreath product of two finite monoids. Let M be a monoid, G a group and let $\eta : A^* \to M \circ G$ be a monoid morphism. We denote by $\pi : M \circ G \to G$ the monoid morphism defined by $\pi(f, g) = g$ and we put $\varphi = \pi \circ \eta$. Thus φ is a monoid morphism from A^* into G. Let $B = G \times A$ and $\sigma : A^* \to B^*$ be the map (which is not a morphism!) defined by

$$\sigma(a_1 a_2 \cdots a_n) = (1, a_1)(\varphi(a_1), a_2) \cdots (\varphi(a_1 a_2 \cdots a_{n-1}), a_n).$$

Theorem 6.3 *If a language L is recognized by $\eta : A^* \to M \circ G$, then L is a finite boolean combination of languages of the form $X \cap \sigma^{-1}(Y)$, where $Y \subset B^*$ is recognized by M and where $X \subset A^*$ is recognized by G.*

We are now ready to prove the inclusions announced above.

Proposition 6.4 *$\mathbf{J} * \mathbf{G}$ is contained in $\Diamond\mathbf{G}$.*

Proof. Let $L \subset A^*$ be a language of $\mathcal{V}_1(A^*)$. Then L is recognized by a wreath product of the form $M \circ G$, where M is $\mathcal{J}$-trivial and G is a group. With the notation of Theorem 6.3, L is a finite boolean combination of languages of the form $X \cap \sigma^{-1}(Y)$, where $Y \subset B^*$ is recognized by M and $X \subset A^*$ is recognized by G. Since X is recognized by a group, $X \in \mathcal{V}_2(A^*)$. Thus it suffices to verify that $\sigma^{-1}(Y) \in \mathcal{V}_2(A^*)$. But since Y is recognized by a $\mathcal{J}$-trivial monoid, Simon's theorem (see section 7 of [28]) implies that Y is a boolean combination of languages of the form $B^*b_1B^*b_2B^* \cdots b_nB^*$, where the b_i's are letters of B. Now since σ^{-1} commutes with boolean operations, it remains to verify that the languages of the form $\sigma^{-1}(B^*b_1B^*b_2B^* \cdots b_nB^*)$ are in $\mathcal{V}_2(A^*)$. Set, for $1 \le i \le n$, $b_i = (g_i, a_i)$. Then we have

$$\begin{aligned}
\sigma^{-1}(B^*b_1B^*b_2B^* \cdots b_nB^*) &= \{u \in A^* \mid \sigma(u) \in B^*b_1B^*b_2B^* \cdots b_nB^*\} \\
&= \{u_0a_1u_1 \cdots a_nu_n \mid b_1 = (\varphi(u_0), a_1),\ b_2 = (\varphi(u_0a_1u_1), a_2),\ \ldots, \\
&\qquad b_n = (\varphi(u_0a_1u_1 \cdots a_{n-1}u_{n-1}), a_n)\} \\
&= \{u_0a_1u_1 \cdots a_nu_n \mid \varphi(u_0) = g_1,\ \varphi(u_0a_1u_1) = g_2,\ \ldots, \\
&\qquad \varphi(u_0a_1u_1 \cdots a_{n-1}u_{n-1}) = g_n\}
\end{aligned}$$

Setting $h_0 = g_1$, $h_1 = (g_1\varphi(a_1))^{-1}g_2, \ldots, h_{n-1} = (g_1\varphi(a_1)g_2\cdots g_{n-1}\varphi(a_{n-1}))^{-1}g_n$, and $H_i = \varphi^{-1}(h_i)$ for $0 \le i \le n$, one gets finally

$$\begin{aligned}
&\sigma^{-1}(B^*b_1B^*b_2B^*\cdots b_nB^*)\\
&\quad = \{u_0a_1u_1\cdots a_nu_n \mid \varphi(u_0) = h_0,\ \varphi(u_1) = h_1,\ \ldots,\ \varphi(u_{n-1}) = h_{n-1}\}\\
&\quad = H_0a_1H_1a_2\cdots H_{n-1}a_nA^*
\end{aligned}$$

Since the H_i and A^* are group languages, it follows that $\sigma^{-1}(B^*b_1B^*b_2B^*\cdots b_nB^*)$ belongs to $\mathcal{V}_2(A^*)$. □

Proposition 6.5 *$\Diamond G$ is contained in PG.*

Proof. It suffices to prove that every language of the form $L = L_0a_1L_1 \cdots a_nL_n$, where the a_i's are letters and the L_i's are group languages of A^*, can be written as $\varphi(L)$ where L is a group language of B^* and $\varphi : B^* \to A^*$ is a length preserving morphism. Let $\bar{A}$ be a copy of A and let $\varphi : (A \cup \bar{A})^* \to A^*$ be the length preserving morphism defined by $\varphi(a) = \varphi(\bar{a}) = a$ for every $a \in A$. Set, for $1 \le i \le n$, $K_i = \varphi^{-1}(L_i)$. Finally, fix a prime number p and let K be the set of all words $u \in (A \cup \bar{A})^*$ whose number of factorizations of the form $u_0\bar{a}_1u_1 \cdots \bar{a}_nu_n$ — where $u_0 \in K_0, \ldots, u_n \in K_n$ — is congruent to 1 modulo p. One can show [26] that K is a group language. We claim that $\varphi(K) = L$. Indeed K is contained in $K_0\bar{a}_1K_1 \cdots \bar{a}_nK_n$ by construction and thus $\varphi(K) \subset \varphi(K_0\bar{a}_1K_1 \cdots \bar{a}_nK_n) \subset L_0a_1L_1 \cdots a_nL_n = L$. Conversely, let $u \in L$. Then u admits a factorization of the form $u = u_0a_1u_1 \cdots a_nu_n$ where $u_0 \in K_0, \ldots, u_n \in K_n$. Let $v = u_0\bar{a}_1u_1 \cdots \bar{a}_nu_n$. Then $v \in K_0\bar{a}_1K_1 \cdots \bar{a}_nK_n$ and $\varphi(v) = u$. Furthermore, $u_0\bar{a}_1u_1 \cdots \bar{a}_nu_n$ is the unique factorization of v in $K_0\bar{a}_1K_1 \cdots \bar{a}_nK_n$ and thus $v \in K$. Therefore $L \subset \varphi(K)$ and thus L is equal to $\varphi(K)$, proving the claim and the theorem. □

Let us summarize the results obtained so far.

Corollary 6.6 *The following formulæ hold:* $\boldsymbol{PG} = \boldsymbol{J} * \boldsymbol{G} = \Diamond \boldsymbol{G} \subset \boldsymbol{BG} = \boldsymbol{EJ}$.

7. Malcev products

Given a variety of semigroups **V**, a monoid M is called a ***V**-extension of a group* G if there exists a surjective morphism $\pi : M \to G$ such that $\pi^{-1}(1) \in \mathbf{V}$. Given a variety of monoids **V**, the Malcev product $\mathbf{V} \circledM \mathbf{G}$ is the variety of monoids generated by **V**-extensions of groups. One can define alternatively the Malcev product by using *relational morphisms*. Recall that a relational morphism between monoids M and N is a function $\tau : M \to \mathcal{P}(N)$ such that:
(1) $\tau(m)\tau(n) \subset \tau(mn)$ for all $m, n \in M$,
(2) $\tau(m)$ is non-empty for all $m \in M$,
(3) $1 \in \tau(1)$.
We will be only interested in relational morphisms into groups in this paper. Note that if $\tau : M \to G$ is a relational morphism into a group G, then $\tau^{-1}(1)$ is a submonoid of M. Let **V** be a variety. Then a monoid M belongs to the Malcev

product $\mathbf{V} ⓜ \mathbf{G}$ if and only if there exists a relational morphism τ from M into a group G such that $\tau^{-1}(1) \in \mathbf{V}$.

The Malcev product is usually a more powerful operation than the semidirect product.

Theorem 7.1 *Let* $\mathbf{V}$ *be any variety. Then* $\mathbf{V} * \mathbf{G} \subset \mathbf{V} ⓜ \mathbf{G}$.

Proof. Let $M \in \mathbf{V} * \mathbf{G}$. Then M divides a monoid of the form $N * G$ where G is a group and $N \in \mathbf{V}$. Let $\pi : N * G \to G$ be the projection. Then $\pi^{-1}(1)$ is isomorphic to N and thus belongs to $\mathbf{V}$. Therefore $N * G$ is in $\mathbf{V} ⓜ \mathbf{G}$ and so is M, since M divides $N * G$. □

Corollary 7.2 $\mathbf{J} * \mathbf{G}$ *is contained in* $\mathbf{J} ⓜ \mathbf{G}$.

It is not difficult to see that a $\mathcal{J}$-trivial extension of a group is a block group.

Proposition 7.3 $\mathbf{J} ⓜ \mathbf{G}$ *is contained in* $\mathbf{BG}$.

Proof. Let M be a **J**-extension of a group G and let $\pi : M \to G$ be a surjective morphism such that $\pi^{-1}(1)$ is $\mathcal{J}$ trivial. Now since $\pi(\langle E(M)\rangle) = \langle 1\rangle = 1$, $\langle E(M)\rangle$ is contained in $\pi^{-1}(1)$ and hence it is $\mathcal{J}$-trivial. Thus, by Theorem 3.3, M is a block group. □

It is slightly more difficult to establish the opposite inclusion.

Proposition 7.4 $\mathbf{BG}$ *is contained in* $\mathbf{J} ⓜ \mathbf{G}$.

Proof. Let M be a block group and let $x \in M$. We claim that the partial action of x on M defined by

$$s \cdot x = \begin{cases} sx & \text{if } s \,\mathcal{R}\, sx \\ \text{undefined} & \text{otherwise} \end{cases}$$

is one-to-one. Indeed, suppose that $rx = sx = t$ for some $r, s, t \in M$. Then $r \,\mathcal{R}\, s \,\mathcal{R}\, t$ and there exist $y, z \in M$ such that $ty = r$ and $tz = s$. Let $Stab(r) = \{u \in M \mid ru = r\}$ be the right stabilizer of r and let G be its minimal ideal. Since $M \in \mathbf{BG}$, $Stab(r) \in \mathbf{BG}$ and thus G is a group. Let e be the identity of G. Then $re = r$ and hence

(1) $r \cdot xye = t \cdot ye = r \cdot e = r$,

(2) $r \cdot ezxye = r \cdot zxye = s \cdot xye = t \cdot ye = r \cdot e = r$.

Therefore, xye and $ezxye$ are in G and thus $e \,\mathcal{R}\, ez \,\mathcal{R}\, ezxye$. It follows that the $\mathcal{H}$-class of ez contains an idempotent. But since M is a block group, the unique idempotent in the $\mathcal{R}$-class of ez is e and hence ez is in G. This means that $r \cdot ez = r$, proving the claim, since $r \cdot ez = r \cdot z = s$.

Furthermore, $s \,\mathcal{R}\, sxy$ if and only if $s \,\mathcal{R}\, sx$ and $sx \,\mathcal{R}\, sxy$. Therefore,

(1) $$s \cdot xy = (s \cdot x) \cdot y \quad \text{for all } s, x, y \in M$$

Let now $\mathfrak{S}(M)$ be the symmetric group on M and let $\tau : M \to \mathfrak{S}(M)$ be the relation that associates to any $x \in M$ the set $\tau(x)$ of all permutations on M that extend the partial permutation on M defined by x. Then formula 1 shows that τ is a relational morphism.

It suffices now to verify that $K = \tau^{-1}(1)$ is a $\mathcal{J}$-trivial monoid. By definition, an element of K defines a partial identity on M. Let $x, y \in K$ and let $s = (xy)^\omega$. Then $s \,\mathcal{R}\, sx$, and since x defines a partial identity, it follows $sx = s$, that is $(xy)^\omega = (xy)^\omega x$. Thus K is an $\mathcal{R}$-trivial block group, and hence a $\mathcal{J}$-trivial monoid. □

The results presented so far were obtained by Margolis and the author in 1984 [16]. The last remaining inclusion, $\mathbf{J} \circledM \mathbf{G} \subset \mathbf{J} * \mathbf{G}$ was established seven years later by Henckell and Rhodes [10, 9]

Theorem 7.5 *$J \circledM G$ is contained in $J * G$.*

The proof relies on several deep results of semigroup theory: the theory of derived categories, Knast's result on graphs and Ash's proof of the Rhodes "cover" conjecture. It is detailed in [11] and thus we shall just give an outline.

Let M be a monoid in $\mathbf{J} \circledM \mathbf{G}$. By the results mentioned above, there exists a relational morphism τ from M into a group G such that $\tau^{-1}(1) \in \mathbf{J}$. One would like to keep $\tau^{-1}(1)$ as small as possible. Define the *kernel* $K(M)$ of M to be the intersection of the submonoids $\tau^{-1}(1)$ over all relational morphisms $\tau : M \to G$ into a group. A compactness argument shows the existence of a relational morphism τ from M into a group G such that $\tau^{-1}(1) = K(M)$. Thus, for this relational morphism, $\tau^{-1}(1)$ is clearly minimal. Actually a stronger result holds: one can show the existence of a relational morphism $\tau : M \to G$ such that, for all $g \in G$, $\tau^{-1}(g)$ is in some sense minimal. To make this definition precise, define a subset X of M to be *pointlike for* τ if there is a $g \in G$ such that $X \subset \tau^{-1}(g)$. A subset of M is called simply *pointlike* if it is pointlike for all relational morphisms from M into a group. Finally, call a relational morphism $\tau : M \to G$ *universal* if
(1) $\tau^{-1}(1) = K(M)$,
(2) any pointlike subset for τ is pointlike.
Again, the existence of universal morphisms for any monoid M is ensured by a compactness argument. Ash's theorems, first conjectured by Rhodes, give some important properties of these universal relational morphisms. The first theorem of Ash [3, 4] gives a precise description of the kernel of M.

An element $\bar{s}$ of a monoid M is a *weak inverse* of s if $\bar{s}s\bar{s} = \bar{s}$. A submonoid K of M is *closed under weak conjugacy* if for every $u \in K$ and every $\bar{s}, s \in M$ such that $\bar{s}$ is a weak inverse of s, $\bar{s}us \in K$ and $su\bar{s} \in K$. Given a monoid M, denote by $D(M)$ the smallest submonoid of M closed under weak conjugacy.

Theorem 7.6 *For every monoid M, $K(M) = D(M)$.*

The second theorem of Ash [3, 4] gives a deep characterization of pointlike sets. For $m \in M$, set

$$m^{(1)} = \{m\} \quad \text{and} \quad m^{(-1)} = \{\bar{m} \mid \bar{m} \text{ is a weak inverse of } m\}.$$

Theorem 7.7 *A subset X of a monoid M is pointlike if and only if there are elements $m_1, m_2, \dots m_n$ of M such that*

$$X \subset D(M)m_1^{(\varepsilon_1)}D(M)m_2^{(\varepsilon_2)}D(M)\cdots D(M)m_n^{(\varepsilon_n)}D(M)$$

where $\varepsilon_i \in \{1, -1\}$ for $1 \le i \le n$.

We now recall the definition of a derived category [18, 39]. Let $\tau : M \to G$ be a relational morphism into a group G. The *derived category* of τ is the (small) category $D(\tau)$ whose objects are the elements of G and whose arrows are the triples $(h, (m, g), hg)$ such that $g \in \tau(m)$. The *exponent of a finite category C* is the smallest integer ω such that m^ω is idempotent for every loop $m \in C$. The following result combines two results of Tilson [39] and Knast [13].

Theorem 7.8 *A monoid M belongs to $\mathbf{J} * \mathbf{G}$ if and only if there is a relational morphism $\tau : M \to G$ onto a finite group G whose derived category $D(\tau)$ satisfies the path identity $(ab)^\omega ad(cd)^\omega = (ab)^\omega(cd)^\omega$ for every subcategory of $D(\tau)$ of the form*

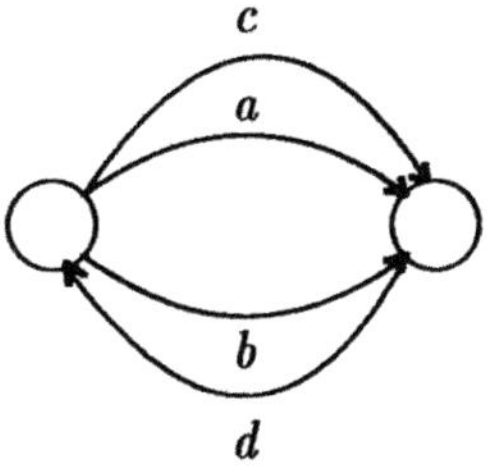

Proof. We have seen that if $M \in \mathbf{J} \circledM \mathbf{G}$, there exists a universal relational morphism $\tau : M \to G$ into a group G such that $\tau^{-1}(1) \in \mathbf{J}$. By Theorem 7.5 it suffices now to show that τ satisfies the path identity $(ab)^\omega ad(cd)^\omega = (ab)^\omega(cd)^\omega$. To verify this identity, consider a subgraph of $D(\tau)$ of the form shown in Figure 7.1

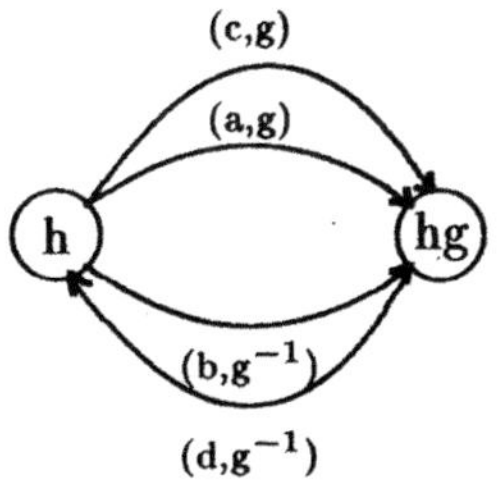

Figure 7.1.

In particular, we have $\{a, c\} \subset \tau^{-1}(g)$ and since τ is universal for M, $\{a, c\}$ is a pointlike subset of M. By Theorem 7.7, there exist elements $d_0, \dots, d_k, d'_0, \dots, d'_k$

of $D(M)$, elements $m_1, \dots, m_k$ of M and integers $\varepsilon_1, \dots, \varepsilon_k \in \{-1, 1\}$ such that

$$a \in d_0 m_1^{(\varepsilon_1)} d_1 m_2^{(\varepsilon_2)} \cdots d_{k-1} m_k^{(\varepsilon_k)} d_k \tag{2}$$

$$c \in d'_0 m_1^{(\varepsilon_1)} d'_1 m_2^{(\varepsilon_2)} \cdots d'_{k-1} m_k^{(\varepsilon_k)} d'_k \tag{3}$$

Thus there is a very loose relation between a and c. The rest of the proof, which is omitted here, amounts to showing that this relation suffices to establish the path identity. □

8. Block groups and ordered monoids

A consequence of the equality $\mathbf{BG} = \mathbf{PG}$ related to ordered monoids is presented in this section. Recall that an *ordered monoid* is a monoid equipped with a partial order relation $\leq$ such that $s \leq t$ implies $sx \leq tx$ and $xs \leq xt$ for all $x, s, t \in M$. An ordered monoid $(M, \leq)$ is called *E-ordered* if, for all $e \in E(M)$, $e \leq 1$. This definition is motivated by the following property of E-ordered monoids.

Proposition 8.1 *Every E-ordered monoid is a block-group.*

Proof. Let $(M, \leq)$ be an E-ordered monoid and let e and f be two $\mathcal{R}$-equivalent idempotents of M. Then $ef = f$ and $fe = e$. Now $e \leq 1$ implies $e = fe \leq f$ and $f \leq 1$ implies $f = ef \leq e$. Thus $e = f$. A dual argument would show that any pair of $\mathcal{L}$-equivalent idempotents are equal. Therefore M is a block group. □

The converse of Proposition 8.1 is false: there exist some block groups that cannot be E-ordered. However, a partial converse holds.

Theorem 8.2 *Every block group is the quotient of some E-ordered monoid.*

Proof. It is easy to verify that the monoids that are quotients of an E-ordered monoid form a variety $\mathbf{V}$. Now if G is a group, then $(\mathcal{P}'(G), \supset)$ is E-ordered. Indeed, by Proposition 4.2 (1), the idempotents of $\mathcal{P}'(G)$ are the subgroups of G, and they contain $\{1\}$, the identity of $\mathcal{P}'(G)$. It follows that $\mathbf{V}$ contains $\mathbf{PG}$ and thus $\mathbf{BG}$, since $\mathbf{PG} = \mathbf{BG}$. □

9. A topological consequence

The progroup topology on A^* is the smallest topology such that every monoid morphism from A^* onto a finite group G is continuous. Equivalently, the group languages form a basis for this topology. The progroup topology was first considered for the free group by M. Hall [8] and by Reutenauer for the free monoid [31, 33]. It is also connected to the study of implicit operations (see the article by J.Almeida and P.Weil elsewhere in this volume).

Some years ago, the author discovered [21, 26, 23] a curious connection between this topology and the Rhodes conjecture $K(M) = D(M)$. The main point was that the Rhodes conjecture reduces to computing the closure of a given recognizable language in this topology. Thus it became a natural and important question to characterize the recognizable closed (resp. open) languages. Several equivalent topological conjectures were presented in [26] and they were later shown to be all equivalent to the Rhodes conjecture [19].

In [29], Reutenauer and the author proposed another conjecture on the progroup topology of the free group and showed that this conjecture also implies the Rhodes conjecture. Again, it was later shown [11] that the Rhodes conjecture and the topological conjecture on the free group were equivalent. The interested reader is referred to the original articles and to the survey [11] for more details.

Finally, when Ash proved the Rhodes conjecture [3, 4], all these topological conjectures became theorems. One year later, a direct proof of the topological conjecture for the free group was obtained by Ribes and Zalesskii [35], giving in turn a new proof of Ash's theorem! The proof of Ribes and Zalesskii uses profinite groups acting on Boolean graphs.

We now come back to the original topological problem: characterize the recognizable closed (resp. open) languages. The author observed that the syntactic monoid of a recognizable closed (resp. open) language is a block group [21, 26]. The characterization of the closed and open recognizable sets is the following. Let L be a subset of A^*, let $\eta : A^* \to M(L)$ be the syntactic morphism of L and let $P = \eta(L)$ be the syntactic image of L. Then

Theorem 9.1 *(1) L is closed if and only if for every $s, t \in M$ and for every $e \in E(M)$, $set \in P$ implies $st \in P$.*
(2) L is open if and only if for every $s, t \in M$ and for every $e \in E(M)$, $st \in P$ implies $set \in P$.

The author recently obtained a more combinatorial description of the open sets [27].

Theorem 9.2 *A recognizable language is open if and only if it is a finite union of languages of the form $L_0a_1L_1 \cdots a_kL_k$ where the a_i's are letters and the L_i's are group languages.*

10. Conclusion and open problems

The sequence of equalities

$$\Diamond\mathbf{G} = \mathbf{PG} = \mathbf{J} * \mathbf{G} = \mathbf{J} \textcircled{\scriptsize M} \mathbf{G} = \mathbf{BG} = \mathbf{EJ}$$

is reminiscent of another sequence of equalities

$$\Diamond_2\mathbf{G} = \mathbf{Inv} = \mathbf{J}_1 * \mathbf{G} = \mathbf{J}_1 \textcircled{\scriptsize M} \mathbf{G} = \mathbf{Ecom} = \mathbf{EJ}_1$$

where $\Diamond_2\mathbf{G}$ denotes the variety generated by Schützenberger products of two groups, $\mathbf{J}_1$ the variety of idempotent and commutative monoids (or semilattices), **Ecom** the variety of monoids whose idempotent commute and **Inv** the variety generated by inverse monoids. Not only does the second sequence have several similarities with the first one, but its success story is also curiously analogous. Indeed, all but one of these equalities were obtained by Margolis and the author [18] and the last missing inclusion $\mathbf{Ecom} \subset \mathbf{J}_1 \textcircled{\scriptsize M} \mathbf{G}$ was obtained by Ash [1, 2].

Let us come back to the equality $\mathbf{BG} = \mathbf{PG}$. It implies in particular that any block group M divides a power group $\mathcal{P}(G)$. It would be interesting to find an upper bound on the size of G as a function of $|M|$. Of course, one could trace the proof of $\mathbf{BG} = \mathbf{PG}$, but with its detours into language theory, and the difficult theorem of Ash, it would give an enormous upper bound.

Another interesting problem is to obtain similar results when the variety $\mathbf{G}$ is replaced by a variety $\mathbf{H}$ of groups. The most interesting cases seem to be the varieties of commutative groups, p-groups (for a given prime p), nilpotent groups and solvable groups. It was for instance proved in [18] that the equalities $\mathbf{J}_1 * \mathbf{H} = \Diamond_2\mathbf{H} = \mathbf{J}_1 \circledM \mathbf{H}$ hold for any non trivial variety of groups, but it seems to be a very difficult problem to find an effective characterization of $\mathbf{PG}$ or to find out the varieties of groups for which the equality $\mathbf{J} * \mathbf{H} = \mathbf{J} \circledM \mathbf{H}$ holds. The topological approach seems to be the more promising way to solve this type of problem.

Acknowledgements

I would like to thank Pascal Weil for his comments on a preliminary version of this article.

References

1. C. J. Ash, *Finite idempotent-commuting semigroups*, in "Semigroups and their applications" (S. Goberstein and P. Higgins eds.), Reidel (Dordrecht, Holland), (1987) 13–23.
2. C. J. Ash, *Finite semigroups with commuting idempotents*, J. Austral. Math. Soc. (Series A) **43**, (1987) 81–90.
3. C. J. Ash, *Inevitable sequences and a proof of the type II conjecture*, in "Proceedings of the Monash Conference on Semigroup Theory", World Scientific, Singapore, (1991) 31–42.
4. C. J. Ash, *Inevitable Graphs: A proof of the type II conjecture and some related decision procedures*, Int. Jour. Alg. and Comp. **1** (1991) 127–146.
5. A. H. Clifford and G. B. Preston, *The Algebraic Theory of Semigroups*, Mathematical Surveys No. 7, American Mathematical Society, Providence, RI, Vol. 1, 1961; Vol. 2, 1967.
6. S. Eilenberg, *Automata, Languages and Machines*, Vol. A, Academic Press, New York, 1974. Vol. B, Academic Press, New York, 1976.
7. R. L. Graham, *On finite 0-simple semigroups and graph theory*, Math. Syst. Theory **2**, (1968) 325–339.
8. M. Hall Jr., *A topology for free groups and related groups*, Ann. of Maths **52**, (1950) 127–139.
9. K. Henckell, *Blockgroups = Power Groups*, in "Proceedings of the Monash Conference on Semigroup Theory", World Scientific, Singapore, (1991) 117–134.
10. K. Henckell and J. Rhodes, *The theorem of Knast, the PG = BG and Type II Conjectures*, in "Monoids and Semigroups with Applications" (J. Rhodes ed.), World Scientific, (1991) 453–463
11. K. Henckell, S. W. Margolis, J.-E. Pin and J. Rhodes, *Ash's Type II Theorem, Profinite Topology and Malcev Products*, International Journal of Algebra and Computation **1** (1991) 411–436.
12. J. M. Howie, *An introduction to Semigroup Theory*, Academic Press, London, 1976.
13. R. Knast, *Some theorems on graph congruences*, RAIRO Inform. Théor. **17** (1983) 331–342.
14. G. Lallement, *Semigroups and combinatorial applications*, Wiley, New York, 1979.
15. S. W. Margolis, *Kernels and expansions: an historical and technical perspective*, in "Monoids and Semigroups with Applications" (J. Rhodes ed.), World Scientific, (1991) 3–30.
16. S. W. Margolis and J.-E. Pin, *Varieties of finite monoids and topology for the free monoid*, in "Proceedings of the 1984 Marquette Conference on Semigroups" (K. Byleen, P. Jones and F. Pastijn eds.), Marquette University (1984) 113–130.

17. S. W. Margolis and J.-E. Pin, *Products of group languages*, FCT Conference, Lecture Notes in Computer Science **199** (1985) 285–299.
18. S. W. Margolis and J.-E. Pin, *Inverse semigroups and varieties of finite semigroups*, J. Algebra **110** (1987) 306–323.
19. S. W. Margolis and J.-E. Pin, *New results on the conjecture of Rhodes and on the topological conjecture*, J. Pure Appl. Algebra **80**, (1992) 305–313.
20. J.-E. Pin, *Variétés de langages et monoïde des parties*, Semigroup Forum **20** (1980) 11–47.
21. J.-E. Pin, *Finite group topology and p-adic topology for free monoids*, in "12th ICALP", Lecture Notes in Computer Science **194**, Springer, Berlin, (1985) 445–455.
22. J.-E. Pin, *Variétés de langages formels*, 160 p., Masson, Paris (1984). *Varieties of formal languages*, 138 p., North Oxford Academic (London), 1986 and Plenum (New York), 1986.
23. J.-E. Pin, *A topological approach to a conjecture of Rhodes*, Bulletin of the Australian Mathematical Society **38** (1988) 421–431.
24. J.-E. Pin, *Relational morphisms, transductions and operations on languages*, in "Formal properties of finite automata and applications" (J.-E. Pin ed.), Lect. Notes in Comp. Sci. **386**, Springer, (1989) 120–137.
25. J.-E. Pin, *On a conjecture of Rhodes*, Semigroup Forum **39** (1989) 1–15.
26. J.-E. Pin, *Topologies for the free monoid*, J. Algebra **137** (1991) 297–337.
27. J.E. Pin, *Polynomial closure of group languages and open sets of the Hall topology*, submitted.
28. J.-E. Pin, *Finite semigroups and recognizable languages: an introduction*, this volume.
29. J.-E. Pin and C. Reutenauer, *A conjecture on the Hall topology for the free group*, Bull. London Math. Soc. **23**, (1991) 356–362.
30. J. Reiterman, *The Birkhoff theorem for finite algebras*, Algebra Universalis, **14** (1982), 1–10.
31. Ch. Reutenauer, *Une topologie du monoïde libre*, Semigroup Forum **18** (1979) 33–49.
32. Ch. Reutenauer, *Sur les variétés de langages et de monoïdes*, Lecture Notes in Computer Science **67**, Springer, Berlin, (1979) 260–265.
33. Ch. Reutenauer, *Sur mon article "Une topologie du monoïde libre"*, Semigroup Forum **22** (1981) 93–95.
34. J. Rhodes, *Global structure theorems for arbitrary semigroups*, in "Proceedings of the 1984 Marquette Conference on Semigroups" (K. Byleen, P. Jones and F. Pastijn eds.) Marquette University (1984) 197–228.
35. L. Ribes and P. A. Zalesskii, *On the profinite topology on a free group*, Bull. London Math. Soc. **25** (1993) 37–43.
36. H. Straubing, *Recognizable sets and power sets of finite semigroups*, Semigroup Forum, **18** (1979) 331–340.
37. B. Tilson, *Monoid Kernels, Semidirect Products and their Adjoint Relationship*, in "Monoids and Semigroups with Applications" (J. Rhodes ed.), World Scientific, (1991) 31–54.
38. B. Tilson, *Type II redux*, in "Semigroups and Their Applications" (S. M. Goberstein and P. M. Higgins eds.), Reidel, Dordrecht, (1987) 201–205.
39. B. Tilson, *Categories as algebras: an essential ingredient in the theory of finite monoids*, J. Pure Appl. Algebra **48** (1987), 83–198.

SEMIGROUPS AND AUTOMATA ON INFINITE WORDS

DOMINIQUE PERRIN
Institut Gaspard Monge, Universté de Marne la Vallée, 93166 Noisy-le-Grand, France.

and

JEAN-ERIC PIN
LITP/IBP, Université Paris VI et CNRS, Tour 55-65, 4 Place Jussieu, 75252 Paris Cedex 05, France.

1. Introduction

This paper is an introduction to the algebraic theory of infinite words. Infinite words are widely used in computer science, in particular to model the behaviour of programs or circuits. From a mathematical point of view, they have a rich structure, at the cross-roads of logic, topology and algebra. This paper emphasizes the combinatorial and algebraic aspects of this theory but the interested reader is referred to the survey articles [32, 43] or to the report [30] for more information on the other aspects. In particular, the important topic of the complexity of the algorithms on infinite words is not treated in this paper.

The paper is written from the viewpoint of generalizing the results on recognizable sets of finite words to infinite words. This does not exactly follow the historical development of the theory, but it gives a good idea of the type of problems that occur in this field. Some of these problems are still open, or have been solved quite recently so that the definitions and results presented below may not be as yet finalized.

The first result to be generalized is the equivalence between finite automata, finite deterministic automata and rational expressions. If one adds infinite iteration (the "omega" operation) to the standard rational operations, union, product and star, one gets a natural definition of the ω-rational sets of infinite words that extends the definition of rational sets of finite words. Büchi [5] was the first to propose a definition of finite automata acting on infinite words. This definition suffices to extend Kleene's theorem to infinite words: the sets of infinite words recognized by finite Büchi automata are exactly the ω-rational sets. This result is now known as Büchi's theorem.

However, Büchi's definition is not totally satisfying since deterministic Büchi automata are not equivalent to non-deterministic ones. The connection between deterministic and non-deterministic Büchi automata was enlightened by a deep theorem of McNaughton: a set of infinite words is recognized by a non-deterministic Büchi automaton if and only if it is a finite boolean combination of sets recognized

J. Fountain (ed.), Semigroups, Formal Languages and Groups, 49–72.

by deterministic Büchi automata. This result was made possible by a suitable definition for automata on infinite words given by Muller [22]. These automata have the same power as the Büchi automata, but this time, non-deterministic automata are equivalent to deterministic ones.

It is a well known fact that finite semigroups can be viewed as a two-sided algebraic counterpart of finite automata that recognize finite words. Several attempts have been made to find an algebraic counterpart of finite automata that recognize infinite words. Since the notion of infinite word is asymmetrical, finite semigroups are not suitable any more. They can be replaced by ω-semigroups, which are, roughly speaking, semigroups equipped with an infinite product. The basic definitions are quite promising, but a technical difficulty arises almost immediately. Indeed, in order to design algorithms on finite ω-semigroups one needs a finite representation for them, but the definition of the infinite product apparently requires an infinite table, even for a 2-element ω-semigroup. However, a Ramsey-type argument shows that the structure of a finite ω-semigroup is totally determined by three operations of finite signature. That is, finite ω-semigroups are equivalent to certain finite algebras of finite signature, the Wilke algebras. Now, the definitions of a recognizable set, of a syntactic congruence, etc., become natural and most results valid for finite words can be adapted to infinite words. Carrying on the work of Arnold [1], Pécuchet [23, 24] and the first author [25, 26, 27], Wilke [44, 45] has pushed the analogy with the theory for finite words sufficiently far to obtain a counterpart of Eilenberg's variety theorem for finite or infinite words. This is approximately the point reached by the algebraic approach today, although current research may already have passed beyond. In any case, this is the place where our article finishes.

The paper is organized as follows: the basic definitions on words and ω-rational sets are given in sections 2 and 3. Büchi automata are defined in section 4, deterministic Büchi automata in section 5 and Muller automata in section 6. The equivalence between finite ω-semigroups and finite Wilke algebras is established in section 7 and their connections with automata are presented in sections 8, 9 and 10. Syntactic ω-semigroups are introduced in section 11 and the variety theorem and its consequences are discussed in section 12. Section 13 presents the conclusion of the article.

2. Words

Let A be a finite set called an *alphabet*, whose elements are *letters*. A *finite word* is a finite sequence of letters, that is, a function u from a finite set of the form $\{0, 1, 2, \dots, n\}$ into A. If one puts $u(i) = a_i$ for $0 \le i \le n$, the word u is usually denoted by $a_0a_1 \cdots a_n$, and the integer $|u| = n + 1$ is the *length* of u. The unique word of length 0 is the empty word, denoted by 1. An *infinite word* is a function u from N into A, usually denoted by $a_0a_1a_2 \cdots$, where $u(i) = a_i$ for all $i \in$ N. A *word* is either a finite word or an infinite word.

Intuitively, the *concatenation* or *product* of two words u and v is the word uv obtained by writing u followed by v. More precisely, if u is finite and v is finite or infinite, then uv is the word defined by

$$(uv)(i) = \begin{cases} u(i) & \text{if } i < |u| \\ v(i - |u|) & \text{if } i \geq |u| \end{cases}$$

We denote respectively by A^*, A^+, $A^{\mathbb{N}}$ the set of all finite words, finite non-empty words, and infinite words, respectively. We also denote by $A^\infty = A^+ \cup A^{\mathbb{N}}$ the set of all non empty words. A word x is a factor of a word w if there exist two words u and v (possibly empty) such that $w = uxv$.

3. Rational sets

The ***rational operations*** are the four operations union, product, plus and star, defined on the set of subsets of A^* as follows

(1) Union: $L_1 \cup L_2 = \{u \mid u \in L_1 \text{ or } u \in L_2\}$

(2) Product: $L_1 L_2 = \{u_1 u_2 \mid u_1 \in L_1 \text{ and } u_2 \in L_2\}$

(3) Plus: $L^+ = \{u_1 \cdots u_n \mid n > 0 \text{ and } u_1, \ldots, u_n \in L\}$

(3) Star: $L^* = \{u_1 \cdots u_n \mid n \geq 0 \text{ and } u_1, \ldots, u_n \in L\}$.

Thus we have the relations

$$L^+ = LL^* = L^*L \quad \text{and} \quad L^* = L^+ \cup \{1\}.$$

The set of rational subsets of A^* is the smallest set of subsets of A^* containing the finite sets and closed under finite union, product and star. For instance, $(a \cup ab)^* ab \cup (ba^*b)^*$ denotes a rational set.

Similarly, the set of rational subsets of A^+ is the smallest set of subsets of A^+ containing the finite sets and closed under finite union, product and plus. It is not difficult to verify that the rational subsets of A^+ are exactly the rational subsets of A^* that do not contain the empty word.

It is possible to generalize the concept of rational sets to infinite words as follows. First, the product can be extended, by setting, for $X \subset A^*$ and $Y \subset A^{\mathbb{N}}$,

$$XY = \{xy \mid x \in X \text{ and } y \in Y\}.$$

Next, we define an infinite iteration ω by setting, for every subset X of A^+

$$X^\omega = \{x_0 x_1 \cdots \mid \text{ for all } i \geq 0, x_i \in X\}$$

Equivalently, X^ω is the set of infinite words obtained by concatenating an infinite sequence of words of X. In particular, if $u = a_0 a_1 \cdots a_n$, we set

$$u^\omega = a_0 a_1 \cdots a_n a_0 a_1 \cdots a_n a_0 a_1 \cdots a_n a_0 a_1 \cdots$$

The set $\mathcal{R}at(A^\infty)$ of ω-***rational*** subsets of A^∞ is the smallest set $\mathcal{R}$ of subsets of A^∞ such that

(a) $\emptyset \in \mathcal{R}$ and for all $a \in A$, $\{a\} \in \mathcal{R}$,

(b) $\mathcal{R}$ is closed under finite union,

(c) For every subset X of A^+ and for every subset Y of A^∞, $X \in \mathcal{R}$ and $Y \in \mathcal{R}$ imply $XY \in \mathcal{R}$,

(d) For every subset X of A^+, $X \in \mathcal{R}$ implies $X^+ \in \mathcal{R}$ and $X^\omega \in \mathcal{R}$.

In other words, the set of ω-rational subsets of A^∞ is the smallest set of subsets of A^∞ containing the finite sets of A^+ and closed under finite union, finite product, plus and omega. The ω-rational sets which are contained in $A^{\mathbb{N}}$ are called, by abuse of language, the ω-rational subsets of $A^{\mathbb{N}}$. There is a very simple characterization of these sets, which is often used as a definition.

Proposition 3.1 *A subset of $A^{\mathbb{N}}$ is ω-rational if and only if it is a finite union of subsets of the form XY^ω where X and Y are non-empty rational subsets of A^+.*

Example 3.1 The set of infinite words on the alphabet $\{a, b\}$ having only a finite number of b's is given by the expression $\{a, b\}^* a^\omega$.

4. Automata

A finite (non-deterministic) automaton is a triple $\mathcal{A} = (Q, A, E)$ where Q is a finite set (the set of *states*), A is an alphabet, and E is a subset of $Q \times A \times Q$, called the set of *transitions*. Two transitions (p, a, q) and (p', a', q') are *consecutive* if $q = p'$. A *path* in $\mathcal{A}$ is a finite sequence of consecutive transitions

$$e_0 = (q_0, a_0, q_1), \quad e_1 = (q_1, a_1, q_2), \quad \ldots \quad , \quad e_{n-1} = (q_{n-1}, a_{n-1}, q_n)$$

also denoted

$$q_0 \xrightarrow{a_0} q_1 \xrightarrow{a_1} q_2 \quad \cdots \quad q_{n-1} \xrightarrow{a_{n-1}} q_n$$

The state q_0 is the *origin* of the path, the state q_n is its *end*, and the word $x = a_0 a_1 \cdots a_{n-1}$ is its *label*.

An *infinite path* in $\mathcal{A}$ is a sequence p of consecutive transitions indexed by $\mathbb{N}$.

$$e_0 = (q_0, a_0, q_1), \quad e_1 = (q_1, a_1, q_2), \quad \ldots$$

also denoted

$$q_0 \xrightarrow{a_0} q_1 \xrightarrow{a_1} q_2 \quad \cdots$$

The state q_0 is the *origin* of the infinite path and the infinite word $a_0 a_1 \cdots$ is its *label*. A state q occurs infinitely often in p if $q_n = q$ for infinitely many n.

Example 4.1 Let $\mathcal{A} = (Q, A, E)$ be the automaton represented in Figure 4.1. Then $Q = \{1, 2\}$, $A = \{a, b\}$ and $E = \{(1, a, 1), (2, b, 1), (1, a, 2), (2, b, 2)\}$ and

$$(1, a, 2)(2, b, 2)(2, b, 1)(1, a, 2)(2, b, 2)(2, b, 1)(1, a, 2)(2, b, 2)(2, b, 1)(1, a, 2)\cdots$$

is an infinite path of $\mathcal{A}$.

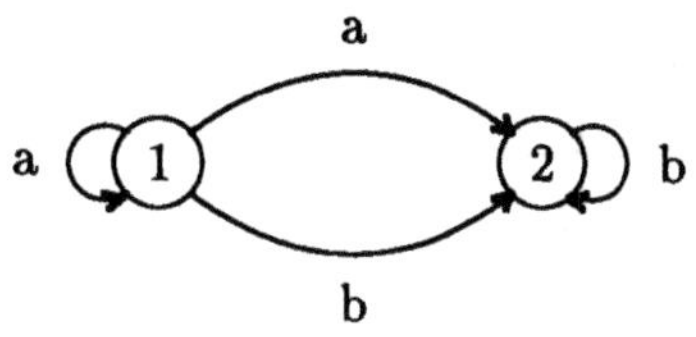

Figure 4.1. A finite automaton.

A finite *Büchi automaton* is a quintuple $\mathcal{A} = (Q, A, E, I, F)$ where
(1) (Q, A, E) is a finite automaton,
(2) I and F are subsets of Q, called the set of *initial* and *final* states, respectively. A finite path in $\mathcal{A}$ is *successful* if its origin is in I and its end is in F. An infinite path p is *successful* if its origin is in I and if some state of F occurs infinitely often in p.

The set of finite (respectively infinite) words *recognized* by $\mathcal{A}$ is the set, denoted $L^+(\mathcal{A})$ (respectively $L^\omega(\mathcal{A})$), of the labels of all successful finite (respectively infinite) paths of $\mathcal{A}$. A set of finite (respectively infinite) words X is *recognizable* if there exists a finite Büchi automaton $\mathcal{A}$ such that $X = L^+(\mathcal{A})$ (respectively $X = L^\omega(\mathcal{A})$).

Example 4.2 Let $\mathcal{A}$ be the Büchi automaton obtained from example 4.1 by taking $I = \{1\}$ and $F = \{2\}$. Initial states are represented by an incoming arrow and final states by an arrow going out. Then $L^+(\mathcal{A}) = a\{a,b\}^*$ is the set of all finite words

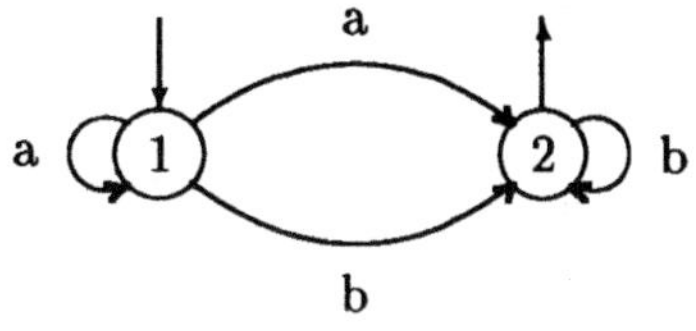

Figure 4.2. A Büchi automaton.

whose first letter is an a, $L^\omega(\mathcal{A}) = a(a^*b)^\omega$ is the set of infinite words whose first letter is an a and containing an infinite number of b's.

Example 4.3 Let $\mathcal{A}$ be the Büchi automaton represented in Figure 4.3. Then $L^\omega(\mathcal{A}) = A^*a^\omega$ is the set of all infinite words containing a finite number of b's. Let $\mathcal{A}'$ be the automaton obtained from $\mathcal{A}$ by taking only 1 as initial state. Then $L^\omega(\mathcal{A}') = A^*ba^\omega$.

Example 4.4 Let $\mathcal{A}$ be the Büchi automaton represented in Figure 4.4. Then $L^\omega(\mathcal{A}) = (a\{b,c\}^* \cup \{b\})^\omega$.

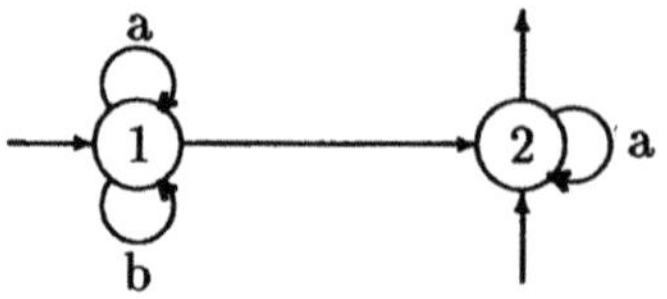

Figure 4.3. The automaton $\mathcal{A}$ recognizing A^*a^ω.

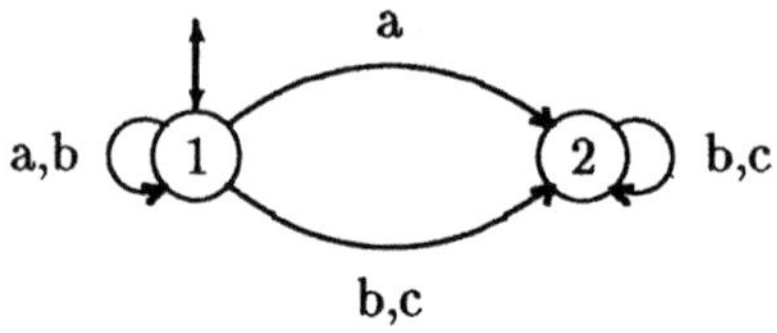

Figure 4.4.

The relationship between rational and recognizable sets of finite words is given by the famous theorem of Kleene.

Theorem 4.1 *A subset of A^* is rational if and only if it is recognizable.*

The counterpart of Kleene's theorem for infinite words is due to Büchi [4].

Theorem 4.2 *A subset of $A^{\mathbb{N}}$ is ω-rational if and only if it is recognizable.*

Sketch of the proof.
(1) Let $\mathcal{A} = (Q, A, E, I, F)$ be a finite Büchi automaton. For $p, q \in Q$, denote by $L^+(E, p, q)$ the set of non empty finite words recognized by the automaton $(Q, A, E, \{p\}, \{q\})$. By Theorem 4.1, all these sets are rational. Furthermore, it is not difficult to see that

$$X = L^\omega(\mathcal{A}) = \bigcup_{i \in I} \bigcup_{f \in F} L^+(E, i, f)(L^+(E, f, f))^\omega.$$

It follows that $L^\omega(\mathcal{A})$ is ω-rational.

(2) It is easy to see that recognizable sets are closed under finite union. Indeed, the disjoint union of two automata $\mathcal{A}_1$ and $\mathcal{A}_2$ recognizes exactly the union of the sets recognized respectively by $\mathcal{A}_1$ and $\mathcal{A}_2$. In order to prove that every ω-rational set is recognizable, it remains to show that every set of the form XX'^ω, where X and X' are non-empty rational subsets of A^+, is recognizable. A short combinatorial argument shows that every rational set can be recognized by a *normal* automaton, that is, an automaton with exactly one initial state i, exactly one final state f and no transition starting from f or ending in i. Now, if $\mathcal{A}$ and $\mathcal{A}'$ are normal automata recognizing X and X', then the automaton obtained by identifying f, i' and f', as indicated in Figure 4.5, recognizes XX'^ω.

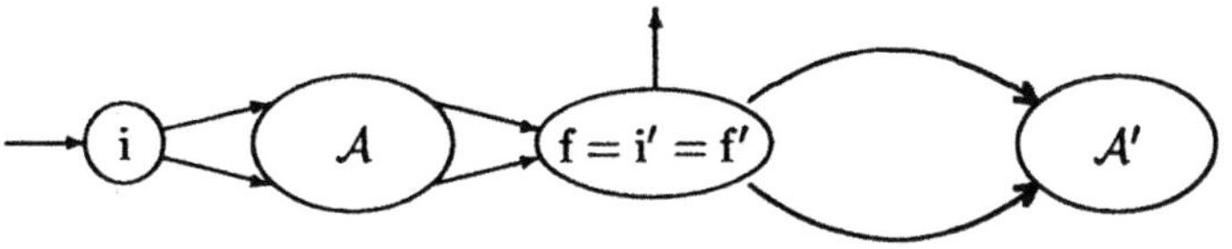

Figure 4.5. An automaton recognizing $X(X')^\omega$.

5. Deterministic Büchi automata

A Büchi automaton $\mathcal{A} = (Q, A, E, I, F)$ is said to be *deterministic* if I is a singleton and if E contains no pair of transitions of the form $(q, a, q_1), (q, a, q_2)$ with $q_1 \neq q_2$. In other words, given a state q and a letter a, there is at most one state q' such that $(q, a, q') \in E$. In particular, each word u is the label of at most one path starting from the initial state. A finite word u is accepted if the end of this unique path is a final state. And an infinite word is accepted if the unique path visits infinitely often a final state. It follows that an infinite word is accepted by $\mathcal{A}$ if and only if it has infinitely many prefixes accepted by $\mathcal{A}$. This leads to the following definition: for any subset L of A^*, put

$$\overrightarrow{L} = \{u \in A^{\mathbb{N}} \mid u \text{ has infinitely many prefixes in } L\}.$$

Then one can state

Proposition 5.1 *If $\mathcal{A}$ is a finite deterministic Büchi automaton, then* $L^\omega(\mathcal{A}) = \overrightarrow{L^+(\mathcal{A})}$.

Example 5.1 If $A = \{a, b\}$ and $L = a^*b$, then $\overrightarrow{L} = \emptyset$.

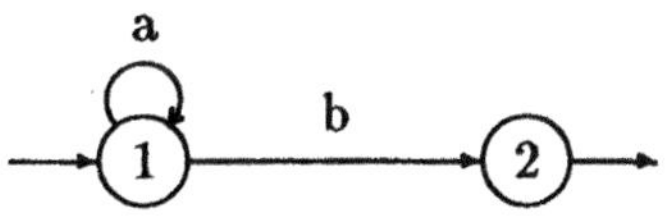

Figure 5.1. $L^+(\mathcal{A}) = a^*b$ and $L^\omega(\mathcal{A}) = \emptyset$.

Consider now the set L of all finite words of the form $a^n b$. Then $\overrightarrow{L} = (a^*b)^\omega$, the set of infinite words containing infinitely many b's (see Figure 5.2).

A set of infinite words is *deterministic* if it is recognized by a finite deterministic Büchi automaton. Landweber [14] has shown the existence of non-deterministic recognizable sets.

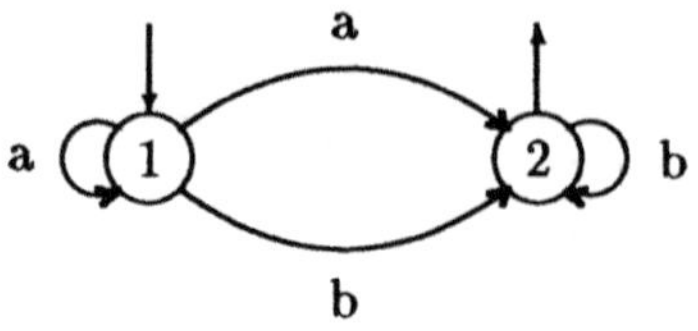

Figure 5.2. $L^+(\mathcal{A}) = (a^*b)^+ = \{a,b\}^*b$ and $L^\omega(\mathcal{A}) = (a^*b)^\omega$.

Proposition 5.2 *The set $\{a,b\}^*a^\omega$ is not deterministic.*

Proof. Let $X = \{a,b\}^*a^\omega$ and suppose that $X = \overrightarrow{L}$ for some subset L of A^*. Then ba^ω has a prefix $u_1 = ba^{n_1}$ in L and similarly, $ba^{n_1}ba^\omega$ has a prefix $u_2 = ba^{n_1}ba^{n_2}$ in L, etc. so that the infinite word $u = ba^{n_1}ba^{n_2}ba^{n_3}\cdots$ has infinitely many prefixes in L. Thus $u \in \overrightarrow{L}$, a contradiction, since u has an infinite number of b's. □

6. Muller automata

We have seen that non-deterministic Büchi automata are not equivalent to deterministic ones. This led Muller [22] to introduce a more satisfying definition.

A finite *Muller automaton* is a quintuple $\mathcal{A} = (Q, A, E, I, \mathcal{T})$ where
(1) (Q, A, E) is a finite automaton,
(2) I is a subset of Q, called the set of *initial* states,
(3) $\mathcal{T} = \{T_1, \dots, T_n\}$ (the state table) is a set of subsets $T_1, \dots, T_n$ of Q.
A path p is *successful* in $\mathcal{A}$ if Infinite(p) $\in \mathcal{T}$, where Infinite(p) is the set of states that are visited infinitely often by p. The set $L^\omega(\mathcal{A})$ (also denoted $L^\omega(E, I, \mathcal{T})$) of infinite words accepted by $\mathcal{A}$ is the set of labels of all successful paths.

Example 6.1 Let $A = \{a,b\}$. The automaton below, with the state table $\mathcal{T} = \{\{2\}\}$, recognizes the set A^*b^ω of all infinite words containing a finite number of a's. The automaton $\mathcal{A}'$ obtained by taking as state table $\mathcal{T}' = \{\{1\},\{2\}\}$, recognizes the

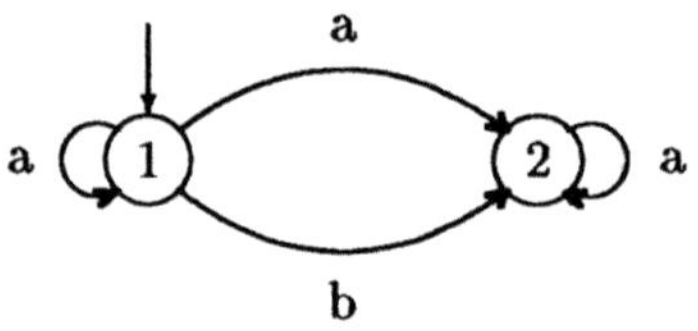

Figure 6.1. A Muller automaton.

set $A^*a^\omega \cup A^*b^\omega$.

A finite Muller automaton $\mathcal{A} = (Q, A, E, I, \mathcal{T})$ is *deterministic* if I is a singleton and if E contains no pair of transitions of the form $(q, a, q_1), (q, a, q_2)$ with $q_1 \neq q_2$.

This time, deterministic automata are equivalent to non-deterministic ones. The following result, mainly due to McNaughton [19], summarizes several equivalences.

Theorem 6.1 *Let X be a subset of $A^{\mathbb{N}}$. The following conditions are equivalent:*
(1) X is ω-rational,
(2) X is recognized by a finite Muller automaton,
(3) X is recognized by a finite deterministic Muller automaton,
(4) X is recognized by a finite Büchi automaton,
(5) X is a boolean combination of sets recognized by finite deterministic Büchi automata.

Sketch of the proof. (3) implies (2) is clear.
(2) implies (1). Let $\mathcal{A} = (Q, A, E, I, \mathcal{T})$ be a Muller automaton with $\mathcal{T} = \{T_1, \ldots, T_n\}$. Then it is easy to see that

$$L^\omega(\mathcal{A}) = \bigcup_{1 \leq j \leq n, i \in I} L^\omega(E, \{i\}, \{T_j\})$$

In other words, we may assume that the set of initial states is a singleton $\{i\}$ and that the table $\mathcal{T}$ contains only one set of states $T = \{t_0, \ldots, t_{k-1}\}$. Let $X' = L^+(E, i, t_0)$ and, for $0 \leq j \leq k-1$, let X_j be the set of labels of all finite paths from t_j to t_{j+1} visiting only states of $\mathcal{T}$ (where the indices j are calculated modulo k). Now a path p is successful if and only if it starts at i and Infinite(p) $= T$. In this case, p can be factorized as follows: a (finite) path from i to t_0, one from t_0 to t_1, one from t_1 to t_2, ..., one from t_{k-2} to t_{k-1}, one from t_{k-1} to t_0, one from t_0 to t_1, etc. Therefore

$$L^\omega(E, \{i\}, \{T\}) = X'(X_0 X_1 \cdots X_{k-1})^\omega$$

and thus $L^\omega(E, \{i\}, \{T\})$ is ω-rational.
(1) implies (4) follows from Theorem 4.2.
(4) implies (3) is the most difficult part of the proof. We shall present in section 10 a semigroup theoretic proof of this result. The best known algorithm to convert a Büchi automaton into a deterministic Muller automaton is Safra's construction, which, given an n-state Büchi automaton, produces an equivalent deterministic Muller automaton with at most n^n states. The reader is referred to the original article of Safra [36] or to [30] for a presentation of this difficult algorithm.

Thus conditions (1) to (4) are equivalent. We conclude by showing that (2) implies (5) and (5) implies (1).
(3) implies (5). Let $\mathcal{A} = (Q, A, E, \{i\}, \mathcal{T})$ be a deterministic Muller automaton. As above, we may assume that $\mathcal{T} = \{T\}$. Then, for an infinite path p, the condition Infinite(p) $= T$ is equivalent to the conjunction of the two conditions
(1) p visits every $t \in T$ infinitely often,
(2) p visits every $t \notin T$ finitely often.
This can be converted into the following formula, which gives (5)

$$L^\omega(E, \{i\}, \{T\}) = \bigcap_{t \in T} L^\omega(E, \{i\}, t) \setminus \bigcap_{t \notin T} L^\omega(E, \{i\}, t).$$

(5) implies (1). By Theorem 4.2, every set recognized by a finite deterministic Büchi automaton is ω-rational. Furthermore, since (1) and (3) are equivalent, ω-rational sets are closed under boolean operations.

Corollary 6.2 *The recognizable sets of $A^{\mathbb{N}}$ are closed under finite boolean operations, morphisms and inverse morphisms between free semigroups.*

7. ω-semigroups

As is shown elsewhere in this volume [33], one can use finite semigroups in place of automata to define the recognizable sets of finite words. It is possible to extend this idea to infinite words by replacing semigroups by ω-semigroups, which are, basically, algebras in which infinite products are defined. Although these algebras do not have a finitary signature, standard results on algebras still hold. In particular, A^∞ behaves like the free algebra on the set A and recognizable sets can be defined, as before, as the sets recognized by finite algebras. However, a problem arises since these finite algebras have an infinitary signature and thus are not finite objects. This problem can be solved by a Ramsey type argument showing that the structure of these finite algebras is totally determined by only three operations of finite signature. This defines a new type of algebras of finite signature, the Wilke algebras, that suffice to deal with infinite products. [1]

We now come to the precise definitions. An *ω-semigroup* is a two-sorted algebra $S = (S_f, S_\omega)$ equipped with the following operations:

(a) A binary operation defined on S_f and denoted multiplicatively,

(b) A mapping $S_f \times S_\omega \to S_\omega$, called *mixed product*, that associates to each pair $(s,t) \in S_f \times S_\omega$ an element of S_ω denoted st,

(c) A mapping $\pi : S_f^{\mathbb{N}} \to S_\omega$, called *infinite product*.

These three operations satisfy the following properties :
(1) S_f, equipped with the binary operation, is a semigroup,
(2) for every $s,t \in S_f$ and for every $u \in S_\omega$, $s(tu) = (st)u$,
(3) for every increasing sequence $(k_n)_{n>0}$ and for all $(s_n)_{n\in\mathbb{N}} \in S_f^{\mathbb{N}}$,

$$\pi(s_0 s_1 \cdots s_{k_1-1}, s_{k_1} s_{k_1+1} \cdots s_{k_2-1}, \ldots) = \pi(s_0, s_1, s_2, \ldots),$$

(4) for every $s \in S_f$ and for every $(s_n)_{n\in\mathbb{N}} \in S_f^{\mathbb{N}}$

$$s\pi(s_0, s_1, s_2, \ldots) = \pi(s, s_0, s_1, s_2, \ldots).$$

Conditions (1) and (2) can be thought of as an extension of associativity. Conditions (3) and (4) show that one can replace the notation $\pi(s_0, s_1, s_2, \ldots)$ by the notation

[1] Actually, the chronology is a little bit different. Ramsey type arguments have been used for a long time in semigroup theory [18, 39, 10], the Wilke algebras were introduced by Wilke in [44] under the name of *binoids* to clarify the approach of Arnold [1], Pécuchet [24] and Perrin [25] and the idea of using infinite products on semigroups came last [30].

$s_0s_1s_2\cdots$ without ambiguity . We shall use this simplified notation in the sequel. Intuitively, an ω-semigroup is a sort of semigroup in which infinite products are defined.

A ω-semigroup $S = (S_f, S_\omega)$ is *complete* if every element of S_ω can be written as an infinite product of elements of S_f. It is a harmless hypothesis to assume that all the ω-semigroups considered in this paper are complete.

Example 7.1 We denote by A^∞ the ω-semigroup $(A^+, A^{\mathbb{N}})$ equipped with the usual concatenation product.

Given two ω-semigroups $S = (S_f, S_\omega)$ and $T = (T_f, T_\omega)$, a *morphism of ω-semigroups* S is a pair $\varphi = (\varphi_f, \varphi_\omega)$ consisting of a semigroup morphism $\varphi_f : S_f \to T_f$ and of a mapping $\varphi_\omega : S_\omega \to T_\omega$ preserving the mixed product and the infinite product: for every sequence $(s_n)_{n\in\mathbb{N}}$ of elements of S_f,

$$\varphi_\omega(s_0s_1s_2\cdots) = \varphi_f(s_0)\varphi_f(s_1)\varphi_f(s_2)\cdots$$

and for every $s \in S_f$, $t \in S_\omega$,

$$\varphi_f(s)\varphi_\omega(t) = \varphi_\omega(st).$$

In the sequel, we shall often omit the subscripts, and use the simplified notation φ instead of φ_f and φ_ω.

Algebraic concepts like ω-subsemigroup, quotient and division are easily adapted to ω-semigroups. The semigroup A^+ is called the *free semigroup* on the set A because it satisfies the following property (which defines free objects in the general setting of category theory): every map from A into a semigroup S can be extended *in a unique way* into a semigroup morphism from A^+ into S. Similarly, it is not difficult to see that the free ω-semigroup on $(A, \emptyset)$ is the ω-semigroup A^∞.

A key result is that when S is finite, the infinite product is totally determined by the elements of the form $s^\omega = sss\cdots$, according to the following result of Wilke [45].

Theorem 7.1 *Let S_f be a finite semigroup and let S_ω be a finite set. Suppose that there exists a mixed product $S_f \times S_\omega \to S_\omega$ and a map from S_f into S_ω, denoted $s \to s^\omega$, satisfying, for every $s, t \in S_f$, the equations*

$$s(ts)^\omega = (st)^\omega$$
$$(s^n)^\omega = s^\omega \quad \textit{for every } n > 0.$$

Then the pair $S = (S_f, S_\omega)$ can be equipped, in a unique way, with the structure of an ω-semigroup such that for every $s \in S$, the product $sss\cdots$ is equal to s^ω.

This is a non trivial result, based on a consequence of Ramsey's theorem which is worth mentioning:

Theorem 7.2 *Let $\varphi : A^+ \to S$ be a morphism from A^+ into a finite semigroup S. For every infinite word $u \in A^{\mathbb{N}}$, there exist a pair (s, e) of elements of S such that $se = s$, $e^2 = e$, and a factorization $u = u_0u_1\cdots$ of u as a product of words of A^+ such that $\varphi(u_0) = s$ and $\varphi(u_n) = e$ for every $n > 0$.*

This motivates the following definition: a ***linked pair*** in a finite semigroup S is a pair $(s, e) \in S \times S$ such that e is idempotent and $se = s$. Two linked pairs (s, e) and (s', e') are *conjugate* (notation $(s, e) \sim (s', e')$) if there exist $x, y \in S^1$ such that $e = xy$, $e' = yx$ and $s' = sx$. Note that these conditions also imply $s = s'y$, since $s'y = sxy = se = s$.

A ***Wilke algebra*** is a two-sorted algebra (S_f, S_ω) equipped with three operations: an associative product on S_f, a mixed product $S_f \times S_\omega \to S_\omega$ and a map from S_f into S_ω, denoted $s \to s^\omega$, satisfying, for every $s, t \in S_f$, the equations

$$s(ts)^\omega = (st)^\omega$$
$$(s^n)^\omega = s^\omega \quad \text{for every } n > 0.$$

A Wilke algebra is *complete* if every element of S_ω can be written as st^ω for some $s, t \in S_f$. Theorem 7.1 states that every finite ω-semigroup is equivalent to a finite Wilke algebra.

One can attach to every finite sèmigroup S a finite complete ω-semigroup $\bar{S} = (S, S_\omega)$, constructed as follows. Denote by π the exponent of S, that is, the smallest positive integer n such that s^n is idempotent for all $s \in S$. Let $[s, e]$ be the conjugacy class of a linked pair (s, e). One defines S_ω as the set of conjugacy classes of linked pairs of S. The pair $\bar{S}$ is equipped with the structure of a Wilke algebra by setting, for all $[s, e] \in S_\omega$ and $t \in S$,

$$t[s, e] = [ts, e] \quad \text{and} \quad t^\omega = [t^\pi, t^\pi].$$

The first operation is well defined, since if (s', e') and (s, e) are conjugate linked pairs, then (ts', e') and (ts, e) are conjugate linked pairs. One can prove that $\bar{S}$ is a complete Wilke algebra such that every semigroup morphism $\varphi : A^+ \to S$ can be extended into a morphism of ω-semigroups $\varphi : A^\infty \to \bar{S}$ defined by $\bar{\varphi}_f = \varphi$ and $\bar{\varphi}_\omega(u) = [s, e]$, where (s, e) is a linked pair associated with u.

The ω-semigroup $\bar{S}$ has the following universal property.

Proposition 7.3 *Let $T = (T_f, T_\omega)$ be a finite ω-semigroup and let S be a finite semigroup. For every semigroup morphism φ from S into T_f, there exists a unique morphism of ω-semigroups $\bar{\varphi} : \bar{S} \to T$ that extends φ. Furthermore, if $\varphi(S)$ generates T as an ω-semigroup, then $\bar{\varphi}$ is onto.*

In particular, every finite ω-semigroup is a quotient of an ω-semigroup of the form $\bar{S}$.

Corollary 7.4 *Every finite complete ω-semigroup $S = (S_f, S_\omega)$ is a quotient of $\bar{S}_f$.*

Example 7.2 Let $S = \{a, b\}$ be the finite semigroup given by the multiplication table $aa = a$, $ab = a$, $ba = b$ and $bb = b$. There are four linked pairs and two conjugacy classes: $(a, a) \sim (a, b)$ and $(b, a) \sim (b, b)$. Thus, $S_\omega = \{[a, a], [b, b]\}$ and the ω-power and the mixed product are given in the following tables

$$a^\omega = [a, a] \qquad\qquad b^\omega = [b, b]$$
$$a[a, a] = a[b, b] = [a, a] \qquad b[a, a] = b[b, b] = [b, b].$$

Example 7.3 Let S be the five element Brandt semigroup: S is the semigroup with zero generated by the set $\{a, b\}$ with relations $a^2 = 0$, $b^2 = 0$, $aba = a$ and $bab = b$. Thus $S = \{a, b, ab, ba, 0\}$ and the $\mathcal{D}$-class structure of S is represented in the figure below:

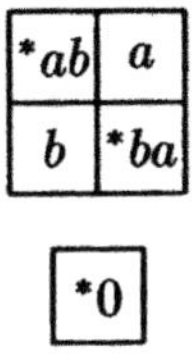

There are seven linked pairs and four conjugacy classes: $(0, ab) \sim (0, ba)$, $(a, ba) \sim (ab, ab)$, $(b, ab) \sim (ba, ba)$ and $(0, 0)$. Thus, $S_\omega = \{[ab, ab], [ba, ba], [0, ab], [0, 0]\}$ and the ω-power and the mixed product are given in the following tables

$$(ab)^\omega = [ab, ab] \qquad (ba)^\omega = [ba, ba] \qquad a^\omega = b^\omega = [0, 0]$$

$$\begin{array}{llll}
a[ab,ab] = [0,ab] & a[ba,ba] = [ab,ab] & a[0,ab] = [0,ab] & a[0,0] = [0,0] \\
b[ab,ab] = [ba,ba] & b[ba,ba] = [0,ab] & b[0,ab] = [0,ab] & b[0,0] = [0,0] \\
ab[ab,ab] = [ab,ab] & ab[ba,ba] = [0,ab] & ab[0,ab] = [0,ab] & ab[0,0] = [0,0] \\
ba[ab,ab] = [0,ab] & ba[ba,ba] = [ba,ba] & ba[0,ab] = [0,ab] & ba[0,0] = [0,0] \\
0[ab,ab] = [0,ab] & 0[ba,ba] = [0,ab] & 0[0,ab] = [0,ab] & 0[0,0] = [0,0].
\end{array}$$

A surjective morphism of ω-semigroups $\varphi : A^\infty \to S$ *recognizes a subset* X of $A^{\mathbb{N}}$ if there exists a subset P of S_ω such that $X = \varphi^{-1}(P)$. By extension, an ω-semigroup S recognizes X if there exists a surjective morphism of ω-semigroups $\varphi : A^\infty \to S$ that recognizes X. This definition can be extended to subsets of A^∞ as follows. A morphism of ω-semigroups $\varphi : A^\infty \to S$ *recognizes a subset* X of A^∞ if and only if $\varphi^{-1}\varphi(X) = X$, that is, if there exists a subset $P = (P_f, P_\omega)$ of (S_f, S_ω) such that $X \cap A^+ = \varphi_f^{-1}(P_f)$ and $X \cap A^\infty = \varphi_\omega^{-1}(P_\omega)$. A consequence of Corollary 7.4 is that every subset of A^∞ recognized by a finite ω-semigroup is recognized by an ω-semigroup of the form $\bar{S}$.

Proposition 7.5 *Let $S = (S_f, S_\omega)$ be a finite ω-semigroup recognizing a subset X of A^∞. Then $\bar{S}_f$ also recognizes X.*

Example 7.4 Let $S = (\{a, b\}, \{a^\omega, b^\omega\})$, where the finite product is given by the equalities $aa = a$, $ab = a$, $ba = b$, $bb = b$ and the infinite product is given by the rules $ax_1x_2\cdots = a^\omega$ and $bx_1x_2\cdots = b^\omega$. One recognizes here the ω-semigroup of Example 7.2. Let $\varphi : A^\infty \to S$ be the morphism of ω-semigroups defined by $\varphi(a) = a$ and $\varphi(b) = b$. Then $\varphi^{-1}(a^\omega) = aA^\omega$.

As for finite words, the following theorem holds.

Theorem 7.6 *A subset of $A^{\mathbb{N}}$ is recognizable if and only if it is recognized by a finite ω-semigroup.*

In other words, finite ω-semigroups can be converted into automata and vice versa. It is not difficult to see that a subset of $A^{\mathbb{N}}$ recognized by a finite ω-semigroup is ω-rational and thus, by Theorem 4.2, recognizable. Indeed, if $\varphi : A^\infty \to S$ recognizes X, then X is a finite union of sets of the form $\varphi^{-1}(s)\varphi^{-1}(e)^\omega$. The algorithm to pass from a finite ω-semigroup to a Büchi automaton presented in section 9 gives a direct proof of this result, without using Theorem 4.2. In the opposite direction, an algorithm to pass from Büchi automata to ω-semigroups is presented in section 8. Finally, the algorithm to pass from a finite ω-semigroup to a Muller automaton given in section 10 gives a proof of McNaughton's theorem.

8. From Büchi automata to ω-semigroups

We give in this section the construction which allows us to pass from a finite (Büchi) automaton to a finite ω-semigroup. This construction is much more involved than the corresponding construction for finite words.

Given a finite Büchi automaton $\mathcal{A} = (Q, A, E, I, F)$ recognizing a subset X of $A^{\mathbb{N}}$, we would like to obtain a finite ω-semigroup recognizing X. Our construction makes use of the semiring $k = \{-\infty, 0, 1\}$ in which addition is the maximum under the ordering $-\infty < 0 < 1$ and multiplication, which extends the boolean addition, is given in the following table

	$-\infty$	0	1
$-\infty$	$-\infty$	$-\infty$	$-\infty$
0	$-\infty$	0	1
1	$-\infty$	1	1

To each letter $a \in A$ is associated a matrix $\mu(a)$ with entries in k defined by

$$\mu(a)_{p,q} = \begin{cases} -\infty & \text{if } (p,a,q) \notin E \\ 0 & \text{if } (p,a,q) \in E \text{ and } p \notin F \text{ and } q \notin F \\ 1 & \text{if } (p,a,q) \in E \text{ and } (p \in F \text{ or } q \in F). \end{cases}$$

We have already used a similar technique to encode automata, but now we want to keep track of the visits to final states. We would like to extend μ into a morphism of ω-semigroups. It is easy to extend μ to a semigroup morphism from A^+ to the multiplicative semigroup of $Q \times Q$-matrices over k. If u is a finite word, one gets

$$\mu(u)_{p,q} = \begin{cases} -\infty & \text{if there exists no path of label } u \text{ from } p \text{ to } q, \\ 1 & \text{if there exists a path from } p \text{ to } q \text{ with label } u \\ & \text{going through a final state,} \\ 0 & \text{if there exists a path from } p \text{ to } q \text{ with label } u \\ & \text{but no such path goes through a final state.} \end{cases}$$

However, trouble arises when one tries to equip $k^{Q \times Q}$ with the structure of an ω-semigroup. The solution consists in coding infinite paths not by square matrices, but by column matrices, in such a way that each coefficient $\mu(u)_p$ codes the existence of an infinite path of label u starting at p.

Let $S = (S_f, S_\omega)$ where $S_f = k^{Q \times Q}$ is the set of square matrices of size $Card\, Q$ with entries in k and $S_\omega = k^Q$ is the set of column matrices with entries in $\{-\infty, 1\}$.

In order to define the operation ω on square matrices, we need the following definition. If s is a matrix of S_f, we call a sequence $p = p_0, p_1, \ldots$ of elements of Q an *infinite s-path starting at p* if for $0 \leq i \leq n-1$, $s_{p_i,p_{i+1}} \neq -\infty$.

The s-path p is *successful* if $s_{p_i,p_{i+1}} = 1$ for an infinite number of coefficients. Then s^ω is the element of S_ω defined, for every $p \in Q$, by

$$s^\omega_p = \begin{cases} 1 & \text{if there exists a successful } s\text{-path of origin } p, \\ -\infty & \text{otherwise.} \end{cases}$$

Note that the coefficients of this matrix can be effectively computed. Indeed, computing s^ω_p amounts to checking the existence of circuits containing a given edge in a finite graph. Then one can verify that S, equipped with these operations, is an ω-semigroup. The morphism μ can now be extended in a unique way to a morphism of ω-semigroups from A^∞ into S. Furthermore, we have the following result.

Proposition 8.1 *The morphism of ω-semigroups from A^∞ into S induced by μ recognizes the set $L^\omega(\mathcal{A})$.*

The ω-semigroup $\mu(A^\infty)$ is called the ω-semigroup *associated with* $\mathcal{A}$.

Example 8.1 Let $X = (a\{b,c\}^* \cup \{b\})^\omega$. This set is recognized by the Büchi automaton represented in Figure 8.1. The ω-semigroup associated with this automa-

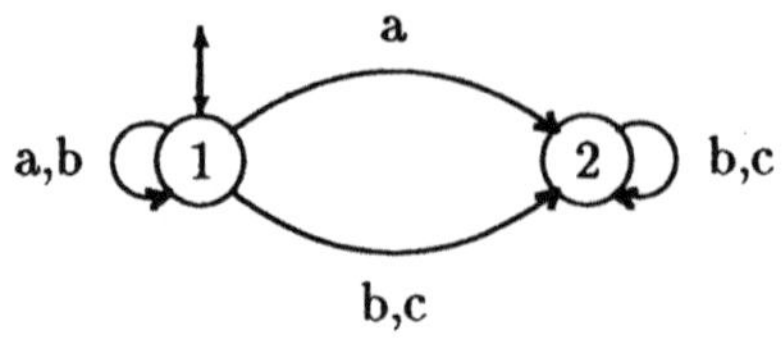

Figure 8.1.

ton contains 9 elements

$$a = \begin{pmatrix} 1 & 1 \\ -\infty & -\infty \end{pmatrix} \qquad b = \begin{pmatrix} 1 & -\infty \\ 1 & 0 \end{pmatrix} \qquad c = \begin{pmatrix} -\infty & -\infty \\ 1 & 0 \end{pmatrix}$$

$$ba = \begin{pmatrix} 1 & 1 \\ 1 & 1 \end{pmatrix} \qquad ca = \begin{pmatrix} -\infty & -\infty \\ 1 & 1 \end{pmatrix} \qquad a^\omega = \begin{pmatrix} 1 \\ -\infty \end{pmatrix}$$

$$b^\omega = \begin{pmatrix} 1 \\ 1 \end{pmatrix} \qquad c^\omega = \begin{pmatrix} -\infty \\ -\infty \end{pmatrix} \qquad (ca)^\omega = \begin{pmatrix} -\infty \\ 1 \end{pmatrix}$$

It is defined by the following relations

$$\begin{array}{lllll} a^2 = a & ab = a & ac = a & b^2 = b & bc = c \\ cb = c & c^2 = c & c^\omega = 0 & (ba)^\omega = b^\omega & aa^\omega = a^\omega \\ ab^\omega = a^\omega & a(ca)^\omega = a^\omega & ba^\omega = b^\omega & bb^\omega = b^\omega & b(ca)^\omega = (ca)^\omega \\ ca^\omega = (ca)^\omega & cb^\omega = (ca)^\omega & c(ca)^\omega = (ca)^\omega. & & \end{array}$$

9. From Wilke algebras to Büchi automata

Given a morphism of ω-semigroups recognizing a subset X of $A^{\mathbb{N}}$, we construct a Büchi automaton that recognizes X. By Proposition 7.5, we may suppose that the ω-semigroup is of the form $\bar{S} = (S, S_\omega)$. Thus, let $\varphi : A^\infty \to \bar{S}$ be a morphism of ω-semigroups recognizing X and let $P = \varphi(X)$. The construction of a Büchi automaton that recognizes X relies on the following result of Pécuchet [23] which gives, for each idempotent e of S, a representation of S by relations in S^1. Here, S^1 is the monoid equal to S if S has an identity and to $S \cup \{1\}$ otherwise.

Lemma 9.1 *Let S be a finite semigroup and e an idempotent of S. The map $\varphi_e : S \to \mathbb{B}^{S^1 \times S^1}$ defined, for all $s \in S$ and for all $p, q \in S^1$ by*

$$\varphi_e(s)_{p,q} = \begin{cases} 1 & \text{if } sq = p \text{ or if } sq = pe \\ 0 & \text{otherwise} \end{cases}$$

is a semigroup morphism.

Let us consider, for every idempotent e of S, the Büchi automaton

$$\mathcal{A}_e = (S^1, A, E_e, I_e, \{1\}) \quad \text{with } I_e = \{s \in S \mid (s, e) \in P\}$$

and

$$E_e = \{(p, a, q) \in S^1 \times A \times S^1 \mid \varphi(a)q = p \text{ or } \varphi(a)q = pe\}.$$

Then the Büchi automaton $\mathcal{A}$ that recognizes X is the disjoint union of the automata $\mathcal{A}_e$, where e ranges over the set of idempotents of S.

Example 9.1 Let $\bar{S} = (\{a, b\}, \{a^\omega, b^\omega\})$ be the ω-semigroup considered in Example 7.2 and let $\varphi : A^\infty \to \bar{S}$ be the morphism of ω-semigroups defined by $\varphi(a) = a$ and $\varphi(b) = b$. The automaton $\mathcal{A}$ is represented in the figure below.

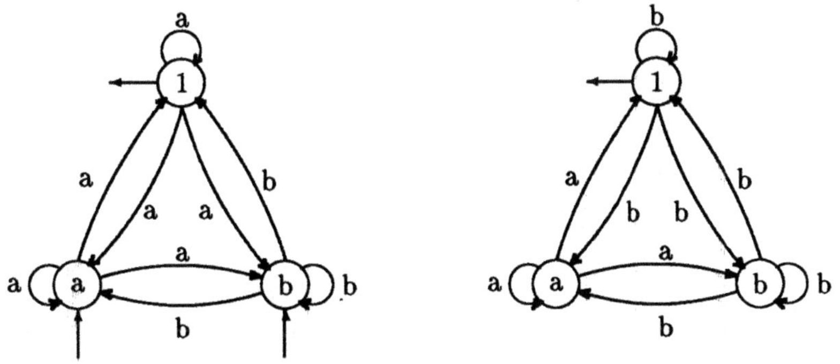

Figure 9.1. The automata $\mathcal{A}_a$ (on the left) and $\mathcal{A}_b$ (on the right).

10. From Wilke algebras to Muller automata

Given a semigroup morphism $\varphi : A^+ \to S$ recognizing a subset X of A^+, it is easy to build a deterministic automaton recognizing X. Take the right representation of A^+ on S^1 defined by $s.a = s\varphi(a)$. This defines a deterministic automaton $\mathcal{A}_S = (S^1, A, E, \{1\}, P\}$, where $E = \{(s, a, s.a) \mid s \in S^1, a \in A\}$, which recognizes L. It is tempting to do the same construction to obtain a deterministic Muller automaton, given a finite ω-semigroup $\bar{S} = (S, S_\omega)$ recognizing a subset X of $A^{\mathbb{N}}$. B. Le Saec [15] has observed that such a construction is possible if the right stabilizers of S are bands. Given a finite semigroup S and an element s of S recall that the right stabilizer of s is the subsemigroup $Stab(s)$ of S defined by $Stab(s) = \{t \mid st = s\}$.

Theorem 10.1 *Let $\bar{S} = (S, S_\omega)$ be a finite ω-semigroup such that the right stabilizers of S are idempotent semigroups. Let $\varphi : A^\infty \to S$ be a morphism of ω-semigroups* *recognizing a subset X of $A^{\mathbb{N}}$. Then the automaton $\mathcal{A}_S = (S^1, A, E, \{1\}, \mathcal{T})$, where $E = \{(s, a, s.a) \mid s \in S^1, a \in A\}$ and $\mathcal{T} = \{\mathit{Infinite}(u) \mid u \in X\}$ is a Muller automaton recognizing X.*

Example 10.1 Let $\bar{S} = (\{a, b\}, \{a^\omega, b^\omega\})$ be the ω-semigroup considered in Examples 7.2 and 9.1 and let $\varphi : A^\infty \to \bar{S}$ be the morphism of ω-semigroups defined by $\varphi(a) = a$ and $\varphi(b) = b$. Let $P = \{a^\omega\}$, so that $\varphi^{-1}(a^\omega) = aA^\omega$. Then the transitions of the automaton $\mathcal{A}_S$ are represented below and its table is $\mathcal{T} = \{\{a\}\}$.

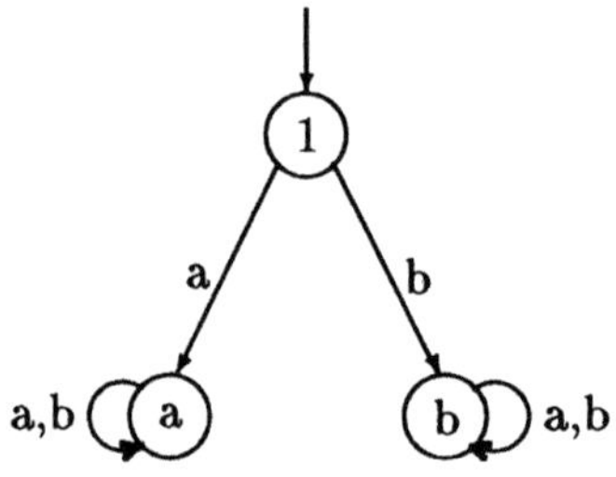

Figure 10.1. A deterministic Muller automaton recognizing aA^ω.

Unfortunately, the right stabilizer of an element of a finite semigroup is not always an idempotent semigroup. However, it is shown in [16, 17] that every finite semigroup is a quotient of a finite semigroup in which the right stabilizer of every element is an idempotent semigroup. In fact, a slightly stronger result holds.

Theorem 10.2 *Every finite semigroup is a quotient of a finite semigroup in which the right stabilizer of every element is a semigroup which satisfies the identities $x = x^2$ and $xy = xyx$.*

Therefore, one can associate with every morphism from A^∞ onto a finite ω-semigroup S a Muller automaton as follows: first compute a finite semigroup T whose right stabilizers are idempotent and satisfy the identity $xy = xyx$ such that S_f is a quotient of T and then apply the result of Le Saec. This gives the proof of (4) implies (3) in Theorem 6.1. To sum up, we have obtained the following result.

Theorem 10.3 *Let X be a set of infinite words. The following conditions are equivalent:*

(1) X is recognizable,

(2) X is recognized by a finite ω-semigroup,

(3) X is ω-rational.

11. Syntactic ω-semigroups

Let $S = (S_f, S_\omega)$ be an ω-semigroup. An *ω-semigroup congruence* on S is a pair $\sim\ = (\sim_f, \sim_\omega)$ where $\sim_f$ is a semigroup congruence on S_f and $\sim_\omega$ is an equivalence on S_ω such that

(1) for all $s, s' \in S_f$ and for all $t, t' \in S_\omega$, $s \sim_f s'$ and $t \sim_\omega t'$ imply $st \sim_\omega s't'$

(2) for all infinite sequences $s_0, s_1, s_2, \ldots$ and $s'_0, s'_1, s'_2, \ldots$ of elements of S_f such that $s_i \sim_f s'_i$ for all i, $s_0 s_1 \cdots \sim_\omega s'_0 s'_1 \cdots$

If S is a finite ω-semigroup, then one may replace (2) by the weaker condition (2')

(2') for all $s, s' \in S_f$, $s \sim_f s'$ implies $s_\omega \sim_\omega s'_\omega$.

In the sequel, we shall often omit the subscripts, and use the simplified notation $\sim$ instead of $\sim_f$ and $\sim_\omega$.

Let X be a subset of A^∞. An ω-semigroup congruence $\sim$ on A^∞ *recognizes* X if the ω-semigroup morphism $\varphi : A^\infty \to A^\infty/\sim$ recognizes X. Contrary to the case for finite words, the lower bound of the congruences that recognize X does not always recognize X. If this lower bound still recognizes X, it is called the *syntactic congruence* of X and is denoted $\sim_X$.

Arnold has shown that the syntactic congruence always exists if X is recognized by a finite ω-semigroup [1]. The reason is that a finite ω-semigroup can be considered as a finite Wilke algebra and that syntactic congruences can be defined on Wilke algebras as on any algebra of finite signature. More precisely, a *congruence* on a Wilke algebra S is a pair $\sim\ = (\sim_f, \sim_\omega)$ where $\sim_f$ is a semigroup congruence on S_f and $\sim_\omega$ is an equivalence relation on S_ω that is compatible with the ω-power and the mixed product. In practice, it is convenient to omit the subscripts f and ω and write $\sim$ for both $\sim_f$ and $\sim_\omega$.

A congruence $\sim$ on S *saturates* a subset $P = (P_f, P_\omega)$ of S if for all $x, y \in S_f$ (resp. S_ω), $x \sim y$ and $x \in P_f$ (resp. P_ω) implies $y \in P_f$ (resp. P_ω). Now the lower bound of the congruences that saturate P also saturates P and is called the *syntactic congruence* of P. It is the congruence $\sim_P$ defined on S_f by $u \sim_P v$ if and only if, for every $x, y \in S_f^1$ and for every $z \in S_f$,

$$(1) \qquad \begin{aligned} xuyz^\omega \in P_\omega &\iff xvyz^\omega \in P_\omega \\ x(uy)^\omega \in P_\omega &\iff x(vy)^\omega \in P_\omega \end{aligned}$$

and on S_ω by $u \sim_P v$ if and only if, for every $x \in S_f^1$,

$$(2) \qquad xu \in P_\omega \iff xv \in P_\omega$$

The *syntactic Wilke algebra* of P is the quotient of S by the syntactic congruence of P.

Now, to compute the syntactic ω-semigroup of a set X recognized by a finite Büchi automaton $\mathcal{A}$, one first computes the finite ω-semigroup S associated with $\mathcal{A}$ and the image P of X in S and then one computes the syntactic Wilke algebra of P in S. This provides an algorithm to compute the syntactic ω-semigroup of a recognizable set.

The syntactic congruence $\sim_X$ of a recognizable set X of A^∞ can also be defined directly as follows. On A^+, $u \sim_X v$ if and only if, for every $x, y \in A^*$ and for every $z \in A^+$,

$$\text{(3)}\qquad \begin{aligned} xuyz^\omega \in X &\iff xvyz^\omega \in X \\ x(uy)^\omega \in X &\iff x(vy)^\omega \in X \end{aligned}$$

and on $A^{\mathbb{N}}$, $u \sim_X v$ if and only if, for every $x \in A^*$,

$$\text{(4)}\qquad xu \in X \iff xv \in X$$

The ***syntactic ω-semigroup*** of X is the quotient of A^∞ under the ω-semigroup congruence $\sim_X$. This is also the minimal (with respect to the quotient ordering) complete finite ω-semigroup recognizing X.

Example 11.1 We come back to example 8.1. The set $X = (a\{b,c\}^* \cup \{b\})^\omega$ is recognized by the ω-semigroup $S = \{a, b, c, ba, ca, a^\omega, b^\omega, (ca)^\omega, 0\}$, defined by the following relations:

$$\begin{array}{lllll} a^2 = a & ab = a & ac = a & b^2 = b & bc = c \\ cb = c & c^2 = c & c^\omega = 0 & (ba)^\omega = b^\omega & aa^\omega = a^\omega \\ ab^\omega = a^\omega & a(ca)^\omega = a^\omega & ba^\omega = b^\omega & bb^\omega = b^\omega b(ca)^\omega = (ca)^\omega & \\ ca^\omega = (ca)^\omega & cb^\omega = (ca)^\omega c(ca)^\omega = (ca)^\omega. & & & \end{array}$$

Since $P = \{a^\omega, b^\omega\}$, the congruence $\sim_P$ is defined by

$$a \sim_P ba \qquad \text{and} \qquad a^\omega \sim_P b^\omega$$

Therefore, the syntactic ω-semigroup is $S(X) = \{a, b, c, ca, a^\omega, (ca)^\omega, 0\}$, with the following relations

$$\begin{array}{lllll} a^2 = a & ab = a & ac = a & ba = a & b^2 = b \\ bc = c & cb = c & c^2 = c & b^\omega = a^\omega & c^\omega = 0 \\ aa^\omega = a^\omega & a(ca)^\omega = a^\omega & ba^\omega = a^\omega & b(ca)^\omega = (ca)^\omega & ca^\omega = (ca)^\omega \\ c(ca)^\omega = (ca)^\omega. & & & & \end{array}$$

12. Varieties

So far, we have extended to infinite words the notions of finite automata, of a semigroup recognizing a set of words and finally of a syntactic semigroup. The next natural step in this process is to extend the variety theorem. Pécuchet [23, 24] made a first attempt in this direction, but the result of Wilke [44] is more satisfactory.

A ***variety of finite Wilke algebras*** is a class of finite Wilke algebras closed under the taking of finite direct products, quotients and complete subalgebras.

Let $X \subset A^\infty$ and let $u \in A^+$. We set

$$u^{-1}X = \{v \in A^\infty \mid uv \in X\}$$
$$Xu^{-\omega} = \{v \in A^+ \mid (vu)^\omega \in X\}$$
$$Xu^{-1} = \{v \in A^+ \mid vu \in X\}.$$

An ∞-*variety* $\mathcal{V}$ associates with every alphabet A a set $\mathcal{V}(A^\infty)$ of recognizable sets of A^∞ such that

(1) for every alphabet A, $\mathcal{V}(A^\infty)$ contains $\emptyset$, A^+, $A^{\mathbb{N}}$ and A^∞ and is closed under finite boolean operations (finite union and complement),
(2) for every semigroup morphism $\varphi : A^+ \to B^+$, $X \in \mathcal{V}(B^\infty)$ implies $\varphi^{-1}(X) \in \mathcal{V}(A^\infty)$,
(3) if $X \in \mathcal{V}(A^\infty)$ and $u \in A^*$, then $u^{-1}X \in \mathcal{V}(A^\infty)$, $Xu^{-1} \in \mathcal{V}(A^\infty)$ and $Xu^{-\omega} \in \mathcal{V}(A^\infty)$.

It is important to notice that one works with A^∞ (and not with $A^{\mathbb{N}}$) in this definition. In other words, the elements of a ∞-variety are sets of finite or infinite words.

If $\mathbf{V}$ is a variety of finite Wilke algebras, we denote by $\mathcal{V}(A^\infty)$ the set of recognizable sets of A^∞ whose syntactic ω-semigroup belongs to $\mathbf{V}$. This is also the set of sets of A^∞ recognized by an ω-semigroup of $\mathbf{V}$.

An ∞-*class* of recognizable sets is a correspondence which associates with every finite alphabet A, a set $\mathcal{C}(A^\infty)$ of recognizable sets of A^∞. In particular, the correspondence $\mathbf{V} \to \mathcal{V}$ associates with every variety of finite Wilke algebras a ∞-class of recognizable sets.

The theorem of Wilke, which extends the corresponding theorem of Eilenberg, can now be stated as follows.

Theorem 12.1 *The correspondence $\mathbf{V} \to \mathcal{V}$ defines a bijection between the varieties of finite Wilke algebras and the ∞-varieties.*

We conclude by giving four examples of correspondence between varieties of finite Wilke algebras and ∞-varieties. The first and the second one extend to infinite words well known results on finite words (the characterization of star-free and locally testable languages), but the last two examples are less familiar since they concern two topological classes.

We refer to our article [33] in this volume for the definitions of star-free and locally testable languages. The set of *star-free subsets of* A^∞ is the smallest set $\mathcal{S}$ of subsets of A^∞ containing the star-free languages of A^+ which is closed under finite boolean operations and such that if X is a star-free subset of A^+ and $Y \in \mathcal{S}$, then $XY \in \mathcal{S}$. Equivalent characterizations of the star-free subsets of $A^{\mathbb{N}}$ were proposed by Thomas [40, 41].

Theorem 12.2 *Let X be a subset of $A^{\mathbb{N}}$. The following conditions are equivalent:*
(1) X is star-free,
(2) X is a finite union of sets of the form UV^ω where U and V are star-free subsets of A^+ and $V^2 \subset V$,
(3) X is a boolean combination of sets of the form $\overrightarrow{L}$ where L is a star-free language.

The syntactic characterization is given in [26]. A finite ω-semigroup (S_f, S_ω) is *aperiodic* if the semigroup S_f is aperiodic.

Theorem 12.3 *A recognizable set of A^∞ is star-free if and only if its syntactic ω-semigroup is aperiodic.*

It follows in particular that the star-free sets form a ∞-variety. Similar results hold for the locally testable sets. A subset of A^∞ is *locally testable* if and only if it is a finite boolean combination of sets of the form uA^*, A^*u, A^*uA^*, uA^ω, A^*uA^ω and $A^*(uA^*)^\omega$, for all $u \in A^+$. The following results are due to Pécuchet [23, 24].

Theorem 12.4 *Let X be a subset of $A^{\mathbb{N}}$. The following conditions are equivalent:*
(1) X is locally testable,
(2) X is a finite union of sets of the form UV^ω where U and V are locally testable subsets of A^+ and $V^2 \subset V$,
(3) X is a boolean combination of sets of the form $\overrightarrow{L}$ where L is a locally testable language.

A finite ω-semigroup (S_f, S_ω) is *locally idempotent and commutative* if the semigroup S_f is locally idempotent and commutative.

Theorem 12.5 *A recognizable set of A^∞ is locally testable if and only if its syntactic ω-semigroup is locally idempotent and commutative.*

One can define a topology on $A^{\mathbb{N}}$ by considering A as a discrete space and by taking the product topology. The open sets of $A^{\mathbb{N}}$ are the sets of the form $XA^{\mathbb{N}}$ for some $X \subset A^+$. A set is closed if its complement is open. The sets of Δ_1 are the clopen sets, sets which are at the same time closed and open. The sets of Δ_2 are at the same time countable unions of closed sets and countable intersection of open sets. One can show that the recognizable sets of Δ_2 are those sets X such that both X and its complement are accepted by a deterministic Büchi automaton. The characterization of these two classes in terms of ω-semigroups is due to Wilke [45]. In these statements, x^π denotes the unique idempotent of the semigroup generated by x.

Theorem 12.6 *The class Δ_1 forms a ∞-variety. The corresponding variety of finite Wilke algebras is the class of algebras defined by the equation $x^\pi yz^\omega = x^\pi y'z'^\omega$.*

Theorem 12.7 *The class Δ_2 forms a ∞-variety. The corresponding variety of finite Wilke algebras is the class of algebras defined by the equation $(x^\pi y^\pi)^\pi x^\omega = (x^\pi y^\pi)^\pi y^\omega$.*

13. Conclusion

Most known results on automata on finite words have found their counterpart on infinite words. However, the complexity of the solutions is apparently increased by an order of magnitude. This is true for instance for the equivalence between deterministic and non deterministic automata and for the variety theorem. Furthermore, several questions which are solved for finite words are still open for infinite words, but one can be reasonably optimistic about them.

Acknowledgements

The authors would like to thank Olivier Carton, John Fountain, Victoria Gould and Pascal Weil for their comments on a preliminary version of this article.

References

1. A. Arnold, *A syntactic congruence for rational ω-languages*, Theoret. Comput. Sci. **39**, (1985) 333–335.
2. D. Beauquier and J.-E. Pin, *Factors of words*, in "Automata, Languages and Programming" (G. Ausiello, M. Dezani-Ciancaglini and S. Ronchi Della Rocca, eds.), Lecture Notes in Computer Science, **372**, Springer, (1989) 63–79.
3. D. Beauquier and J.-E. Pin, *Languages and scanners*, Theoret. Comput. Sci. **84**, (1991) 3–21.
4. J. R. Büchi, *Weak second-order arithmetic and finite automata*, Z. Math. Logik und Grundl. Math. **6**, (1960) 66–92.
5. J. R. Büchi, *On a decision method in restricted second-order arithmetic*, in Proc. 1960 Int. Congr. for Logic, Methodology and Philosophy of Science, Stanford Univ. Press, Stanford, (1962) 1–11.
6. J. R. Büchi and L.H. Landweber, *Definability in the monadic second-order theory of successors*, J. Symbolic Logic **34**, (1969) 166–170.
7. S. Eilenberg, *Automata, languages and machines*, Vol. A, Academic Press, New York (1974).
8. S. Eilenberg, *Automata, languages and machines*, Vol. B, Academic Press, New York (1976).
9. S. C. Kleene, *Representation of events in nerve nets and finite automata*, in "Automata Studies" (C. E. Shannon and J. McCarthy eds.), Princeton University Press, Princeton, New Jersey, (1956) 3–42.
10. J. Justin and G. Pirillo, *On a natural extension of Jacob's rank*, J. Comb. Theory **43**, (1986) 205–218.
11. R. Ladner, *Application of model theoretic games to discrete linear orders and finite automata*, Information and Control, **33**, (1977) 281–303.
12. G. Lallement, *Semigroups and combinatorial applications*, Wiley, New York (1979).
13. H. Landweber, *Finite state games - A solvability algorithm for restricted second-order arithmetic*, Notices Amer. Math. Soc. **14**, (1967) 129–130.
14. H. Landweber, *Decision problems for ω-automata*, Math. Syst. Theor. **3**, (1969) 376–384.
15. B. Le Saec, *Saturating right congruences*, Informatique Théorique et Applications **24**, (1990) 545–560.
16. B. Le Saec, J.-E. Pin and P. Weil, *A purely algebraic proof of McNaughton's theorem on infinite words*, in "Foundations of Software Technology and Theoretical Computer Science" (S. Biswas and K. V. Nori eds.), Lecture Notes in Computer Science **560**, Springer, (1991) 141–151.
17. B. Le Saec, J.-E. Pin and P. Weil, *Finite semigroups whose stabilizers are idempotent semigroups*, International Journal of Algebra and Computation **1**, (1991) 291–314.
18. M. Lothaire, *Combinatorics on Words*, Cambridge University Press, (1982).
19. R. McNaughton, *Testing and generating infinite sequences by a finite automaton*, Information and Control **9**, (1966) 521–530.
20. R. McNaughton, *Algebraic decision procedures for local testability*, Math. Syst. Theor. **8**, (1974) 60–76.
21. R. McNaughton and S. Papert, *Counter-free Automata*, MIT Press, (1971).
22. D. Muller, *Infinite sequences and finite machines*, in "Switching Theory and Logical Design", Proc. Fourth Annual Symp. IEEE, (1963) 3–16.
23. J.-P. Pécuchet, *Variétés de semigroupes et mots infinis*, in (B. Monien and G. Vidal-Naquet eds.), STACS 86, Lecture Notes in Computer Science **210**, Springer, (1986) 180–191.
24. J.-P. Pécuchet, *Étude syntaxique des parties reconnaissables de mots infinis*, in Proc. 13th ICALP, (L. Kott ed.), Lecture Notes in Computer Science **226**, (1986) 294–303.

25. D. Perrin, *Variétés de semigroupes et mots infinis*, C. R. Acad. Sci. Paris **295**, (1982) 595–598.
26. D. Perrin, *Recent results on automata and infinite words*, in "Mathematical Foundations of Computer Science", Lecture Notes in Computer Science **176** Springer-Verlag, New-York/Berlin, (1984) 134–148.
27. D. Perrin, *An introduction to automata on infinite words*, in "Automata on Infinite Words" (Nivat, M. ed.), Lecture Notes in Computer Science **192**, Springer, (1984) 2–17.
28. D. Perrin, *Automata*, in "Handbook of Theoretical Computer Science" (Van Leeuwen, J. ed.), Vol B : Formal Models and Semantics, Elsevier (1990).
29. D. Perrin and J.-E. Pin, *First order logic and star-free sets*, J. Comput. System Sci. **32**, (1986) 393–406.
30. D. Perrin and J.-E. Pin, *Mots infinis*, to appear (LITP report 93–40).
31. D. Perrin and P. E. Schupp, *Automata on the integers, recurrence distinguishability and the equivalence and decidability of monadic theories*, Proc. 1st IEEE Symp. on Logic in Computer Science, (1986) 301–304.
32. J.-E. Pin, *Logic, Semigroups and Automata on words*, to appear.
33. J.-E. Pin, *Finite semigroups and recognizable languages: an introduction*, this volume.
34. J.-E. Pin, *Variétés de langages formels*, Masson, Paris (1984); English translation: *Varieties of formal languages*, Plenum, New-York (1986).
35. F. D. Ramsey, *On a problem of formal logic*, Proc. London Math. Soc. **30**, (1929) 338–384.
36. S. Safra, *On the complexity of the ω-automata*, in: Proc. 29th Ann. IEEE Symp. on Foundations of Computer Science, (1988) 319–327.
37. M. P. Schützenberger, *On finite monoids having only trivial subgroups*, Information and Control **8**, (1965) 190–194.
38. I. Simon, *Piecewise testable events*, Proc. 2nd GI Conf., Lecture Notes in Computer Science **33**, Springer, Berlin, (1975) 214–222.
39. I. Simon, *Word Ramsey theorems*, in "Graph Theory and Combinatorics" (B. Bollobas ed.), Academic Press, (1984).
40. W. Thomas, *Star-free regular sets of ω-sequences*, Information and Control **42**, (1979) 148–156.
41. W. Thomas, *A combinatorial approach to the theory of ω-automata*, Information and Control **48**, (1981) 261–283.
42. W. Thomas, *Classifying regular events in symbolic logic*, J. Comput. Syst. Sci. **25**, (1982) 360–375.
43. W. Thomas, *Automata on infinite objects*, in "Handbook of Theoretical Computer Science", vol B, Formal models and semantics, Elsevier, (1990) 135–191.
44. T. Wilke, *An Eilenberg theorem for ∞-languages*, in "Automata, Languages and Programming" Proc. of 18th ICALP Conference, **510**, Springer, Berlin, (1991) 588–599.
45. T. Wilke, *An algebraic theory for regular languages of finite and infinite words*, International Journal of Algebra and Computation, **3**, (1993) 447–489.
46. T. Wilke, *Locally threshold testable languages of infinite words*, in STACS 93, (P. Enjalbert, A. Finkel, K. W. Wagner eds.), Lecture Notes in Computer Science **665**, Springer, Berlin, (1993) 607–616.

RELATIVELY FREE PROFINITE MONOIDS: AN INTRODUCTION AND EXAMPLES*

J. ALMEIDA
Grupo de Matemática Pura
Faculdade de Ciências do Porto
P. Gomes Teixeira
4000 Porto, Portugal
E-mail: jalmeida@fc.up.pt

and

P. WEIL
LITP
Institut Blaise Pascal
4 place Jussieu
75252 Paris Cedex 05, France
E-mail: weil@litp.ibp.fr

Introduction

One of the main strengths of the concept of variety for the study of classes of algebras, say, of monoids, is the dual viewpoint which it allows: algebraic, or semantic, on one hand — a class of monoids closed under taking submonoids, homomorphic images and (arbitrary) direct products —; syntactic on the other hand — a class defined by a set of identities. Connecting the two points of view is the construction of the free object of a variety $\mathcal{V}$ over a set A as a subdirect product of all the A-generated elements of $\mathcal{V}$, or as the quotient of the free monoid A^* by the fully invariant congruence generated by the identities of $\mathcal{V}$ in $|A|$ variables.

When one wants to study pseudovarieties of finite monoids (algebraic characterization, solution of the membership problem, combinatorial description of the associated classes of recognizable languages), the symmetry of this double viewpoint is somewhat modified. Of course, as an intermediate case, if a pseudovariety of finite monoids $\mathbf{V}$ admits a free object over a finite set A, the understanding of this free object gives considerable information on the A-generated elements of $\mathbf{V}$: syntactic information on the identities satisfied by these monoids, and semantic-combinatorial information on the languages of A^* they recognize. In effect, the free object of $\mathbf{V}$ over A is a finite monoid which contains all the information on all A-generated elements of $\mathbf{V}$. However the existence of such free objects is far from being the general case, since it occurs if and only if $\mathbf{V}$ has only finitely many A-generated elements

* Both authors gratefully acknowledge support from ESPRIT-BRA Working Group 6317 "ASMICS". The first author was also supported in part by JNICT through the project SAL (PBIC/C/CEN/1021/92). The second author was also supported in part by the Projet de Recherche Coordonnée "Mathématiques et Informatique".

J. Fountain (ed.), Semigroups, Formal Languages and Groups, 73–117.

(up to isomorphism).

The theory of implicit operations and free profinite monoids presented here aims at dealing with the general case by constructing a possibly infinite topological monoid containing the desired information on the infinitely many A-generated elements of $\mathbf{V}$. This monoid, denoted $\overline{\Omega}_A\mathbf{V}$ and called the free pro-$\mathbf{V}$ monoid over A, is constructed by taking the projective limit of the A-generated elements of $\mathbf{V}$. It is always compact, so that it inherits many important properties of finite monoids (e.g. Green's relations $\mathcal{J}$ and $\mathcal{D}$ coincide). Moreover, the submonoid of $\overline{\Omega}_A\mathbf{V}$ generated by A is dense in it. We will see that the elements of $\mathbf{V}$ are exactly the continuous homomorphic images of the $\overline{\Omega}_A\mathbf{V}$, that the collection of the $\overline{\Omega}_A\mathbf{V}$ (A finite) characterizes the pseudovariety $\mathbf{V}$ (Reiterman's theorem), and that the languages of A^* which are recognized by monoids of $\mathbf{V}$ can be characterized in terms of the topology of $\overline{\Omega}_A\mathbf{V}$. These fundamental definitions and properties are covered in section 1.

So the program for understanding a given pseudovariety $\mathbf{V}$ can be viewed as computing the free pro-$\mathbf{V}$ monoids $\overline{\Omega}_A\mathbf{V}$, or at least learning something about their structure. Not surprisingly, this study is quite difficult in general. We will however discuss two classes of examples, namely the sub-pseudovarieties of $\mathbf{DS}$ (section 2) and the pseudovarieties of groups (section 3), for which some interesting results are known.

Recall that $\mathbf{DS}$ is the pseudovariety of all monoids in which each regular $\mathcal{D}$-class is a subsemigroup (or equivalently, in which each regular element is a group element). The key fact regarding the sub-pseudovarieties $\mathbf{V}$ of $\mathbf{DS}$ is that the elements of free pro-$\mathbf{V}$ monoids can be factored as products of elements of A and of regular elements. Moreover, one can give interesting combinatorial characterizations of these regular elements in terms of multiplicity of subwords. We will see that, for certain sub-pseudovarieties of $\mathbf{DS}$, more can be said on the regular elements (e.g. sufficient conditions for equality), and in favorable cases, one can give unique factorization theorems for the elements of $\overline{\Omega}_A\mathbf{V}$.

The case of groups is quite different. Profinite groups have been studied a lot, notably in the framework of number theory and algebraic geometry, and we can benefit from this work. The results presented here are of a geometrical nature. They deal with certain properties of the Cayley graphs of free profinite groups and with their consequences for the so-called reduced factorizations of elements of free profinite groups.

The paper closes with a case study which uses simultaneously properties of sub-pseudovarieties of $\mathbf{DS}$ and those of pseudovarieties of groups. We compute the free pro-$\mathbf{V}$ monoids when $\mathbf{V} = \mathbf{DG} \cap \mathbf{ECom}$ (the class of monoids in which the idempotents commute and the regular $\mathcal{D}$-classes are groups), and we provide a join decomposition for this pseudovariety.

Historically, the theory of free pro-$\mathbf{V}$ monoids and of implicit operations received its major impetus from Reiterman's theorem [20]. Much of the work thereafter was accomplished by Almeida ([3], [5], etc.) and, for what concerns the sub-pseudovarieties of $\mathbf{DS}$, by Azevedo [13] and by Almeida and Azevedo [8]. Almeida's book [6] offers a comprehensive treatment of the theory. The major connection with the classical study of free profinite groups, and in particular with the introduction of the geometrical ideas involving the Cayley graphs of these groups, arose through work of Ribes and Zalesskiĭ [23] who offered a direct proof of the topological version of the Rhodes type-II conjecture. (Independently and a year earlier, Ash [11] had proved the original statement of the Rhodes conjecture.) The results of sections 3 and 4, which make use of this geometric intuition, are due to Almeida and Weil [10].

In this respect, there is a difference between the first two sections of this paper and the last two. The first two offer material which is not so recent, which has been presented in conferences several times, has been published in journals, and now also partly appears in book form. The last two sections, on the contrary, deal with very recent results, which are presented here for the first time in a conference, and are still "to appear". This may account to some extent for stylistic differences between the two sets of sections, the first one being "slower" and assuming a more "pedagogical" tone. However, throughout the paper, we have tried to give a proof of (almost) all the statements, so as to keep it as self-contained as possible.

Finally, let us note that most statements in this paper are given in terms of monoids and pseudovarieties of monoids. Unless otherwise indicated, they also hold for semigroups and pseudovarieties of semigroups, up to obvious changes (e.g. replace free monoids by free semigroups). Examples will be drawn freely from the realm of monoids as well as from that of semigroups.

Prerequisites

We assume the reader to be familiar with some basic notions of semigroup theory: idempotents (the set of idempotents of a semigroup S is denoted $E(S)$), regular elements, Green's relations.

We recall that a pseudovariety of monoids (resp. semigroups) is a class of finite monoids (resp. semigroups) which is closed under taking submonoids (resp. subsemigroups), homomorphic images and finitary direct products.

For certain sections, we also assume familiarity with the notion of languages recognized by a monoid or a semigroup, and with the notion of variety of languages. For details on these notions, the reader is referred to Pin's book [19], or to Pin's contribution to this volume.

From the language setting, we borrow the following: an alphabet always means a finite set. If A is an alphabet, its elements are called letters. The elements of the free monoid A^* (or of the free semigroup A^+) over A can be seen as finite sequences of letters, and are called words. We will use freely this combinatorial terminology when dealing with free monoids.

1. Free profinite monoids and implicit operations

Projective limits constitute an essential tool for our purpose: to each pseudovariety $\mathbf{V}$ and to each finite set A, we associate the projective limit of the A-generated elements of $\mathbf{V}$, called the free pro-$\mathbf{V}$ monoid over A.

Projective limits can be defined for any type of algebras (e.g. groups, rings, modules) or partial algebras (e.g. graphs). In this section we will be interested only in monoids and semigroups but modules and graphs will play an important role in section 3. For this reason, section 1.1, which deals with the basic properties of projective limits, is written in terms of algebras in general.

Section 1.2 introduces relatively free profinite monoids, their topology and their elementary properties. Examples are computed in section 1.3. In section 1.4, we consider the monoids of implicit operations on a pseudovariety **V**, and show that they are naturally isomorphic to the free pro-**V** monoids. Section 1.5 is devoted to the statement and proof of Reiterman's theorem on the definition of pseudovarieties by sets of pseudoidentities (formal equalities of implicit operations). Finally, section 1.6 deals with the characterization of the class of languages recognized by **V** in terms of the topology of the free pro-**V** monoids.

A general reference for the contents of this section is Almeida's book [6]. Other expository papers on the topic are Almeida's [2] and Weil's [26]. On the elementary properties of projective limits, we have used the first pages of Serre's [25] and of Fried and Jarden's [15]. In fact, the presentation given here, which introduces first the relatively free profinite monoids, and next the monoids of implicit operations, is in the reverse order of that adopted in [2], [6] and [26], but in keeping with that of Almeida's thorough treatment in [3]. Sections 1.1 to 1.5 are an expanded version of the first sections of a forthcoming paper by Almeida and Weil about implicit operations on semidirect products [10].

1.1. Projective limits

Recall that a partially ordered set (poset) I is *directed* if any two elements of I admit a common upper bound. A *directed system of algebras* $(S_i)_{i\in I}$ is a family of algebras indexed by a directed poset I such that, for each pair (i,j) of elements of I satisfying $i \geq j$, there exists a homomorphism $\varphi_{i,j}\colon S_i \to S_j$, and such that, for all $i \geq j \geq k$ in I, we have $\varphi_{i,i} = \mathrm{id}_{S_i}$ and $\varphi_{j,k} \circ \varphi_{i,j} = \varphi_{i,k}$.

By definition, the *projective limit* of a directed system of algebras $(S_i)_{i\in I}$ is the following subalgebra of the direct product $\prod_{i\in I} S_i$:

$$\varprojlim (S_i)_{i\in I} = \{(x_i)_{i\in I} \in \prod_{i\in I} S_i : \forall i \geq j,\ \varphi_{i,j}x_i = x_j\}.$$

Example. Let $I = \mathbb{N}$ be the set of natural numbers. For each $n \geq 0$, let $S_n = \langle a; a^n = a^{n+1}\rangle = \{1, a, \ldots, a^n\}$ with

$$a^i a^j = \begin{cases} a^{i+j} & \text{if } i+j \leq n \\ a^n & \text{if } i+j > n. \end{cases}$$

For each $n, k \geq 0$, let $\varphi_{n+k,n}\colon S_{n+k} \to S_n$ be the morphism given by $\varphi_{n+k,n}a^i = a^{\min(i,n)}$. The morphisms $\varphi_{n+k,n}$ make $(S_n)_n$ into a directed system of monoids, and $\varprojlim (S_n)_n$ is isomorphic to the free monoid generated by a with a zero added, that is, $\varprojlim (S_n)_n$ is the set $\{1, a, a^2, \ldots, a^n, \ldots\} \cup \{0\}$.

$$\begin{array}{ll}
S_1 & 1 - a \\
S_2 & 1 - a - a^2 \\
\vdots & \vdots \\
S_n & 1 - a - a^2 \quad \cdots \quad - a^n \\
\vdots & \vdots \\
\varprojlim(S_n)_n & 1 - a - a^2 \quad \cdots \quad - a^n \quad \cdots \qquad\qquad 0
\end{array}$$

The *canonical morphisms* $\pi_i \colon \varprojlim(S_i)_{i \in I} \to S_i$ are defined to be the restrictions of the coordinate projections to $\varprojlim(S_i)_i$. Note that these morphisms are not necessarily onto.

Example. Let $I = \mathbb{N}$. For each $n, k \geq 0$, let $S_n = \mathbb{Z}$, and let $\varphi_{n+k,n} \colon \mathbb{Z} \to \mathbb{Z}$ be given by $\varphi_{n+k,n} x = 2^k x$. These data define a projective system and $\varprojlim(S_n)_n = \{0\}$.

A *directed system of morphisms* defined on an algebra S is a family of morphisms $(\rho_i \colon S \to S_i)_{i \in I}$ indexed by a directed poset I such that:

- there exist morphisms $\varphi_{i,j} \colon S_i \to S_j$ for all pairs (i, j) of elements of I verifying $i \geq j$, and these morphisms make $(S_i)_i$ a directed system of algebras,
- for all $i \geq j$ in I, we have $\rho_j = \varphi_{i,j} \circ \rho_i$.

It is not difficult to verify the following universal property of projective limits.

Proposition 1.1 *Let $(\rho_i \colon S \to S_i)_{i \in I}$ be a directed system of morphisms defined on an algebra S. Then there exists a unique morphism $\rho \colon S \to \varprojlim(S_i)_i$ satisfying $\rho_i = \pi_i \circ \rho$ for all i. Moreover ρ is defined by $\rho s = (\rho_i s)_i$.*

This property defines $\varprojlim(S_i)_i$ up to an isomorphism.

We say that the morphism ρ described above is *induced* by the directed system $(\rho_i)_i$.

Let us now assume that the algebras S_i are topological algebras, and that the morphisms $\varphi_{i,j}$ are continuous. The direct product $\prod_i S_i$ is equipped with the product topology. By definition, a basis of open sets for this topology is given by the sets of the form

$$V_{i_1} \times \cdots \times V_{i_n} \times \prod_{\substack{j \in I \\ j \neq i_1, \ldots, i_n}} S_j,$$

where $n \geq 1$, $i_1, \ldots, i_n$ are distinct elements of I and V_{i_h} is an open subset of S_{i_h}. Then one verifies that $\varprojlim(S_i)_i$ is a closed subalgebra of $\prod_i S_i$. Thus the canonical morphisms π_i are continuous. In particular, if the S_i are compact, then so is their direct product (by the classical Tikhonov's theorem), and hence so is their projective limit.

Note. Using the fact that the poset I is directed, one may verify that a basis of open subsets of $\varprojlim(S_i)_{i\in I}$ is given by the sets of the form $\pi_i^{-1}(V_i)$ where $i \in I$ and V_i is an open subset of S_i.

The case where the S_i are finite algebras is particularly important for our purpose. We consider each finite algebra to be equipped with the discrete topology: each of them is compact, and any morphism between them is continuous. In this situation the topological algebra $\varprojlim(S_i)_i$ is compact and totally disconnected.

We say that an algebra is *profinite* if it is the projective limit of a directed system of finite algebras. In fact, for the most usual kinds of algebras (like groups, rings, lattices, semigroups, monoids, sets; in general for those in which syntactic congruences are characterized by a finite number of terms), one can verify that an algebra is profinite if and only if it is compact and totally disconnected (cf. [4]). We do not know, for instance, if this property holds for groupoids.

More generally, for each class C of algebras, we say that an algebra is *pro-C* if it is a projective limit of algebras in C.

Example. If we consider algebras with empty signature, that is, sets, it follows that profinite sets are compact totally disconnected topological spaces. Conversely, one can verify that any such topological space is a projective limit of finite sets, i.e., it is profinite.

Let us note the following important properties of projective limits.

Lemma 1.1 *Let $(\rho_i\colon S \to S_i)_{i\in I}$ be a directed system of morphisms defined on S. Let $\rho\colon S \to \varprojlim(S_i)_i$ be the induced morphism.*

a) ρ is one-to-one if and only if, for each $s \in S$, $\{s\} = \bigcap_{i\in I} \rho_i^{-1}(\rho_i s)$.
b) If S is compact and if, for each $i \in I$, ρ_i is onto, then ρ is onto.

Proof. (a) is immediate by definition of projective limits.

In order to prove (b), for each element $(s_i)_i$ of $\varprojlim(S_i)_i$, we consider the family of closed sets $\left(\rho_i^{-1}(s_i)\right)_i$. Since the system of morphisms $(\rho_i)_i$ is directed, each finite intersection of $\rho_i^{-1}(s_i)$ is non-empty. But S is compact, so $\bigcap_{i\in I} \rho_i^{-1}(s_i)$ is non-empty, which concludes the proof. □

The following lemma is an immediate but useful observation. It shows that, when we are given a pro-C algebra S where C is a class closed under taking closed subalgebras, we may always assume that S is the projective limit of a directed system $(S_i)_{i\in I}$ of algebras in C in such a way that the canonical morphisms $\pi_i\colon S \to S_i$ are onto.

Lemma 1.2 *Let $S = \varprojlim(S_i)_i$ and let $\pi_i\colon S \to S_i$ be the canonical morphisms. For each i, $\pi_i S$ is a subalgebra of S_i and $S = \varprojlim(\pi_i S)_i$.* □

The next lemma shows that the finite continuous images of a projective limit $\varprojlim(S_i)_i$ of compact algebras are exactly the images of the S_i.

Lemma 1.3 *Let $S = \varprojlim(S_i)_{i\in I}$ be a projective limit of compact algebras and let $\pi_i \colon S \to S_i$ $(i \in I)$ be the canonical morphisms, which we assume to be onto. Let T be a finite algebra and let $\varphi \colon S \to T$ be a continuous morphism. There exists $i \in I$ and a continuous morphism $\varphi_i \colon S_i \to T$ such that $\varphi = \varphi_i \circ \pi_i$.*

Proof. The $\pi_i^{-1}(V)$ ($i \in I$, V open subset of S_i) constitute a basis of open sets for the topology of S. Since φ is continuous, for each $s \in S$, there exists an element i_s of I and an open neighborhood V_s of $\pi_{i_s} s$ in S_{i_s} such that $s \in \pi_{i_s}^{-1}(V_s)$ and $\pi_{i_s}^{-1}(V_s) \subseteq \varphi^{-1}(\varphi s)$. Now S is compact and $S = \bigcup_s \pi_{i_s}^{-1}(V_s)$. So there exists an integer n, elements $s_1, \ldots, s_n \in S$, elements $i_1, \ldots, i_n \in I$ and open sets $V_1, \ldots, V_n$ respectively of $S_{i_1}, \ldots, S_{i_n}$ such that

$$S = \bigcup_{h=1}^{n} \pi_{i_h}^{-1}(V_h) \qquad \text{with } s_h \in \pi_{i_h}^{-1}(V_h) \text{ and } \pi_{i_h}^{-1}(V_h) \subseteq \varphi^{-1}(\varphi s_h) \text{ for all } h.$$

By hypothesis, I is a directed poset, so we may consider an element i of I which is greater than or equal to $i_1, \ldots, i_n$. For each $s \in S$, there exists $1 \leq h \leq n$ such that $s \in \pi_{i_h}^{-1}(V_h)$. Therefore $\pi_i^{-1}(\pi_i s) \subseteq \pi_{i_h}^{-1}(\pi_{i_h} s) \subseteq \pi_{i_h}^{-1}(V_h) \subseteq \varphi^{-1}(\varphi s_h)$. In particular $\varphi s_h = \varphi s$, so

$$\text{for each } s \in S, \quad \pi_i^{-1}(\pi_i s) \subseteq \varphi^{-1}(\varphi s).$$

Thus, letting $\varphi_i(\pi_i s) = \varphi s$ defines a mapping φ_i from S_i into T which is trivially a morphism and satisfies $\varphi = \varphi_i \circ \pi_i$.

It remains to observe that φ_i is continuous. For this it suffices to verify that $\varphi_i^{-1}(t)$ is closed in S_i for each $t \in T$. By hypothesis, $\varphi^{-1}(t)$ is closed in S, and hence is compact. Since π_i is continuous, it follows that $\varphi_i^{-1}(t) = \pi_i(\varphi^{-1}(t))$ is compact, which concludes the proof. □

To conclude this section, we observe that compact monoids and semigroups, and among them profinite ones, have many useful properties in common with finite ones. In particular, we will use freely the following elementary properties of compact semigroups.

Proposition 1.2 *Let S be a compact semigroup. Then*

a) S contains an idempotent;

b) if $s \leq_{\mathcal{J}} t$ in S and $t \leq_{\mathcal{R}} s$ (resp. $t \leq_{\mathcal{L}} s$, $t \leq_{\mathcal{H}} s$), then $s \mathrel{\mathcal{R}} t$ (resp. $s \mathrel{\mathcal{L}} t$; $s \mathrel{\mathcal{H}} t$);

c) the Green relations $\mathcal{J}$ and $\mathcal{D}$ coincide.

Proof. Let $s \in S$ and let M_s be the closure of the subsemigroup generated by s. Note that M_s is compact and commutative. We claim that M_s admits a unique non-empty minimal closed ideal, namely the intersection of all its closed ideals. Indeed any finite intersection of closed ideals is non-empty, and the claim follows by compactness. Let I_s be this minimal ideal. If $x \in I_s$, then $xI_s = I_s$ by commutativity and minimality of I_s, so I_s is a group, and hence I_s contains an idempotent, thus proving (a).

Now let $s, t \in S$ be such that $s \leq_{\mathcal{J}} t$ and $t \leq_{\mathcal{R}} s$. Then $s = utv$ and $t = sx$ for some $u, v, x \in S$. So, for each $n \geq 0$, $s = u^n s(xv)^n$. By (a), there exists a sequence $\left((xv)^{k_n}\right)_n$ which converges to an idempotent e. Then $s = \lim_n u^{k_n} s(xv)^{2k_n}$. But, for each n, $u^{k_n} s(xv)^{2k_n} = sxv(xv)^{k_n - 1} = tv(xv)^{k_n - 1} \leq_{\mathcal{R}} t$, so $s \leq_{\mathcal{R}} t$ by passing to the limit. This proves (b).

Finally, let $s, t \in S$ with $s \mathcal{J} t$. Then $s = utv$ for some $u, v \in S$. We have $ut \leq_{\mathcal{J}} t \mathcal{J} s$, so $ut \leq_{\mathcal{J}} s \leq_{\mathcal{R}} ut$. By (b), this implies $s \mathcal{R} ut$. Now, $t \mathcal{J} s \mathcal{J} ut$ and $ut \leq_{\mathcal{L}} t$, so $t \mathcal{L} ut$ by (b) again. Thus $s \mathcal{D} t$. This completes the proof of the proposition. □

1.2. Relatively free profinite monoids and semigroups

Let A be a finite set and let $\mathbf{V}$ be a pseudovariety. An *A-generated profinite (resp. pro-$\mathbf{V}$) monoid* is a mapping $\mu\colon A \to M$ into a profinite (resp. pro-$\mathbf{V}$) monoid M such that the submonoid generated by μA is dense in M. By extension, we will say that the profinite (resp. pro-$\mathbf{V}$) monoid M is A-generated if there exists a mapping $\mu\colon A \to M$ as above.

A-generated profinite monoids are the objects of a category whose morphisms are defined as follows: Let $\mu\colon A \to M$ and $\nu\colon A \to N$ be A-generated profinite monoids. A morphism of profinite monoids (that is, a monoid morphism which is continuous) $\varphi\colon M \to N$ is a *morphism of A-generated profinite monoids* from μ to ν if $\nu = \varphi \circ \mu$. In particular, φ must be onto.

Let $\mathbf{V}$ be a pseudovariety of finite monoids and let A be a finite set. The class $\mathbf{V}_A$ of all A-generated profinite monoids in $\mathbf{V}$, together with all the morphisms of A-generated profinite monoids between them, constitutes a category and a directed system. By definition, we let $\overline{\Omega}_A\mathbf{V}$ be its projective limit.

For each A-generated profinite monoid $\mu\colon A \to M$ with $M \in \mathbf{V}$, we may consider the natural extension of μ to a morphism $\mu\colon A^* \to M$. Since M is finite, we have $\mu A^* = M$. By the universal property of projective limits, these morphisms μ induce a morphism $\iota\colon A^* \to \overline{\Omega}_A\mathbf{V}$, which we call the *natural morphism* from A^* into $\overline{\Omega}_A\mathbf{V}$. We let ι_A be the restriction of ι to A. Note that ι need not be injective. However, if $\mathbf{V}$ is not trivial, then ι_A is one-to-one.

Example. Let $\mathbf{V} = \mathbf{Com}$ be the pseudovariety of all commutative monoids. Let A be a finite set, and let a, b be distinct elements of A. Since $\mu(ab) = \mu(ba)$ for each morphism $\mu\colon A^* \to M$ such that $M \in \mathbf{Com}$, we have $\iota(ab) = \iota(ba)$. So ι is not injective. In general, ι is injective if and only if $\mathbf{V}$ satisfies no nontrivial identity on $|A|$ variables.

Note. More precisely, one can verify that ιA^* is the free object of the variety generated by the pseudovariety $\mathbf{V}$. In the sequel, we will often identify the isomorphic monoids ιA^* and $A^*/\ker \iota$. In particular, when $\mathbf{V}$ satisfies no nontrivial identity, we will view A^* as a submonoid of $\overline{\Omega}_A\mathbf{V}$.

Proposition 1.3 *Let A be a finite set and let $\mathbf{V}$ be a pseudovariety of monoids. Then $\overline{\Omega}_A\mathbf{V}$ (together with the mapping ι_A) is the free pro-$\mathbf{V}$ monoid over the finite set A.*

Furthermore, an A-generated finite monoid is in $\mathbf{V}$ if and only if it is a continuous homomorphic image of $\overline{\Omega}_A\mathbf{V}$.

Proof. We first verify that ιA^* is dense in $\overline{\Omega}_A\mathbf{V}$. For each A-generated profinite monoid in $\mathbf{V}$, that is, for each onto morphism $\mu\colon A^* \to M$ with $M \in \mathbf{V}$, let $\pi_\mu\colon \overline{\Omega}_A\mathbf{V} \to M$ be the corresponding canonical morphism. Then $\pi_\mu \circ \iota = \mu$. By definition of the topology on $\overline{\Omega}_A\mathbf{V}$, a basis of neighborhoods of an element $x \in \overline{\Omega}_A\mathbf{V}$ consists of all $\pi_\mu^{-1}(\pi_\mu x)$. But $\mu A^* = M$ for each μ, so if $\pi_\mu x = \mu u$ for some $u \in A^*$, then $\iota u \in \pi_\mu^{-1}(\pi_\mu x)$. Thus ιA^* is dense in $\overline{\Omega}_A\mathbf{V}$.

Let now S be an A-generated pro-$\mathbf{V}$ monoid: there exists a continuous mapping $\sigma\colon A \to S$ such that the submonoid generated by σA is dense in S, and there exists a directed system $(S_i)_i$ of monoids in $\mathbf{V}$ such that $S = \varprojlim(S_i)_i$. Let $\pi_i\colon S \to S_i$ $(i \in I)$ be the corresponding canonical morphisms, which, by Lemma 1.2, can be assumed to be onto. By definition of projective limit, each element x of S is of the form $x = (\pi_i x)_i$. Let $\sigma_i = \pi_i \circ \sigma\colon A \to S_i$. Then the σ_i are A-generated profinite monoids in $\mathbf{V}$. Therefore, considering the canonical morphisms associated with the projective limit defining $\overline{\Omega}_A\mathbf{V}$, we find that there exist continuous morphisms $\varphi_i\colon \overline{\Omega}_A\mathbf{V} \to S_i$ such that $\varphi_i \circ \iota_A = \sigma_i = \pi_i \circ \sigma$. We use the universal property of projective limits once more to see that these morphisms induce an onto continuous morphism $\varphi\colon \overline{\Omega}_A\mathbf{V} \to S$ such that $\pi_i \circ \varphi = \varphi_i$. So $\pi_i \circ \varphi \circ \iota_A = \pi_i \circ \sigma$ for each i, and hence $\varphi \circ \iota_A = \sigma$.

The proof of the last statement of the proposition is an immediate consequence of the definition of $\overline{\Omega}_A\mathbf{V}$ and of Lemma 1.3. □

The following corollary of Proposition 1.3 will turn out to be rather useful.

Corollary 1.4 *Let* $\mathbf{V}$ *and* $\mathbf{W}$ *be pseudovarieties of monoids such that* $\mathbf{W} \subseteq \mathbf{V}$, *let* A *and* B *be finite sets, let* ι *be the natural map from* A^* *into* $\overline{\Omega}_A\mathbf{V}$, *and let* ι_A *be the restriction of* ι *to* A. *If* $\varphi\colon A \to \overline{\Omega}_B\mathbf{W}$ *is a mapping, then* φ *induces a unique continuous morphism* $\bar\varphi\colon \overline{\Omega}_A\mathbf{V} \to \overline{\Omega}_B\mathbf{W}$ *such that* $\bar\varphi \circ \iota_A = \varphi$.

In particular, the identity function of A *induces an onto continuous morphism from* $\overline{\Omega}_A\mathbf{V}$ *onto* $\overline{\Omega}_A\mathbf{W}$. □

The construction of the free pro-$\mathbf{V}$ monoid over a finite set A and of its topology can also be approached from a different angle. We first define an écart d on A^* as follows. For each pair of distinct words u and v of A^*, we let

$$\begin{aligned} r(u,v) &= \min\big\{|S| : S \in \mathbf{V}, \exists \mu\colon A^* \to S, \mu u \neq \mu v\big\} \\ d(u,v) &= 2^{-r(u,v)}. \end{aligned}$$

We also let $d(u,u) = 0$. Then we have

$$d(u,v) = d(v,u) \qquad \text{and} \qquad d(u,v) \leq \max\big(d(u,w), d(w,v)\big)$$

for all $u, v, w \in A^*$. That is, d is an ultrametric écart and as such, d defines a uniform structure on A^*. The resulting topology is called the *pro-*$\mathbf{V}$ *topology* of A^*. One easily verifies that d is a distance (that is, $d(u,v) = 0$ if and only if $u = v$) if and only if $\mathbf{V}$ satisfies no nontrivial identity on $|A|$ variables, and hence if and only if the morphism $\iota\colon A^* \to \overline{\Omega}_A\mathbf{V}$ is injective.

More precisely, $d(u,v) = 0$ if and only if $\iota u = \iota v$. Thus d induces an ultrametric distance on the quotient set $A^*/\ker\iota$, called the *pro-*$\mathbf{V}$ *distance*. If one identifies the isomorphic monoids ιA^* and $A^*/\ker\iota$, one can also verify that the topology thus defined on $A^*/\ker\iota$ coincides with that induced by $\overline{\Omega}_A\mathbf{V}$ on ιA^*. For this reason and since $\overline{\Omega}_A\mathbf{V}$ is compact, $\overline{\Omega}_A\mathbf{V}$ is sometimes called the *pro-*$\mathbf{V}$ *completion* of A^*.

1.3. Examples

The situation is particularly simple when $\mathbf{V}$ admits a finite free object $F_A(\mathbf{V})$ over the set A. This is the case for instance when $\mathbf{V}$ is generated by a finite number of monoids. Then, the category $\mathbf{V}_A$ of A-generated elements of $\mathbf{V}$ admits only finitely many elements, up to isomorphism, and $F_A(\mathbf{V})$ is its initial object by definition. Therefore, we have:

Proposition 1.5 *Let A be a finite set and let $\mathbf{V}$ be a pseudovariety. If $\mathbf{V}$ admits a free object $F_A(\mathbf{V})$ over A, then $\overline{\Omega}_A\mathbf{V} = F_A(\mathbf{V})$. In particular, $\overline{\Omega}_A\mathbf{V}$ is finite and $\overline{\Omega}_A\mathbf{V}$ is an A-generated element of $\mathbf{V}$.* □

A very important example of this situation is given by the pseudovariety $\mathbf{J}_1$ of finite semilattices, or equivalently, of finite idempotent and commutative monoids. It is well-known that $\mathbf{J}_1$ is generated by the 2-element monoid $\{0,1\}$, and that the free object of $\mathbf{J}_1$ over the set A is the set $\mathcal{P}(A)$ of all subsets of A, under union. Let $c\colon A^* \to \mathcal{P}(A)$ be the natural morphism from A^* into $\overline{\Omega}_A\mathbf{J}_1 = \mathcal{P}(A)$. This morphism is called the *content morphism*. For each word $u \in A^*$, the content of u is the set $c(u)$ of all letters of A which occur in u.

Let now $\mathbf{V}$ be a pseudovariety containing $\mathbf{J}_1$. By Corollary 1.4, the identity function of A induces an onto continuous morphism $c_{\mathbf{V}}\colon \overline{\Omega}_A\mathbf{V} \to \mathcal{P}(A)$. Let $\iota\colon A^* \to \overline{\Omega}_A\mathbf{V}$ be the natural morphism. Then the composed morphism $c_{\mathbf{V}} \circ \iota$ is equal to c. Thus, we have been able to extend the notion of content of a word to the elements of $\overline{\Omega}_A\mathbf{V}$. The morphism $c_{\mathbf{V}}$ is also called *content morphism*, and when there is no risk of ambiguity, we write c for $c_{\mathbf{V}}$.

Some information regarding the passage from the special case of a pseudovariety with free objects to the general one is given in the following proposition. Recall that each pseudovariety $\mathbf{V}$ is the union of an increasing sequence of pseudovarieties admitting free objects over each finite alphabet. Indeed, if we identify isomorphic monoids, then $\mathbf{V}$ is at most countable, so that we can enumerate its elements,

$$\mathbf{V} = \{S_0, S_1, \ldots, S_n, \ldots\}.$$

For each $k \geq 0$, let $\mathbf{V}_k$ be the pseudovariety generated by $S_0, \ldots, S_k$. Then $\mathbf{V}_k$ admits free objects over each finite set, and $\mathbf{V} = \bigcup_{k\geq 0}\mathbf{V}_k$.

Proposition 1.6 *Let A be a finite set and let $(\mathbf{V}_k)_{k\geq 0}$ be a sequence of pseudovarieties such that $\mathbf{V}_k \subseteq \mathbf{V}_{k+1}$ for all $k \geq 0$. Let $\mathbf{V} = \bigcup_k \mathbf{V}_k$. Then $\overline{\Omega}_A\mathbf{V} = \varprojlim(\overline{\Omega}_A\mathbf{V}_k)_k$.*

Proof. According to Corollary 1.4, we may consider the continuous morphisms π_k from $\overline{\Omega}_A\mathbf{V}$ onto $\overline{\Omega}_A\mathbf{V}_k$ induced by the identity mapping of A^*. These morphisms are easily seen to constitute a directed system.

By Lemma 1.1, it suffices to show that, for all $x, y \in \overline{\Omega}_A\mathbf{V}$, if $\pi_k x = \pi_k y$ for all k, then $x = y$. But, for each A-generated monoid M in $\mathbf{V}$, there exists $k \geq 0$ such that $M \in \mathbf{V}_k$. So the canonical morphism π_M from $\overline{\Omega}_A\mathbf{V}$ onto M factors through π_k, and hence $\pi_k x = \pi_k y$ implies $\pi_M x = \pi_M y$. Since this holds for each M in $\mathbf{V}$, and since $\overline{\Omega}_A\mathbf{V}$ is the projective limit of the A-generated elements of $\mathbf{V}$, this implies $x = y$. □

We now consider a few simple, but nontrivial, examples, namely the pseudovarieties of semigroups **Nil**, **K**, $\mathbf{K}^{\mathrm{r}}$ and **LI**. These pseudovarieties are defined by

$$\begin{aligned}
S \in \mathbf{Nil} &\iff \exists k,\ \forall s_1, \ldots, s_k, t \in S,\ s_1 \cdots s_k t = s_1 \cdots s_k = t s_1 \cdots s_k,\\
S \in \mathbf{K} &\iff \exists k,\ \forall s_1, \ldots, s_k, t \in S,\ s_1 \cdots s_k t = s_1 \cdots s_k,\\
S \in \mathbf{K}^{\mathrm{r}} &\iff \exists k,\ \forall s_1, \ldots, s_k, t \in S,\ t s_1 \cdots s_k = s_1 \cdots s_k,\\
S \in \mathbf{LI} &\iff \exists k,\ \forall s_1, \ldots, s_k, t \in S,\ s_1 \cdots s_k t s_1 \cdots s_k = s_1 \cdots s_k.
\end{aligned}$$

The semigroups in **Nil** are called *nilpotent*, and those in **LI** are called *locally trivial*.

Let A be a fixed finite alphabet. For each $k \geq 1$, let I_k be the ideal of A^+ consisting of all words of length at least k. One can verify that each $S_k = A^+/I_k$ is a nilpotent semigroup. Moreover, each morphism from A^+ onto a nilpotent semigroup can be factored through some S_k. Therefore, $\overline{\Omega}_A\mathbf{Nil} = \varprojlim(S_k)_{k\geq 0}$. Note that the semigroup S_k is the set of all words of length less than k, with a zero added and with product given, for all words u and v of length less than k by

$$uv = \begin{cases} uv & \text{if } uv \text{ is of length less than } k \\ 0 & \text{otherwise.} \end{cases}$$

In particular, this means that the natural morphism $\iota\colon A^+ \to \overline{\Omega}_A\mathbf{Nil}$ is injective.

As in the first example given in section 1.1, it is not difficult to verify that the semigroup $\overline{\Omega}_A\mathbf{Nil}$ is obtained from A^+ by adding a zero 0. A basis of clopen sets for its topology is the family of all $\{w\}$ $(w \in A^+)$ and the sets $I_k \cup \{0\}$ $(k \geq 1)$. So $\overline{\Omega}_A\mathbf{Nil}$ is the one-point compactification of A^+, where A^+ is endowed with the discrete topology. Moreover, one can verify that a sequence of words $(u_n)_n$ of A^+ converges in $\overline{\Omega}_A\mathbf{Nil}$ (if we identify A^+ and ιA^+) if and only if, either it is ultimately constant, or the lengths of the u_n tend towards infinity. In this last case, the sequence $(u_n)_n$ tends to 0.

We now turn to **K**. Again one can define a sequence of A-generated semigroups S_k in **K** such that $\overline{\Omega}_A\mathbf{K} = \varprojlim(S_k)_{k\geq 0}$, namely, by letting S_k be defined by the set of generators A and by the relation $y_1 \cdots y_k z = y_1 \cdots y_k$. Informally speaking, the projection of A^+ onto S_k tests prefixes of length k, and S_k can be identified with the set of all words of length less than or equal to k, with product

$$uv = \begin{cases} uv & \text{if } uv \text{ is of length less than or equal to } k \\ \text{prefix of length } k \text{ of } uv & \text{otherwise.} \end{cases}$$

Again, the natural morphism $\iota\colon A^+ \to \overline{\Omega}_A\mathbf{K}$ is injective. Moreover, one can check that a convergent sequence of words $(u_n)_n$ is either ultimately constant, or it satisfies

$$\forall k \quad \exists n_k \quad \forall n, m \geq n_k \qquad u_n \text{ and } u_m \text{ have identical prefixes of length } k.$$

So $\overline{\Omega}_A\mathbf{K}$ can be identified with the set $A^\infty = A^+ \cup A^\omega$, where A^ω denotes the set of (right) infinite words over A. In particular, $\overline{\Omega}_A\mathbf{K}$ is uncountable when $|A| \geq 2$. The product in A^∞ is extended from the product in A^+ by letting $ww' = w$ if w is an infinite word. A basis of clopen sets for the topology of $\overline{\Omega}_A\mathbf{K} = \varprojlim(S_k)_{k\geq 0}$ is the family of all $\{w\}$ and wA^∞ ($w \in A^+$). Note that wA^∞ is the family of all finite and infinite words with prefix w. In particular, the topology on A^+ is discrete, while the topology on $A^\omega = A^{\mathbb{N}}$ is the product topology of the discrete topology on A, that is, A^ω is a Cantor set. Thus, $\overline{\Omega}_A\mathbf{K}$ can be sketchily described as a "nilpotent extension of a Cantor set".

Dually $\overline{\Omega}_A\mathbf{K}^{\mathrm{r}}$ can be identified with ${}^\infty A = A^+ \cup {}^\omega A$, where ${}^\omega A$ is the set of left infinite words over A.

Finally, similar considerations allow the computation of $\overline{\Omega}_A\mathbf{LI}$. One can show that $\overline{\Omega}_A\mathbf{LI}$ is a nilpotent extension of an uncountable rectangular band. More precisely, $\overline{\Omega}_A\mathbf{LI} = A^+ \cup (A^\omega \times {}^\omega A)$. A^+ has the discrete topology and $A^\omega \times {}^\omega A$ the product topology. Finally $A^\omega \times {}^\omega A$ is a rectangular band and if $w \in A^+$, $w_1 \in A^\omega$ and $w \in {}^\omega A$, then $w(w_1, w_2) = (ww_1, w_2)$ and $(w_1, w_2)w = (w_1, w_2w)$. A basis of clopen sets for this topology is the family of all $\{w\}$ and $(wA^\omega \times {}^\omega Aw') \cup wA^+w'$, where w and w' are finite words.

We will refer in section 2 another example of complete computation of a free pro-$\mathbf{V}$ monoid, namely the free pro-$\mathbf{J}$ monoid ($\mathbf{J}$ is the pseudovariety of $\mathcal{J}$-trivial monoids). The description of the free pro-$\mathbf{J}$ monoid will be obtained by a rather different approach.

1.4. Implicit operations and pro-V monoids

The theory of implicit operations constitutes yet another way to approach the profinite completions of the finitely generated free monoids induced by a pseudovariety $\mathbf{V}$. Informally speaking, an implicit operation on $\mathbf{V}$ is any operation that makes sense on each element of $\mathbf{V}$ and is preserved under morphisms between elements of $\mathbf{V}$. More precisely, if A is a finite set, an *A-ary implicit operation π on* $\mathbf{V}$ is a family $\pi = (\pi_S)$ indexed by the elements S of $\mathbf{V}$, such that π_S is a mapping from S^A into S, and such that, for each morphism $\varphi\colon S \to T$ between elements of $\mathbf{V}$, the following diagram is commutative,

$$\begin{array}{ccc} S^A & \xrightarrow{\pi_S} & S \\ \big\downarrow{\scriptstyle \varphi^A} & & \big\downarrow{\scriptstyle \varphi} \\ T^A & \xrightarrow{\pi_T} & T \end{array}$$

that is, such that $\varphi\pi_S\big((s_a)_a\big) = \pi_T\big((\varphi s_a)_a\big)$ for all $(s_a)_{a\in A}$ in S^A.

An unsophisticated, yet fundamental, class of examples is given by the words on some fixed countable alphabet $\{x_1, \ldots, x_n, \ldots\}$. For instance, the word $u = x_1x_2^2x_3x_1$ defines a 3-ary operation u_S on any monoid S, by letting, for all s_1, s_2 and s_3 in S, $u_S(s_1, s_2, s_3) = s_1s_2^2s_3s_1$. It is not difficult to verify that, for any pseudovariety $\mathbf{V}$, $u = (u_S)_{S\in\mathbf{V}}$ is an implicit operation. In general, a word on the alphabet A defines an A-ary operation. Such implicit operations, which are defined by words, are called *explicit operations.*

In general, there are implicit operations which are not explicit. The easiest to describe is the ω-power. Let S be a finite monoid. For each $s \in S$, let s^ω be the unique idempotent in the subsemigroup of S generated by s. It is not difficult to verify that, for any morphism $\varphi\colon S \to T$ and any $s \in S$, we have $\varphi(s^\omega) = (\varphi s)^\omega$. So, for any pseudovariety $\mathbf{V}$, $(\ )^\omega$ is a unary implicit operation on $\mathbf{V}$. For $\mathbf{V} = \mathbf{M}$, the pseudovariety of all finite monoids, or for $\mathbf{V} = \mathbf{Com}$, this operation is not explicit. Of course, in other pseudovarieties, it may be explicit. For instance in any nontrivial pseudovariety of finite idempotent monoids, it is also defined by the word x_1. In the pseudovariety $(\mathbb{Z}/2\mathbb{Z})$ generated by the cyclic group of order 2, it is defined by x_1^2.

For each finite set A, pointwise multiplication makes the set of implicit operations a monoid: If π and π' are A-ary implicit operations on $\mathbf{V}$, the implicit operation $\pi\pi'$ is defined, for each $S \in \mathbf{V}$ and each $(s_a)_{a\in A} \in S^A$, by $(\pi\pi')_S\big((s_a)_a\big) = \pi_S\big((s_a)_a\big)\ \pi'_S\big((s_a)_a\big)$. We denote this monoid by $I_A(\mathbf{V})$. Its unit element is the explicit operation induced by the empty word.

Proposition 1.7 *Let* $\mathbf{V}$ *be a pseudovariety and let* A *be a finite set. Then the monoids* $I_A(\mathbf{V})$ *and* $\overline{\Omega}_A\mathbf{V}$ *are naturally isomorphic.*

Proof. For each A-generated profinite monoid $\mu\colon A \to M$ with $M \in \mathbf{V}$, we define a mapping φ_μ, from $I_A(\mathbf{V})$ into M, by letting $\varphi_\mu(\pi) = \pi_M\big((\mu a)_{a\in A}\big)$. This mapping is easily seen to be a morphism, and the φ_μ constitute a directed system of morphisms. By Lemma 1.1, the induced morphism $\varphi\colon I_A(\mathbf{V}) \to \overline{\Omega}_A\mathbf{V}$ is one-to-one.

The reciprocal morphism ψ, from $\overline{\Omega}_A\mathbf{V}$ into $I_A(\mathbf{V})$, is constructed as follows. For each $S \in \mathbf{V}$ and for each A-tuple $s = (s_a)_a \in S^A$, we let μ_s be the mapping from A into S which maps each element a of A to s_a. Let $\hat\mu_s\colon \overline{\Omega}_A\mathbf{V} \to S$ be the corresponding canonical morphism. In particular, $\hat\mu_s \circ \iota_A = \mu_s$. For each element x of $\overline{\Omega}_A\mathbf{V}$, we define $(\psi x)_S$ by

$$\begin{aligned}(\psi x)_S\colon S^A &\longrightarrow S\\ s = (s_a)_a &\longmapsto \hat\mu_s x.\end{aligned}$$

It is not difficult to verify that $\psi x = (\psi x)_S$ defines an implicit operation on $\mathbf{V}$ and that ψ is the reciprocal morphism of φ. □

In the sequel, we drop the notation $I_A(\mathbf{V})$, and we identify the monoid of A-ary implicit operations on $\mathbf{V}$ with $\overline{\Omega}_A\mathbf{V}$. It will sometimes be useful to view the elements of the free pro-$\mathbf{V}$ monoid as implicit operations on $\mathbf{V}$. In this context, the results of section 1.2 translate as follows.

- The A-ary explicit operations on $\mathbf{V}$ form a dense submonoid of $\overline{\Omega}_A\mathbf{V}$. In the isomorphism between the monoid of implicit operations and $\overline{\Omega}_A\mathbf{V}$, this submonoid is mapped onto ιA^*, where ι is the natural morphism from A^* into $\overline{\Omega}_A\mathbf{V}$.
- A sequence $(\pi_n)_{n\geq 0}$ of A-ary implicit operations on $\mathbf{V}$ converges to an A-ary implicit operation π if, for each $S \in \mathbf{V}$, $\pi_S = (\pi_n)_S$ for all n large enough. That is, if and only if the sequence $(\pi_n)_n$ eventually coincides with π on each element S of $\mathbf{V}$.

- The distance on $A^*/\ker\iota$, that is, on the monoid of A-ary explicit operations on $\mathbf{V}$, which was defined at the end of section 1.2 can be extended in a natural way to a distance on the monoid of all A-ary implicit operations on $\mathbf{V}$ by letting, for each pair (π, π') of distinct A-ary implicit operations

$$
\begin{aligned}
r(\pi, \pi') &= \min \{|S| : S \in \mathbf{V}, \pi_S \neq \pi'_S\} \\
d(\pi, \pi') &= 2^{-r(\pi,\pi')},
\end{aligned}
$$

and by letting $d(\pi, \pi) = 0$. This distance defines the topology of $\overline{\Omega}_A\mathbf{V}$.

Example. The ω-power $(\ )^\omega$ is the limit of the sequence of explicit operations $(a^{n!})_n$. Indeed, for each monoid S and each element s of S, the idempotent s^ω is equal to $s^{n!}$ for all $n \geq |S|$.

1.5. Reiterman's theorem

Let us recall the following important consequence of Corollary 1.4: if $\mathbf{V}$ and $\mathbf{W}$ are pseudovarieties such that $\mathbf{W} \subseteq \mathbf{V}$, and if A is a finite set, then id_{A^*} induces a continuous morphism from $\overline{\Omega}_A\mathbf{V}$ onto $\overline{\Omega}_A\mathbf{W}$, called the *natural projection* of $\overline{\Omega}_A\mathbf{V}$ onto $\overline{\Omega}_A\mathbf{W}$.

We will see that the natural projections of $\overline{\Omega}_A\mathbf{V}$ onto $\overline{\Omega}_A\mathbf{W}$ are injective for all finite sets A if and only if $\mathbf{V} = \mathbf{W}$. This very useful result will in fact appear as a consequence of a more precise statement, due to Reiterman, which is stated in terms of pseudoidentities.

Let $\mathbf{V}$ be a pseudovariety and let A be a finite set. A *pseudoidentity* on $\mathbf{V}$ on the set A (or, in $|A|$ variables) is a pair of elements (u, v) of the free pro-$\mathbf{V}$ monoid $\overline{\Omega}_A\mathbf{V}$ (we also say *the pseudoidentity* $u = v$). It is said to be nontrivial if the elements u and v of $\overline{\Omega}_A\mathbf{V}$ are distinct. Recall that an *identity* in $|A|$ variables is a pair (u, v) of elements of the free monoid A^* for some finite set A. Since each element of $\overline{\Omega}_A\mathbf{V}$ is the limit of a sequence of words, or rather of a sequence of explicit operations, we may view the pseudoidentity (u, v) as the limit of a sequence of identities.

Let S be a pro-$\mathbf{V}$ monoid. Recall that each mapping $\mu \colon A \to S$ induces a continuous morphism $\mu \colon \overline{\Omega}_A\mathbf{V} \to S$. We say that S *satisfies the pseudoidentity* (u, v), and we write $S \models u = v$ if, for each map $\mu \colon A \to S$, we have $\mu u = \mu v$. This generalizes the notion of S satisfying an identity. Note that the set of pseudoidentities of $\mathbf{V}$ on the set A satisfied by a given pro-$\mathbf{V}$ monoid S is a closed subset of $\overline{\Omega}_A\mathbf{V} \times \overline{\Omega}_A\mathbf{V}$.

Example. If $S = \varprojlim(S_i)_{i \in I}$, then S satisfies all pseudoidentities which are satisfied by the S_i. In particular, the free pro-$\mathbf{V}$ monoid $\overline{\Omega}_A\mathbf{V}$ satisfies all pseudoidentities which are satisfied by the A-generated elements of $\mathbf{V}$.

Let now Σ be a set of pseudoidentities on $\mathbf{V}$ (not necessarily involving a bounded number of variables). We let $[\![\Sigma]\!]_\mathbf{V}$ be the class of all elements of $\mathbf{V}$ satisfying all pseudoidentities in Σ. For instance, if $\mathbf{S}$, $\mathbf{M}$, $\mathbf{G}$ and $\mathbf{Ab}$ denote respectively the pseudovarieties of all semigroups, monoids, groups and abelian groups, then

$$\begin{aligned}\mathbf{J}_1 &= [\![xy = yx, x^2 = x]\!]_{\mathbf{M}},\\ \mathbf{LI} &= [\![x^\omega y x^\omega = x^\omega]\!]_{\mathbf{S}},\\ \mathbf{K} &= [\![x^\omega y = x^\omega]\!]_{\mathbf{S}},\\ \mathbf{Nil} &= [\![x^\omega y = x^\omega, yx^\omega = x^\omega]\!]_{\mathbf{S}} = [\![yx^\omega = x^\omega]\!]_{\mathbf{K}},\\ \mathbf{G} &= [\![x^\omega y = y, yx^\omega = y]\!]_{\mathbf{M}},\\ \mathbf{Ab} &= [\![xy = yx]\!]_{\mathbf{G}}.\end{aligned}$$

We can now state Reiterman's theorem.

Theorem 1.8 *Let* $\mathbf{V}$ *be a pseudovariety and let* $\mathbf{W} \subseteq \mathbf{V}$ *be a subclass of* $\mathbf{V}$. *Then* $\mathbf{W}$ *is a sub-pseudovariety if and only if there exists a class* Σ *of pseudoidentities on* $\mathbf{V}$ *such that* $\mathbf{W} = [\![\Sigma]\!]_{\mathbf{V}}$.

Proof. It is not difficult to verify that, if Σ is a class of pseudoidentities on $\mathbf{V}$, then $[\![\Sigma]\!]_{\mathbf{V}}$ is closed under taking submonoids, homomorphic images and finitary direct products. That is, $[\![\Sigma]\!]_{\mathbf{V}}$ is a sub-pseudovariety of $\mathbf{V}$.

We now consider a sub-pseudovariety $\mathbf{W}$ of $\mathbf{V}$, and the class Σ of all pseudoidentities on $\mathbf{V}$ which are satisfied by all elements of $\mathbf{W}$. Let S be a finite monoid in $\mathbf{V}$, satisfying all pseudoidentities in Σ. We need to show that $S \in \mathbf{W}$.

Since S is finite, there exists a finite set A and an onto continuous morphism $\sigma\colon \overline{\Omega}_A\mathbf{V} \to S$. Let also π be the natural projection of $\overline{\Omega}_A\mathbf{V}$ onto $\overline{\Omega}_A\mathbf{W}$. By definition, $\ker\pi$ is exactly the subset of Σ consisting of all pseudoidentities on $\mathbf{V}$ on the set A which are satisfied by the elements of $\mathbf{W}$. Thus, the hypothesis made on S implies that $\ker\pi \subseteq \ker\sigma$. Therefore σ factors through π, that is, there exists an onto morphism $\tau\colon \overline{\Omega}_A\mathbf{W} \to S$ such that $\sigma = \tau\circ\pi$. Note that, for each $s \in S$, $\sigma^{-1}(s)$ is closed and hence compact, so that, by continuity of π, $\tau^{-1}(s) = \pi(\sigma^{-1}(s))$ is closed. Therefore, τ is continuous. By Proposition 1.3, this implies $S \in \mathbf{W}$. □

We will often use the following corollary of Reiterman's theorem.

Corollary 1.9 *Let* $\mathbf{V}$ *and* $\mathbf{W}$ *be pseudovarieties such that* $\mathbf{W} \subseteq \mathbf{V}$. *The natural projections of* $\overline{\Omega}_A\mathbf{V}$ *onto* $\overline{\Omega}_A\mathbf{W}$ *are injective for all finite sets* A *if and only if* $\mathbf{V} = \mathbf{W}$.

Proof. It suffices to notice that the projection of $\overline{\Omega}_A\mathbf{V}$ onto $\overline{\Omega}_A\mathbf{W}$ is injective if and only if $\mathbf{W}$ satisfies no nontrivial pseudoidentity of $\mathbf{V}$ on the set A. □

Note. Recall that we say that a monoid S ultimately satisfies the sequence of identities $(u_n = v_n)_n$ if it satisfies the identities $u_n = v_n$ for all large enough n. A class of finite monoids is ultimately defined by a sequence of identities $(u_n = v_n)_n$ if it consists exactly of all monoids which ultimately satisfy this sequence. The Eilenberg-Schützenberger theorem states that a class of finite monoids is a pseudovariety if and only if it is ultimately defined by a sequence of identities (see for instance Pin's contribution to this volume). Using the fact that a pseudoidentity can be viewed as the limit of a sequence of identities, it is not difficult to verify that Reiterman's theorem above implies the Eilenberg-Schützenberger theorem. On the other hand, Higgins gave a proof of Reiterman's theorem from the Eilenberg-Schützenberger theorem [17]. Unlike the proof given here, Higgins's proof explicitly avoids the use of the axiom of choice. This axiom intervenes in our treatment in the form of Tikhonov's theorem.

1.6. Languages recognized by a pseudovariety $\mathbf{V}$

There is a very interesting connection between the topology of $\overline{\Omega}_A\mathbf{V}$ and the rational languages of A^* which are recognized by the elements of $\mathbf{V}$, which we call $\mathbf{V}$-*recognizable*. We refer the reader to [19] and Pin's contribution to this volume for background on recognizable languages and varieties of recognizable languages.

In this section, we consider only pseudovarieties $\mathbf{V}$ which satisfy no nontrivial identity, that is, for which the natural morphism $\iota\colon A^* \to \overline{\Omega}_A\mathbf{V}$ is injective for each finite set A. In particular, we identify the free monoid A^* with a submonoid of the free pro-$\mathbf{V}$ monoid $\overline{\Omega}_A\mathbf{V}$. Furthermore, under this hypothesis, each morphism from A^* into a monoid S of $\mathbf{V}$ can be extended to a uniquely defined continuous morphism from $\overline{\Omega}_A\mathbf{V}$ into S.

The main observation in this section, is that the $\mathbf{V}$-recognizable languages over the alphabet A can be characterized in terms of the clopen subsets of $\overline{\Omega}_A\mathbf{V}$.

First we notice that, since A^* is dense in $\overline{\Omega}_A\mathbf{V}$, we have $K = \overline{K \cap A^*}$ for each clopen subset K of $\overline{\Omega}_A\mathbf{V}$. Next, one can verify that the following are equivalent for each language L in A^*.

i) L is $\mathbf{V}$-recognizable.
ii) $L = K \cap A^*$ for some clopen subset K of $\overline{\Omega}_A\mathbf{V}$.
iii) The closure $\overline{L}$ of L in $\overline{\Omega}_A\mathbf{V}$ is clopen and $L = \overline{L} \cap A^*$.

As a consequence, the closures of the $\mathbf{V}$-recognizable languages in A^* constitute exactly the set of clopen sets of $\overline{\Omega}_A\mathbf{V}$.

It is often an important question to characterize the $\mathbf{V}$-recognizable languages for a given peudovariety $\mathbf{V}$. These languages form a variety of languages, and in particular, on each finite alphabet, they form a Boolean algebra. In this context, we look for simple generating sets of this Boolean algebra, and the following observation turns out to be very useful.

We say that a family $\mathcal{X}$ of subsets of $\overline{\Omega}_A\mathbf{V}$ *separates* the points of $\overline{\Omega}_A\mathbf{V}$ if, for each pair of distinct elements x and y in $\overline{\Omega}_A\mathbf{V}$, there exists an element X of $\mathcal{X}$ such that either $x \in X$ and $y \notin X$, or $x \notin X$ and $y \in X$.

Proposition 1.10 *Let A be a finite alphabet, let $\mathbf{V}$ be a pseudovariety satisfying no nontrivial identity, and let $A^*\mathcal{V}$ be the Boolean algebra of languages over A recognized by $\mathbf{V}$. Let $\mathcal{L}$ be a subset of $A^*\mathcal{V}$ and let $\overline{\mathcal{L}}$ be the set of the topological closures in $\overline{\Omega}_A\mathbf{V}$ of the elements of $\mathcal{L}$.*

The Boolean algebra $A^*\mathcal{V}$ is generated by $\mathcal{L}$ if and only if the points of $\overline{\Omega}_A\mathbf{V}$ are separated by $\overline{\mathcal{L}}$.

Proof. Let $\mathcal{B}$ be the Boolean algebra generated by $\overline{\mathcal{L}}$. By the above remarks, $\mathcal{B}$ consists only of clopen subsets of $\overline{\Omega}_A\mathbf{V}$. Furthermore, the Boolean algebra $A^*\mathcal{V}$ is generated by $\mathcal{L}$ if and only if $\mathcal{B}$ is the class of all clopen subsets of $\overline{\Omega}_A\mathbf{V}$.

Let us first assume that $\overline{\mathcal{L}}$ separates the points of $\overline{\Omega}_A\mathbf{V}$ and let $\mathcal{L}^C$ be the class of all complements in A^* of the languages of $\mathcal{L}$. Let K be a clopen subset of $\overline{\Omega}_A\mathbf{V}$. For each $x \in \overline{\Omega}_A\mathbf{V} \setminus K$ and for each $y \in K$, there exists a language $L_{x,y}$ in $\mathcal{L}$ or $\mathcal{L}^C$, such that $\overline{L_{x,y}}$ contains y and not x. By compactness, one can select a finite collection of points of K, say, $y_1, \dots, y_n$ such that $x \notin \bigcup_{i=1}^n \overline{L_{x,y_i}}$ and $K \subseteq \bigcup_{i=1}^n \overline{L_{x,y_i}}$. Therefore, for each $x \in \overline{\Omega}_A\mathbf{V} \setminus K$, there exists a language L_x such that $x \notin \overline{L_x}$, $K \subseteq \overline{L_x}$ and $\overline{L_x} \in \mathcal{B}$. In particular, $K = \bigcap_{x \notin K} \overline{L_x}$, and by compactness again, there exists a finite collection of points, say $x_1, \dots, x_m$, such that $K = \bigcap_{j=1}^m \overline{L_{x_j}}$. Thus, $K \in \mathcal{B}$.

In order to prove the converse, we assume that $\mathcal{B}$ is the class of all clopen subsets of $\overline{\Omega}_A\mathbf{V}$, and we consider two distinct elements x and y of $\overline{\Omega}_A\mathbf{V}$. The topology of $\overline{\Omega}_A\mathbf{V}$ is compact and totally disconnected, so the clopen sets form a basis for this topology. Therefore, there exists $X \in \mathcal{B}$ such that $x \in X$ and $y \notin X$. But X can be expressed as a finite union of finite intersections of closures of languages in $\mathcal{L}$ and their complements. Note that the complement of a $\mathbf{V}$-recognizable language L is $\mathbf{V}$-recognizable, and that $\overline{A^* \setminus L} = \overline{\Omega}_A\mathbf{V} \setminus \overline{L}$. So there exist finitely many $\mathbf{V}$-recognizable languages $L_1, \dots, L_k$ in $\mathcal{L}$ or $\mathcal{L}^C$ such that $x \in \bigcap_{i=1}^k \overline{L_i}$ and $y \notin \bigcap_{i=1}^k \overline{L_i}$. Therefore, $x \in \overline{L_i}$ and $y \notin \overline{L_i}$ for some $1 \leq i \leq k$. This shows that $\overline{\mathcal{L}}$ separates the points of $\overline{\Omega}_A\mathbf{V}$, which concludes the proof. □

Example. Using the examples developed in section 1.3, we recover the well-known characterizations of the varieties of languages associated with **Nil**, **K**, $\mathbf{K}^{\mathrm{r}}$ and **LI**. For each finite alphabet A, the following hold.

- The Boolean algebra of all **Nil**-recognizable languages of A^+ is generated by the singleton languages.
- The Boolean algebra of all **K**- (resp. $\mathbf{K}^{\mathrm{r}}$-) recognizable languages of A^+ is generated by the languages of the form wA^* (resp. A^*w), where $w \in A^+$.
- The Boolean algebra of all **LI**-recognizable languages of A^+ is generated by the languages of the form wA^* and A^*w, where $w \in A^+$. This last result provides a proof of the equality $\mathbf{LI} = \mathbf{K} \vee \mathbf{K}^{\mathrm{r}}$.

2. Multiplicity of subwords and implicit operations on DS

This section is devoted to the study of implicit operations on sub-pseudovarieties of **DS**, the class of all monoids in which each regular $\mathcal{D}$-class is a subsemigroup. This is a rather large class of pseudovarieties, which contains notably the pseudovariety **J** of $\mathcal{J}$-trivial monoids, and the pseudovariety **G** of all groups.

As we mentioned in the introduction of this paper, the implicit operations on **DS** can be factored as products of explicit and regular operations, and this fact provides sufficient insight to obtain more detailed statements for certain sub-pseudovarieties of **DS**.

Sections 2.1, 2.2 and 2.3 lay the background for this factorization theorem, which is proved in section 2.4. Section 2.5 presents some further results on **DO** and **DG**, the classes of monoids in which the regular $\mathcal{D}$-classes are respectively orthodox semigroups and groups. These results will be used in section 4. Finally, section 2.6 summarizes some results concerning implicit operations on $\mathcal{J}$-trivial monoids. In this case, the monoids of implicit operations are countable, and are completely described.

Apart from section 2.6, this section is self-contained, and again we have not included the bibliographical references in the text. Proposition 2.1 is due to Schützenberger [24]. The results of sections 2.2, 2.3, 2.4, and 2.5 are due to Almeida and Azevedo, and can be found in Azevedo's [13], Almeida and Azevedo's [8], and of course in Almeida's book [6]. Finally, the results of section 2.6 are due to Almeida and can be found in [5] and in [6].

2.1. A FEW WORDS ABOUT **DS**

If **V** is a pseudovariety of semigroups, we let **DV** be the class of all monoids in which each regular $\mathcal{D}$-class is a subsemigroup in **V**. It is not difficult to verify that **DV** is a pseudovariety.

In particular, **DS** is the class of all monoids in which each regular $\mathcal{D}$-class is a subsemigroup, or equivalently, in which each regular element is a group element.

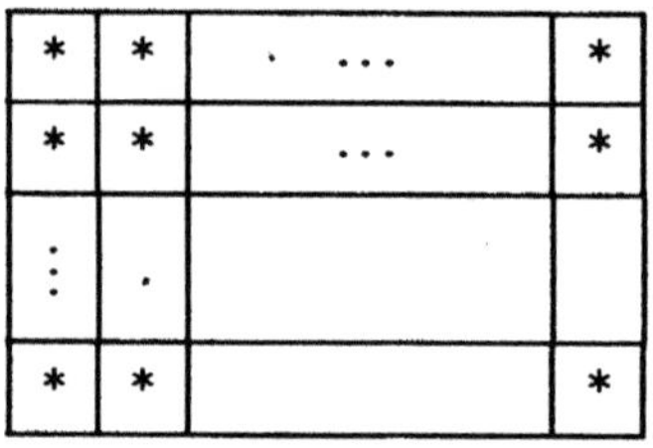

A regular $\mathcal{D}$-class of a monoid in **DS**

This can be expressed by the following equalities.

$$\begin{aligned}\mathbf{DS} &= [\![\big((xy)^\omega (yx)^\omega (xy)^\omega\big)^\omega = (xy)^\omega]\!]_{\mathbf{M}} \\ &= [\![\big((xy)^\omega x\big)^{\omega+1} = (xy)^\omega x]\!]_{\mathbf{M}}.\end{aligned}$$

Before we mention another important characterization of **DS**, let us recall that **LG** is the pseudovariety of all semigroups S such that eSe is a group for each idempotent e of S, or equivalently, the class of all semigroups in which the regular elements form the minimal ideal:

$$\mathbf{LG} = [\![(x^\omega y x^\omega)^\omega = x^\omega]\!]_{\mathbf{S}}.$$

Proposition 2.1 *Let S be a finite monoid. Then the following are equivalent:*

i) $S \in \mathbf{DS}$,

ii) For each $e \in E(S)$, $\{s \in S \mid e \leq_{\mathcal{J}} s\}$ *is a submonoid of* S,

iii) S *is a semilattice of semigroups in* **LG**, *that is, there exists a morphism* $\varphi\colon S \to T$ *with* $T \in \mathbf{J}_1$ *such that* $\varphi^{-1}(t) \in \mathbf{LG}$ *for each* $t \in T$.

Proof. Let us first assume that $S \in \mathbf{DS}$, and let T be the power set $T = \mathcal{P}\big(E(S)\big)$ under intersection. Then T is a monoid in $\mathbf{J}_1$. Also, let $\varphi\colon S \to T$ be defined by

$$\varphi s = \{e \in E(S) : e \leq_{\mathcal{J}} s\}.$$

Let s and s' be elements of S. We want to show that $\varphi(ss') = \varphi s \cap \varphi s'$. If $e \leq_{\mathcal{J}} ss'$, then clearly $e \leq_{\mathcal{J}} s$ and $e \leq_{\mathcal{J}} s'$. Conversely, if $e \leq_{\mathcal{J}} s$ and $e \leq_{\mathcal{J}} s'$, there exists elements u, v, x, y of S such that $e = eusve = exs'ye$. In particular, e, eus and $s'ye$ are $\mathcal{D}$-equivalent regular elements of S. But $S \in \mathbf{DS}$, so $euss'ye$ is also $\mathcal{D}$-equivalent to e, and hence $e \leq_{\mathcal{J}} ss'$. Thus φ is a morphism.

Now let X be an element of T, that is, $X \subseteq E(S)$. Then

$$\varphi^{-1}(X) = \{s \in S : \forall e \in E(S),\ e \leq_{\mathcal{J}} s \Longleftrightarrow e \in X\}.$$

In particular, if $\varphi^{-1}(X) \neq \emptyset$, then X is saturated under the restriction of $\mathcal{J}$ to $E(S)$, and the idempotents of $\varphi^{-1}(X)$ are exactly the $\mathcal{J}$-maximal elements of X. Thus these idempotents are all $\mathcal{D}$-equivalent, which implies $\varphi^{-1}(X) \in \mathbf{LG}$. This proves that (i) implies (iii).

We now verify that (iii) implies (ii). Let $\varphi\colon S \to T$ be a morphism with $T \in \mathbf{J}_1$, and such that $\varphi^{-1}(t) \in \mathbf{LG}$ for each $t \in T$. Let $e \in E(S)$ and let $s, s' \in S$ such that $e \leq_{\mathcal{J}} s, s'$. Then $\varphi e \leq_{\mathcal{J}} \varphi s, \varphi s'$. But $T \in \mathbf{J}_1$, so we have $\varphi e = \varphi(ess')$. Since $\varphi^{-1}(\varphi e) \in \mathbf{LG}$, $e \in E(S)$ and $ess' \leq_{\mathcal{J}} e$, this implies $e \mathrel{\mathcal{D}} ess'$, and hence $e \leq_{\mathcal{J}} ss'$.

Finally, we verify that (ii) implies (i). Let D be a regular $\mathcal{D}$-class of S, let $s, s' \in D$ and let e be an idempotent in D. Then $e \leq_{\mathcal{J}} s, s'$, and by hypothesis, this implies $e \leq_{\mathcal{J}} ss'$. But $ss' \leq_{\mathcal{J}} s \mathrel{\mathcal{J}} e$, so we have $e \mathrel{\mathcal{D}} ss'$, that is, $ss' \in D$: D is a subsemigroup of S. This completes the proof. □

We now list a few important sub-pseudovarieties of **DS**.

- **DO** is the pseudovariety of all monoids in which each regular $\mathcal{D}$-class is an orthodox subsemigroup, that is, a subsemigroup in which a product of idempotents is idempotent.
- **DG** is the pseudovariety of all monoids in which each regular $\mathcal{D}$-class is a group.
- **J** is the variety of $\mathcal{J}$-trivial monoids, that is, the class of aperiodic elements of **DG**, or the class of all the monoids in which each regular $\mathcal{D}$-class is a singleton.

Then we have

$$\mathbf{J}_1 \subset \mathbf{J} \subset \mathbf{DG} \subset \mathbf{DO} \subset \mathbf{DS}$$

and

$$\begin{aligned}
\mathbf{DO} &= [\![(xy)^\omega(yx)^\omega(xy)^\omega = (xy)^\omega]\!]_{\mathbf{M}},\\
\mathbf{DG} &= [\![(xy)^\omega = (yx)^\omega]\!]_{\mathbf{M}},\\
\mathbf{J} &= [\![x^{\omega+1} = x^\omega]\!]_{\mathbf{DG}}\\
&= [\![(xy)^\omega = (yx)^\omega,\ x^{\omega+1} = x^\omega]\!]_{\mathbf{M}}\\
&= [\![(xy)^\omega x = (xy)^\omega,\ y(xy)^\omega = (xy)^\omega]\!]_{\mathbf{M}}.
\end{aligned}$$

2.2. Implicit operations on DS

Since the semigroups of **Nil** are $\mathcal{J}$-trivial and since **Nil** satisfies no nontrivial identity (see section 1.3), it follows that the same holds for **J**. In particular, this means that, for each finite set A, the free monoid A^* is a dense submonoid of $\overline{\Omega}_A\mathbf{J}$, and more generally of $\overline{\Omega}_A\mathbf{V}$ for each pseudovariety $\mathbf{V}$ containing **J**.

The characterization of **DS** given in the previous section has interesting consequences for the free pro-**DS** monoids $\overline{\Omega}_A\mathbf{DS}$. Indeed, by allowing us to consider the content of elements, the passage to these projective limits makes it possible to rewrite the third statement of Proposition 2.1 as follows. Recall that $c\colon \overline{\Omega}_A\mathbf{DS} \to \mathcal{P}(A) = \overline{\Omega}_A\mathbf{J}_1$ denotes the content morphism.

Proposition 2.2 *Let A be a finite set and let $x, y \in \overline{\Omega}_A\mathbf{DS}$. Then*

$$c(y) \subseteq c(x) \quad\Longrightarrow\quad (x^\omega y x^\omega)^\omega = x^\omega.$$

In particular, the content morphism c is such that, for each $t \in \mathcal{P}(A)$, the semigroup $c^{-1}(t)$ satisfies $(x^\omega y x^\omega)^\omega = x^\omega$: its regular elements form a $\mathcal{J}$-class which is its minimal ideal.

Proof. It suffices to show that, for each morphism $\mu\colon A^* \to S$ with $S \in \mathbf{DS}$ and for each $x, y \in \overline{\Omega}_A\mathbf{DS}$ such that $c(y) \subseteq c(x)$, we have $\pi(x^\omega y x^\omega)^\omega = \pi x^\omega$. Here π is the canonical morphism $\pi\colon \overline{\Omega}_A\mathbf{DS} \to S$ extending μ.

By Proposition 2.1, there exists a morphism $\varphi\colon S \to T$ into a monoid $T \in \mathbf{J}_1$ such that $\varphi^{-1}(t) \in \mathbf{LG}$ for each $t \in T$. Since $T \in \mathbf{J}_1$, the morphism $\varphi \circ \pi$ factors through the content morphism c, that is, there exists a morphism $\psi\colon \overline{\Omega}_A\mathbf{J}_1 \to T$ such that the following diagram commutes.

$$\begin{array}{ccc} \overline{\Omega}_A\mathbf{DS} & \xrightarrow{\ \pi\ } & S \\ {\scriptstyle c}\big\downarrow & & \big\downarrow{\scriptstyle \varphi} \\ \overline{\Omega}_A\mathbf{J}_1 & \xrightarrow{\ \psi\ } & T \end{array}$$

By hypothesis, $c(x) = c(x^\omega y x^\omega)$, so $\psi(c(x)) = \psi(c(x^\omega y x^\omega))$ and hence $\varphi(\pi(x)) = \varphi(\pi(x^\omega y x^\omega))$. But $\varphi^{-1}(\varphi(\pi x)) \in \mathbf{LG}$, so $\big((\pi x)^\omega \pi(x^\omega y x^\omega)(\pi x)^\omega\big)^\omega = (\pi x)^\omega$. Now, $(\pi z)^\omega = \pi z^\omega$ for all $z \in \overline{\Omega}_A\mathbf{DS}$, so we have $\pi(x^\omega y x^\omega)^\omega = \pi x^\omega$, thus concluding the proof. □

This result yields a complete description of the relation $\mathcal{J}$ for regular elements.

Corollary 2.3 *Let A be a finite set and let $\mathbf{V}$ be a pseudovariety containing $\mathbf{J}_1$. Then $\mathbf{V} \subseteq \mathbf{DS}$ if and only if*

(1) *for all $x, y \in \overline{\Omega}_A\mathbf{V}$, if x and y are regular, then*

$$x \mathrel{\mathcal{J}} y \quad\Longleftrightarrow\quad c(x) = c(y)$$

For such a pseudovariety, $\overline{\Omega}_A\mathbf{V}$ has $2^{|A|}$ regular $\mathcal{D}$-classes.

Proof. Let us first assume that $\mathbf{V} \subseteq \mathbf{DS}$. Then $\overline{\Omega}_A\mathbf{V}$ is a continuous homomorphic image of $\overline{\Omega}_A\mathbf{DS}$, so the implication of Proposition 2.2 holds in $\overline{\Omega}_A\mathbf{V}$ as well. Let x and y be regular elements of $\overline{\Omega}_A\mathbf{V}$. If $x \mathrel{\mathcal{J}} y$, then $c(x) \mathrel{\mathcal{J}} c(y)$ and hence $c(x) = c(y)$.

Conversely, if $c(x) = c(y)$, then $(x^\omega y x^\omega)^\omega = x^\omega$ by Proposition 2.2. Moreover, by regularity, x can be written as $x = (xz)^\omega x$ for some z. But $\mathbf{DS}$ satisfies $\big((uv)^\omega u\big)^{\omega+1} = (uv)^\omega u$, so $x^{\omega+1} = x$. Thus $x \leq_{\mathcal{J}} y$ and, by symmetry, $x \mathrel{\mathcal{J}} y$.

Let us now assume that $\mathbf{V}$ satisfies (1), and let $x, y \in \overline{\Omega}_A\mathbf{V}$. Then $(xy)^\omega$ and $\big((xy)^\omega(yx)^\omega(xy)^\omega\big)^\omega$ are idempotents with equal content, and hence they are $\mathcal{J}$-equivalent. But $\big((xy)^\omega(yx)^\omega(xy)^\omega\big)^\omega \leq_{\mathcal{H}} (xy)^\omega$, so

$$\mathbf{V} \models \big((xy)^\omega(yx)^\omega(xy)^\omega\big)^\omega = (xy)^\omega$$

and hence $\mathbf{V} \subseteq \mathbf{DS}$. □

2.3. Multiplicity of subwords and regular elements of $\overline{\Omega}_A\mathbf{DS}$

Let us now pay closer attention to the regular elements of $\overline{\Omega}_A\mathbf{DS}$. We first notice the following general property of regular elements of relatively free profinite monoids.

Lemma 2.1 *Let* $\mathbf{V}$ *be a pseudovariety, let* A *be a finite set, and let* $x \in \overline{\Omega}_A\mathbf{V}$. *Then* x *is regular if and only if, for each* A*-generated monoid* $\mu\colon A \to S$ *in* $\mathbf{V}_A$, πx *is a regular element of* S, *where* $\pi\colon \overline{\Omega}_A\mathbf{V} \to S$ *is the corresponding canonical morphism.*

Proof. If x is regular, then $x = xyx$ for some $y \in \overline{\Omega}_A\mathbf{V}$. Therefore, if $\pi\colon \overline{\Omega}_A\mathbf{V} \to S$ is a morphism, then $\pi x = \pi x\ \pi y\ \pi x$, so πx is regular.

We now prove the converse implication. Up to the identification of isomorphic elements, $\mathbf{V}_A$ has only countably many elements $\mu_n\colon A \to S_n$ $(n \geq 0)$. For each $n \geq 0$, we let $\mu'_n\colon A \to S_1 \times \cdots \times S_n$ be the direct product of $\mu_1, \ldots, \mu_n$, we let $\pi_n\colon \overline{\Omega}_A\mathbf{V} \to S_1 \times \cdots \times S_n$ be the corresponding canonical morphism, and we pick $y_n \in \overline{\Omega}_A\mathbf{V}$ such that $\pi_n(xy_nx) = \pi_n x$. By compactness, the sequence $(y_n)_{n\geq 0}$ admits a convergent subsequence $(y_{k_n})_n$. Let $y = \lim_n y_{k_n}$. Then, for each $n \geq 0$, $\pi_{k_n} y = \pi_{k_n} y_m$ for all m large enough, and hence $\pi_{k_n}(xyx) = \pi_{k_n} x$. But the definition of the π_n then implies that, for each $n \geq 0$, $\pi_n(xyx) = \pi_n x$. Therefore $xyx = x$ and x is regular. □

Recall that a word $u = a_1 \cdots a_r \in A^+$ $(a_i \in A)$ is a subword of a word $v \in A^*$ if v can be factored as $v = v_0a_1v_1 \cdots a_rv_r$ $(v_i \in A^*)$. Thus, if we consider words as finite sequences of letters, the subwords of a word v are given by its subsequences.

If $u, v \in A^*$ and $u \neq 1$, we let

$$\begin{bmatrix} v \\ u \end{bmatrix} = \max\left\{r \geq 0 : u^r \text{ is a subword of } v\right\}.$$

Example. $\left[\begin{smallmatrix} abaab \\ a \end{smallmatrix}\right] = 3$ and $\left[\begin{smallmatrix} abaab \\ ab \end{smallmatrix}\right] = 2$.

For each word $u = a_1 \cdots a_k \in A^*$ $(a_i \in A)$, we let $L(u)$ be the language $L(u) = A^*a_1A^* \cdots a_kA^*$ of all words admitting u for a subword. By definition of $\left[\begin{smallmatrix} v \\ u \end{smallmatrix}\right]$, we have

$$\begin{bmatrix} v \\ u \end{bmatrix} \geq r \iff v \in L(u^r).$$

We will use the following observation.

Lemma 2.2 *For each finite set* A *and for each word* $u \in A^*$, *the language* $L(u)$ *is* $\mathbf{J}$*-recognizable.*

Proof. Let $u \in A^+$ and let k be the length of u. Let also $\sim$ be the syntactic congruence of $L(u)$ and let $x, y \in A^*$. To prove that the syntactic monoid of $L(u)$ is in $\mathbf{J}$, it suffices to prove that $x^{k+1} \sim x^k$ and $(xy)^k \sim (yx)^k$.

Let $r, s \in A^*$. Since rx^ks is a subword of $rx^{k+1}s$, $rx^ks \in L(u)$ implies $rx^{k+1}s \in L(u)$. Conversely, if $rx^{k+1}s \in L(u)$, then we can view u as a subsequence of $rx^{k+1}s$ using at most one letter of each of the $k+1$ factors x, and leaving one unused. Therefore, u is also a subword of rx^ks, and $rx^ks \in L(u)$. Thus, $x^{k+1} \sim x^k$.

Similarly, if $r(xy)^k s \in L(u)$, we can view u as a subsequence of $r(xy)^k s$ using at most one letter in each factor xy. This letter can then be found in yx as well, and hence u is a factor of $r(yx)^k s$ too. By symmetry, this shows that $r(xy)^k s \in L(u)$ if and only if $r(yx)^k s \in L(u)$, and hence $(xy)^k \sim (yx)^k$. □

The computation of multiplicity of subwords is directly relevant for the determination of the regular elements of monoids in **DS**.

Proposition 2.4 *Let A be a finite set and let $u, v \in A^+$ be words with equal content. If $S \in$ **DS** and $\left[\begin{smallmatrix} v \\ u \end{smallmatrix}\right] > |S|$, then, for each morphism $\mu\colon A^* \to S$, μv is regular. That is, $S \models v^{\omega+1} = v$.*

Proof. It is possible to factor v as $v = v_1 \cdots v_k$ with $k = \left[\begin{smallmatrix} v \\ u \end{smallmatrix}\right]$ and $\left[\begin{smallmatrix} v_i \\ u \end{smallmatrix}\right] = 1$ for each $1 \leq i \leq k$. For each morphism $\mu\colon A^* \to S$, there exist $0 < m < n \leq k$ such that $\mu(v_1 \cdots v_m) = \mu(v_1 \cdots v_n)$. Therefore $\mu(v_1 \cdots v_m) = \mu(v_1 \cdots v_m)\, \mu(v_{m+1} \cdots v_n)^\ell$ for all $\ell \geq 0$, and hence

$$\begin{aligned} \mu(v_1 \cdots v_m) &= \mu(v_1 \cdots v_m)\, \mu(v_{m+1} \cdots v_n)^{\omega+1} \\ \mu v &= \mu(v_1 \cdots v_m)\, \mu(v_{m+1} \cdots v_n)^{\omega+1}\, \mu(v_{n+1} \cdots v_k). \end{aligned}$$

But the v_i all have the same content, equal to $c(u)$, and $\mu(v_{m+1} \cdots v_n)^{\omega+1}$ is regular. Since $\mu v \leq_{\mathcal{J}} \mu(v_{m+1} \cdots v_n)^{\omega+1}$, it follows from Proposition 2.2 that μv is regular. □

Moreover, for each non-empty word u, it is possible to extend the definition of $\left[\begin{smallmatrix} x \\ u \end{smallmatrix}\right]$ to all implicit operations x on **M**.

Proposition 2.5 *Let* **V** *be a pseudovariety containing* **J**. *For each finite set A and for each $u \in A^+$, the mapping $v \mapsto \left[\begin{smallmatrix} v \\ u \end{smallmatrix}\right]$ is uniformly continuous for the pro-*
V *topology of A^*, and hence can be extended in a unique fashion to a continuous mapping*

$$\begin{aligned} \overline{\Omega}_A \mathbf{V} &\longrightarrow \mathbb{N} \cup \{\infty\} \\ x &\longmapsto \begin{bmatrix} x \\ u \end{bmatrix} \end{aligned}$$

Proof. It suffices to show that, for each $r \geq 0$, there exists $\varepsilon > 0$ such that, for all $v, v' \in A^*$,

$$d(v, v') < \varepsilon \quad \Longrightarrow \quad \begin{bmatrix} v \\ u \end{bmatrix} = \begin{bmatrix} v' \\ u \end{bmatrix} \text{ or } \begin{bmatrix} v \\ u \end{bmatrix}, \begin{bmatrix} v' \\ u \end{bmatrix} \geq r,$$

where d denotes the pro-**V** distance on A^* (see section 1.2).

For each $r \geq 0$, let n_r be the cardinality of the syntactic monoid of $L(u^r)$. Recall that this monoid is in **J**. Therefore, if $d(v, v') < 2^{n_r}$, then v and v' are syntactically equivalent vis-à-vis $L(u^r)$, and hence $\left[\begin{smallmatrix} v \\ u \end{smallmatrix}\right], \left[\begin{smallmatrix} v' \\ u \end{smallmatrix}\right] \geq r$ or $\left[\begin{smallmatrix} v \\ u \end{smallmatrix}\right], \left[\begin{smallmatrix} v' \\ u \end{smallmatrix}\right] < r$. It is now easy to conclude the proof. □

Note. Let **V** and **W** be pseudovarieties such that $\mathbf{J} \subseteq \mathbf{V} \subset \mathbf{W}$, and let π be the natural projection from $\overline{\Omega}_A \mathbf{W}$ onto $\overline{\Omega}_A \mathbf{V}$. Let $u \in A^+$. Both mappings $x \mapsto \left[\begin{smallmatrix} x \\ u \end{smallmatrix}\right]$, defined respectively on $\overline{\Omega}_A \mathbf{V}$ and $\overline{\Omega}_A \mathbf{W}$, are continuous extensions of the same mapping defined on A^*. So for all $x \in \overline{\Omega}_A \mathbf{W}$, $\left[\begin{smallmatrix} x \\ u \end{smallmatrix}\right] = \left[\begin{smallmatrix} \pi x \\ u \end{smallmatrix}\right]$.

This leads to the following important characterization of the regular elements of $\overline{\Omega}_A\mathbf{J}$ and $\overline{\Omega}_A\mathbf{DS}$. We say that a word in A^* is *linear* if no letter occurs more than once in it. There are only finitely many linear words in A^*.

Corollary 2.6 *Let* $\mathbf{V}$ *be a pseudovariety containing* $\mathbf{J}$ *and contained in* $\mathbf{DS}$. *Let* A *be a finite set and let* $x \in \overline{\Omega}_A\mathbf{V}$. *Then the following are equivalent:*

i) x *is regular;*

ii) for each word $u \in A^+$, $\left[\begin{smallmatrix} x \\ u \end{smallmatrix}\right] \in \{0, \infty\}$;

iii) for each linear word $u \in A^+$, $\left[\begin{smallmatrix} x \\ u \end{smallmatrix}\right] \in \{0, \infty\}$.

Furthermore, if x *is regular and* $u \in A^+$, *then* $\left[\begin{smallmatrix} x \\ u \end{smallmatrix}\right] = \infty$ *if and only if* $c(u) \subseteq c(x)$.

Proof. Let x be a regular element of $\overline{\Omega}_A\mathbf{V}$. Then $x = xyx$ for some y, so that $x = (xy)^k x$ for all $k \geq 0$. Let $u \in A^+$ and let $(u_n)_n$ and $(v_n)_n$ be sequences of words converging respectively to x and y in $\overline{\Omega}_A\mathbf{V}$. By continuity, we have

$$\begin{bmatrix} x \\ u \end{bmatrix} = \begin{bmatrix} (xy)^k x \\ u \end{bmatrix} = \lim_n \begin{bmatrix} (u_n v_n)^k u_n \\ u \end{bmatrix} \quad \text{for all } k.$$

But it is not difficult to see that $\left[\begin{smallmatrix} (u_n v_n)^k u_n \\ u \end{smallmatrix}\right] \geq (k+1)\left[\begin{smallmatrix} u_n \\ u \end{smallmatrix}\right]$, so $\left[\begin{smallmatrix} x \\ u \end{smallmatrix}\right] \geq k\left[\begin{smallmatrix} x \\ u \end{smallmatrix}\right]$ for all k, and hence either $\left[\begin{smallmatrix} x \\ u \end{smallmatrix}\right] = 0$ or $\left[\begin{smallmatrix} x \\ u \end{smallmatrix}\right] = \infty$. This proves that (i) implies (ii). Similarly, if $c(u) \subseteq c(x)$ and u is of length k, then $\left[\begin{smallmatrix} (u_n v_n)^{k-1} u_n \\ u \end{smallmatrix}\right] \geq 1$ for all sufficiently large n, and hence $\left[\begin{smallmatrix} x \\ u \end{smallmatrix}\right] = \infty$.

It is clear that (ii) implies (iii). Let us now assume that $\left[\begin{smallmatrix} x \\ u \end{smallmatrix}\right] \in \{0, \infty\}$ for each linear word u in A^+, and let $(v_n)_{n \geq 0}$ be a sequence of words such that $x = \lim_n v_n$. We may assume the v_n to have all the same content as x. Then, up to extracting a subsequence of $(v_n)_n$, we can find a linear word u with the same content as x which is a subword of all the v_n. Indeed, there are only finitely many such linear words, so one of them is a subword of infinitely many v_n. For this word u, $\left[\begin{smallmatrix} x \\ u \end{smallmatrix}\right] = \lim_n \left[\begin{smallmatrix} v_n \\ u \end{smallmatrix}\right] \geq 1$, so $\left[\begin{smallmatrix} x \\ u \end{smallmatrix}\right] = \infty$.

Now let $\mu\colon A^* \to S$ be a morphism with $S \in \mathbf{V}$. Since $\mathbf{J} \subseteq \mathbf{V}$, A^* is a dense submonoid of $\overline{\Omega}_A\mathbf{V}$. We write also μ for the induced morphism $\mu\colon \overline{\Omega}_A\mathbf{V} \to S$. For all n large enough, $\left[\begin{smallmatrix} v_n \\ u \end{smallmatrix}\right] > |S|$ and $\mu v_n = \mu x$, so μx is regular by Proposition 2.4. This implies that x is regular in $\mathbf{DS}$ by Lemma 2.1. □

2.4. Factorization of implicit operations on DS

We are now ready to prove that each implicit operation on $\mathbf{DS}$ is a product of explicit operations and of regular implicit operations: that is, for each finite set A, the monoid $\overline{\Omega}_A\mathbf{DS}$ is generated by its regular elements and by the letters of A. It is immediate that this property is inherited by the monoids $\overline{\Omega}_A\mathbf{V}$ with $\mathbf{V} \subset \mathbf{DS}$, since these monoids are homomorphic images of $\overline{\Omega}_A\mathbf{DS}$.

Let A be a fixed alphabet and let π be the natural projection of $\overline{\Omega}_A\mathbf{M}$ onto $\overline{\Omega}_A\mathbf{DS}$. In terms of implicit operations, if $x \in \overline{\Omega}_A\mathbf{M}$, then πx is the restriction of x to $\mathbf{DS}$. For each $x \in \overline{\Omega}_A\mathbf{M}$, we let

$$\varphi x = \text{Card}\left\{u \in A^* : u \text{ linear}, \begin{bmatrix} x \\ u \end{bmatrix} \neq 0\right\}.$$

First we notice that, for each $a \in c(x)$, $\left[{x \atop a}\right] \geq 1$. Indeed, by continuity of the content morphism c, there exists a sequence $(v_n)_n$ of words converging to x in $\overline{\Omega}_A\mathbf{M}$ and such that $c(v_n) = c(x)$ for all n. For each n, $\left[{v_n \atop a}\right] \geq 1$, and by continuity, $\left[{x \atop a}\right] \geq 1$. Therefore, $\varphi x = 0$ if and only if $c(x) = \emptyset$, that is, if and only if $x = 1$, in which case x is regular.

Let us now consider $n \geq 1$, let us assume that if $x \in \overline{\Omega}_A\mathbf{M}$ and $\varphi x < n$, then x is a product of letters and of implicit operations whose restriction to **DS** is regular, and let us consider $x \in \overline{\Omega}_A\mathbf{M}$ such that $\varphi x = n$. Two cases arise. If $\left[{x \atop u}\right] = \infty$ for each linear word u such that $\left[{x \atop u}\right] \neq 0$, then πx is regular by Corollary 2.6. Otherwise, there exists a linear word u such that $r = \left[{x \atop u}\right] \notin \{0, \infty\}$. By continuity of the mapping $x \mapsto \left[{x \atop u}\right]$, there exists a sequence of words $(v_n)_{n \geq 0}$ converging to x in $\overline{\Omega}_A\mathbf{M}$, and such that $\left[{v_n \atop u}\right] = r$ for all $n \geq 0$. Thus, if $u^r = a_1a_2 \cdots a_k$ $(a_i \in A)$, each v_n can be factored as

$$v_n = v_{n,0}a_1v_{n,1} \cdots a_kv_{n,k} \qquad (v_{n,i} \in A^*).$$

Furthermore, this factorization can be chosen to show the leftmost occurrence of $a_1 \cdots a_k$ as a subword of v_n, that is, we can assume that, for each $1 \leq i \leq k$, the letter a_i does not occur in $v_{n,i-1}$. Now, by compactness and up to extracting a subsequence of $(v_n)_n$, we may assume that each sequence $(v_{n,i})_n$ converges in $\overline{\Omega}_A\mathbf{M}$: let x_i be its limit. Then

$$x = \lim_n v_n = x_0a_1x_1 \cdots a_kx_k.$$

By construction, the word u is not a subword of any of the $v_{n,i}$ $(0 \leq i < k)$ since one of its letters does not occur in $v_{n,i}$. It is not a subword of $v_{n,k}$ either since $\left[{v_n \atop u}\right] = r$. Therefore, for each $0 \leq i \leq k$, $\left[{x_i \atop u}\right] = 0$. On the other hand, let u' be a linear word such that $\left[{x_i \atop u'}\right] \neq 0$. Then u' is a subword of almost all the $v_{n,i}$, and hence of almost all the v_n. Therefore $\left[{x \atop u'}\right] \neq 0$. Thus, for each $0 \leq i \leq k$, $\varphi x_i < \varphi x$.

So, using induction on φx, we can show that x is a product of letters and of implicit operations whose restriction to **DS** is regular.

We can make this statement more precise by noticing the following. Let y be a regular element of $\overline{\Omega}_A\mathbf{DS}$ and let z be an element of $\overline{\Omega}_A\mathbf{DS}$ whose content is contained in that of y. Then yz, zy and y have the same content and, by Proposition 2.2, yz and zy are regular as well.

Theorem 2.7 *Let A be a finite set. Then each element of $\overline{\Omega}_A\mathbf{M}$ admits a factorization of the form*

$$x = u_0x_1u_1 \cdots x_ru_r$$

where

- *for each $0 \leq i \leq r$, u_i is in A^*,*
- *for each $1 \leq i \leq r$, the restriction of x_i to* **DS** *is regular,*
- *for each $1 \leq i \leq r$ (resp. $0 \leq i < r$) such that u_i is not the empty word, the first (resp. last) letter of u_i does not lie in the content of x_i (resp. u_{i+1}),*
- *for each $1 \leq i \leq r$ such that $u_i = 1$, the contents of x_{i-1} and x_i are $\subseteq$-incomparable.* □

Corollary 2.8 *Let* $\mathbf{V}$ *be a pseudovariety contained in* $\mathbf{DS}$ *and let* A *be a finite set. Then* $\overline{\Omega}_A\mathbf{V}$ *is generated by its regular elements and by the letters of* A. □

2.5. Implicit operations on DO

In the case of sub-pseudovarieties of $\mathbf{DO}$, we can go further in the description of the monoids of implicit operations, and in particular of their regular elements. We show that these elements are determined by their idempotent power and by their restriction to groups.

Lemma 2.3 *Let* $S \in \mathbf{DO}$, *let* $e \in E(S)$, *and let* $s, t \in S$ *such that* $e \leq_{\mathcal{J}} s, t$. *Then* $este = esete$.

Proof. By Proposition 2.1, es and te are group elements in the $\mathcal{J}$-class of e. So $este = es(es)^\omega(te)^\omega te$. But $(es)^\omega(te)^\omega$ is idempotent since $S \in \mathbf{DO}$ and $(es)^\omega(te)^\omega \mathcal{H} e$. So $(es)^\omega(te)^\omega = e$, and $este = esete$. □

Proposition 2.9 *Let* $\mathbf{V}$ *be a pseudovariety contained in* $\mathbf{DO}$ *and let* $\mathbf{H} = \mathbf{V} \cap \mathbf{G}$. *Let* A *be a finite set and let* $x, y \in \overline{\Omega}_A\mathbf{V}$. *If* x *and* y *are regular, then* $x = y$ *if and only if* $x^\omega = y^\omega$ *and* $\mathbf{H} \models x = y$.

Proof. If $\mathbf{V} \subseteq \mathbf{G}$, then $\mathbf{V} = \mathbf{H}$ and the proposition is trivial. We now suppose that $\mathbf{V}$ is not a pseudovariety of groups, and hence, that $\mathbf{J}_1 \subseteq \mathbf{V}$. Let $x \in \overline{\Omega}_A\mathbf{V}$ and let $S \in \mathbf{V}$. There exists a word $u \in A^*$ with content $c(u) = c(x)$, and such that x coincides with the explicit operation u on S, that is, $x_S = u_S$.

Let $s = (s_a)_{a \in A} \in S^A$. Then $x_S(s) = u_S(s)$ is a regular element of S: let $e = x_S^\omega(s)$ and let H be its $\mathcal{H}$-class. In particular, $x_S(s) = e\, x_S(s)\, e = e\, u_S(s)\, e \in H$. Moreover, e is equal to a certain product of the s_a ($a \in c(x)$), so $e \leq_{\mathcal{J}} s_a$ for each $a \in c(x)$.

Let $s' = (s'_a)_{a \in A} \in S^A$ be given by $s'_a = es_ae$ if $a \in c(x)$ and $s'_a = e$ otherwise. Using the fact that u is an explicit operation of content $c(x)$, Lemma 2.3 implies $e\, u_S(s)\, e = u_S(s')$, and hence $x_S(s) = x_S(s')$. Now we observe that $es_ae \in H$ for each $a \in c(x)$, so that $x_S(s') = x_H(s')$. Thus, x_S is determined by x_S^ω and by the restrictions of x to the subgroups of S. □

In section 4, we will be interested in certain sub-pseudovarieties of $\mathbf{DG}$. For these pseudovarieties, the above proposition yields the following.

Corollary 2.10 *Let* $\mathbf{V}$ *be a pseudovariety containing* $\mathbf{J}_1$ *and contained in* $\mathbf{DG}$, *and let* $\mathbf{H} = \mathbf{V} \cap \mathbf{G}$. *Let* A *be a finite set and let* $x, y \in \overline{\Omega}_A\mathbf{V}$. *If* x *and* y *are regular, then* $x = y$ *if and only if* $c(x) = c(y)$ *and* $\mathbf{H} \models x = y$.

Proof. This is immediate in view of Proposition 2.9 above, since $x^\omega = y^\omega$ if and only if $c(x) = c(y)$ (Corollary 2.3). □

2.6. Implicit operations on J

In the case of aperiodic elements of **DG**, that is, of $\mathcal{J}$–trivial monoids, the regular elements are all idempotent, so Corollary 2.10 (or, alternately, Corollary 2.3) implies that the idempotents are determined by their contents. In particular, each idempotent is the ω-power of a linear word (in fact, of any word with the appropriate content). Moreover, since each element of $\overline{\Omega}_A\mathbf{J}$ is a product of letters and of idempotents by Theorem 2.7, it follows that $\overline{\Omega}_A\mathbf{J}$ is countable.

In fact, we can state a unique factorization result for the free pro-**J** monoids. Its proof is too long to be included here, and we refer the reader to Almeida [6] or [5].

Theorem 2.11 *Let A be a finite set.*
a) If $x, y \in \overline{\Omega}_A\mathbf{J}$, then $x = y$ if and only if $\left[\begin{smallmatrix} x \\ u \end{smallmatrix}\right] = \left[\begin{smallmatrix} y \\ u \end{smallmatrix}\right]$ for each word $u \in A^+$.
b) If $x \in \overline{\Omega}_A\mathbf{M}$ and if $x = u_0x_1u_1 \cdots x_ru_r$ and $x = v_0y_1v_1 \cdots y_sv_s$ are factorizations as in Theorem 2.7 (with respect to $\mathbf{J}$), then $r = s$, $u_i = v_i$ for all $1 \le i \le r$, and $\mathbf{J} \models x_i = y_i$, that is, $c(x_i) = c(y_i)$ for all $0 \le i \le r$.
c) $\overline{\Omega}_A\mathbf{J}$ is the free monoid with a unary operation $x \mapsto x^\omega$ over A in the variety defined by the identities $(xy)^\omega = (yx)^\omega = (x^\omega y^\omega)^\omega$ and $x^\omega x = xx^\omega = (x^\omega)^\omega = x^\omega$. Moreover, the word problem in $\overline{\Omega}_A\mathbf{J}$ (viewed as such a (2,1)-algebra) is decidable.

As a consequence of this remarkable result, we obtain Simon's theorem on the combinatorial description of **J**-recognizable languages. Recall that, for a word $u = a_1 \cdots a_n \in A^*$, the language $L(u) = A^*a_1A^* \cdots a_nA^*$ is **J**-recognizable (Lemma 2.2).

Theorem 2.12 *Let A be a finite set. The Boolean algebra of all $\mathbf{J}$-recognizable languages in A^* is generated by the $L(u)$ ($u \in A^+$).*

Proof. Recall that a word $v \in A^*$ lies in $L(u)$ ($u \in A^+$) if and only if $\left[\begin{smallmatrix} v \\ u \end{smallmatrix}\right] \ge 1$. By continuity of the mapping $x \mapsto \left[\begin{smallmatrix} x \\ u \end{smallmatrix}\right]$ on $\overline{\Omega}_A\mathbf{J}$, it follows that an element x of $\overline{\Omega}_A\mathbf{J}$ lies in $\overline{L(u)}$ if and only if $\left[\begin{smallmatrix} x \\ u \end{smallmatrix}\right] \ge 1$. Therefore statement (a) of Theorem 2.11 above implies that the points of $\overline{\Omega}_A\mathbf{J}$ are separated by the closures of the languages $L(u)$ ($u \in A^+$). By Proposition 1.10, this suffices to prove the theorem. □

3. Relatively free profinite groups

Cayley graphs provide a convenient tool for working with relatively free profinite groups such as the groups $\overline{\Omega}_A\mathbf{H}$ of implicit operations on a pseudovariety **H** of groups. Since not only the algebraic structure of $\overline{\Omega}_A\mathbf{H}$ is important, we will need to deal with some additional structure besides the usual combinatorial Cayley graphs. We therefore start by reviewing some basic notions and results of the theory of profinite graphs and then pursue to study specifically profinite Cayley graphs of profinite groups.

Section 3.1 introduces profinite graphs in general. Lemmas 3.2 and 3.3 are taken from Zalesskiĭ and Mel'nikov [27] and Ribes and Zalesskiĭ [23], respectively, where (homologically defined) tree-like properties of profinite graphs were investigated. Cayley graphs of some **H**-free profinite groups were shown by Gildenhuys and Ribes

[16] (see also Ribes [22]) to be $\mathbb{A}$-trees for a naturally associated profinite ring (of homological coefficients). The complete characterization of the pseudovarieties of groups $\mathbf{H}$ with this property, due to Almeida and Weil [9], is the theme of section 3.2. Such pseudovarieties of groups, called arborescent, turn out to be particularly amenable to syntactical arguments since certain "reduced factorizations" have a kind of uniqueness property in this case, a topic which is considered in detail in section 3.3.

The main result of section 3.3 (Theorem 3.6) and its applications (Propositions 3.7 and 3.8) were first proved in Almeida and Weil [9]; the proof presented here, in terms of cuts, which is developed in sections 3.1 and 3.3, reformulates in a more geometrical language — and somewhat simplifies — the original proof.

3.1. Profinite graphs

By a *graph* we mean a pair $G = (V, E)$ of disjoint sets endowed with two mappings $\alpha, \omega : E \to V$. The elements of V are called *vertices* and the elements of E are called *edges*. We often write $G = V \cup E$ when the context makes clear which elements are vertices and which are edges. For $e \in E$, αe and ωe are called, respectively, the *beginning* and the *end* vertices of the edge e and we often draw the picture

$$\alpha e \cdot \xrightarrow{e} \cdot\, \omega e$$

to represent this situation. The vertices αe and ωe are said to be the *extremities* of e and to be *joined* by e. Gluing all these pictures via the identification of vertices with the same name, we obtain the usual graphical representation of G.

A *morphism of graphs* $\varphi : (V, E) \to (V', E')$ is given by two functions $\varphi_1 : V \to V'$ and $\varphi_2 : E \to E'$ which respect the operations in the following sense: for each $e \in E$, $\chi\varphi_2 e = \varphi_1 \chi e$ ($\chi \in \{\alpha, \omega\}$). A *quotient* of a graph G is a morphic image of G. Just as in section 1.1, where total algebras were considered, profinite graphs are defined as projective limits of finite graphs. In particular, in a profinite graph, the vertices and the edges each constitute topological spaces which are compact and zero-dimensional and the mappings α and ω are continuous. Whenever we consider graph (quotient) morphisms defined on profinite graphs, we will be referring to continuous graph morphisms.

Let $\mathbb{A}$ be a profinite commutative ring with 1. Viewing $\mathbb{A}$-modules as (universal) algebras of a suitable type, we may also consider profinite $\mathbb{A}$-modules, i.e., projective limits of finite $\mathbb{A}$-modules, as just another case of profinite (universal) algebras. We denote by $\mathbb{A}[S]$ the free profinite $\mathbb{A}$-module over the profinite set S. In particular, if S is finite, then $\mathbb{A}[S]$ is a direct sum of $|S|$ copies of $\mathbb{A}$. In general, if $S = \varprojlim S_i$, then $\mathbb{A}[S] = \varprojlim \mathbb{A}[S_i]$ and the submodule of $\mathbb{A}[S]$ generated by S, which is the free $\mathbb{A}$-module freely generated by S, is only a dense submodule of $\mathbb{A}[S]$. Thus, the elements of $\mathbb{A}[S]$ are limits of (Cauchy) sequences of formal linear combinations of elements of S with coefficients in $\mathbb{A}$.

Now let $G = (V, E)$ be a profinite graph. We define two morphisms of profinite $\mathbb{A}$-modules by

$$\begin{aligned} \partial : \mathbb{A}[E] &\longrightarrow \mathbb{A}[V] \\ e \in E &\longmapsto \omega e - \alpha e \end{aligned} \qquad \text{and} \qquad \begin{aligned} \varepsilon : \mathbb{A}[V] &\longrightarrow \mathbb{A} \\ v \in V &\longmapsto 1 \end{aligned}$$

and we consider the sequence

$$\mathbb{A}[E] \xrightarrow{\partial} \mathbb{A}[V] \xrightarrow{\varepsilon} \mathbb{A}$$

where we note that $\operatorname{Im}\partial \subseteq \operatorname{Ker}\varepsilon$. The corresponding homology $\mathbb{A}$-modules are

$$H_0(G,\mathbb{A}) = \operatorname{Ker}\varepsilon/\operatorname{Im}\partial \quad \text{and} \quad H_1(G,\mathbb{A}) = \operatorname{Ker}\partial.$$

If $\varphi : G \to G'$ is a morphism of profinite graphs then there are associated morphisms of profinite $\mathbb{A}$-modules $\mathbb{A}[E] \to \mathbb{A}[E']$ and $\mathbb{A}[V] \to \mathbb{A}[V']$ and the following diagram commutes

$$\begin{array}{ccccc} \mathbb{A}[E] & \xrightarrow{\partial_G} & \mathbb{A}[V] & \xrightarrow{\varepsilon_G} & \mathbb{A} \\ \downarrow & & \downarrow & & \| \\ \mathbb{A}[E'] & \xrightarrow{\partial_{G'}} & \mathbb{A}[V'] & \xrightarrow{\varepsilon_{G'}} & \mathbb{A} \end{array}$$

It is then just an easy diagram chasing exercise to show that there are also induced morphisms $H_0(G,\mathbb{A}) \to H_0(G',\mathbb{A})$ and $H_1(G,\mathbb{A}) \to H_1(G',\mathbb{A})$, that the first one is onto if φ is onto, that the second one is one-to-one if φ is one-to-one, and that the functors $H_0(\_,\mathbb{A})$ and $H_1(\_,\mathbb{A})$ preserve projective limits.

Now, for a finite graph G, we claim that $H_0(G,\mathbb{A}) = 0$ if and only if G is connected in the usual (undirected) sense. Indeed, if G is not connected, then G has a quotient graph G' with two vertices and no edges except possibly loops; whence, for G', $\operatorname{Im}\partial = 0 \neq \operatorname{Ker}\varepsilon$ and, since the associated mapping $H_0(G,\mathbb{A}) \to H_0(G',\mathbb{A})$ is onto, we deduce that $H_0(G,\mathbb{A}) \neq 0$. Conversely, if G is connected, fixing a vertex v_0, we may define a distance function $d(v)$ on V as the length of the shortest undirected path from v_0 to v. If $\sum_{v\in V}\lambda_v = 0$, i.e., $\sum_{v\in V}\lambda_v v \in \operatorname{Ker}\varepsilon$, and $v_1 \in V$ is such that $d(v_1)$ is maximal among the vertices v for which $\lambda_v \neq 0$, then there is a vertex v_1' at smaller distance from v_0 and an edge e_1 with $\{\alpha e_1, \omega e_1\} = \{v_1, v_1'\}$. Then $\pm\partial(\lambda_{v_1}e_1) = \lambda_{v_1}v_1 - \lambda_{v_1}v_1'$ and so $\sum_{v\in V}\lambda_v v - \big(\pm\partial(\lambda_{v_1}e_1)\big)$ is an element of $\operatorname{Ker}\varepsilon$ in which the coefficient of v_1 is 0. An easy inductive scheme then shows that $\sum_{v\in V}\lambda_v v \in \operatorname{Im}\partial$ and proves the above claim. Since the functor $H_0(\_,\mathbb{A})$ transforms onto morphisms into onto morphisms, it follows that, for a profinite graph G, $H_0(G,\mathbb{A}) = 0$ if and only if all finite quotients of G are connected. We then say that G is *profinitely connected*.

On the other hand, for an arbitrary graph G, if $(e_1,\dots,e_n)$ is an undirected path from v_1 to v_2, then $\partial(\sum_{i=1}^{n}\pm e_i) = v_2 - v_1$ where an edge is given a minus sign if it appears in the wrong direction in the path. So, if G has (finite) cycles, then $H_1(G,\mathbb{A}) \neq 0$. Conversely, if $\sum_{e\in E}\lambda_e e \in \operatorname{Ker}\partial \setminus \{0\}$ and $\lambda_{e_1} \neq 0$, then, as $\sum_{e\in E}\lambda_e(\omega e - \alpha e) = 0$, there is some $e_2 \in E\setminus\{e_1\}$ such that $\lambda_{e_2} \neq 0$ and e_2 has one of its ends equal to, say, ωe_1. Again an appropriate induction scheme allows us to construct arbitrarily long undirected paths in G. Thus, in case G is finite, it follows that G has cycles. Hence a finite graph G is acyclic if and only if $H_1(G,\mathbb{A}) = 0$. Since $H_1(\_,\mathbb{A})$ preserves projective limits, if G is a projective limit of acyclic finite graphs, then $H_1(G,\mathbb{A}) = 0$. However, the converse is not true as the examples in the next section will show. Basically, this is because a quotient of an acyclic graph is not necessarily acyclic.

We say that a profinite graph G is $\mathbb{A}$-*acyclic* if $H_1(G,\mathbb{A}) = 0$ and that G is an $\mathbb{A}$-*tree* if G is profinitely connected and $\mathbb{A}$-acyclic. Since $H_1(\_,\mathbb{A})$ transforms one-to-one morphisms into one-to-one morphisms, a closed subgraph of an $\mathbb{A}$-acyclic graph is also $\mathbb{A}$-acyclic.

The *degree* of a vertex v of a graph G is the number of edges of G with an extremity at v. The *(vertex) degree* of a graph G is the supremum of the degrees of its vertices in case all these degrees are finite, and is ∞ otherwise. We next use more or less standard compactness arguments to show that profinitely connected graphs of finite degree must have long paths.

Lemma 3.1 *Let $G = \varprojlim G_n$ be the projective limit of a projective sequence $G_1 \leftarrow G_2 \leftarrow \cdots$ of finite graphs in which the projection morphisms are onto. Then the following conditions are equivalent:*

i) G has finite degree;

ii) all the vertices of G have finite degree;

iii) there is a bound for the degrees of the G_n $(n \geq 1)$.

Proof. Clearly (i)$\Longrightarrow$(ii) by definition of degree of a graph.

If G has a vertex of degree n, then some G_k must also have a vertex of degree n since the edges of G are separated by the morphisms onto the G_k. Hence (iii)$\Longrightarrow$(i).

Finally, suppose there is no bound for the degrees of the vertices of the G_n $(n \geq 1)$. Then we may assume that, for each $n \geq 1$, G_n has some vertex v_n of degree $\geq n$. Let $\psi_n : G \to G_n$ be the canonical projection and let $\varphi_{mn} : G_m \to G_n$ $(m > n)$ be the onto morphism associated with the projective sequence, so that $\psi_n = \varphi_{mn} \circ \psi_m$. Then, for each $n \geq 1$, there are n distinct edges $e_{n1}, \ldots, e_{nn}$ of G such that, for $i = 1, \ldots, n$, $\psi_n e_{ni}$ has v_n as one of its extremities.

The sequence $(e_{n1})_{n\geq 1}$ of edges of the profinite graph G has some convergent subsequence $(e_{n_k 1})_{k\geq 1}$, say with limit e_1. In turn, the sequence $(e_{n_k 2})_{k\geq 2}$ has some convergent subsequence $(e_{n_{k_l} 2})_{l\geq 1}$, say with limit e_2. Inductively, for each positive integer p, there is a strictly increasing sequence $(n_r)_{r\geq 1}$ of indices such that each of the sequences $(e_{n_r i})_{r\geq i}$ $(i = 1, \ldots, p)$ converges, say to the respective edges e_i. Moreover, this choice can be made so that, at each step, the already chosen edges $e_1, \ldots, e_{p-1}$ are preserved and so we obtain an infinite sequence $e_1, e_2, \ldots$ of edges.

For a fixed p, the convergence of the above sequences implies that, for every k, there is r such that, for each $i \in \{1, \ldots, p\}$, $\psi_k e_{n_r i} = \psi_k e_i$, whence $\psi_k e_i = \psi_k e_{n_r i} = \varphi_{n_r k} \psi_{n_r} e_{n_r i}$. Since the edges $\psi_{n_r} e_{n_r i}$ $(i = 1, \ldots, p)$ have the common extremity v_{n_r}, the edges $\psi_k e_i$ $(i = 1, \ldots, p)$ also have a common extremity. Being true for all $k \geq 1$, this forces $e_1, \ldots, e_p$ to have a common extremity. Since p is arbitrary, this shows that G has infinite degree, thereby proving (ii)$\Longrightarrow$(iii). $\square$

The *diameter* of a graph is the supremum of the lengths of finite paths with no repeated vertices.

Proposition 3.1 *Let G be an infinite profinitely connected graph of finite degree. Then G contains some infinite path without repeated vertices.*

Proof. Let $G = \varprojlim G_n$, a projective limit of a projective sequence of finite connected graphs in which the projection morphisms are onto. Let ψ_n and φ_{mn} be morphisms as in the proof of Lemma 3.1. Note that, since G is infinite, the sizes of the graphs G_n ($n \geq 1$) are not bounded. By Lemma 3.1, the G_n ($n \geq 1$) have bounded degrees and, therefore, have unbounded diameters and so we may assume that each G_n has diameter $\geq n$. For each $n \geq 1$, choose edges $e_{n1}, \ldots, e_{nn}$ of G such that $(\psi_n e_{n1}, \ldots, \psi_n e_{nn})$ is a path without repeated vertices.

Proceeding as in the proof of Lemma 3.1, we conclude that, for each $p \geq 1$, there is a strictly increasing sequence $(n_k)_{k\geq 1}$ of indices such that each of the sequences $(e_{n_r i})_{r\geq 1}$ ($i = 1, \ldots, p$) converges to an edge e_i. Moreover, the argument is such that, having done so for p, the choice for $p+1$ preserves the choice of the first p edges $e_1, \ldots, e_p$, and so we obtain an infinite sequence $e_1, e_2, \ldots$ of edges. Again the argument of the proof of Lemma 3.1 shows that $(e_1, \ldots, e_p)$ is a path without repeated vertices. Hence $(e_1, e_2, \ldots)$ is an infinite path without repeated vertices. □

We proceed with a couple of useful properties of $\mathbb{A}$-trees.

Lemma 3.2 *Let G be a profinite graph.*

a) Let H_1 and H_2 be profinitely connected subgraphs of G. If $H_1 \cap H_2 \neq \emptyset$, then $H_1 \cup H_2$ is profinitely connected.

b) If G is an $\mathbb{A}$-tree and $(H_i)_i$ is a collection of profinitely connected subgraphs of G such that $\bigcap_i H_i \neq \emptyset$, then $\bigcap_i H_i$ is an $\mathbb{A}$-tree.

Proof. (a) In each finite quotient H' of $H = H_1 \cup H_2$, the images of H_1 and H_2 are two connected subgraphs with at least one vertex in common and so H' is connected. Hence H is profinitely connected.

(b) Let $H = \bigcap_i H_i$. Since H is a closed subgraph of G, it suffices to show that H is profinitely connected. Let $H_i = (V_i, E_i)$ for each i and let $H = (V, E)$. Then $\mathbb{A}[E] = \bigcap_i \mathbb{A}[E_i]$ and $\mathbb{A}[V] = \bigcap_i \mathbb{A}[V_i]$. Note that

$$\operatorname{Ker} \varepsilon_H = \mathbb{A}[V] \cap \operatorname{Ker} \varepsilon_G = \bigcap_i \operatorname{Ker} \varepsilon_{H_i} \quad \text{and} \quad \operatorname{Im} \partial_H \subseteq \bigcap_i \operatorname{Im} \partial_{H_i}.$$

Since G is an $\mathbb{A}$-tree, the mapping ∂_G is one-to-one and so the inclusion above must be an equality. Finally, since $\operatorname{Ker} \varepsilon_{H_i} = \operatorname{Im} \partial_{H_i}$ for all i, we conclude that H is profinitely connected. □

In view of Lemma 3.2, for any two vertices u and v of an $\mathbb{A}$-tree G, the intersection of all profinitely connected subgraphs of G which contain u and v is an $\mathbb{A}$-tree and the smallest profinitely connected subgraph containing u and v; it is called the *segment joining u and v* and is denoted by $[u, v]$.

Lemma 3.3 *Let G be an $\mathbb{A}$-tree, let u, v and w be vertices of G, and let H and K be two profinitely connected subgraphs of G. Then:*

a) $[u, v] \subseteq [u, w] \cup [w, v]$;

b) $[u, v] = [u, w] \cup [w, v]$ if and only if $w \in [u, v]$;

c) if $H \cap K \neq \emptyset$, $u \in H$ and $v \in K$, then $[u, v] \cap H \cap K \neq \emptyset$;

d) if $[u,w] \cap [w,v] = \{w\}$, *then* $[u,v] = [u,w] \cup [w,v]$.

Proof. (a) By Lemma 3.2(a), $[u,w] \cup [w,v]$ is a profinitely connected subgraph of G containing u and v and so it contains $[u,v]$.

(b) is immediate in view of (a).

(c) Again by Lemma 3.2(a), $H \cup K$ is a profinitely connected subgraph of G containing u and v and so it contains the segment $[u,v]$. Hence

$$[u,v] = ([u,v] \cap H) \cup ([u,v] \cap K)$$

is a decomposition of $[u,v]$ into closed subgraphs. If these two graphs are disjoint, then $[u,v]$ has a quotient which has only two vertices and no edges joining them, which contradicts profinite connectedness of $[u,v]$. Hence $[u,v] \cap H \cap K \neq \emptyset$.

(d) Take $H = [u,w]$, $K = [w,v]$ and apply (c) to conclude that $w \in [u,v]$. The result follows from (b). □

A vertex v is said to be a *cut* of an $\mathbb{A}$-tree G if G is the union of two nontrivial profinitely connected subgraphs with no edges in common and whose only vertex in common is v.

For the remainder of this section, G denotes an $\mathbb{A}$-tree.

Lemma 3.4 *Let u and v be two vertices of G. Then a vertex w is a cut of the segment $[u,v]$ if and only if $w \neq u,v$ and $[u,w] \cap [w,v] = \{w\}$.*

Proof. Suppose first that $[u,w] \cap [w,v] = \{w\}$, so that $[u,v] = [u,w] \cup [w,v]$ by Lemma 3.3(d). If $w \neq u,v$, then the segments $[u,w]$ and $[w,v]$ are nontrivial and so, being profinitely connected with only the vertex w in common and having union $[u,v]$, it follows that w is a cut of $[u,v]$.

Conversely, suppose that w is a cut of $[u,v]$ and let H and K be two nontrivial profinitely connected subgraphs with only the vertex w in common whose union is $[u,v]$. Without loss of generality, we may assume that $u \in H$ and $v \in K$ since $[u,v]$ cannot be entirely contained in one of H or K. Consider the segments $[u,w]$ and $[w,v]$. By definition of segment, since $\{u,w\} \subseteq H$ and $\{w,v\} \subseteq K$, we then have $[u,w] \cap [w,v] \subseteq H \cap K = \{w\}$. □

Lemma 3.5 *Let u and v be two vertices of G and let w_1 and w_2 be two distinct cuts of the segment $[u,v]$ such that $w_1 \in [u,w_2]$. Then w_1 is a cut of $[u,w_2]$ and w_2 is a cut of $[w_1,v]$.*

Proof. Since $w_1 \in [u,w_2]$, if $w_2 \in [u,w_1]$, then the two segments $[u,w_1]$ and $[u,w_2]$ coincide. Then, as $[u,v] = [u,w_1] \cup [w_1,v] = [u,w_2] \cup [w_2,v]$ and $[u,w_i] \cap [w_i,v] = \{w_i\}$ for $i = 1,2$, we must also have $[w_1,v] = [w_2,v]$, whence $w_1 = w_2$, in contradiction with the hypothesis. Hence $w_2 \notin [u,w_1]$ and so $w_2 \in [w_1,v]$. To conclude the proof, by symmetry, it suffices to show that w_1 is a cut of $[u,w_2]$. Indeed, since $w_2 \in [w_1,v]$, $[u,w_1] \cap [w_1,w_2] \subseteq [u,w_1] \cap [w_1,v] = \{w_1\}$. □

In the situation of Lemma 3.5, we write $w_1 \prec_{[u,v]} w_2$. More generally, for two cuts w_1 and w_2 of the segment $[u,v]$ (**from** u **to** v), we write $w_1 \preceq_{[u,v]} w_2$ if $w_1 \in [u,w_2]$ or, equivalently, if $w_2 \in [w_1,v]$.

Proposition 3.2 *The cuts of a segment $[u,v]$ of an $\mathbb{A}$-tree are linearly ordered by the relation $\preceq_{[u,v]}$.* □

Example. There may be infinitely many cuts in a segment. For example, consider the projective limit G of the sequence of graphs

in which each morphism identifies the two rightmost edges. Then G is a projective limit of trees and therefore, it is an $\mathbb{A}$-tree for any $\mathbb{A}$. The graph G looks as follows:

The segment $[u,v_1]=[u,v_2]$ is the whole graph and has infinitely many cuts, namely all vertices but u, v_1 and v_2 are cuts. This example also shows that the set of all cuts of a segment, together with the two extremities, may not be closed.

Lemma 3.6 *Let w_1 be a cut of $[u,v]$ and let w_2 be a cut of $[w_1,v]$. Then w_2 is a cut of $[u,v]$.*

Proof. Since $[u,w_2]\subseteq[u,w_1]\cup[w_1,w_2]$ by Lemma 3.3(a),

$$[u,w_2]\cap[w_2,v]\subseteq\big([u,w_1]\cap[w_2,v]\big)\cup\big([w_1,w_2]\cap[w_2,v]\big)=\{w_2\}$$

because $[u,w_1]\cap[w_2,v]\subseteq[u,w_1]\cap[w_1,v]=\{w_1\}$ and $w_1\notin[w_2,v]$. □

3.2. Profinite Cayley graphs

Let G be a profinite A-generated group, i.e., A is a profinite set and a continuous mapping $\varphi: A\to G$ is fixed such that the subgroup of G generated by φA is dense. Although we do not restrict φ to be one-to-one, we usually denote the image of $a\in A$ under φ by a. The *profinite Cayley graph of G (with respect to φ)* is the graph $\Gamma(G)=(G,G\times A)$ with $\alpha(g,a)=g$ and $\omega(g,a)=ga$. For an edge (g,a), we call the letter a its *label*. We will only be interested here in the case where A is finite. If G is finite, then $\Gamma(G)$ is the usual Cayley graph of G. Since $\Gamma(\_)$ preserves projective limits, in general $\Gamma(G)$ is the projective limit of the Cayley graphs of the finite quotients of G. In particular, $\Gamma(G)$ is profinitely connected. On the other hand, the Cayley graph of any nontrivial finite group has nontrivial cycles. Nevertheless, under certain conditions, the profinite Cayley graph of a profinite group may be $\mathbb{A}$-acyclic for suitable $\mathbb{A}$. We proceed to examine an important case in which this situation arises.

Fix a nontrivial pseudovariety $\mathbf{H}$ of groups. For a finite set A, the profinite Cayley graph of $\overline{\Omega}_A\mathbf{H}$ with respect to the natural mapping $A\to\overline{\Omega}_A\mathbf{H}$ is denoted by $\Gamma_{\mathbf{H}}(A)$. The finite quotients of the group $\overline{\Omega}_1\mathbf{H}$ have a natural ring structure which is respected by morphisms of 1-generated groups. Hence, we may view $\overline{\Omega}_1\mathbf{H}$ as a profinite ring. We say that $\Gamma_{\mathbf{H}}(A)$ is an **H**-*tree* if $\Gamma_{\mathbf{H}}(A)$ is $\overline{\Omega}_1\mathbf{H}$-acyclic. The nontrivial pseudovariety **H** is said to be ***arborescent*** if $\Gamma_{\mathbf{H}}(A)$ is an **H**-tree for every finite alphabet A.

For example, the pseudovariety **Ab** of all finite abelian groups is not arborescent since, if the alphabet A contains at least two distinct letters a and b, then the edges $(1,a)$, (a,b), (b,a) and $(1,b)$ constitute a cycle in $\Gamma_{\mathbf{Ab}}(A)$. On the other hand, $\Gamma_{\mathbf{Ab}}(\{a\}) = \Gamma_{\mathbf{G}}(\{a\})$ and so $\Gamma_{\mathbf{Ab}}(\{a\})$ is an **Ab**-tree (i.e., a **G**-tree) by Theorem 3.3 below, even though no finite nontrivial quotient of $\Gamma_{\mathbf{Ab}}(\{a\})$ is acyclic.

Let us try to reformulate the condition that $\Gamma_{\mathbf{H}}(A)$ is an **H**-tree. Since $\Gamma_{\mathbf{H}}(A)$ is profinitely connected, i.e., $H_0(\Gamma_{\mathbf{H}}(A), \mathbb{A}) = 0$ for $\mathbb{A} = \overline{\Omega}_1\mathbf{H}$, $\Gamma_{\mathbf{H}}(A)$ is an **H**-tree if and only if the mapping

$$\partial : \mathbb{A}[\overline{\Omega}_A\mathbf{H} \times A] \to \mathbb{A}[\overline{\Omega}_A\mathbf{H}]$$

defines an isomorphism between $\mathbb{A}[\overline{\Omega}_A\mathbf{H} \times A]$ and the profinite ring $I = \operatorname{Im}\partial = \operatorname{Ker}\varepsilon$, which is often called the ***augmentation ideal***. Indeed, note that the $\mathbb{A}$-module $\mathbb{A}[\overline{\Omega}_A\mathbf{H}]$ is in fact an $\mathbb{A}$-algebra where the multiplication continuously and linearly extends the multiplication of $\overline{\Omega}_A\mathbf{H}$. (In case $\overline{\Omega}_A\mathbf{H}$ is finite, $\mathbb{A}[\overline{\Omega}_A\mathbf{H}]$ is just the usual group ring.) This ring $\mathbb{A}[\overline{\Omega}_A\mathbf{H}]$ acts naturally on $\mathbb{A}[\overline{\Omega}_A\mathbf{H} \times A]$ by the continuous linear extension of the action of $\overline{\Omega}_A\mathbf{H}$ on $\overline{\Omega}_A\mathbf{H} \times A$ given by $g(h,a) = (gh,a)$. It is then an easy exercise on universal properties to show that $\mathbb{A}[\overline{\Omega}_A\mathbf{H} \times A]$ is the $\mathbb{A}[\overline{\Omega}_A\mathbf{H}]$-free module over $\{1\} \times A$. Note also that ∂ respects the action of $\mathbb{A}[\overline{\Omega}_A\mathbf{H}]$. We therefore conclude that $\Gamma_{\mathbf{H}}(A)$ is an **H**-tree if and only if the augmentation ideal I is freely generated by the $\partial(1,a) = a - 1$ ($a \in A$) as a profinite $\mathbb{A}[\overline{\Omega}_A\mathbf{H}]$-module.

We claim that the elements $a - 1$ ($a \in A$) always generate I as a profinite $\mathbb{A}[\overline{\Omega}_A\mathbf{H}]$-module, i.e., the $\mathbb{A}[\overline{\Omega}_A\mathbf{H}]$-module generated by these elements is dense in I. Indeed, the latter assertion is equivalent to the augmentation ideal I_G of $\mathbb{A}[G]$ being generated by the $a - 1$ ($a \in A$) for each A-generated G in **H**. By induction on the number of nonzero coefficients λ_g in an element $\sum_{g \in G} \lambda_g g$ of I_G, it is easy to show that the elements $g - 1$ ($g \in G$) generate I_G as an $\mathbb{A}$-module. But, if $g = a_1 \ldots a_n$ with $a_i \in A$, then $g - 1 = \sum_{i=1}^{n} a_1 \ldots a_{i-1}(a_i - 1)$ lies in the $\mathbb{A}[G]$-module generated by the $a - 1$ ($a \in A$), which proves the claim.

Recall that, for two semigroups S and T and a monoid homomorphism $T^1 \to \operatorname{End} S$ with the image of $s \in S$ under the endomorphism corresponding to $t \in T^1$ denoted by ts, the associated semidirect product is the semigroup $S*T$ with universe the cartesian product $S \times T$ and operation given by $(s_1,t_1)(s_2,t_2) = (s_1 + t_1 s_2, t_1 t_2)$ where additive notation is adopted for S. For two pseudovarieties **V** and **W**, $\mathbf{V} * \mathbf{W}$ denotes the pseudovariety generated by all semidirect products $S * T$ with $S \in \mathbf{V}$ and $T \in \mathbf{W}$.

Consider now a profinite $\mathbb{A}[\overline{\Omega}_A\mathbf{H}]$-module M and a continuous mapping $r : A \to M$. The action of $\mathbb{A}[\overline{\Omega}_A\mathbf{H}]$ on M yields, in particular, a semidirect product $M * \overline{\Omega}_A\mathbf{H}$. Suppose that the mapping $r' : A \to M * \overline{\Omega}_A\mathbf{H}$ defined by $r'(a) = \big(r(a), a\big)$ extends to a continuous group homomorphism $\widehat{r'} : \overline{\Omega}_A\mathbf{H} \to M * \overline{\Omega}_A\mathbf{H}$. Composing with the projection on the first component, we obtain a continuous mapping $d : \overline{\Omega}_A\mathbf{H} \to M$ such that $\widehat{r'}(g) = \big(d(g), g\big)$ for all $g \in \overline{\Omega}_A\mathbf{H}$. Although d is not in general a group homomorphism, it is a derivation in the sense that

$$d(gh) = d(g) + g \cdot d(h) \quad \text{for all } g, h \in \overline{\Omega}_A \mathbf{H}. \tag{2}$$

We extend d to a morphism of profinite $\mathbb{A}$-modules $\widehat{d} : \mathbb{A}[\overline{\Omega}_A \mathbf{H}] \to M$ by using the universal property of $\mathbb{A}[\overline{\Omega}_A \mathbf{H}]$. By continuity and $\mathbb{A}$-linearity of $\widehat{d}$, from (2) we deduce that

$$\widehat{d}(xy) = \varepsilon(y) \cdot d(x) + x \cdot d(y) \quad \text{for all } x, y \in \mathbb{A}[\overline{\Omega}_A \mathbf{H}].$$

In particular, the restriction ρ of $\widehat{d}$ to the augmentation ideal I is a morphism of profinite $\mathbb{A}[\overline{\Omega}_A \mathbf{H}]$-modules satisfying, for each $a \in A$,

$$\rho(a-1) = \widehat{d}(a-1) = d(a) - d(1) = d(a) = r(a).$$

The various mappings involved in this argument are depicted in the following diagram:

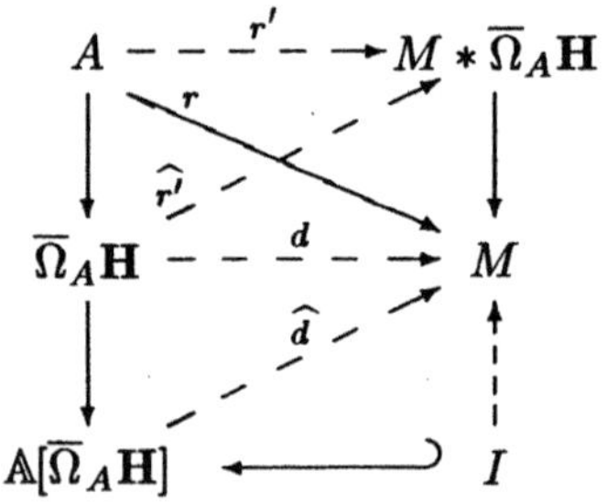

Finally, note that, if the abelian group $M * \overline{\Omega}_A \mathbf{H}$ is pro-$\mathbf{H}$, then the mapping r' does extend to $\widehat{r'}$ as indicated above. We thus have established half of the following theorem.

Theorem 3.3 *A nontrivial pseudovariety* $\mathbf{H}$ *of groups is arborescent if and only if* $(\mathbf{H} \cap \mathbf{Ab}) * \mathbf{H} = \mathbf{H}$.

Proof. It remains to be shown that, if $\mathbf{H}$ is arborescent, S and T are groups in $\mathbf{H}$ with S abelian, and $S * T$ is a semidirect product, then $S * T \in \mathbf{H}$.

Let A be a finite set and let $a \mapsto (s_a, t_a)$ define a mapping $A \to S * T$ whose image generates $S * T$. Since $T \in \mathbf{H}$, the correspondence $a \mapsto t_a$ defines a unique continuous homomorphism $\psi : \overline{\Omega}_A \mathbf{H} \to T$. If $\varphi : T \to \operatorname{End} S$ defines the action of T on S, the composite $\varphi \circ \psi$ gives an action of $\overline{\Omega}_A \mathbf{H}$ on S and a semidirect product $S * \overline{\Omega}_A \mathbf{H}$. On the other hand, since $S \in \mathbf{H} \cap \mathbf{Ab}$, S is a direct product of cyclic groups in $\mathbf{H}$ on which $\mathbb{A} = \overline{\Omega}_1 \mathbf{H}$ acts naturally. Hence, S is a (finite) $\mathbb{A}[\overline{\Omega}_A \mathbf{H}]$-module.

Consider the mapping $r : A \to S$ given by $a \mapsto s_a$. Since $\mathbf{H}$ is arborescent, there is a morphism of profinite $\mathbb{A}[\overline{\Omega}_A \mathbf{H}]$-modules $\rho : I \to S$ such that $\rho(a-1) = r(a)$ $(a \in A)$. Then $g \mapsto \rho(g-1)$ defines a continuous derivation $d : \overline{\Omega}_A \mathbf{H} \to S$ and so the mapping $\sigma : \overline{\Omega}_A \mathbf{H} \to S * \overline{\Omega}_A \mathbf{H}$ is a continuous homomorphism. Composing with the homomorphism $(s, g) \mapsto (s, \psi(g))$ $(s \in S,\ g \in \overline{\Omega}_A \mathbf{H})$, we obtain a continuous homomorphism $\overline{\Omega}_A \mathbf{H} \to S * T$ which is onto since the $(s_a, t_a) = (d(a), \psi(a))$ $(a \in A)$ generate $S * T$. Hence $S * T \in \mathbf{H}$ by Proposition 1.3. $\square$

3.3. Reduced factorizations in relatively free profinite groups

It is well known that the Cayley graph of a free group with respect to a set A of free generators is a tree in the usual combinatorial sense. Reading the second components of the edges (g, a) from the shortest path from 1 to h, taking a^{-1} instead of a if the edge appears in the wrong direction, we obtain a so-called *reduced word* $b_1 \ldots b_n$ on the alphabet $A \cup \overline{A}$ where $\overline{A} = \{a^{-1} : a \in A\}$ in the sense that no factor of length two is of any of the forms aa^{-1} or $a^{-1}a$ with $a \in A$.

Now, suppose that $\mathbf{H}$ is an arborescent pseudovariety of groups and A is a finite alphabet. Then the profinite Cayley graph $\Gamma_{\mathbf{H}}(A)$ is an $\mathbf{H}$-tree so that, as was already observed in section 3.1, for each $g \in \overline{\Omega}_A\mathbf{H}$ we have a segment $[1, g]$. This segment is not necessarily a finite path and so we may not be able to apply the procedure described above for the free group. For example, if $g \in \overline{\Omega}_{\{a\}}\mathbf{H}$ is such that both g and g^{-1} are non-explicit (monoid) operations, then no finite path connects 1 with g in $\Gamma_{\mathbf{H}}(\{a\})$. In $\overline{\Omega}_{\{a\}}\mathbf{H}$, each vertex h has exactly one edge leading to it, namely (ha^{-1}, a), and one edge leaving from it, namely (h, a). By Proposition 3.1, it follows that, if $[1, g]$ is not finite, then it contains an infinite sequence $h, ha, ha^2, \ldots$ or an infinite sequence $h, ha^{-1}, ha^{-2}, \ldots$ and so all vertices of $\Gamma_{\mathbf{H}}(\{a\})$ (and, therefore, also all edges). These arguments, although formulated for the case when $\mathbf{H}$ is arborescent, prove the following general result.

Proposition 3.4 *For any pseudovariety* $\mathbf{H}$ *of groups, the only possibly infinite profinitely connected subgraph of* $\Gamma_{\mathbf{H}}(\{a\})$ *is* $\Gamma_{\mathbf{H}}(\{a\})$ *itself.* □

Another way to look at reduced words $b_1 \ldots b_n$ in free groups is to note that the property of having no mutually inverse consecutive letters amounts to saying that the corresponding path repeats no edges. We use an appropriate abstraction of this property to define reduced factorizations in the profinite setting.

Throughout the remainder of this section, $\mathbf{H}$ denotes a fixed arborescent pseudovariety of groups.

For $\pi, \rho \in \overline{\Omega}_A\mathbf{H}\setminus\{1\}$, we say that the product $\pi\rho$ is *reduced* if $[1, \pi] \cap [\pi, \pi\rho] = \{\pi\}$. More generally, a product $\pi_1 \cdots \pi_n$ of $n \geq 0$ factors $\pi_1, \ldots, \pi_n \in \overline{\Omega}_A\mathbf{H}\setminus\{1\}$ is *reduced* if each product $\pi_i\pi_{i+1}$ $(1 \leq i < n)$ is reduced; we then also say that $\pi_1 \cdots \pi_n$ is a *reduced factorization* of $\pi = \pi_1 \cdots \pi_n$. For example, by Proposition 3.4, the only reduced products in $\overline{\Omega}_{\{a\}}\mathbf{H}$ of length greater than 1 are those of the forms $\underbrace{a \cdots a}_{n}$ and $\underbrace{a^{-1} \cdots a^{-1}}_{n}$ $(n > 1)$. In particular, since $\overline{\Omega}_{\{a\}}\mathbf{H}$ is uncountable, most of its elements do not admit nontrivial reduced factorizations. Sufficient conditions for nontrivial reduced factorizability are given below but they are obviously not necessary.

Note that the action of $\overline{\Omega}_A\mathbf{H}$ on the edges of $\Gamma_{\mathbf{H}}(A)$ considered in section 3.2 is in fact part of a graph action in the sense that each $g \in \overline{\Omega}_A\mathbf{H}$ gives a continuous endomorphism of the graph $\Gamma_{\mathbf{H}}(A)$ by acting on the edges as described in section 3.2 and on vertices by left multiplication. It follows that, for any $\pi, \rho, \sigma \in \overline{\Omega}_A\mathbf{H}$, $\pi[\rho, \sigma] = [\pi\rho, \pi\sigma]$.

Lemma 3.7 *Let* $\pi_1, \ldots, \pi_n \in \overline{\Omega}_A\mathbf{H} \setminus \{1\}$. *If the product* $\pi_1 \cdots \pi_n$ *is reduced then* $[1, \pi_1] \cap [\pi_1, \pi_1 \cdots \pi_n] = \{\pi_1\}$.

Proof. We proceed by induction on n. The case $n = 1$ is obvious and the case $n = 2$ follows directly from the definition of reduced product. So, assume that $n > 2$ and the result is valid for $n - 1$.

Since the product $\pi_2 \cdots \pi_n$ is reduced, by the induction hypothesis $[1, \pi_2] \cap [\pi_2, \pi_2 \cdots \pi_n] = \{\pi_2\}$ and so $[\pi_1, \pi_1\pi_2] \cap [\pi_1\pi_2, \pi_1 \cdots \pi_n] = \{\pi_1\pi_2\}$. By Lemma 3.3(d), it follows that

$$[\pi_1, \pi_1 \cdots \pi_n] = [\pi_1, \pi_1\pi_2] \cup [\pi_1\pi_2, \pi_1 \cdots \pi_n]. \tag{3}$$

Now, suppose that σ is a vertex different from π_1 in $[1, \pi_1] \cap [\pi_1, \pi_1 \cdots \pi_n]$. By (3), since $[1, \pi_1] \cap [\pi_1, \pi_1\pi_2] = \{\pi_1\}$ by Lemma 3.6, we must also have $\sigma \in [1, \pi_1] \cap [\pi_1\pi_2, \pi_1 \cdots \pi_n]$. Then

$$[\pi_1, \pi_1\pi_2] \subseteq [\pi_1, \sigma] \cup [\sigma, \pi_1\pi_2] \subseteq [1, \pi_1] \cup [\pi_1\pi_2, \pi_1 \cdots \pi_n]$$

and so

$$[\pi_1, \pi_1\pi_2] \subseteq ([1, \pi_1] \cap [\pi_1, \pi_1\pi_2]) \cup ([\pi_1\pi_2, \pi_1 \cdots \pi_n] \cap [\pi_1, \pi_1\pi_2]) = \{\pi_1, \pi_1\pi_2\}$$

yielding $[\pi_1, \pi_1\pi_2] = \{\pi_1, \pi_1\pi_2\}$, which is a contradiction since $\pi_2 \neq 1$ and so the finite graph $\{\pi_1, \pi_1\pi_2\}$ is not connected. This shows that there is no such vertex σ and completes the induction step. □

The next proposition gives a "geometrical" interpretation of the definition of reduced factorization.

Proposition 3.5 *Let* $\pi_1, \ldots, \pi_n \in \overline{\Omega}_A\mathbf{H}$. *Then the product* $\pi_1 \cdots \pi_n$ *is reduced if and only if the vertices* $\pi_1, \pi_1\pi_2, \ldots, \pi_1 \cdots \pi_{n-1}$ *constitute a strictly increasing chain of cuts of the segment* $[1, \pi_1 \cdots \pi_n]$.

Proof. Suppose first that the product $\pi_1 \cdots \pi_n$ is reduced. Proceeding by induction on n, we may assume that the result is valid for $n-1$. Then, by Lemma 3.7, π_1 is a cut of $[1, \pi_1 \cdots \pi_n]$ and, by the induction hypothesis, the vertices $\pi_1\pi_2, \ldots, \pi_1 \cdots \pi_{n-1}$ are cuts of the segment $[\pi_1, \pi_1 \cdots \pi_n]$, whence also of the segment $[1, \pi_1 \cdots \pi_n]$ by Lemma 3.6.

Conversely, suppose that $\pi_1, \pi_1\pi_2, \ldots, \pi_1 \cdots \pi_{n-1}$ constitute a strictly increasing chain of cuts of $[1, \pi_1 \cdots \pi_n]$. Then, for any $i \in \{1, \ldots, n-1\}$, denoting $\pi_1 \cdots \pi_j$ by σ_j, we have

$$[\sigma_{i-1}, \sigma_i] \cap [\sigma_i, \sigma_{i+1}] \subseteq [\sigma_1, \sigma_i] \cap [\sigma_i, \sigma_n]$$

by the definition of the ordering of cuts and so $[\sigma_{i-1}, \sigma_i] \cap [\sigma_i, \sigma_{i+1}] = \{\sigma_i\}$ because σ_i is a cut of $[1, \sigma_n]$. Hence $[1, \pi_i] \cap [\pi_1, \pi_{i+1}] = \{\pi_i\}$ and so the product $\pi_1 \cdots \pi_n$ is reduced. □

Given $\pi, \rho \in \overline{\Omega}_A\mathbf{H}$, we write $\pi \preceq \rho$ if $\pi = 1$, $\pi = \rho$ or π is a cut of the segment $[1, \rho]$ of $\Gamma_{\mathbf{H}}(A)$. This defines a partial order on $\overline{\Omega}_A\mathbf{H}$ which, by Proposition 3.2, is in fact a tree ordering, in the sense that elements with a common upper bound are comparable.

We say that a factorization $\pi = \pi_1 \cdots \pi_m$ *refines* a factorization $\pi = \rho_1 \cdots \rho_n$ of the same element if there are indices i_j $(j = 0, \ldots, n)$ such that $0 = i_0 < i_1 < \cdots < i_n = m$ and $\rho_j = \pi_{i_{j-1}+1} \ldots \pi_{i_j}$ $(j = 1, \ldots, n)$.

Lemma 3.8 *Let $\pi = \pi_1 \cdots \pi_m$ and $\pi = \rho_1 \cdots \rho_n$ be two reduced factorizations in $\overline{\Omega}_A\mathbf{H}$ of the same element. Then $\pi_1 \cdots \pi_m$ refines $\rho_1 \cdots \rho_n$ if and only if*

$$\{\pi_1, \pi_1\pi_2, \ldots, \pi_1 \cdots \pi_{m-1}\} \supseteq \{\rho_1, \rho_1\rho_2, \ldots, \rho_1 \cdots \rho_{n-1}\}.$$

Proof. Since necessity is obvious, it suffices to show that the inclusion in the statement is sufficient for $\pi_1 \cdots \pi_m$ to refine $\rho_1 \cdots \rho_n$. Indeed, for each $j \in \{1, \ldots, n-1\}$, there is $i_j \in \{1, \ldots, m-1\}$ such that $\rho_1 \cdots \rho_j = \pi_1 \cdots \pi_{i_j}$ and, since both sequences $(\pi_1, \pi_1\pi_2, \ldots, \pi_1 \cdots \pi_{m-1})$ and $(\rho_1, \rho_1\rho_2, \ldots, \rho_1 \cdots \rho_{n-1})$ are strictly increasing under $\preceq$, the sequence $0 = i_0, i_1, \ldots, i_n = m$ is also strictly increasing and $\rho_j = \pi_{i_{j-1}+1} \cdots \pi_{i_j}$ for $j = 1, \ldots, n$. □

We may now establish the confluence of refinement of reduced factorizations.

Theorem 3.6 *Any two reduced factorizations in $\overline{\Omega}_A\mathbf{H}$ of the same element have a common refinement.*

Proof. Let $\pi = \pi_1 \cdots \pi_m$ and $\pi = \rho_1 \cdots \rho_n$ be two reduced factorizations. Then, by Proposition 3.2, the union of the two chains $\pi_1 \prec \pi_1\pi_2 \prec \cdots \prec \pi_1 \cdots \pi_{m-1}$ and $\rho_1 \prec \rho_1\rho_2 \prec \cdots \prec \rho_1 \cdots \rho_{n-1}$ of cuts of $[1, \pi]$ is also a chain of cuts which we may write in the form $\xi_1 \prec \xi_1\xi_2 \prec \cdots \prec \xi_1 \cdots \xi_{r-1}$. Taking $\xi_r = (\xi_1 \cdots \xi_{r-1})^{-1}\pi$, we obtain a reduced factorization $\pi = \xi_1 \cdots \xi_r$ by Proposition 3.5 which refines the two given factorizations by Lemma 4.1. □

Note that the order $\preceq$ is not stable under multiplication. For example, if $a \in A$, then $a \preceq a^3$ but $a^{-1} = a^{-2}a \not\preceq a^{-2}a^3 = a$ since $[1, a] = \{1, a, (1, a)\}$. To our best knowledge, no example has yet been found of an element π of $\overline{\Omega}_A\mathbf{H}$ such that $[1, \pi]$ has infinitely many cuts. If no such example exists, then by Theorem 3.6 every element π of $\overline{\Omega}_A\mathbf{H}$ has a "canonical" reduced factorization, namely that which uses all cuts of $[1, \pi]$ according to Proposition 3.5. But, not even the weaker property that the set $\{\rho : \rho \preceq \pi\}$ is closed has yet been established, and perhaps both properties fail. (Compare with the example near the end of section 3.1.)

We conclude this section with two applications of Theorem 3.6 which will be useful in the next section. We sketch only the proof of the first and refer the reader to [9] for details.

Proposition 3.7 *Let $\mathbf{H}$ be an arborescent pseudovariety of groups and let A be an alphabet. For $1 \leq i \leq n$, let A_i and B_i be subsets of A, let $\pi_i \in \overline{\Omega}_{A_i}\mathbf{H} \setminus \{1\}$ and $\rho_i \in \overline{\Omega}_{B_i}\mathbf{H} \setminus \{1\}$, and let u_j $(1 \leq j \leq n-1)$ be reduced words over $A \cup \overline{A}$*

such that the first letter of u_j is not in $A_j \cup \overline{A}_j \cup B_j \cup \overline{B}_j$, while its last letter is not in $A_{j+1} \cup \overline{A}_{j+1} \cup B_{j+1} \cup \overline{B}_{j+1}$. Then the factorizations $\pi_1 u_1 \pi_2 \dots u_{n-1}\pi_n$ and $\rho_1 u_1 \rho_2 \dots u_{n-1}\rho_n$ are reduced.

If $\pi_1 u_1 \pi_2 \dots u_{n-1}\pi_n = \rho_1 u_1 \rho_2 \dots u_{n-1}\rho_n$ in $\overline{\Omega}_A \mathbf{H}$, then $\pi_i = \rho_i$ for all $1 \leq i \leq n$.

Proof. The two factorizations $\pi_1 u_1 \pi_2 \dots u_{n-1}\pi_n$ and $\rho_1 u_1 \rho_2 \dots u_{n-1}\rho_n$ are reduced since the $\Gamma_{\mathbf{H}}(A_j)$ and $\Gamma_{\mathbf{H}}(B_j)$ are profinitely connected subgraphs of $\Gamma_{\mathbf{H}}(A)$ which do not include any edge whose label is either the last letter of u_{j-1} or the first letter of u_{j+1} (or their inverses). By Theorem 3.6, the two factorizations have a common refinement $\xi_1 \cdots \xi_r$. Let $\pi_1 = \xi_1 \cdots \xi_i$ and $\rho_1 = \xi_1 \cdots \xi_j$ and suppose that $i < j$. Arguing by induction, it suffices to show this leads to a contradiction. Indeed, we have $u_1 = \xi_{i+1} \cdots \xi_k$ for some $k > i$. Hence, if $a \in A$ is such that a or a^{-1} is the first letter of u_1, then the segment $[1, \xi_{i+1}]$ has an edge labeled a while $[1, \rho_1] \subseteq \Gamma_{\mathbf{H}}(B_1)$ includes no such edge. □

Proposition 3.8 *Let $\mathbf{H}$ be an arborescent pseudovariety of groups, let A be an alphabet and let B and C be disjoint non-empty subsets of A. Let $\pi_1, \dots, \pi_r, \rho_1, \dots, \rho_s$ be elements of $\overline{\Omega}_B \mathbf{H} \setminus \{1\}$, and let $\sigma_0, \dots, \sigma_r$ and $\tau_0, \dots, \tau_s$ be elements of $\overline{\Omega}_C \mathbf{H}$ such that $\sigma_i \neq 1$ if $1 \leq i < r$ and $\tau_j \neq 1$ if $1 \leq j < s$. Then the factorizations $\sigma_0 \pi_1 \sigma_1 \cdots \pi_r \sigma_r$ and $\tau_0 \rho_1 \tau_1 \cdots \rho_s \tau_s$ are reduced.*

If $\sigma_0 \pi_1 \sigma_1 \cdots \pi_r \sigma_r = \tau_0 \rho_1 \tau_1 \cdots \rho_s \tau_s$ in $\overline{\Omega}_A \mathbf{H}$, then $r = s$, $\sigma_i = \tau_i$ for $0 \leq i \leq r$ and $\pi_i = \rho_i$ for $1 \leq i \leq r$.

4. A case study: DS with commuting idempotents

Let **ECom** and **ZE** denote, respectively, the pseudovarieties of all finite monoids in which idempotents commute (among themselves) and in which idempotents are central (i.e., commute with every element). The aim of this section is to give a canonical factorization for implicit operations on the pseudovariety $\mathbf{DS} \cap \mathbf{ECom} = \mathbf{DG} \cap \mathbf{ECom}$ and to deduce some join decompositions for certain subpseudovarieties.

Section 4.1 presents the study of canonical factorizations on free profinite monoids over pseudovarieties of the forms $\mathbf{DH} \cap \mathbf{ECom}$ and $\mathbf{ZE} \cap \overline{\mathbf{H}}$ where $\mathbf{H}$ is a pseudovariety of groups. Such factorizations are obtained by simplifying the general factorizations over **DS** considered in section 2.4. To establish their uniqueness, suitable finite automata are considered. As an application of these unique factorization statements and the results of section 3.3, via Reiterman's Theorem, join decompositions are given for $\mathbf{DH} \cap \mathbf{Ecom}$ and $\mathbf{ZE} \cap \overline{\mathbf{H}}$ for the case of an arborescent pseudovariety $\mathbf{H}$ of groups. We also present examples to show that such join decompositions are not valid for arbitrary **H** and to indicate that the arborescence of **H** is not a necessary condition for them to hold. Finally, section 4.3 gives the counterpart of these results for varieties of languages via Eilenberg's correspondence between these entities and pseudovarieties of monoids [14].

All results of this section are taken from Almeida and Weil [9]. Theorems 4.5 and 4.7 extend the particular cases where **H** is the pseudovariety of all finite groups which had already been obtained by Almeida and Azevedo [7] and Almeida [1], respectively. Theorem 4.9 first appeared in Ash, Hall and Pin [12]. For related problems concerning join decompositions, see Rhodes [21] and Almeida [6].

4.1. Some canonical factorizations

We start by observing that the commutation of idempotents in a monoid implies that the regular elements constitute a submonoid. This allows us to simplify the factorization of implicit operations on subpseudovarieties of **DG** $\cap$ **ECom** given by Theorem 2.7.

Let $\mathbf{V} \subseteq \mathbf{DG} \cap \mathbf{ECom}$ and let $\pi \in \overline{\Omega}_A\mathbf{M}$. We say that a factorization of π of the form

$$\pi = u_0\pi_1u_1 \dots \pi_r u_r$$

is *normal* (*with respect to* **V**) if

- $u_i \in A^*$ $(0 \leq i \leq r)$;
- $u_i \neq 1$ $(0 < i < r)$;
- the restriction of π_i to **V** is regular $(1 \leq i \leq r)$;
- for each $i < r$, the last letter of the word u_i (if it exists) does not belong to the content of π_{i+1};
- for each $i > 0$, the first letter of the word u_i (if it exists) does not belong to the content of π_i.

Note that, if $\mathbf{W} \subseteq \mathbf{V}$ and a factorization is normal with respect to **V**, then it is also normal with respect to **W**.

Proposition 4.1 *Let* **V** *be a subpseudovariety of* **DG** $\cap$ **ECom**. *Then every* $\pi \in \overline{\Omega}_A\mathbf{M}$ *admits a normal factorization with respect to* **V**. □

For our analysis of implicit operations on **DG** $\cap$ **ECom**, we need some test monoids to separate distinct factorizations. They are constructed as transition monoids of suitable automata.

For $r \geq 0$, let $u_0, \dots, u_r \in A^*$ and let $A_1, \dots, A_r \subseteq A$ be such that $A_i \neq \emptyset$ $(1 \leq i \leq r)$, $u_i \neq 1$ $(0 < i < r)$, the last letter of each u_i does not belong to A_{i+1}, and the first letter of each u_i does not belong to A_i. For each $i \in \{1, \dots, r\}$, let $\mathcal{A}_i$ be a permutation automaton on the alphabet A_i with set of states Q_i and let $p_i, q_i \in Q_i$. The automaton $\mathcal{A} = \mathcal{A}(u_0; \mathcal{A}_1, p_1, q_1; u_1; \dots; \mathcal{A}_r, p_r, q_r; u_r)$ is the following

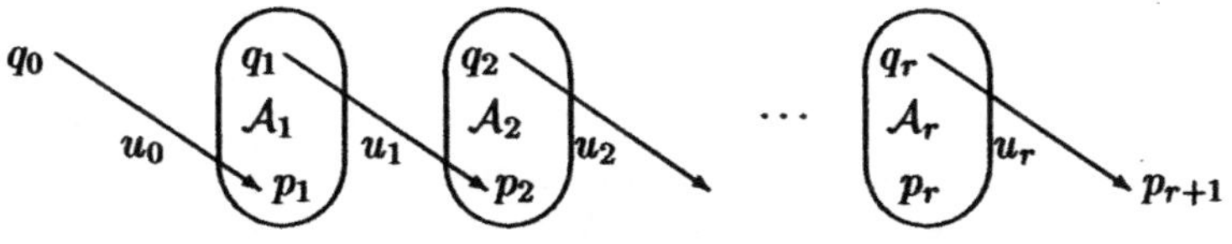

The condition on the extreme letters of the words u_i guarantees that $\mathcal{A}$ is a deterministic and co-deterministic automaton, meaning that each letter defines a partial one-to-one transformation of the set Q of states. Hence all words define partial one-to-one transformations of the set Q. In particular, the transition monoid $M_\mathcal{A}$ of the automaton $\mathcal{A}$ belongs to **ECom**. A slightly more detailed analysis of the structure of $\mathcal{A}$ proves the following lemma.

Lemma 4.1 *The monoid* $M_\mathcal{A}$ *lies in* **DG** $\cap$ **ECom** *and its subgroups lie in the pseudovariety generated by the permutation groups* $M_{\mathcal{A}_i}$. *If* $A_1 = \dots = A_r$, $A_1 \cap$

*$c(u_0 \dots u_r) = \emptyset$, $u_0 = u_r = 1$, and $|u_i| = 1$ $(i = 1, \dots, r-1)$, then $M_{\mathcal{A}} \in$ **ZE**. Furthermore, if $w \in A^+$, $k > |u_0 u_1 \dots u_r| + r$ and $p, q \in Q$ are such that $pw^k = q$, then there is some $i \in \{1, \dots, r\}$ such that $p, q \in Q_i$ and $w \in A_i^+$.*

Theorem 4.2 *Let* **H** *be a pseudovariety of groups and let* $\mathbf{V} = \mathbf{DH} \cap \mathbf{ECom}$. *Let $\pi, \rho \in \overline{\Omega}_A\mathbf{M}$ and let $\pi = u_0 \pi_1 u_1 \dots \pi_r u_r$ and $\rho = v_0 \rho_1 v_1 \dots \rho_s v_s$ be factorizations in normal form with respect to* **V**. *Then* **V** *satisfies $\pi = \rho$ if and only if $r = s$, $u_i = v_i$ $(0 \le i \le r)$ and* **V** *satisfies $\pi_i = \rho_i$ $(1 \le i \le r)$ (or, equivalently, $c(\pi_i) = c(\rho_i)$ and $\mathbf{H} = \mathbf{V} \cap \mathbf{G}$ satisfies $\pi_i = \rho_i$).*

Proof. Let $A_i = c(\pi_i)$ $(1 \le i \le r)$ and let $B_j = c(\rho_j)$ $(1 \le j \le s)$. Consider the following automaton $\mathcal{A}$:

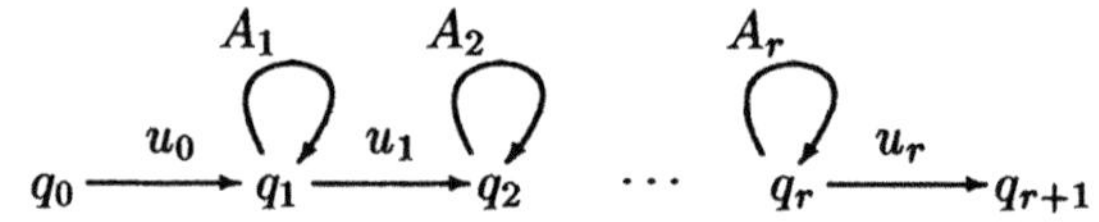

and its transition homomorphism $\mu : A^* \to M$. Let $\hat{\mu} : \overline{\Omega}_A\mathbf{M} \to M$ be the unique continuous homomorphic extension of μ. By Lemma 4.1, $M \in \mathbf{J} \cap \mathbf{ECom}$ and so $M \in \mathbf{V}$. Let $k > \max\{|u_0 \dots u_r| + r, |v_0 \dots v_s| + s\}$.

Since the factors π_i and ρ_j are regular on **V**, their images in M, $\hat{\mu}(\pi_i)$ and $\hat{\mu}(\rho_j)$, are idempotents. By density of A^* in $\overline{\Omega}_A\mathbf{M}$, there are words w_i such that $c(w_i) = A_i$ and $\hat{\mu}(\pi_i) = \mu(w_i^k)$. By Lemma 4.1, $\mu(w_i^k)$ is a partial identity function whose domain consists of all q_j such that $c(w_i) = A_i \subseteq A_j$.

We claim that $\hat{\mu}(\pi) = \mu(u_0 w_1^k u_1 \dots w_r^k u_r)$ is the function $\{q_0\} \to \{q_{r+1}\}$. Clearly $q_0\,\hat{\mu}(\pi) = q_{r+1}$. Suppose that a state $q \ne q_0$ belongs to the domain of $\hat{\mu}(\pi)$. Let $p = q\,u_0 w_1^k u_1 \dots u_{i-1} w_i^k$. Since $\mu(w_i^k)$ is idempotent, $p w_i^k = p$ and so $p = q_{j(i)}$ for some $j(i)$ such that $c(w_i) = A_i \subseteq A_{j(i)}$. Then either $u_0 = 1$ and $q_0 = q_1$, and so $j(1) > 1$ because there is no way back from right to left in the automaton $\mathcal{A}$; or $u_0 \ne 1$ and so $j(1) = 1$ would imply that the last letter of u_0 belongs to $c(\pi_1) = A_1$, contrary to the assumption that the product $u_0 w_1^k u_1 \dots w_r^k u_r$ is in canonical form. This shows that $j(1) > 1$. Suppose next that i is such that $j(i) > i$. Then $j(j(i)) = j(i)$ implies $c(u_i w_{i+1}^k u_{i+1} \dots u_{j(i)-1} w_{j(i)}^k) \subseteq A_{j(i)}$ which again contradicts the assumption that we are dealing with a normal factorization. Hence $j(j(i)) > j(i)$. Moreover, this argument applies whenever $i < j(i) \le r$. Whence $1, j(1), j(j(1)), \dots$ is a strictly increasing sequence of integers which is bounded above by r, which is absurd. Hence $q = q_0$ and the claim is established.

Consider now words w_i' $(1 \le i \le s)$ such that $c(w_i') = B_i$ and $\hat{\mu}(\rho_i) = \mu(w_i'^k)$. Again by Lemma 4.1, $\mu(w_i'^k)$ is a partial identity with domain the set of all q_j such that $B_i \subseteq A_j$. Since $\mathbf{V} \models \pi = \rho$, we have $\hat{\mu}(\pi) = \hat{\mu}(\rho)$ and so $q_0 v_0 w_1'^k v_1 \dots w_s'^k v_s = q_{r+1}$. As above, $q_0\, v_0 w_1'^k = q_i$ for some $i \in \{1, \dots, r\}$. This implies that $q_0\, v_0 = q_i$ and so u_0 is a prefix of v_0. By symmetry, $v_0 = u_0$ so that $i = 1$ and $c(w_1') \subseteq A_1 = c(w_1)$. Again by symmetry, it follows that $c(\pi_1) = c(\rho_1)$. Iterating the above argument, we deduce that $r = s$, $u_i = v_i$ $(0 \le i \le r)$ and $c(\pi_i) = c(\rho_i)$ $(1 \le i \le r)$.

Finally, to establish that $\mathbf{V} \models \pi_i = \rho_i$, since π_i and ρ_i are regular on $\mathbf{V}$ and have the same content, by Corollary 2.10, it suffices to show that $\mathbf{H} \models \pi_i = \rho_i$. Let G be an A_i-generated group in $\mathbf{H}$ and let $\mathcal{A}_i$ be its Cayley graph with respect to the set of generators A_i. Note that the transition monoid of $\mathcal{A}_i$ is precisely G. Let $A_i = \{a_{i1}, \ldots, a_{in_i}\}$ and consider the vertices $p_i = 1$ and $q_i = (\pi_i)_G(a_{i1}, \ldots, a_{in_i})$. We have an associated automaton $\mathcal{A}$ defined as follows:

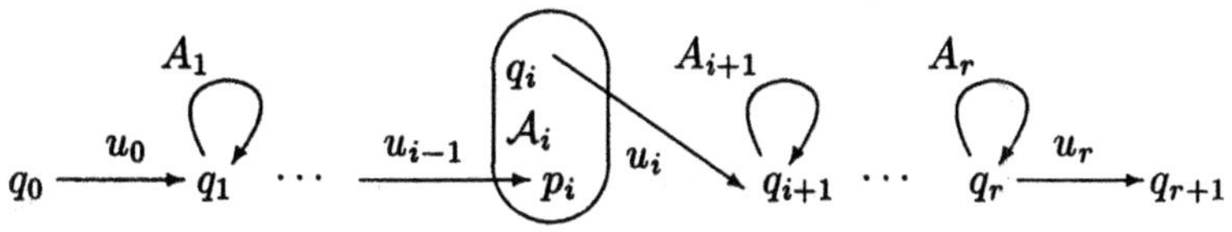

(4)

We again denote by μ and $\hat{\mu}$ respectively the transition homomorphism of $\mathcal{A}_i$ and its continuous homomorphic extension to $\overline{\Omega}_A\mathbf{M}$. By the first part of the proof, the mappings $\hat{\mu}(u_0\rho_1u_1\ldots\rho_{i-1}u_{i-1})$, $\hat{\mu}(u_i\rho_{i+1}u_{i+1}\ldots\rho_ru_r)$ and $\hat{\mu}(u_0\pi_1u_1\ldots\pi_ru_r)$ send, respectively, q_0 to p_i, q_i to q_{r+1}, and q_0 to q_{r+1}. Hence, since $\hat{\mu}(u_0\pi_1u_1\ldots\pi_ru_r) = \hat{\mu}(u_0\rho_1u_1\ldots\rho_ru_r)$ by Lemma 4.1 and we are dealing with partial one-to-one mappings, the equality $(\rho_i)_G(a_{i1}, \ldots, a_{in_i}) = (\pi_i)_G(a_{i1}, \ldots, a_{in_i})$ holds, which shows that $\mathbf{H} \models \pi_i = \rho_i$. □

The above proof shows that the transition monoids of the automata of the form (4) suffice to separate any two distinct normal products with respect to $\mathbf{DH} \cap \mathbf{ECom}$.

Corollary 4.3 *Let* $\mathbf{H}$ *be a pseudovariety of groups. Then the pseudovariety* $\mathbf{DH} \cap \mathbf{ECom}$ *is generated by the transition monoids of the automata of the form (4) with* $\mathcal{A}_i$ *a permutation automaton over the alphabet* A_i *such that* $M_{\mathcal{A}_i} \in \mathbf{H}$*, the last letter of each* u_i *not in* A_{i+1}*, and the first letter of each* u_i *not in* A_i. □

Corollary 4.4 *An implicit operation* π *on* M *is regular on* $\mathbf{DG} \cap \mathbf{ECom}$ *if and only if it is regular on* $\mathbf{J} \cap \mathbf{ECom}$*. In particular, a factorization is normal with respect to any pseudovariety in the interval* $[\mathbf{J} \cap \mathbf{ECom}, \mathbf{DG} \cap \mathbf{ECom}]$ *if and only if it is normal with respect to* $\mathbf{J} \cap \mathbf{ECom}$.

Proof. Since $\overline{\Omega}_A\mathbf{J} \cap \mathbf{ECom}$ is a quotient of $\overline{\Omega}_A\mathbf{DG} \cap \mathbf{ECom}$, clearly regularity on $\mathbf{DG} \cap \mathbf{ECom}$ implies regularity on $\mathbf{J} \cap \mathbf{ECom}$. For the converse, just note that a transition function on an automaton $\mathcal{A}$ of the form (4) is regular in the transition monoid $M_{\mathcal{A}_i}$ if and only if the corresponding function on the quotient automaton obtained by identifying all states of $\mathcal{A}_i$ is regular. □

A similar but somewhat simpler situation arises when we consider the pseudovarieties $\mathbf{DH} \cap \mathbf{ZE} = \overline{\mathbf{H}} \cap \mathbf{ZE}$, where $\overline{\mathbf{H}}$ denotes the pseudovariety of all finite monoids whose subgroups lie in $\mathbf{H}$.

Theorem 4.5 *Let* $\mathbf{H}$ *be a pseudovariety of groups and let* $\mathbf{V} = \mathbf{ZE} \cap \overline{\mathbf{H}}$*. Each element* π *of* $\overline{\Omega}_A\mathbf{V}$ *admits a factorization of the form* $\pi = \pi_1u_1\pi_2\ldots u_{r-1}\pi_r$*, where* $u_i \in A^+$ $(1 < i < r)$*,* π_i *is regular* $(1 \le i \le r)$*,* $\pi_i \ne \pi_i^\omega$ $(1 \le i < r)$*,* $\pi_r = \pi_r^\omega$ *implies* $\pi_1 = \pi_1^\omega$*,* $c(\pi_1) = c(\pi_2) = \cdots = c(\pi_r)$ *and* $c(u_1 \cdots u_{r-1}) \cap c(\pi_1) = \emptyset$.

Furthermore, if $\pi \in \overline{\Omega}_A\mathbf{V}$ *and* $\pi = \pi_1 u_1 \pi_2 \dots u_{r-1}\pi_r$ *and* $\pi = \rho_1 v_1 \rho_2 \dots v_{s-1}\rho_s$ *are two such factorizations, then* $r = s$, $u_i = v_i$ $(1 \leq i < r)$, $c(\pi_1) = c(\rho_1)$ *and* $\mathbf{H}$ *satisfies* $\pi_i = \rho_i$ $(1 \leq i \leq r)$.

4.2. Some join decompositions

Putting together the results of section 4.1 and Propositions 3.7 and 3.8, we now give some join decompositions of pseudovarieties.

Theorem 4.6 *Let* $\mathbf{H}$ *be an arborescent pseudovariety of groups. Then*

$$(\mathbf{J} \cap \mathbf{ECom}) \vee \mathbf{H} = \mathbf{DH} \cap \mathbf{ECom}.$$

Proof. The inclusion from left to right is obvious. By Reiterman's Theorem, to prove the reverse inclusion, it suffices to show that, for any $\pi, \rho \in \overline{\Omega}_A\mathbf{M}$,

$$(\mathbf{J} \cap \mathbf{ECom}) \vee \mathbf{H} \models \pi = \rho \quad \Longrightarrow \quad \mathbf{DH} \cap \mathbf{ECom} \models \pi = \rho.$$

Let $\pi = u_0\pi_1 u_1 \dots \pi_r u_r$ and $\rho = v_0 \rho_1 v_1 \dots \rho_s v_s$ be normal factorizations with respect to $\mathbf{J}\cap\mathbf{ECom}$ (and so also with respect to $\mathbf{DH}\cap\mathbf{ECom}$ by Corollary 4.4) and suppose that the pseudoidentity $\pi = \rho$ is valid in $\mathbf{J} \cap \mathbf{ECom}$ and in $\mathbf{H}$. Then, by Theorem 4.2, $r = s$, $u_i = v_i$ $(0 \leq i \leq r)$ and $c(\pi_i) = c(\rho_i)$ $(1 \leq i \leq r)$. On the other hand, by Proposition 3.7, $\mathbf{H} \models \pi_i = \rho_i$ $(i = 1, \dots, r)$. Hence $\pi = \rho$ is valid in $\mathbf{DH} \cap \mathbf{ECom}$ by Corollary 2.10. □

Theorem 4.7 *Let* $\mathbf{H}$ *be an arborescent pseudovariety of groups. Then*

$$(\mathbf{A} \cap \mathbf{Com}) \vee \mathbf{H} = \mathbf{ZE} \cap \overline{\mathbf{H}}.$$

Proof. As in the proof of Theorem 4.6, suppose that both $\mathbf{A} \cap \mathbf{Com}$ and $\mathbf{H}$ satisfy the pseudoidentity $\pi = \rho$ and let $\pi = \pi_1 u_1 \pi_2 \dots u_{r-1}\pi_r$ and $\rho = \rho_1 v_1 \rho_2 \dots v_{s-1}\rho_s$ be factorizations according to Theorem 4.5. We start by observing that, by substituting 1 for every variable except for a given variable a, we obtain a pseudoidentity $|\pi|_a = |\rho|_a$ which is valid in all finite cyclic monoids (which generate $\mathbf{A}\cap\mathbf{Com}$) if and only if it is trivial. But $|\pi|_a$ is a word if and only if $a \notin c(\pi_1)$. Hence, from $\mathbf{A}\cap\mathbf{Com} \models \pi = \rho$ we deduce that $c(\pi_1) = c(\rho_1)$. Then a direct application of Proposition 3.8 shows that $r = s$, $u_i = v_i$ $(i = 1, \dots, r-1)$ and $\mathbf{H} \models \pi_i = \rho_i$ $(i = 1, \dots, r)$. Finally, by Corollary 2.10, $\pi = \rho$. □

Note that the above proof of Theorem 4.7 does not really use Theorem 4.5. We next present an example to show that the decompositions of Theorems 4.6 and 4.7 are not valid in general. For the second decomposition, we may already observe that any $\mathbf{H} \subseteq \mathbf{Ab}$ gives a counter-example since $\mathbf{ZE} \cap \overline{\mathbf{Ab}}$ contains non-commutative monoids, namely all monoids of the form S^1 where S is a nilpotent semigroup (in fact, these monoids generate $\mathbf{ZE} \cap \overline{\mathbf{I}} = \mathbf{ZE} \cap \mathbf{A}$).

Example. For a group G, let $G_0 = G$ and let G_{n+1} be the subgroup generated by the commutators $[g, h] = ghg^{-1}h^{-1}$ with $g \in G_n$ and $h \in G$. Recall that the group G is said to be *nilpotent of class n* if $G_n = \{e\}$. We denote by $\mathbf{G}_{nil,n}$ the pseudovariety of all finite nilpotent groups of class n.

Let $A_\infty = \{a_1, a_2, \dots\}$ be an infinite alphabet. Define, recursively, the words $\alpha_0 = a_1$, $\beta_0 = 1$, $\alpha_{n+1} = \alpha_n a_{n+1} \beta_n$ and $\beta_{n+1} = \beta_n a_{n+1} \alpha_n$. It is well-known that $\mathbf{G}_{nil,n}$ is the subpseudovariety of $\mathbf{G}$ defined by the identity $\alpha_n = \beta_n$ and that $\mathbf{G}_{nil,n}$ is strictly contained in $\mathbf{G}_{nil,n+1}$ [18].

Consider the pseudoidentity

$$\alpha_n(\alpha_n\beta_n)^\omega a_{n+1}\beta_n(\alpha_n\beta_n)^\omega = \beta_n(\alpha_n\beta_n)^\omega a_{n+1}\alpha_n(\alpha_n\beta_n)^\omega.$$

It is valid in $\mathbf{J} \cap \mathbf{ECom}$ and in $\mathbf{G}_{nil,n+1}$. However, since both its sides are in normal form, it is not valid in $\mathbf{DG}_{nil,n+1} \cap \mathbf{ECom}$. Hence $(\mathbf{J} \cap \mathbf{ECom}) \vee \mathbf{G}_{nil,n+1} \neq \mathbf{DG}_{nil,n+1} \cap \mathbf{ECom}$.

On the other hand, the pseudovariety $\mathbf{G}_{nil}$ of all finite nilpotent groups is generated by a union of arborescent pseudovarieties of groups, namely the pseudovarieties $\mathbf{G}_p$ of all finite p-groups, where p is a prime. This implies that $\mathbf{G}_{nil}$ satisfies a pseudoidentity $\pi = \rho$ if and only if $\pi = \rho$ is valid in all $\mathbf{G}_p$ (p prime). Therefore, adopting the same notation as in the proof of Theorem 4.6, by Proposition 3.7 we have

$$\mathbf{G}_{nil} \models u_0\pi_1u_1 \dots \pi_r u_r = u_0\rho_1u_1 \dots \rho_r u_r \quad \Longrightarrow \quad \mathbf{G}_{nil} \models \pi_i = \rho_i \text{ for } 1 \leq i \leq r.$$

Hence $(\mathbf{J} \cap \mathbf{ECom}) \vee \mathbf{G}_{nil} = \mathbf{DG}_{nil} \cap \mathbf{ECom}$. Similarly, $(\mathbf{A} \cap \mathbf{Com}) \vee \mathbf{G}_{nil} = \mathbf{ZE} \cap \overline{\mathbf{G}_{nil}}$.

4.3. The corresponding varieties of languages

The results of the preceding sections may now be easily translated into corresponding results on varieties of languages according to Eilenberg's correspondence.

For a pseudovariety $\mathbf{H}$ of groups, denote by $\mathcal{L}(\mathbf{H})$ the class of all languages of the form $u_0L_1u_1 \dots L_ru_r$ such that $r \geq 0$, $u_i \in A^*$ and:

- $u_i \neq 1$ for $0 < i < r$;
- each L_i is a group language over the non-empty sub-alphabet A_i of A, whose syntactic monoid lies in $\mathbf{H}$ and whose minimal automaton has only one terminal state;
- the first letter of each u_i is not in A_i;
- the last letter of each u_i is not in A_{i+1}.

Note that, under the notation of section 4.1, if the minimal automaton of L_i is $\mathcal{A}_i$, then the minimal automaton of such a language $u_0L_1u_1 \dots L_ru_r$ is the automaton $\mathcal{A} = \mathcal{A}(u_0; \mathcal{A}_1, p_1, q_1; u_1; \dots; \mathcal{A}_r, p_r, q_r; u_r)$ where the initial and terminal states are, respectively, q_0 and p_{r+1}. Hence, by Lemma 4.1, the languages of $\mathcal{L}(\mathbf{H})$ are $\mathbf{DH} \cap \mathbf{ECom}$-recognizable. If $L \in \mathcal{L}(\mathbf{H})$, then the closure $\overline{L}$ of L in $\overline{\Omega}_A(\mathbf{DH} \cap \mathbf{ECom})$ consists precisely of those elements of $\mathbf{DH} \cap \mathbf{ECom}$ whose canonical images in the transition monoid of $\mathcal{A}$ map q_0 to p_{r+1}. Finally, note that, in case $\mathbf{H} = \mathbf{I}$, the languages L_i are of the form A_i^*.

Theorem 4.8 *For any pseudovariety* $\mathbf{H}$ *of groups, the class of languages in* A^* *which are recognized by monoids in* $\mathbf{DH} \cap \mathbf{ECom}$ *is the Boolean algebra generated by* $\mathcal{L}(\mathbf{H})$.

Proof. In view of the above remarks, by Proposition 1.10, it suffices to observe that the closures of the languages in $\mathcal{L}(\mathbf{H})$ separate the points of $\overline{\Omega}_A(\mathbf{DH} \cap \mathbf{ECom})$ and this was done in the proof of Theorem 4.2. □

Take for $\mathcal{L}'(\mathbf{H})$ the class of all languages of the form $L_1 a_1 L_2 \ldots a_{r-1} L_r$ such that $r \geq 0$ and:

- each L_i is a group language over a fixed subset B of A, whose syntactic monoid (over B) is in $\mathbf{H}$ and whose minimal automaton has only one terminal state;
- each $a_i \in A \setminus B$.

Then, just as above, we have the following result.

Theorem 4.9 *For any pseudovariety* $\mathbf{H}$ *of groups, the class of languages in* A^* *which are recognized by monoids in* $\mathbf{ZE} \cap \overline{\mathbf{H}}$ *is the Boolean algebra generated by* $\mathcal{L}'(\mathbf{H})$. □

Finally, in case $\mathbf{H}$ is arborescent, the results of section 4.2 allow us to restrict even further the generators needed to obtain the above varieties of languages.

Theorem 4.10 *Let* $\mathbf{H}$ *be an arborescent pseudovariety of groups. The class of languages in* A^* *which are recognized by monoids in* $\mathbf{DH} \cap \mathbf{ECom}$ *is the Boolean algebra generated by the group languages with syntactic monoid in* $\mathbf{H}$ *and the languages of the form* $u_0 A_1^* u_1 \cdots A_r^* u_r$, *where* $r \geq 0$, $\emptyset \neq A_i \subseteq A$, *each* u_i *is in* A^* *with* $|u_i| \geq 1$ *for* $0 < i < r$, *the first letter of each* u_i *is not in* A_i *and the last letter of each* u_i *is not in* A_{i+1}. □

Theorem 4.11 *Let* $\mathbf{H}$ *be an arborescent pseudovariety of groups. The class of languages in* A^* *which are recognized by monoids in* $\mathbf{ZE} \cap \overline{\mathbf{H}}$ *is the Boolean algebra generated by the group languages with syntactic monoid in* $\mathbf{H}$ *and the languages of the form* $L(a, k, n) = \{u \in A^* : |u|_a = k\}$, *where* $0 \leq k < n$ *and* $a \in A$. □

Of course, Theorems 4.10 and 4.11 fail exactly for those pseudovarieties for which Theorems 4.6 and 4.7, respectively, fail. So, in particular, they fail for each $\mathbf{H} = \mathbf{G}_{nil,n}$ with $n \geq 1$.

References

1. J. Almeida, *Some pseudovariety joins involving the pseudovariety of finite groups*, Semigroup Forum **37** (1988) 53–57.
2. ———, *Equations for pseudovarieties*, in J.-E. Pin (Ed.), *Formal properties of finite automata and applications*, Lect. Notes in Comp. Sci. **386**, Springer, 1989, 148–164.
3. ———, *The algebra of implicit operations*, Algebra Universalis **26** (1989) 16–32.
4. ———, *Residually finite congruences and quasi-regular subsets in uniform algebras*, Portugaliæ Mathematica **46** (1989) 313–328.
5. ———, *Implicit operations on finite $\mathcal{J}$-trivial semigroups and a conjecture of I. Simon*, J. Pure and Applied Algebra **69** (1990) 205–218.
6. ———, *Semigrupos Finitos e Álgebra Universal*, Publicações do Instituto de Matemática e Estatística da Universidade de São Paulo, 1992. English translation to appear.
7. J. Almeida and A. Azevedo, *Implicit operations on certain classes of semigroups*, in S. Goberstein and P. Higgins (Eds.), *Semigroups and their Applications*, D. Reidel, 1987, 1–11.

8. ——————, *On regular implicit operations*, Portugaliæ Mathematica **50** (1993) 35–61.
9. J. Almeida and P. Weil, *Reduced factorizations in free profinite groups and join decompositions of pseudovarieties*, Int. J. Algebra and Computation **4** (1994) 375–403.
10. ——————, *Free profinite semigroups over semidirect products*, Izv. VUZ Matem. (to appear).
11. C. Ash, *Inevitable graphs: a proof of the type-II conjecture and some related decision procedures*, Int. J. Algebra and Computation **1** (1991) 127–146.
12. C. J. Ash, T. E. Hall, and J.-E. Pin, *On varieties of languages associated with some varieties of finite monoids with commuting idempotents*, Inform. and Control **86** (1990) 32–42.
13. A. Azevedo, *Operações implícitas sobre pseudovariedades de semigrupos. Aplicações*, Doctoral dissertation, University of Porto, 1989.
14. S. Eilenberg, *Automata, Languages and Machines*, vol. B, Academic Press, New York, 1976.
15. M. Fried and M. Jarden, *Field Arithmetic*, Ergeb. der Math. **11**, Springer, 1986.
16. D. Gildenhuys and L. Ribes, *Profinite groups and Boolean graphs*, J. Pure and Applied Algebra **12** (1978) 21–47.
17. P. Higgins, *An algebraic proof that pseudovarieties are defined by pseudoidentities*, Algebra Universalis **27** (1990) 597–599.
18. A. Mal'cev, *Nilpotent semigroups*, Uch. Zap. Ivanov Gos. Ped. Inst. **5** (1953) 107–111. In Russian.
19. J.-E. Pin, *Variétés de Langages Formels*, Masson, Paris, 1984. English translation: *Varieties of Formal Languages*, North Oxford, London and Plenum, New York, 1986.
20. J. Reiterman, *The Birkhoff theorem for finite algebras*, Algebra Universalis **14** (1982) 1–10.
21. J. Rhodes, *New techniques in global semigroup theory*, *in* S. Goberstein and P. Higgins (Eds.), *Semigroups and their Applications*, Dordrecht, 1987, D. Reidel, 168–181.
22. L. Ribes, *Grupos profinitos y grafos topológicos*, Publicacions de la Secció de Matemàtiques **4**, Universitat Autònoma de Barcelona, 1977, 1–64.
23. L. Ribes and P. Zalesskiĭ, *On the profinite topology on a free group*, Bulletin London Math. Soc. **25** (1993) 37–43.
24. M.-P. Schützenberger, *Sur le produit de concaténation non ambigu*, Semigroup Forum **13** 1976, 47–76.
25. J.-P. Serre, *Cohomologie galoisienne*, Lect. Notes in Mathematics **5**, Springer, 1965.
26. P. Weil, *Implicit operations on pseudovarieties: an introduction*, *in* J. Rhodes (Ed.), *Monoids and semigroups with applications*, World Scientific, Singapore, 1991.
27. P. Zalesskiĭ and O. Mel'nikov, *Subgroups of profinite groups acting on trees*, Math. USSR Sbornik **135** (1989) 405–424. English translation.

LOGICS FOR REGULAR LANGUAGES, FINITE MONOIDS, AND CIRCUIT COMPLEXITY

H. STRAUBING
Computer Science Department
Boston College
Chestnut Hill, Mass.
U.S.A. 02167

D. THÉRIEN
School of Computer Science
McGill University
Montréal, Québec
Canada H3A 2K6

and

W. THOMAS
Institut für Informatik und Praktische Mathematik
Christian-Albrechts-Universität zu Kiel
D-24098 Kiel
Germany

1. Introduction

In the sixties, the theory of regular languages developed from two principal sources: First, the notion of *syntactic monoid* opened the possibility of studying regular languages and their properties in the framework of semigroup theory. A basis of this approach is Eilenberg's Variety Theorem which sets up a correspondence between classes of regular languages (respecting certain closure properties) and pseudovarieties of finite monoids. A second source was the view that regular languages are the behaviors of finite automata and that these behaviors can be defined naturally in certain systems of symbolic logic. Büchi and Elgot showed that the system of *monadic second-order logic* characterizes the expressive power of finite automata (and thus the class of regular languages). Both aspects were connected in a line of research opened by Schützenberger and McNaughton, and it was subsequently shown that several sublogics of monadic second-order logic, notably systems of first-order logic, correspond to special properties of syntactic monoids.

In the past ten years, surprising connections between this field and the complexity theory of boolean circuits have been discovered. The principal aim in this theory is to evaluate the computational power of boolean circuits in terms of their size, depth, and the types of gates used in their construction. Circuit families of particular interest are those with circuits of small depth and of polynomial size (i.e., with polynomially many gates in the number of input ports), consisting of *AND*- and *OR*-gates and possibly some generalizations of these gates. A family of circuits

J. Fountain (ed.), Semigroups, Formal Languages and Groups, 119–146.

which contains one circuit for each possible number of input ports can be viewed as a definition of a formal language (to which a 0-1-word belongs if it causes output 1 in the circuit with the appropriate number of input ports). Restrictions in the size and depth of circuits thus constitute corresponding classes of formal languages, and these in turn have been characterized by suitable classes of finite monoids or certain logical systems. As a result, all the different aspects of definability—the algebraic, the logical, and the complexity theoretic one—have now been merged into a unified and appealing theory.

The present lecture notes give an introduction to this rapidly developing field and survey the most important results. In Section 2, we shall present the logical framework for the definition of regular languages and introduce some important specializations. Section 3 is devoted to the characterization of classes of regular languages in terms of the corresponding (varieties of) syntactic monoids. In Section 4, a short introduction to circuit complexity theory is given, together with a survey of the algebraic description of circuit complexity classes (based on "polynomial length programs over finite monoids"). Finally, in Section 5, the logical approach is taken up again; here extensions of the logical systems of the first section are used to describe circuit complexity classes.

While the necessary terminology from formal language theory, logic and circuit theory will be briefly recalled, the reader is assumed to have basic knowledge of semigroup theory.

For a more detailed exposition of the material see the forthcoming monograph [31].

2. Logical Description of Regular Languages

2.1. Regular Languages and Monadic Second-Order Logic

Let us first recall the definiton of regular languages in terms of regular expressions and finite automata. ***Regular expressions*** over the finite alphabet A are the expressions built up from the symbols a in A (denoting the respective singletons $\{a\}$), $\emptyset$ (denoting the empty language), and ϵ (denoting the singleton containing the empty word) by means of the operations $+$ (for union), $\cdot$ (for concatenation of languages), and the Kleene star * (for the iteration of concatenation). Conventions on bracketing and omission of the concatenation dot are assumed as usual, and often we do not distinguish formally between expressions and denoted languages. So, for instance, we speak of the regular language

$$L_0 = a^*ba(a+b)^*$$

over the alphabet $A = \{a, b\}$, which contains all words that start with a sequence of letters a followed by a segment ba.

A ***(nondeterministic) finite automaton*** is a structure $\mathcal{A} = (Q, A, \iota, \Delta, F)$ with a finite set Q of "states", an "initital state" $\iota \in Q$, a "transition relation" $\Delta \subseteq Q \times A \times Q$, and a set $F \subseteq Q$ of "final states". A word $w = a_1 \dots a_n$ is ***accepted*** by $\mathcal{A}$ if there is a state sequence (a "successful run") $p_1, \dots, p_n, p_{n+1}$ such that

- $p_1 = \iota$,

- $(p_i, a_i, p_{i+1}) \in \Delta$ for $i = 1, \ldots, n$,
- $p_{n+1} \in F$.

The language *recognized* by $\mathcal{A}$ consists of the words over A accepted by $\mathcal{A}$. Any finite automaton can be converted to a *deterministic* one (where a function $\delta : Q \times \Sigma \to Q$ replaces the relation Δ) recognizing the same language. By Kleene's Theorem, the languages definable by regular expressions and those recognizable by finite automata coincide. For example, the language L_0 defined by the above mentioned regular expression is recognized by the finite automaton with the states $1, 2, 3$, initial state 1, final state set $\{3\}$, and the transitions $(1, b, 2)$, $(2, a, 3)$ as well as $(1, a, 1)$, $(3, a, 3)$, and $(3, b, 3)$ (see Figure).

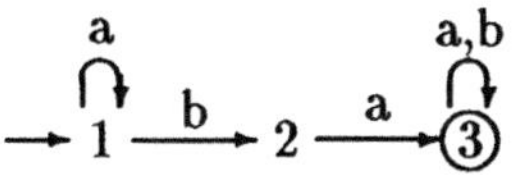

A natural alternative for the description of formal languages is to express properties of words in first-order logic. Here one uses variables $x, y, \ldots$ that range over positions of letters in a word, and uses formulas such as $Q_a x$ and xSy to express the conditions "position x carries letter a", "position x has position y as successor", respectively. Thus, the language L_0 is defined over the alphabet $A = \{a, b\}$ by the formula

$$\varphi_0 : \; \exists x \exists y (Q_b x \wedge xSy \wedge Q_a y \wedge \forall z (z < x \to Q_a z)).$$

More precisely, we identify a word $w = a_1 \ldots a_n$ over A with the corresponding *word model*, namely the relational structure

$$\underline{w} = (\{1, \ldots, n\}, S, <, (Q_c)_{c \in A})$$

where $\{1, \ldots, n\}$ is the set of "positions", S and $<$ are the usual successor relation and order relation on $\{1, \ldots, n\}$, and $Q_c = \{i \in \{1, \ldots, n\} \mid a_i = c\}$. The predicates S and $<$, which only depend on the length of w, are called *numerical predicates*, while the unary predicates Q_c are *letter predicates* (coding which positions carry which letters). The corresponding *first-order language* has variables $x, y, z, \ldots, x_1, x_2, \ldots$ ranging over the positions in word models, and is built up from atomic formulas of the form

$$x = y, \; xSy, \; x < y, \; Q_c x \; (for \; a \in A)$$

by means of the usual connectives $\neg, \wedge, \vee, \to, \leftrightarrow$ and the quantifiers $\exists$ and $\forall$. If $r_1, \ldots, r_n$ are positions in $\{1, \ldots, n\}$ and $\varphi(x_1, \ldots, x_n)$ is a formula with at most $x_1, \ldots, x_n$ occurring free in φ, then $(\underline{w}, r_1, \ldots, r_n) \models \varphi(x_1, \ldots, x_n)$ means that φ is satisfied in $\underline{w}$ under the mentioned interpretation for S and $<$ and with $r_1, \ldots, r_n$ as interpretation of $x_1, \ldots, x_n$, respectively. A *sentence* is a formula without free variables. The language *defined by* the sentence φ is

$$L(\varphi) = \{w \in A^* \mid \underline{w} \models \varphi\}.$$

Thus, $L(\varphi_0) = L_0$. Let us denote by $FO[S, <]$ the class of languages definable by first-order sentences of this form. Similarly, $FO[S]$ and $FO[<]$ indicate the classes of languages defined by first-order sentences with S, resp. $<$, as the only numerical predicate besides equality.

In defining word properties, it is often convenient to use constants *min* and *max* for the first and last positions of a word. In the sequel we shall allow these constants in order to abbreviate notation. For example, the formulas $Q_a min$ and $xSmax$ are abbreviations for $\exists z(Q_a z \wedge \neg\exists y\, ySz)$ and $\exists z(xSz \wedge \neg\exists y\, zSy)$, respectively. Also the use of S could be eliminated via a definition in terms of $<$, replacing xSy by

$$x < y \wedge \neg\exists z(x < z \wedge z < y).$$

Hence we have $FO[S,<] = FO[<]$. On the other hand, we shall see later that $FO[S]$ forms a proper subclass of $FO[<]$.

We now extend the logical formalism by second-order variables $X, Y, \ldots X_1, \ldots,$ which range over sets of positions, together with corresponding atomic formulas $Xx, Xy, \ldots$. Since sets are "monadic second-order objects" , in contrast to relations of higher arity that are polyadic second-order objects, the system is called *monadic second-order logic* (over the signature with the relation symbols $S, <, Q_a$). Note that in this logic, the order relation $<$ becomes definable in terms of successor S: We have (over word models) that $x < y$ is satisfied iff

$$\neg x = y \wedge \forall X(Xx \wedge \forall z \forall z'(Xz \wedge zSz' \rightarrow Xz') \rightarrow Xy).$$

Since $x = y$ is also definable by $\forall X(Xx \leftrightarrow Xy)$, the class $MSO[S]$ of languages definable in monadic second-order logic with S as only numerical predicate is the class of languages definable by monadic second-order sentences including $=$ and $<$. Altogether we obtain

$$FO[S] \subseteq FO[<] \subseteq MSO[S].$$

The key result connecting finite automata with logic states that $MSO[S]$ is the class of regular languages.

Theorem 2.1 (Büchi [10], Elgot [14]) *A language is regular iff it belongs to $MSO[S]$.*

Proof. To show the direction from left to right, let $\mathcal{A} = (Q, A, \iota, \Delta, F)$ be a finite automaton. Assume $Q = \{0, \ldots, k\}$ where $\iota = 0$. We have to find a monadic second-order sentence that expresses in any given word model $\underline{w}$ (over A) that $\mathcal{A}$ accepts w. Over a word $w = a_1 \ldots a_n$, the sentence will state the existence of a successful run $p_1, \ldots, p_{n+1}$ of $\mathcal{A}$. We may code such a state sequence up to p_n by a tuple $(X_0, \ldots, X_k)$ of pairwise disjoint subsets of $\{1, \ldots, n\}$ such that X_i contains those positions of w where state i is assumed. From the last state p_n the automaton should be able to reach a final state via the word's last letter a_n. Thus, $\mathcal{A}$ accepts w iff

$$\begin{aligned}\underline{w} \models \exists X_0 \ldots \exists X_k \ \ & (\textstyle\bigwedge_{i \neq j} \neg\exists x(X_i x \wedge X_j x) \\ & \wedge\ X_0 min \\ & \wedge\ \forall x \forall y(xSy \rightarrow \textstyle\bigvee_{(i,a,j)\in\Delta}(X_i x \wedge Q_a x \wedge X_j y) \\ & \wedge\ \textstyle\bigvee_{\exists j \in F:(i,a,j)\in\Delta}(X_i max \wedge Q_a max))\end{aligned}$$

Let us briefly sketch also the proof from right to left. One may proceed in two steps: First we replace first-order variables by a suitable use of second-order variables (which simplifies the logic and hence the translation to automata), and then we construct a finite automaton by induction over these simplified formulas. For the first step, we introduce new types of atomic formulas involving only set variables, namely XSY, $X \subseteq Y$, and $X \subseteq Q_a$ for $a \in A$, with the intended interpretations "X and Y are singletons $\{x\}$ and $\{y\}$ with xSy", "X is subset of Y", "X is subset of Q_a", respectively. It is easy to see that this modified logic, involving only second-order variables and the new atomic formulas, has the same expressive power as the original one. In particular, equality between sets is definable in terms of $\subseteq$, and being a singleton set can be defined by the condition that exactly one proper subset exists. Then first-order variables (and quantifications over them) are captured by variables restricted to singletons and second-order quantifications.

In the second step, we associate to each formula of the modified logic a finite automaton. Since this is done by induction, the case that free set variables occur has to be considered. A formula $\varphi(X_1, \ldots, X_m)$ which has free variables $X_1, \ldots X_m$ defines a set of words over the extended alphabet $A \times \{0,1\}^m$, where the j-th additional components of the letters determine the interpretation of X_j. (More precisely, the set of positions i of a word model where the j-th additional component is 1 is the interpretation of the set variable X_j.) It is easy to present finite automata that recognize the sets defined by atomic formulas X_jSX_k, $X_j \subseteq X_k$, and $X_j \subseteq Q_a$. For the inductive step, it suffices to consider the connectives $\neg, \vee$ and the existential set quantification, since the other connectives and the universal set quantifier are definable in terms of them. This in turn amounts to the proof that the class of regular languages shares certain closure properties, namely closure under complement, under union, and under projection. All these properties are well-known for the class of regular languages. □

From the first half of the proof we obtain that regular languages are definable by monadic second-order sentences of a special form, where a prefix of existential second-order quantifiers is followed by a first-order formula. This fragment is called *existential monadic second-order logic*, and $EMSO[S]$ is the associated class of definable languages. Combining both parts of the above proof, we obtain $MSO[S] = EMSO[S]$.

In showing Theorem 2.1, the original motivation of Büchi [10] and Elgot [14] was to prove that certain monadic second-order theories are decidable, for example the "weak monadic second-order theory of successor" (which is the set of monadic second-order sentences with S as the only predicate symbol which are true over the domain of the natural numbers, when set variables range only over *finite* sets). By the conversion of formulas into finite automata as described in the proof of the theorem, this decision problem is reduced to the (decidable) emptiness problem for finite automata.

In order to be able to treat more powerful theories, including quantifiers over arbitrary sets of natural numbers, or including a second successor relation (which leads to the infinite binary tree as underlying structure), more general models of finite automata were introduced, working over infinite words or infinite node-labelled trees. Büchi and Rabin succeeded in proving analogues to the theorem above for these generalized finite automata and monadic second-order logic (over the successor structure of the natural numbers, respectively over the binary tree). The nontrivial point here was to establish the necessary closure properties (union, complement, projection) for the automata under consideration, which required intricate constructions. By the same technique, many decidability results for modal logics, temporal logics, and logics of programs have been obtained. For more information on this subject see the survey [38].

If more general ways of quantification or more general underlying structures (than words or trees) are considered, the situation changes radically. Let $SO[S]$ be the class of languages that are definable by general second-order sentences (allowing also quantification over relations of higher arity), and denote by $ESO[S]$ the class of languages definable by existential second-order formulas (where only existential relation quantifiers occur, preceding a first-order formula). As shown by Fagin [15], $ESO[S]$ is equal to the complexity class NP (and hence much larger than the class of regular languages), whereas $SO[S]$ exhausts the polynomial time hierarchy. A similar effect arises even within monadic second-order logic when the underlying structures are no longer words or trees, but finite graphs, i.e., when the successor relation S is replaced by an edge relation E. Already in existential monadic second-order logic one may define graph properties that are NP-complete, for example 3-colorability.

A third way of generalizing Büchi's and Elgot's Theorem is to stay within restricted means of quantification (namely, within first-order logic) and within the domain of word models, but to allow more numerical predicates than S and $<$ in defining formulas. If $\mathcal{P}$ is a class of numerical predicates, then one may consider the class $FO[<, \mathcal{P}]$ of languages which are definable with the numerical predicates in $\mathcal{P}$ as well as $<$ and equality. In Section 5 below it will be shown that suitable choices of $\mathcal{P}$ allow us to characterize basic circuit complexity classes.

2.2. First-Order Definability

In this subsection we survey (without proofs) some results on first-order definable languages. We consider the language classes $FO[S]$ and $FO[<]$ and their structure which is derived from a complexity measure of the defining first-order formulas. The measure is based on the prenex normal form of formulas, in which quantifiers of the same type (existential or universal) are grouped into blocks. If there are n such blocks starting with existential quantifiers, we speak of a Σ_n-formula. A Σ_n*-sentence* is thus of the form

$$\exists \overline{x}_1 \forall \overline{x}_2 \ldots \exists/\forall \overline{x}_n \varphi(\overline{x}_1, \ldots, \overline{x}_n)$$

where $\overline{x}_1, \ldots, \overline{x}_n$ are nonempty tuples of variables, the quantifier blocks alternate between existential and universal ones, and φ is quantifier-free. A Π_n*-sentence* has n quantifier blocks starting with a universal one. A boolean combination of Σ_n-formulas will be called a $B(\Sigma_n)$-formula.

In the context of monadic second-order quantifiers and with *first-order* formulas as "kernels" φ, we know from the result $MSO[S] = EMSO[S]$ of the previous section that the Σ_n-form can be reduced to Σ_1-form. Let us consider this question for first-order formulas with numerical predicates S and $<$, respectively.

The situation for first-order logic with successor S is illustrated by the following example sentences, to be interpreted in word models over the alphabet $A = \{a, b\}$:

$$\varphi_1 : \ \forall x \forall y ((xSy \wedge Q_a x \wedge Q_b y) \rightarrow \exists z (ySz \wedge Q_b z))$$

$$\varphi_2 : \ \exists x (Q_a x \wedge \forall y (xSy \rightarrow \exists z (ySz \wedge Q_a z)))$$

Converted to prenex normal forms these sentences yield quantifier prefixes of type Π_2 and Σ_3, respectively. It is easy to see that a word satisfies φ_1 iff it does not end with ab and has no segment aba. Similarly, a word satisfies φ_2 iff it has a segment $a * a$, where $*$ indicates a or b, or (in the case that x has no successor) the word ends with a. Such conditions on the existence of segments, suffixes, or prefixes can be formulated directly by $B(\Sigma_1)$-sentences when we allow the constants *min* and *max*. More general conditions that could be formalized by $B(\Sigma_1)$-sentences involve *multiple* occurrences of segments; for example the condition that at least two segments ab occur is formalizable by

$$\exists x \exists y \exists x' \exists y' (xSy \wedge Q_a x \wedge Q_b y \wedge \neg x = x' \wedge x'Sy' \wedge Q_a x' \wedge Q_b y')$$

As it turns out, *all* first-order sentences in the signature of the successor relation S together with *min* and *max* can be reduced to boolean combinations of Σ_1-sentences.

There is a more combinatorial formulation of this reduction result. For a word $w \in A^*$, a segment $\sigma \in A^+$, and a natural number t let

$$occ_{\sigma,t} := \begin{cases} \text{number of occurrences of } \sigma \text{ in } w & \text{if this number is } < t \\ t & \text{otherwise} \end{cases}$$

Furthermore, let $pref_d(w)$, resp. $suf_d(w)$ be the segment of the first, resp. last, d letters of w (or w itself if its length is $\leq d$). Now define, for $u, v \in A^*$ and $d, t \geq 0$,

$$u \sim_{d,t} v$$

iff for all segments σ of length $\leq d$ we have $occ_{\sigma,t}(u) = occ_{\sigma,t}(v)$, $pref_d(u) = pref_d(v)$, and $suf_d(u) = suf_d(v)$. Then $\sim_{d,t}$ is a congruence of finite index, and we say that a language L is *locally threshold testable* if L is a (finite) union of $\sim_{d,t}$-classes for some fixed d and t. Thus membership of a word in a locally threshold testable language depends only on the prefix and suffix of a given length and on the existence of segments of a given length, counted up to a fixed threshold.

Theorem 2.2 (Thomas [35], [36]) *The following are equivalent:*
(a) L is in FO[S],
(b) L is definable by a $B(\Sigma_1)$-sentence of first-order logic with numerical predicate S and the constants min and max.
(c) L is locally threshold testable.

The implications from (c) to (b) and from (b) to (a) are immediate. For the step from (a) to (c), the "Ehrenfeucht-Fraïssé technique" can be applied. For an introduction see e.g. [39].

From this theorem we obtain languages which belong to $FO[<]$ but not to $FO[S]$. For example, the language

$$a^* b a^* c a^*$$

is clearly first-order definable in terms of $<$ but easily seen to be not locally threshold testable. However, this leaves unsettled the question how to decide *effectively* for a given regular language (presented say by a regular expression) whether it belongs to $FO[S]$ or $FO[<]$. In the next section effective procedures are derived from semigroup theoretical characterizations of $FO[S]$ and $FO[<]$.

We now turn to the class $FO[<]$. As shown by McNaughton and Papert [22], it can be characterized by a variant of the calculus of regular expressions, the "star-free expressions", where the Kleene star * is left out but a symbol $\sim$ for complementation (with respect to A^* for the given alphabet A) is included. As an example over the alphabet $A = \{a, b, c\}$ consider again the language $a^*ba^*ca^*$. It can be defined without use of the star * via a represention of a^* in the form

$$\sim (A^* \cdot b \cdot A^* \ + \ A^* \cdot c \cdot A^*)$$

where A^* stands for the star-free expression $\sim \emptyset$. For technical reasons we will use the concatenation dot only via alphabet letters as in this example, i.e. we allow the formation of an expression $r_0 \cdot a_1 \cdot r_1 \cdots a_n \cdot r_n$ from given expressions $r_0, \ldots, r_n$ and letters $a_1, \ldots, a_n$. Furthermore, the atomic star-free expressions are restricted to $\emptyset$ alone. A ***star-free expression*** is any expression obtained from $\emptyset$ by applications of the boolean operations $\sim$, $+$ and concatenation via letters. A language over a given alphabet is called ***star-free*** if it is definable by a star-free expression. (It is not difficult to show that our version of star-free expressions, originating in Straubing [28] and Thérien [33], leads to the same class of languages as the classical version that starts from the atomic expressions $\emptyset$, ϵ, $a \in A$, and includes the boolean connectives $\sim$ and $+$ as well as concatenation products $r_0 \cdot r_1 \cdots r_n$. For instance, concerning the atomic expressions ϵ and a, note that $\{\epsilon\}$ is obtainable by subtracting all languages A^*aA^* from A^*, i.e., is definable by a boolean combination of expressions $\sim \emptyset$ and $\sim \emptyset \cdot a \cdot \sim \emptyset$, and that a language $\{a\}$ is definable via the representation $\{\epsilon\} \cdot a \cdot \{\epsilon\}$.)

Theorem 2.3 (McNaughton, Papert [22]) *A language is star-free iff it belongs to $FO[<]$.*

There is a tight correspondence between first-order formulas with $<$ and star-free expressions, based on the classification of formulas by number of quantifier alternations and the classification of star-free expressions by their "dot-depth".

Define the ***dot-depth*** $\mathrm{dd}(r)$ of star-free expressions r inductively as follows:

(1) $\mathrm{dd}(\emptyset) = 0$
(2) $\mathrm{dd}(\sim r) = \mathrm{dd}(r)$
(3) $\mathrm{dd}(r + s) = \max\{\mathrm{dd}(r), \mathrm{dd}(s)\}$
(4) $\mathrm{dd}(r_0 \cdot a_1 \cdot r_1 \cdots a_m \cdot r_m) = \max\{\mathrm{dd}(r_0), \ldots, \mathrm{dd}(r_m)\} + 1$

The ***dot-depth of a star-free language*** $L \subseteq A^*$ is defined to be the smallest dot-depth of a star-free expression defining L. The classes $\mathcal{V}_n$ of star-free languages of dot-depth at most n (over a given alphabet A) form the ***dot-depth hierarchy*** (in the sense of Straubing and Thérien); by convention, $\mathcal{V}_0 = \{\emptyset, A^*\}$. The example language $a^*ba^*ca^*$ over $A = \{a, b, c\}$ belongs to $\mathcal{V}_2$; note that the definition of a^* given above is of dot-depth 1.

The notion of dot-depth was originally introduced by Cohen and Brzozowski in [11] for the version of star-free expressions with atomic expressions including ϵ and letters a and with concatenation products of the form $r_0 \cdot r_1 \cdots r_n$. The resulting hierarchy, often called "Brzozowski hierarchy", was shown to be strict by Brzozowski and Knast [9], and a logical characterization was given by Thomas [36]. For the present version of star-free expressions, corresponding results were obtained by Straubing [29], Thomas [37], and Perrin, Pin [23]:

Theorem 2.4 *(1) The dot-depth hierachy is strict, i.e., for $n \geq 0$ the class $\mathcal{V}_n$ is properly included in $\mathcal{V}_{n+1}$.*
(2) For $n \geq 1$, a language belongs to $\mathcal{V}_n$ iff it is definable by a $B(\Sigma_n)$-sentence with $<$ as numerical predicate.

This result shows that, in contrast to the preceding cases $MSO[S]$ and $FO[S]$, a reduction of quantifier alternation depth to existential formulas fails for first-order logic with $<$.

In the next section we shall consider languages which are not star-free (i.e., fall outside $FO[<]$) but are still regular. Such examples can be obtained by conditions which involve counting modulo a fixed number (*modular counting* for short). For example, the set of words of even length is regular but not star-free. This motivates the extension of first-order logic by features which allow us to express conditions on modular counting. Two such extended logics are introduced in Examples 3 and 4 below.

3. Regular Languages and Finite Monoids

In this section we shall discuss further classes of regular languages (besides $FO[S]$ and $FO[<]$) and characterizations of language classes by properties of syntactic monoids. The *syntactic monoid* of a language $L \subseteq A^*$ is the quotient monoid $M(L) = A^*/\sim_L$ where $\sim_L$ is the *syntactic congruence* over A^* defined by

$$w \sim_L w' \quad iff \quad \forall u, v \in A^*(uwv \in L \Leftrightarrow uw'v \in L).$$

The *syntactic morphism* of L is the projection μ_L of A^* onto $M(L)$. It is well known that a language L is regular iff $M(L)$ is finite.

We begin with some examples.

3.1. Examples

Example 1. Let

$$L_1 = (a + b + c)^* aa(a + b + c)^*.$$

which is defined by the first-order sentence

$$\exists x \exists y (xSy \wedge Q_a x \wedge Q_a y).$$

and hence belongs to $FO[S]$. A simple computation shows that the syntactic monoid of this language has six elements, consisting of the identity, a zero (which is the image of a^2 under the syntactic morphism) and a four-element regular $\mathcal{J}$-class with three idempotents (the images of b, ab and ba) and a null element (the image of a).

Example 2. Let

$$L_2 = a^*ba^*ca^*.$$

(considered already after Theorem 2.2) which is defined by the sentence

$$\exists x \exists y((x < y \wedge Q_b x \wedge Q_c y) \wedge \forall z((\neg z = x \wedge \neg z = y) \rightarrow Q_a z)).$$

This sentence uses order and equality as the only numerical predicates; thus $L_2 \in FO[<]$. The syntactic monoid consists of an identity element, which is the image of the letter a, together with a four-element nilpotent semigroup, whose nonzero elements are the images of b, c and bc.

Example 3. Let

$$L_3 = \{w \in \{a,b\}^* : |w| \equiv 0 \pmod 2\}.$$

We can write a first-order sentence that defines L_3 by introducing numerical predicates

$$x \equiv 0 \pmod k,$$

for $k > 1$. This predicate is interpreted to mean, 'x is a position divisible by k', where the positions are numbered from left to right, beginning with 1. L_3 is then defined by the sentence

$$\forall x(\neg \exists y(x < y) \rightarrow (x \equiv 0 \pmod 2)),$$

which says, in effect, that if there is a final position in the string, then that position has an even number. We denote by $FO[<, \{x \equiv 0 \pmod k\}]$ the family of languages defined by first-order sentences in which we admit both the ordering and these new predicates as numerical predicates. Hence $L_3 \in FO[<, \{x \equiv 0 \pmod k\}]$. The syntactic monoid of L_3 is the cyclic group of order 2; all words of even length are mapped by the syntactic morphism to the identity of the group, and all words of odd length are mapped to the generator.

Example 4. Let L_4 be the set of words over the alphabet $\{a, b\}$ in which there are an even number of occurrences of the letter a. As in the preceding example, the syntactic monoid is the group of order 2, however the syntactic morphism is different. In the present example the letter a is mapped to the generator of the group and the letter b to the identity. We define L_4 by the sentence

$$\exists^{(2,0)} x Q_a x.$$

The symbol $\exists^{(2,0)}$ is a *modular quantifier*. It is interpreted to mean 'there are exactly 0 mod 2 positions such that...' Thus the sentence says precisely that the number of positions with the letter a is even. We will use the notation $MOD_k[...]$ analogously to $FO[...]$ to denote the classes of languages defined by sentences with modular quantifiers of modulus k with specified numerical predicates. We also use $(FO + MOD_k)[..]$ to denote the classes defined when both the ordinary quantifiers and modular quantifiers are available. Thus $L_4 \in MOD_2[\emptyset]$, since we did not use any numerical predicates in the defining sentence.

What are the inclusion relations among these language classes, and how can we effectively decide to which classes a given regular language belongs? First, it is straightforward to verify that we have the chain of inclusions

$$FO[S] \subseteq FO[<] \subseteq FO[<, \{x \equiv 0 \pmod k\}] \subseteq (FO + MOD_k)[<] \subseteq MSO[S].$$

To answer the question on the strictness of these inclusions and to obtain effective tests for determining if a given regular language belongs to any of these classes, we shall present semigroup theoretic characterizations of all these classes.

3.2. The Key Theorem

In the next two sections we will outline a proof of the following theorem.

Theorem 3.1 *Let $L \subseteq A^*$. $L \in FO[<]$ if and only if L is regular and $M(L)$ is aperiodic.*

In particular, this theorem implies that the languages L_3 and L_4 of the examples in the preceding section are not in $FO[<]$, since their syntactic monoids are groups.

This theorem is actually a consequence of two different results: One, due to McNaughton and Papert [22] and mentioned above (Theorem 2.3), characterizes $FO[<]$ as the family of ***star-free*** regular languages—those that can be expressed using only boolean operations and concatenation— and the other, due to Schützenberger [26], characterizes the star-free regular languages as those with aperiodic syntactic monoids.

We will use some heavy algebraic machinery to prove this theorem. There are, to be sure, simpler proofs than the one we outline here, but the argument we give has the advantage of generalizing to treat different classes of quantifiers and numerical predicates.

3.3. The Block Product

Let M and N be monoids. The ***block product*** of N and M, denoted $N \square M$, is the set $N^{M\times M} \times M$, together with a multiplication given by

$$(F_1, m_1)(F_2, m_2) = (F, m_1 m_2),$$

where for all $m, m' \in M$,

$$F(m, m') = F_1(m, m_2 m')F_2(m m_1, m').$$

It is straightforward to verify that this multiplication is associative, and that $(I, 1)$, where $I(m, m') = 1$ for all $m, m' \in M$, is the identity element. Thus $N \square M$ is itself a monoid. The map

$$(F, m) \mapsto m,$$

is a homomorphism from the block product onto M. It is not hard to show that if G is a group contained in $N \square M$, then the kernel of the restriction of this homomorphism to G is isomorphic to a group in the direct product of $|M|^2$ copies of N. In particular, *the block product of two aperiodic monoids is aperiodic.* By the way, the block product is *not* an associative operation, and it most certainly is not a commutative operation.

The definition of the block product is a bit obscure, so let us try to provide a way to understand what it's all about. Let us say that a homomorphism $\varphi : A^* \to K$, where K is a monoid, *decomposes* with respect to M and N if there exist homomorphisms

$$\alpha : A^* \to M,$$

and

$$\beta : (M \times A \times M)^* \to N$$

such that the value of φ at a word $a_1 \cdots a_r$ is determined by

$$\alpha(a_1 \cdots a_r)$$

and

$$\beta(1, a_1, \alpha(a_2 \cdots a_r))\beta(\alpha(a_1), a_2, \alpha(a_3 \cdots a_r)) \cdots \beta(\alpha(a_1 \ldots a_{r-1}), a_r, 1).$$

One can then show that if φ decomposes with respect to M and N, then φ factors through a homomorphism $\psi : A^* \to N \;\square\; M$, and that if φ factors through a homomorphism into the block product, then φ decomposes with respect to M and $N^{M \times M}$. It is this property of the block product that we will use in the analysis of logical formulas.

The *Krohn-Rhodes Theorem* is a general decomposition theorem for finite monoids. Although it is usually stated in terms of the wreath product, there is a version for block products that we will use in the next subsection:

Theorem 3.2 *Every homomorphism φ from A^* into a finite monoid K factors through an iterated block product*

$$M_r \;\square\; (M_{r-1} \;\square\; \cdots \;\square\; (M_2 \;\square\; M_1) \cdots),$$

where each M_i is either a simple group that divides K or the monoid $U_1 = \{0, 1\}$.

The block product, along with this version of the Krohn-Rhodes theorem, is described in Rhodes and Tilson [25].

3.4. How the Key Theorem is Proved

We will merely sketch the main points of the argument. If $M(L)$ is aperiodic, then by the Krohn-Rhodes Theorem, the syntactic morphism μ_L factors through a block product

$$U_1 \;\square\; (U_1 \;\square\; \cdots \;\square\; (U_1 \;\square\; U_1) \cdots).$$

Let us denote this iterated block product by $U_1^{[r]}$. It follows from our remarks concerning decomposition, and the fact that U_1 is idempotent and commutative, that there is a homomorphism

$$\alpha : A^* \to U_1^{[r-1]}$$

such that the image of μ_L at $w \in A^*$ is determined by $\alpha(w)$ and by the set of triples

$$\{(\alpha(w'), a, \alpha(w'')) : w = w'aw''\}.$$

This is used to prove by induction on r that for all homomorphisms $\theta : A^* \to U_1^{[r]}$, the sets $\theta^{-1}(k)$, $k \in U_1^{[r]}$ are in $FO[<]$, and thus L, which is a finite union of such sets, is in $FO[<]$.

For the converse, we must show that if $L \in FO[<]$, then $M(L)$ is aperiodic. We will interpret formulas with free variables in $\{x_1, \ldots, x_k\}$ in *word structures*, which we view as words over the extended alphabet $A \times 2^{\{x_1,\ldots,x_k\}}$. A formula θ with free variables thus defines a language over this extended alphabet. For example, if θ is $Q_a x_1$, then L_θ is the language

$$A^*(a, \{x_1\})A^*.$$

If θ is $x_1 < x_2$, then L_θ is

$$A^*(A \times \{\{x_1\}\})A^*(A \times \{\{x_2\}\})A^*.$$

We show by induction on the construction of a formula θ of $FO[<]$ that $M(L_\theta)$ is aperiodic. This is easy to show for the atomic formulas, and it is also easy to show that aperiodicity is preserved under boolean operations. The heart of the argument is the proof that aperiodicity is preserved under quantification: Let θ be a formula with free variables in $\{x, x_1, \ldots, x_k\}$, and suppose that $M(L_\theta)$ is aperiodic. Let ζ be the formula $\exists x\theta$. We denote by μ_θ and μ_ζ the syntactic morphisms of L_θ and L_ζ, respectively. Now μ_ζ decomposes with respect to $M(L_\theta)$ and U_1, since we can take $\alpha = \mu_\theta$, and set

$$\beta(m, (a, X), m') = 0$$

if and only if

$$m \cdot \alpha(a, X \cup \{x\}) \cdot m' \in \mu_\theta(L_\theta).$$

Thus μ_ζ factors through $U_1 \,\square\, M(L_\theta)$. It follows from the inductive hypothesis and our remarks concerning groups in block products that $M(L_\zeta)$ is aperiodic.

3.5. Characterizations of Other Language Classes

The preceding result can be extended in a number of different ways. The first extension characterizes the language classes defined with modular quantifiers and ordering:

Theorem 3.3 *Let $L \subseteq A^*$ be a regular language. Then $L \in MOD[<]$ if and only if $M(L)$ is a solvable group, and $L \in (FO + MOD)[<]$ if and only if every group in $M(L)$ is solvable.*

This result is due to the authors [32]. The argument used to prove this is virtually identical to the one used to prove the characterization of $FO[<]$. The crucial point is that modular quantification with modulus q is equivalent to formation of the block product with Z_q on the left, in the same sense that existential quantification is equivalent to forming the block product with U_1 on the left. The Krohn-Rhodes

Theorem implies that every homomorphism into a monoid containing only solvable groups factors through a block product whose factors are copies of U_1 and cyclic groups, and this is used to show that if $M(L)$ contains only solvable groups then $L \in (FO{+}MOD)[<]$. Conversely, if $L \in (FO{+}MOD)[<]$, then we use the properties of groups contained in block products to conclude that every group in $M(L)$ is solvable. For the characterization of $MOD[<]$ we need the additional facts that the block product of two groups is a group, and that a homomorphism into a group G factors through an iterated block product all of whose factors are the simple composition factors of G.

Quite similar techniques are used to show the following two theorems, due to Barrington, Compton, Straubing and Thérien [3]:

Theorem 3.4 *Let $L \subseteq A^*$ be a regular language. Then*

$$L \in FO[<, \{x \equiv 0 \pmod{k} : k > 1\}]$$

if and only if for all $t > 0$, $\mu_L(A^t)$ (the image of A^t by syntactic morphism μ_L of L) contains no nontrivial group.

This shows that the language L_4 of the examples in 3.1 is not in

$$FO[<, \{x \equiv 0 \pmod{k} : k > 1\}].$$

There is also a modular version of this theorem:

Theorem 3.5 *Let $L \subseteq A^*$ be a regular language. Then*

$$L \in (FO + MOD_q)[<, \{x \equiv 0 \pmod{k} : k > 1\}]$$

if and only if for all $t > 0$, every group in $\mu_L(A^t)$ is solvable, and has cardinality that divides a power of q.

Somewhat different arguments are used to characterize the classes in which successor and equality are the only numerical predicates:

Theorem 3.6 *Let $L \subseteq A^*$ be regular. Then $L \in FO[S]$ if and only if $M(L)$ is aperiodic, and for all $e, s, f, t, u \in \mu_L(A^+)$, with e, f idempotent, $esfteuf = euftesf$.*

This result was first observed by Beauquier and Pin [7] in the context of infinite words. It follows from Theorem 2.2, which characterizes $FO[S]$ as the class of 'locally threshold testable' languages, and the algebraic characterization of this class, which is a result of Thérien and Weiss [34]. A corollary is that the language L_2 (from Example 2 of Section 3.1) is not in $FO[S]$, because the identity element of $M(L_2)$ is an idempotent of $\mu_{L_2}(A^+)$, and we get a contradiction to the condition in the theorem by taking $e = f = t = 1$, $s = \mu_{L_2}(b)$ and $u = \mu_{L_2}(c)$. In particular, the order relation cannot be defined by a first-order formula using only successor and equality.

In fact, there are effective algebraic characterizations for all the classes formed by various combinations of first-order and modular quantifiers and the numerical predicates discussed here. For details, see the forthcoming book by Straubing [31].

3.6. A Question

Up until now the numerical predicates we have considered give rise only to regular languages when used in defining sentences. Of course one can consider sentences with such numerical relations as '$y = 2x$', or 'y is the Gödel number of a Turing machine that halts when started on the binary representation of x'. These will give rise to nonregular languages (in the first instance) and nonrecursive languages (in the second). So let us ask the question posed at the end of 1.1 in this wider setting: Are there *any* numerical predicates we can introduce so that L_4, the set of strings with an even number of occurrences of the letter a, is first-order definable? We shall see in Section 5 that the answer to this very difficult question is 'No', and that this reveals interesting connections between logic, the theory of finite automata, and circuit complexity.

4. Circuit Complexity and Finite Monoids

Boolean circuits are now extensively studied as a model for parallel computations. In recent years, deep connections have been found between certain classes of circuits and computations taking place in finite monoids. In this section, we will present some results in this area: we assume that the reader has more knowledge of algebra than of computational complexity.

4.1. Computational Complexity

For both practical and theoretical reasons, computer scientists are interested in classifying problems according to the amount of computing resources required to get their solution. The classical model for sequential computations is the Turing machine and the usual complexity measures are time and space. A language L is recognizable in time $t(n)$ if there exists a Turing machine that determines membership in L using at most $t(n)$ steps on any input of length n; L is recognizable in space $s(n)$ if there exists a Turing machine that determines membership in L while visiting at most $s(n)$ memory cells on any input of length n. Two typical complexity classes definable in this framework are $LOGSPACE$, the family of languages recognizable in deterministic logarithmic space, and $PTIME$, the family of languages recognizable in deterministic polynomial time: an important open question is to determine if the easily shown inclusion $LOGSPACE \subseteq PTIME$ is strict or not. A standard reference for this material is [18].

The development of parallel technology has led to the introduction of new computing models. A formalism commonly considered for parallel computations is that of boolean circuit. An n-input boolean circuit C_n is given by a vertex-labeled directed acyclic graph where

- vertices have fan-in 0 or 2
- vertices of fan-in 0 (the input gates) have their labels in $\{1, 0, X_1, \ldots, X_n, \overline{X_1}, \ldots, \overline{X_n}\}$
- vertices of fan-in 2 (the internal gates) have their labels in $\{AND, OR\}$
- there is a unique vertex of fan-out 0 (the output gate)

Such an object computes a function $C_n : \{0,1\}^n \to \{0,1\}$ in a natural way. Given $x \in \{0,1\}^n$, an input gate labeled $X_i(\overline{X_i}, 1, 0)$ returns the value of the ith bit of x (the complemented value of the ith bit of x, 1, 0); an AND gate returns 1 iff both edges entering it carry the value 1; an OR gate returns 1 iff at least one edge entering it carries the value 1; finally the value of $C_n(x)$ is the bit returned by the output gate.

To recognize subsets of $\{0,1\}^*$, we consider families of circuits, $C = (C_n)_{n\geq 0}$, and the words of length n accepted by C are precisely those satisfying $C_n(x) = 1$. We say that L is recognized in size $s(n)$ iff there exists a circuit family $C = (C_n)_{n\geq 0}$ accepting L such that the number of vertices in C_n is $s(n)$; we say that L is recognized in depth $d(n)$ iff there exists a circuit family $C = (C_n)_{n\geq 0}$ accepting L such that the length of the longest path in C_n is $d(n)$.

Circuits are meant to simulate parallel computations since all gates at a given level operate simultaneously. Thus circuit depth measures parallel time. Also we identify parallel space with circuit size. It is interesting to note that, with the above definitions, sequential space is polynomially related to parallel time and sequential time is polynomially related to parallel space. For a discussion of this fact and of circuits in general, the reader may consult [12].

We end this introduction with some remarks.

(1) What we have defined here is non-uniform families of circuits, that is, no condition on the algorithmic definability of $C = (C_n)_{n\geq 0}$ is imposed. In particular the model allows non recursively enumerable languages to be recognized. The remedy is to require some condition on how to construct the graph C_n; for example, it may be asked that, given 1^n, some log-space bounded Turing machine is able to produce a description of C_n. Barrington, Immerman and Straubing [4] offer a detailed study of various uniformity criteria; our presentation will concentrate exclusively on the non-uniform model.

(2) One can consider circuits over an arbitrary input alphabet A by allowing input gates to be labeled by "$X_i = a$" for any $a \in A$.

(3) We do not distinguish between a class $\mathcal{C}$ of circuits and the class of languages recognizable by circuits in $\mathcal{C}$. The context makes clear which is intended.

(4) By allowing circuits to have several output gates, we can investigate computation of functions instead of simply recognition of languages.

4.2. NC^1 AND ITS SUBCLASSES

The circuit class NC^1 that we will be interested in consists, by definition, of those circuit families $C = (C_n)_{n\geq 0}$ of $O(\log n)$ depth (we have $d(n) = O(\log n)$ iff there is some constant c such that for all n, $d(n) \leq c \log n$); note that such circuits necessarily have polynomial size since only binary gates are used. It is an easy exercise to show that, with the proper uniformity condition, $NC^1 \subseteq LOGSPACE$; it is an open question to determine if the inclusion is proper or not.

An example of a non-trivial function computable in NC^1 is iterated addition, which is defined as follows:

input: n n-bit integers

output: the $(n + \log n)$-bit sum of the inputs

The trick is to repeatedly replace each group of three integers of length m (say $X = x_1 \ldots x_m, Y = y_1 \ldots y_m$ and $Z = z_1 \ldots z_m$) by two integers of length $m+1$ (say

$C = 0c_1 \dots c_m$ and $D = d_0 \dots d_{m-1}0$) such that $X + Y + Z = C + D$. This can be done by letting

$$c_i = 1 \text{ iff } x_i + y_i + z_i \text{ is odd (for } i = 1, \dots, m)$$
$$d_i = 1 \text{ iff } x_{i+1} + y_{i+1} + z_{i+1} \geq 2 \text{ (for } i = 1, \dots, m)$$

In $O(\log n)$ stages, each of constant depth, we thus produce two integers, whose sum is equal to that of the original n input numbers. We leave as an exercise to show how these two remaining integers can be added up with $O(\log n)$ levels of binary gates. This example implies that the language

$$MAJORITY(x_1, \dots, x_n) = \begin{cases} 1 & \text{if } \sum x_i \geq n/2 \\ 0 & \text{otherwise} \end{cases}$$

and the language

$$MOD_q(x_1, \dots, x_n) = \begin{cases} 1 & \text{if } \sum x_i \equiv 0 \bmod q \\ 0 & \text{otherwise} \end{cases}$$

are in NC^1. (Here we identify a language $L \subseteq \{0,1\}^*$ with its characteristic function, mapping $(x_1, \dots, x_n)$ to 1 iff $x_1 \dots x_n \in L$.)

We next define certain subclasses of NC^1 by restricting the depth of our circuit families $C = (C_n)_{n \geq 0}$ to be constant while allowing more powerful gates than simply binary AND and OR. The class of constant-depth circuit families constructed with binary gates (usually denoted NC^0) is not very interesting: if $C = (C_n)_{n \geq 0}$ is such a circuit family, there is a fixed bound t on the number of bits that can influence the value of $C_n(x)$, and thus C cannot compute the AND function since it depends on all bits.

A more interesting case is obtained by considering circuit families of constant-depth, polynomial size using AND gates and OR gates of arbitrary fan-in. This class of languages is denoted by AC^0. The bound on the size implies that all gates have polynomial fan-in and thus each one can be replaced by an $O(\log n)$-depth tree of binary gates. Hence $AC^0 \subseteq NC^1$. A deep result, independently proved by Furst, Saxe and Sipser [16] and Ajtai [1], says that the inclusion is proper; any MOD_q language belongs to NC^1 but not to AC^0.

In a similar vein, we define, for any $q \geq 2$, $CC^0(q)$ as the class of circuit families $C = (C_n)_{n \geq 0}$ having constant depth, polynomial size, constructed with binary AND and OR gates, and MOD_q gates of arbitrary fan-in. Since MOD_q is computable in NC^1, it follows that $CC^0(q) \subseteq NC^1$. We will write CC^0 for $\bigcup_q CC^0(q)$. Finally $ACC^0(q)$ is similarly obtained by allowing AND, OR and MOD_q gates of arbitrary fan-in: ACC^0 will stand for $\bigcup_q ACC^0(q)$ and we have $ACC^0 \subseteq NC^1$. It can be shown that if p and q have the same prime divisors then $CC^0(p) = CC^0(q)$ and $ACC^0(p) = ACC^0(q)$. If p and q are distinct primes, Smolensky has shown in [27] that $MOD_p \notin ACC^0(q)$, hence that $ACC^0(q) \not\subseteq NC^1$ in that case. The status of the inclusion when q is composite remains an open question.

4.3. Programs over finite monoids

The connection between circuits and monoids is based on the following definition: Let M be a finite monoid; an n-input M-program φ_n (over input alphabet A) is a sequence $\varphi_n = \nu_1 \dots \nu_l$ of instructions, where ν_j has the form (i_j, f_j) for some $1 \leq i_j \leq n, f_j : A \to M$. On input $x = x_1 \dots x_n \in A^n, \varphi_n(x)$ returns the monoid element which is the product of $f_1(x_{i_1}) \dots f_l(x_{i_l})$. As in the circuit case, functions from A^* into M are obtained by considering families of M-programs $\varphi = (\varphi_n)_{n \geq 0}$. Fixing an accepting subset $F \subseteq M$, φ then recognizes the language L, where $L \cap A^n = \{x \in A^n : \varphi_n(x) \in F\}$. We will denote by $\mathcal{B}(M)$ the class of languages that can be thus recognized by families of M-programs $\varphi = (\varphi_n)_{n \geq 0}$, with the added constraint that φ_n has polynomial length, i.e. φ_n contains at most n^c instructions for some constant c. We will consider only the non-uniform version of this definition, but as in the circuit case, uniformity criteria on the algorithmic definability of our programs can be imposed.

Such computing devices were originally introduced, in a different but equivalent form, under the name bounded-width branching program (Borodin et al. [8]). They observed the following

Lemma 4.1 *For any M, $\mathcal{B}(M) \subseteq NC^1$.*

Proof. Suppose we have a family $\varphi = (\varphi_n)_{n \geq 0}$ of M-programs, where φ_n has length $l = n^c$ and $L \cap A^n = \{x : \varphi_n(x) \in F\}$. We define an n-input circuit C_n that recognizes the same set in depth $O(\log n)$ by the following process.

Given $x = x_1 \dots x_n \in A^n$:

- in constant-depth (in fact using wires only) produce for each instruction $\nu_j = (i_j, f_j)$ the binary encoding of $f_j(x_{i_j})$;
- in parallel multiply the l monoid elements two by two, i.e. produce $l/2$ binary encodings corresponding to the products of pairs of elements; this can be done in constant depth;
- repeat the previous step until the product of the l elements, i.e. the value of $\varphi_n(x)$, is obtained; this will require $\log l = O(\log n)$ stages, each of constant depth;
- test, in constant depth, if $\varphi_n(x)$ belongs to F or not.

□

Borodin et al. [8] conjectured that $NC^1 \subseteq \bigcup_M \mathcal{B}(M)$ was false. Indeed, it seems that the fixed-size memory of the program should not allow computing $MAJORITY$, i.e. counting if the number of bits that are one is at least $n/2$. It came as quite a surprise when Barrington disproved the conjecture in [2], using a clever but simple trick available in simple non-abelian groups. The deep relationship between small-depth circuits and computations realized by M-programs was further strengthened by Barrington and Thérien [6] and Barrington, Straubing and Thérien [5] where it was shown that natural subclasses of NC^1 correspond exactly to natural algebraic restrictions on the monoids that can be used.

Before presenting some of these results, we introduce a notion of reducibility between languages which arises naturally in this context and which will make some of the arguments to come easier to describe. Let A, B be finite alphabets and consider n-input programs for which the instructions have the form (i, f) for some

$1 \leq i \leq n, f : A \to B$. The program thus induces a map $\varphi_n : A^n \to B^*$. Let $L \subseteq A^*, K \subseteq B^*$; we say that L is reducible to K, denoted $L \leq K$, iff there is a family $\varphi = (\varphi_n)_{n \geq 0}$ of programs as above such that the length of φ_n is at most n^c for some constant c and $x \in L \cap A^n$ iff $\varphi_n(x) \in K$. It is easily seen that $\leq$ is transitive and that any class $\mathcal{B}(M)$ is closed under this reducibility.

Recall that a language $K \subseteq B^*$ is regular iff there exists a morphism $\varphi : B^* \to M$, where M is some finite monoid, such that $K = \varphi^{-1}\varphi(K)$. The interest of our reducibility notion is that the classes $\mathcal{B}(M)$ are exactly the closure under $\leq$ of the regular languages recognized by M via morphisms.

Lemma 4.2 *$L \in \mathcal{B}(M)$ iff $L \leq K$ for some regular language K recognized by M via a morphism.*

Proof. If $L \in \mathcal{B}(M)$ we can view the family of M-programs $\varphi = (\varphi_n)_{n \geq 0}$ as producing a string in M^* instead of an element of M. Letting $\eta_M : M^* \to M$ be the canonical morphism and $K = \eta_M^{-1}(F)$, we have that $L \leq K$. For the converse, suppose φ is the reduction from L to K and $\psi : B^* \to M$ is the morphism recognizing K. Composing φ and ψ gives us a family of M-programs that recognizes L, using $\psi(K)$ as accepting set. □

4.4. Barrington's theorem

We now prove the theorem of Barrington ([2]). Consider S_5, the group of permutations on five points: let e be the identity and σ be any 5-cycle. We define $\mathcal{B}_\sigma(S_5)$ to be the class of languages L for which there is a polynomial length family of S_5-programs φ such that

$$\varphi(x) = \begin{cases} \sigma & \text{if } x \in L \\ e & \text{otherwise} \end{cases}$$

Lemma 4.3 *For any two 5-cycles σ and τ $\mathcal{B}_\sigma(S_5) = \mathcal{B}_\tau(S_5)$.*

Proof. We have that σ and τ are conjugates in S_5, say $\tau = \theta^{-1}\sigma\theta$. Suppose $\varphi_n = \nu_1 \ldots \nu_l$ is such that $\varphi_n(x) = \sigma$ if $x \in L, e$ otherwise. We modify φ_n as follows: the first instruction is changed to produce $\theta^{-1} f_1(x_{i_1})$ instead of $f(x_{i_1})$ and the last one is changed to produce $f(x_{i_l})\theta$ instead of $f(x_{i_l})$. Then the new program produces $\theta^{-1}\varphi_n(x)\theta$ instead of $\varphi_n(x)$, i.e. it produces τ iff the original one was yielding σ and e otherwise. Note that the modification preserves the length of the program. □

Lemma 4.4 *$\mathcal{B}_\sigma(S_5)$ is closed under complement*

Proof. As in the previous proof, we modify each program by changing the last instruction so that it produces $f_l(x_{i_l})\sigma^{-1}$ instead of $f(x_{i_l})$. The new program produces $\varphi_n(x)\sigma^{-1}$ instead of $\varphi_n(x)$, i.e. it produces σ^{-1} when the original one was yielding e and e when the original one was yielding σ. Hence $\overline{L}$ is in $\mathcal{B}_{\sigma^{-1}}(S_5) = \mathcal{B}_\sigma(S_5)$. Once again, the modification preserves the length. □

Lemma 4.5 *Let $\sigma = (12345), \tau = (13542)$. Then $\theta = \sigma\tau\sigma^{-1}\tau^{-1}$ is a 5-cycle.*

Proof. (12345)(13542)(54321)(24531) = (13254). □

Theorem 4.1 $NC^1 = \bigcup_M \mathcal{B}(M)$.

Proof. The theorem is proved by showing that an n-input circuit, which we can assume to be a tree, of depth d can be simulated by an n-input S_5-program of length 4^d which will yield some 5-cycle if the input is accepted by the circuit and the identity otherwise. By induction on d;

$d = 0$ The circuit has the form $X_i, \overline{X_i}, 1$ or 0. In each case, a single instruction can appropriately simulate the circuit.

$d > 0$ By Lemma 5.2, we can assume that the output gate is an AND, i.e. that $C = AND(C_1, C_2)$ where each of C_1, C_2 have depth $d-1$. By the induction hypothesis, there exist φ_1 via which the language accepted by C_1 is in $\mathcal{B}_\sigma(S_5)$, φ_2 via which the language accepted by C_2 is in $\mathcal{B}_\tau(S_5)$, φ_3 via which the language accepted by C_1 is in $\mathcal{B}_{\sigma^{-1}}(S_5)$, φ_4 via which the language accepted by C_4 is in $\mathcal{B}_{\tau^{-1}}(S_5)$, each of which has length 4^{d-1}. Let $\varphi = \varphi_1\varphi_2\varphi_3\varphi_4$; then for any $x \in \{0,1\}^n$,

$$\varphi(x) = \begin{cases} \sigma\tau\sigma^{-1}\tau^{-1} = \theta & \text{if } C(x) = 1 \\ e & \text{otherwise} \end{cases}$$

□

We add the following remarks.

- The theorem in effect proves that $NC^1 = \mathcal{B}_\sigma(S_5)$. In fact it is true that $NC^1 = \mathcal{B}(M)$ for any monoid M containing a simple non-abelian group.
- In terms of the reducibility notion introduced in the last section, the theorem says that for any language L in NC^1, we have that $L \leq K$ where K is the word problem of S_5.
- An identical trick was used in a different context by Maurer and Rhodes in [20]. Indeed, the theorem of this section follows directly from their result.

4.5. Algebraic descriptions of some subclasses of NC^1

In the last section, it was shown that $\mathcal{B}(M) = NC^1$ for any monoid containing a simple non-abelian group. A natural question is to investigate the computing power of monoids that do not have this property (they are called solvable monoids). One nice feature of the relationship via programs between circuits and monoids is the fact that natural subclasses of NC^1 (such as AC^0, CC^0, ACC^0) can be put in correspondence with natural restrictions on the monoids (respectively aperiodic monoids, solvable groups, solvable monoids).

These results are easier to prove using reductions to regular languages: we then use the nice combinatorial descriptions of the languages recognized via morphisms into aperiodic monoids (Schützenberger [26]), solvable groups (Straubing [28], Thérien [33]) and solvable monoids ([33]).

We will now give in detail the proof for aperiodic monoids. Consider the following hierarchy of languages over some alphabet A:

- $\mathcal{H}_1$ = boolean closure of languages of the form A^*aA^* with $a \in A$

– $\mathcal{H}_k$ = boolean closure of $\mathcal{H}_{k-1}$ and languages of the form $L_0 a_1 L_1 \ldots a_r L_r$ with $L_i \in \mathcal{H}_{k-1}, a_i \in A$.

This is the dot-depth hierarchy as introduced in Section 2.2, modified slightly at the first level $\mathcal{H}_1$. A language L is star-free, that is can be described by a regular expression not using the $*$ operator (but allowing complement) iff it belongs to some $\mathcal{H}_k$. From Sections 2 and 3 we know that L is star-free iff it can be recognized via a morphism into an aperiodic monoid. Note as a consequence of this result that the language MOD_q cannot be described by a star-free expression since any monoid recognizing MOD_q via a morphism must contain the cyclic group of order q.

We are now ready to prove our algebraic characterization of the circuit class AC^0.

Theorem 4.2 (Barrington, Thérien [6]) *$AC^0 = \bigcup_M \mathcal{B}(M)$, where the union ranges over all aperiodic monoids.*

Proof. To show the inclusion from right to left we use the fact that $L \in \mathcal{B}(M)$ for some M aperiodic iff $L \leq K$ for some star-free K. Let $\varphi = (\varphi_n)_{n \geq 0}$ be the reduction from L to K. We construct an appropriate circuit as follows. Given $x = x_1 \ldots x_n \in A^n$:

– in constant depth, in fact using wires only, produce the binary encoding of $w = \varphi_n(x)$;

– determine if $w \in K$; a union operation translates into a binary OR, a complement translates into a negation (which can be pushed to the input level using de Morgan's laws); a concatenation $K_0 b_1 K_1 \ldots b_r K_r$ translates into an OR of fan-in $\binom{n}{r}$ of ANDs of fan-in $2r+1$ (the subcircuit has to check every set of r positions of the input to verify if these positions contain $b_1, \ldots, b_r$ and if the intermediate segments belong to $K_0, \ldots, K_r$).

Note that when K belongs to $\mathcal{H}_k$, the circuit will have k levels of unbounded gates.

For the converse, we assume that our constant depth circuit $C = (C_n)_{n \geq 0}$ is over the binary alphabet and that it has k alternating levels of OR gates and AND gates. We say that a circuit is an OR-circuit (AND-circuit) if the output gate is an OR (AND). We show by induction on k that $L \leq K_k$ for some star-free $K_k \subseteq B_k^* = \{a, b, c_1, \ldots, c_{k-1}\}^*$.

$k = 1$: Suppose $C_n = OR(X_{i_1}, \ldots, X_{i_s}, \overline{X_{j_1}}, \ldots, \overline{X_{j_t}})$. We let $\varphi_n = (i_1, f) \ldots (i_s, f), (j_1, \overline{f}), \ldots, (j_t, \overline{f})$, where $f(0) = \overline{f}(1) = a, f(1) = \overline{f}(0) = b$. Clearly $\varphi_n(x) \in B_1^* b B_1^*$ iff $C_n(x) = 1$. If C_n is an AND gate instead, we have $\varphi_n(x) \in \overline{B_1^* a B_1^*}$ iff $C_n(x) = 1$. Letting $S_1 = B_1^* b B_1^*$ and $P_1 = \overline{B_1^* a B_1^*}$, we notice that both S_1, P_1 are recognized via a morphism into U_1, the 2-element semilattice.

$k > 1$: We suppose inductively that there exist S_{k-1}, P_{k-1}, each recognized via a morphism into an aperiodic monoid U_{k-1}, such that for any n-input OR-circuit C (AND-circuit D) of depth $k-1$ and size n^c, there exists an n-input program φ having the property that $\varphi(x) \in S_{k-1}(P_{k-1})$ iff $C(x) = 1 (D(x) = 1)$ and whose length is a polynomial with the degree depending only on k and c. Let now $C = OR(D_1, \ldots, D_s)$ be an OR-circuit of depth k and size n^c. We let $\varphi = (1, f)\varphi_1(1, f) \ldots \varphi_s(1, f)$, where $f(0) = f(1) = c_{k-1}$, and $\varphi_1, \ldots, \varphi_s$ are the programs recognizing the languages of the AND-circuits $D_1, \ldots, D_s$

respectively. It is easily seen that $\varphi(x) \in B_k^* c_{k-1} P_{k-1} c_{k-1} B_k^*$ iff $C(x) = 1$ since the only way the program can produce a segment in P_{k-1} between two markers c_{k-1} is to have some D_i being 1. Moreover the condition on the length of φ is obviously satisfied. If C is an AND-circuit instead, we see that $\varphi(x) \in (c_{k-1}S_{k-1})^* c_{k-1}$ iff $C(x) = 1$. Let S_k, P_k be the two languages just introduced. Consider U_{k-1} as a transformation monoid and let $\overline{U_{k-1}}$ be obtained from U_{k-1} by adding the constant transformations. Then one can show that S_k, P_k are both recognized by the wreath product $U_k = U_1 \circ \overline{U_{k-1}}$; the idea is that the markers c_{k-1} act as resets to the identity of U_{k-1} in the front monoid and the end copy of U_1 is used to detect if what happens between two consecutive markers belongs to the appropriate set. Note that U_k is aperiodic, hence S_k, P_k are star-free.

□

We have used a different alphabet B_k for circuits of depth k to simplify the proof. A more careful argument can be given that works with only one marker for arbitrary depth. Finally, define AC_k^0 as those constant depth polynomial size families of circuits $C = (C_n)_{n \geq 0}$ satisfying the condition that, for any n, any path in C_n contains at most k gates of fan-in greater than 2 (for an unimportant technical reason, we will suppose that all gates on the first level have fan-in greater than 2; this can always be achieved by duplicating inputs); it is possible to specialize the above theorem to each individual class AC_k^0. This will be stated in full in the next subsection.

Similar characterizations exist for the classes CC^0 and ACC^0 in terms of solvable groups and solvable monoids respectively. One way to prove these is to use the combinatorial descriptions of languages recognized via morphisms into such monoids ([28], [33]), descriptions which are based in part on a modular counting version of concatenation.

Theorem 4.3 *[5] $CC^0 = \bigcup_M \mathcal{B}(M)$, where the union ranges over all solvable groups;*
[6] $ACC^0 = \bigcup_M \mathcal{B}(M)$, where the union ranges over all solvable monoids.

Once again this general theorem specializes to take into consideration the exact depth of the circuits.

4.6. Conclusion

We define a variety to be a class V of finite monoids closed under division (i.e. morphic image of submonoid) and finite direct product. It has clearly emerged over the years that varieties are the proper level at which to classify languages recognized via morphisms into finite monoids (see [13], [24]).

The results of the last subsection show that many interesting subclasses of NC^1 can be algebraically characterized using polynomial length programs. It turns out that varieties seem to play a key role in this setting as well and McKenzie, Péladeau and Thérien [21] explore this point of view in detail. Define $\mathcal{B}(V) = \bigcup_{M \in V} \mathcal{B}(M)$; let M stand for the variety of all monoids, A for the variety of aperiodic monoids, $Gsol$ for the variety of solvable groups and $Msol$ for the variety of solvable monoids.

We thus have $NC^1 = \mathcal{B}(M)$, $AC^0 = \mathcal{B}(A)$, $CC^0 = \mathcal{B}(Gsol)$ and $ACC^0 = \mathcal{B}(Msol)$. The separation of circuit classes is thus, in all those cases, equivalent to separating the computing power of the corresponding varieties. The problem is that contrary to what happens for morphisms, distinct varieties V and W can give rise to the same class $\mathcal{B}(V) = \mathcal{B}(W)$. We give two examples:

Example 1 It follows from Barrington's theorem that for any variety V containing a simple non-abelian group we have $\mathcal{B}(V) = \mathcal{B}(M)$.

Example 2 Let J_1 be the variety of idempotent and commutative monoids, and for $k \geq 2$, let J_k be the variety generated by monoids of the form $U_1 \circ \overline{M_{k-1}}$ where $M_{k-1} \in J_{k-1}$. For $k \geq 1$, let H_k be the smallest variety containing enough monoids to recognize all languages in $\mathcal{H}_k$; we have $H_1 = J_1$ but $J_k \nsubseteq H_k$ for all $k > 1$. On the other hand a careful inspection of the proof of Theorem 4.7 shows that $AC^0_k = \mathcal{B}(J_k) = \mathcal{B}(H_k)$.

In [21], it is shown that $\mathcal{B}(V) = \mathcal{B}(W)$ iff the two classes contain the same regular languages: more precisely $\mathcal{B}(V) \subseteq \mathcal{B}(W)$ iff any word problem of any monoid in V can be computed by a polynomial length family of programs over some monoid in W. Thus the same claim applies to corresponding circuit classes. This paper also offers a general conjecture about the computing power of various varieties contained in $Msol$ which, if proved, would give as corollaries virtually all the known results and conjectures believed to be true about the internal structure of NC^1.

5. Tying in Logic Again

5.1. Circuit Complexity and Logic

Let us return to the question asked at the end of Section 3: Is there *any* numerical predicate we can introduce so that the set of strings over $\{a, b\}$ with an even number of occurrences of a is first-order definable?

The answer is 'No'. This follows from the circuit lower bounds of the previous section, and the next theorem. Let us denote by N the class of all numerical predicates, so that $FO[N]$ denotes the family of languages definable by first-order sentences in which arbitrary numerical predicates are admitted.

Theorem 5.1 $FO[N] = AC^0$.

This theorem appears in Immerman [19] and Gurevich and Lewis [17]. By the results of [1] and [16], cited in Section 4, the language MOD_2 is not in AC^0, so the theorem immediately gives the answer to our question.

Theorem 5.1 is one of a collection of results that characterize computational complexity classes in terms of formal logic. The first of these is a theorem of Fagin [15] that the class NP of languages recognizable in nondeterministic polynomial time are precisely those definable by existential second-order sentences with successor. Immerman [19] gives a large number of such results for various complexity classes, including Theorem 5.1.

The theorem is quite surprising at first glance, because the polynomial size bound in the definition of AC^0 appears nowhere in the definition of $FO[N]$. In fact, the theorem is not all that difficult to prove. We will sketch here a quick proof of the theorem that uses the characterization of AC^0 in terms of programs over finite monoids (see Theorem 4.6). First, suppose L is defined by a first-order sentence φ. Beginning with the outermost quantifier, we replace

$$\exists x\theta(x)$$

by

$$\bigvee_{i=1}^{n} \theta(i),$$

and similarly replace universal quantifiers by AND-gates. When we reach the atomic formulas we replace sentences of the form

$$\rho(i_1, \ldots, i_m),$$

where ρ is a numerical predicate by a constant 1 or 0, depending on whether the sentence is true or false, and $Q_1 i, Q_0 i$ by X_i and $\bar{X}_i$, respectively. The resulting expression is a circuit that recognizes the set of strings of length n in L. The depth of this circuit is equal to the depth k of nesting of the quantifiers in φ, and the size is bounded by a polynomial in n of degree k. Thus $L \in AC^0$.

Conversely, if $L \in AC^0$, then L is recognized by a family of programs of length n^k over a finite aperiodic monoid M. That is, an input string $w = a_1 \cdots a_n$ over $\{0,1\}$ is translated to a string

$$\pi(w) = b_1 \cdots b_{n^k} \in M^*.$$

Since M is aperiodic, it follows from the results cited in Section 3 that w is accepted if and only if $\pi(w)$ satisfies a sentence θ of $FO[<]$. We will use θ to construct a sentence of $FO[N]$ that defines L. We can encode each position in $\pi(w)$ by a k-tuple of positions in w. We then replace each quantifier

$$\exists x\psi(x), \forall x\psi(x),$$

in θ by

$$\exists x_1 \cdots \exists x_k \psi(x_1, \ldots, x_k), \forall x_1 \cdots \forall x_k \psi(x_1, \ldots, x_k).$$

The numerical relation $<$ on positions in $\pi(w)$ is replaced by a $2k$-ary numerical predicate expressing the corresponding relation on positions of w (this will depend on the encoding). The atomic formula $Q_m x$, where $m \in M$ is replaced by

$$(R_m^0(x_1, \ldots, x_k, y, z) \wedge Q_0 y) \vee (R_m^1(x_1, \ldots, x_k, y, z) \wedge Q_1 y),$$

where $R_m^j(x_1, \ldots, x_k, y, z)$ is interpreted to mean: 'in the program for inputs of length z, the xth instruction, where x is encoded by $(x_1, \ldots, x_k)$, reads input bit y and emits m if this bit is equal to j'. This is a numerical predicate that depends on the choice of encoding. When this translation is complete we have the promised sentence of $FO[N]$ that defines L.

The same techniques can be used to give characterizations of other circuit complexity classes in terms of modular quantifiers:

Theorem 5.2 *Let $q > 0$. Then*

$$ACC^0(q) = (FO + MOD_k[N]),$$

and

$$CC^0(q) = MOD_k[N].$$

5.2. Regular Languages in Circuit Complexity Classes

The following theorem, due to Barrington, Compton, Straubing and Thérien [3], characterizes the regular languages in AC^0.

Theorem 5.3 *Let $L \subseteq \{0,1\}^*$ be a regular language. $L \in AC^0$ if and only if*

$$L \in FO[<, \{x \equiv 0 \pmod{k} : k > 1\}].$$

The 'if' direction of this theorem follows immediately from Theorem 5.1. For the only if direction, we use the algebraic characterization of the class $FO[<, \{x \equiv 0 \pmod{k} : k > 1\}]$ from Section 3: If $L \in AC^0$ is regular and $L \notin FO[<, \{x \equiv 0 \pmod{k} : k > 1\}]$, then there exist $t > 0$, $q > 1$ such that $\mu_L(A^t)$ contains a cyclic group of order q. We use this together with the circuits for L to construct an AC^0 family of circuits that counts the number of occurrences of 1 in the input modulo q. This contradicts the results of [1] and [16] cited in Section 4.

5.3. A General Principle?

We will say that a numerial predicate is *regular* if it can be expressed as a first-order formula in the atomic formulas $x < y$ and $x \equiv 0 \pmod{k}$. The terminology is justified by the fact that any second-order monadic sentence that uses these numerical predicates is regular, while any nonregular numerical predicate can be used to define a nonregular language. For example,

$$x + y \equiv z \pmod{7}$$

is a regular numerical predicate, while

$$x + y = z$$

is not. Let us denote by *Reg* the class of all regular numerical predicates, and by $\mathcal{L}_{Reg}$ the class of all regular languages. The last theorem can then be restated as

$$FO[N] \cap \mathcal{L}_{Reg} = FO[Reg].$$

Let us see what this means. It may happen, of course, that a sentence using nonregular numerical predicates defines a regular language. This occurs, for instance, with the sentence

$$\forall x \exists y (y | x \wedge Q_0 y).$$

Here '|' has the usual meaning 'divides'. The language defined is the regular language $0(0+1)^*$. Now in this case we could just as well have written

$$\forall x \exists y (y \leq x \wedge Q_0 y),$$

which uses only regular numerical predicates. The theorem says that this phenomenon is general: Whenever we define a regular language by a first-order sentence using nonregular numerical predicates we can define the same language by a sentence that contains only regular numerical predicates.

Thus Theorem 5.3, which at first glance concerns lower bounds for boolean circuits, is equivalent to a very simple and natural-looking principle about the definability of regular languages in first-order logic. Is there a direct proof of this principle? If we could prove

$$FO[N] \cap \mathcal{L}_{Reg} = FO[Reg]$$

using only algebraic and automaton-theoretic considerations, then we would obtain as a consequence a new proof that the language MOD_2 is not in AC^0.

But there is more: there is considerable evidence that the analogous principle holds for modular quantifiers, and these are provably equivalent to *open* problems about circuit complexity. For example, we have:

Theorem 5.4 *The following are equivalent:*
(a) $ACC^0 \neq NC^1$.
(b) $(FO + MOD)[N] \cap \mathcal{L}_{Reg} = (FO + MOD)[Reg]$.

Let us see why this is so. If $ACC^0 = NC^1$, then ACC^0 contains all regular languages; in particular, ACC^0 contains regular languages with nonsolvable syntactic monoids. Thus

$$ACC^0 \cap \mathcal{L}_{Reg} = (FO + MOD)[N] \cap \mathcal{L}_{Reg}$$

is strictly larger than $(FO + MOD)[Reg]$, which contains only languages with solvable syntactic monoids. Conversely, if $(FO + MOD)[Reg]$ is strictly contained in $(FO+MOD)[N] \cap \mathcal{L}_{Reg}$, then ACC^0 contains a regular language with a nonsolvable syntactic monoid. It follows from the results in Section 4 on complete problems for NC^1 that ACC^0 contains every language in NC^1.

There are analogous equivalences for CC^0 and for the classes $ACC^0(q)$ and $CC(q)$ for a fixed modulus $q > 1$:

Theorem 5.5 *The following are equivalent:*
(a) CC^0 *does not contain the* AND *function.*
(b) $MOD[N] \cap \mathcal{L}_{Reg} = MOD[Reg]$.

Theorem 5.6 *Let* $q > 1$. *The following are equivalent:*
(a) $ACC^0(q)$ *does not contain the language*

$$\{a_1 \cdots a_n \in \{0,1\}^* : \sum_{i=1}^{n} a_i \equiv 0 \pmod p\},$$

where p *is any prime that does not divide* q.
(b) $(FO + MOD_q)[N] \cap \mathcal{L}_{Reg} = (FO + MOD_q)[Reg]$.

Theorem 5.7 *The following are equivalent:*
(a) $CC^0(q)$ contains neither the AND function nor the language

$$\{a_1 \cdots a_n \in \{0,1\}^* : \sum_{i=1}^{n} a_i \equiv 0 \pmod p\},$$

where p is any prime that does not divide q.
(b) $MOD_q[N] \cap \mathcal{L}_{Reg} = MOD_q[Reg]$.

Moreover, in the last two theorems, the pairs of equivalent statements are known to be true when q is prime, because we possess direct proofs of the statements concerning circuits. We regard this as evidence that the equivalent pairs of statements are true in general. We thus conjecture that for any set $\mathcal{Q}$ of ordinary and modular quantifiers, $\mathcal{Q}[N] \cap \mathcal{L}_{Reg} = \mathcal{Q}[Reg]$. Again, this is a general principle concerning the logical definability of regular languages, equivalent to a quite different-looking principle concerning constant-depth circuits. The cases where it is known to be true are proved using the equivalent circuit formulations. We would like to see a direct proof of the general statement, one that brings to the fore the automaton-theoretic and algebraic considerations. When N is replaced by the union of the regular numerical predicates and the class of all monadic numerical predicates, the conjecture is known to be true ([30]).

References

1. M. Ajtai, Σ_1^1 formulae on finite structures, *Annals of Pure and Applied Logic* **24** (1983), 1–48.
2. D.A.M. Barrington, Bounded width branching programs recognize exactly those languages in NC^1, *J. Computer and Systems Science* **38** (1989), 150–164.
3. D. A. M. Barrington, K. Compton, H. Straubing and D. Thérien, Regular languages in NC^1, *J. Comp. Syst. Sci.* **44** (1992), 478-499.
4. D.A.M. Barrington, N. Immerman and H. Straubing, On uniformity within NC^1, *J. Computer and Systems Science* **41** (1990), 274–306.
5. D.A.M. Barrington, H. Straubing and D. Thérien, Non-uniform automata over groups, *Information and Computation* **89 (2)** (1990), 109–132.
6. D.A.M. Barrington and D. Thérien, Finite monoids and the fine structure of NC^1, *J. of the Association for Computing Machinery* **35** (1988), 941–952.
7. D. Beauquier and J. E. Pin, Factors of words, *Proc. 16th ICALP,* Springer Lecture Notes in Computer Science **372** (1989), 63–79.
8. A. Borodin, D. Dolev, F.E. Fich and W. Paul, Bounds for width two branching programs, *Proc. of the 15 ACM Symp. on the Theory of Computing* (1983), 87–93.
9. J.A. Brzozowski, R. Knast, The dot-depth hierarchy of star-free languages is infinite, *J. Comput. System Sci.* **16** (1978), 37–55.
10. J.R. Büchi, Weak second-order arithmetic and finite automata, *Z. Math. Logik Grundl. math.* **6** (1960), 66–91.
11. R.S. Cohen, J.A. Brzozowski, Dot-depth of star-free events, *J. Comput. System Sci.* **5** (1971), 1–15.
12. S.A. Cook, A taxonomy of problems with fast parallel solutions, *Information and Computation* **64** (1985), 2–22 .
13. S. Eilenberg, *Automata, Languages and Machines,* Vol. B, Academic Press, New York 1976.
14. C.C. Elgot, Decision problems of finite automata design and related arithmetics, *Trans. Amer. math. Soc.* **98** (1061), 21–52.
15. R. Fagin, Generalized first-order spectra and polynomial-time recognizable sets, in R. Karp, ed., *The Complexity of Computation,* SIAM-AMS Proceedings, vol. 7, American Mathematical Society, Providence, Rhode Island, 1974.

16. M.L. Furst, J.B. Saxe and M. Sipser, Parity, circuits, and the polynomial-time hierarchy, *Math. Syst. Theory* **17** (1984), 13–27.
17. Y. Gurevich and H. Lewis, A logic for constant-depth circuits, *Information and Control,* **61** (1984), 65–74.
18. J.E. Hopcroft and J.D. Ullman, *Introduction to Automata Theory, Languages, and Computation,* Addison-Wesley (1979).
19. N. Immerman, Languages that capture complexity classes, *SIAM J. Computing* **16** (1987), 760–778.
20. W. Maurer and J. Rhodes, A property of finite simple non-abelian groups, *Proc. Amer. Math. Soc.* **16** (1965), 552–554.
21. P. McKenzie, P. Péladeau and D. Thérien, NC^1: the automata-theoretic viewpoint, *Computational Complexity* **1** (1991), 330–359.
22. R. McNaughton and S. Papert, *Counter-Free Automata,* MIT Press, Cambridge, Mass. 1971.
23. D. Perrin, J.E. Pin, First-order logic and star-free sets, *J. Comput. System Sci.* **32** (1986), 393–406.
24. J.-E. Pin, *Varieties of Formal Languages,* Plenum Press (1986).
25. J. Rhodes and B. Tilson, The kernel of monoid morphisms, *J. Pure and Applied Algebra* **62** (1989), 227–268.
26. M. P. Schützenberger, On finite monoids having only trivial subgroups, *Information and Control* **8** (1965), 190–194.
27. R. Smolensky, Methods in the theory of lower bounds for boolean circuit complexity, *Proc. of the 19 ACM Symp. on the Theory of Computing* (1987), 77–82.
28. H. Straubing, *Varieties of recognizable sets whose syntactic monoids contain solvable groups,* Ph. D. Thesis, UC Berkeley, 1978.
29. H. Straubing, A generalization of the Schützenberger product of finite monoids, *Theoretical Computer Science* **13** (1981), 107–110.
30. H. Straubing Circuit complexity and the expressive power of generalized first-order formulas, in *Proc. 19th ICALP,* Springer Lecture Notes in Computer Science **623** (1992), 16–27.
31. H. Straubing, *Finite Automata, Formal Logic, and Circuit Complexity,* Birkhäuser, Boston, 1994.
32. H. Straubing, D. Thérien and W. Thomas, Regular languages defined with generalized quantifiers, in *Proc. 15th ICALP,* Springer Lecture Notes in Computer Science **317** (1988), 561–575.
33. D. Thérien, Classification of finite monoids: the language approach, *Theoretical Computer Science* **14** (1981), pp. 195–208.
34. D. Thérien and A. Weiss, Graph congruences and the wreath product, *J. Pure and Applied Algebra* **36** (1985), 205–215.
35. W. Thomas, The theory of successor with an extra predicate, *Math. Annalen* **237** (1978), 121–132.
36. W. Thomas, Classifying regular events in symbolic logic, *J. Comput. System Sci.* **25** (1982), 360–376.
37. W. Thomas, An application of the Ehrenfeucht-Fraïssé game in formal language theory, *Bull. Soc. Math. de France* **16** (1984), 11–21.
38. W. Thomas, Automata on infinite objects, in: *Handbook of Theoretical Computer Science* (J. v. Leeuwen, ed.), Vol. B, Elsevier Sci. Publ., Amsterdam 1990, 133–191.
39. W. Thomas, The Ehrenfeucht-Fraïssé game in theoretical computer science, in: TAPSOFT 93 (M.C. Gaudel, J.P. Jouannaud, eds.), Springer Lecture Notes in Computer Science **669** (1993), 559–568.

ALGORITHMIC PROBLEMS IN GROUPS, SEMIGROUPS AND INVERSE SEMIGROUPS

S. MARGOLIS, J. MEAKIN, M. SAPIR
*University of Nebraska at Lincoln**

1. Introduction

Algorithmic problems in modern algebra had their origins in logic and topology - in the works of Thue [124], Tietze [121] and Dehn [25] at the beginning of the century. They showed that the problem of deducibility of relations in associative calculi, the homeomorphism problem for topological manifolds and the homotopy equivalence problem in finite dimensional manifolds all turned out to be equivalent to algebraic problems, namely the word problem for finitely presented semigroups and groups and the isomorphism and conjugacy problems for finitely presented groups.

Let $\mathcal{V}$ be a variety of universal algebras, let F_X be the finitely generated free object in $\mathcal{V}$ with a set X of free generators. Let σ be a congruence on F_X generated by finitely many pairs $R = \{(u_i, v_i) : i = 1, 2, \ldots\}$. Then the factor algebra F_X/σ is called *finitely presented inside* $\mathcal{V}$. We will denote this algebra by $< X : R, \mathcal{V} >$. If it is clear what variety $\mathcal{V}$ we are dealing with, we will omit $\mathcal{V}$ from this notation. We say that the word problem is *decidable (solvable)* in F_X/σ if there exists an algorithm which tells us for every pair of elements (x, y) from F_X if $(x, y) \in \sigma$.

As is well known the unsolvability of the word problem for semigroups was proved in 1947 by A.A.Markov [61] and E.L.Post [92] and significantly more difficult results on the unsolvability of the three group problems: the word problem, the conjugacy problem and the isomorphism problem — were proved by P.S.Novikov [79] at the beginning of the fifties. A few years later W.W.Boone [16] gave another proof of the unsolvability of the word problem in groups.

The investigation of algorithmic problems, stimulated initially from problems in logic and topology, is largely motivated now by the internal needs of algebra. Algorithmic problems lie very often at the heart of difficult algebraic problems. For example, central to the deep results on Burnside problems for periodic groups is the solution to the word and conjugacy problems in the Burnside groups [80], [2], [83], [42], [51].

While the main algorithmic problems were proved to be unsolvable in general, further investigations have been carried out in two directions. The first direction deals with algebras given by defining relations which satisfy certain restrictions, for example groups and semigroups with small cancellation restrictions, hyperbolic

* This research was supported in part by the NSF grant DMS-9203981 and by the Center for Communication and Information Science, University of Nebraska-Lincoln

J. Fountain (ed.), Semigroups, Formal Languages and Groups, 147–214.

groups, automatic groups and semigroups, groups and semigroups given by a small number of relations. There are many deep theorems obtained in this direction (see [50], [34], [29]) and many deep problems. Among these problems one can mention the word problem for one-relator semigroups. It is known that every semigroup given by a relation of the form $w = 1$ has solvable word problem, but the general case is still unsolved in spite of 30 years of continuous attempts. For other partial results see the surveys [3] and [62]. Recall that in the case of groups Magnus' theorem states that the word problem is solvable in every one-relator group [50].

Another direction investigates algorithmic problems in some subclasses of semigroups, groups, etc. Algorithmic problems are important from a general, "philosophical", point of view also. Thus, the solvability of the word problem in a class of algebraic systems usually means that it is possible to study structural properties of systems in this class. Conversely unsolvability of the word problem usually means that there will be major difficulties in the investigation of this class "as a whole". Among all possible subclasses, the most natural and important ones are, of course, varieties – classes given by identities. The remarkable role they play in algebra and the fact that it is very natural to consider algorithmic problems in varieties induced a large and robust interest in this branch of algebra during the last 15-20 years. This study occupied a major part in most surveys devoted to algorithmic problems in algebra (see e.g. [15], [3], [96], [14], [45]).

Varieties are the most intensively studied classes of universal algebras because they combine the most natural operations on both syntactic and semantic levels. Birkhoff's fundamental theorem connecting these two points of view leads to a rich interplay between combinatorics on words and structural methods. We recall the basic definitions here.

Syntactic: a variety of algebras of a given type is a collection of algebras all satisfying a given set of identities.

Semantic: a variety of algebras of a given type is a collection of algebras closed under the operations of taking subalgebras, quotients and direct products.

Birkhoff's theorem then says that these two definitions define the same classes of algebras. In this paper we will be interested in finitely described varieties of groups, semigroups and inverse semigroups. Clearly, there is both a syntactic and semantic notion of the term "finitely described".

Syntactically finitely described: a variety is finitely based if it is defined by a finite set of identities.

Semantically finitely described: a variety is finitely generated if it is the smallest variety containing a fixed finite algebra.

Given a property α of varieties one can ask for an algorithm which decides if a finitely described variety has the property α. If we find such an algorithm then we will say that we have an *algorithmic description* of those varieties that satisfy our property α.

The last ten years have yielded algorithmic descriptions of varieties with many important properties. Among all properties α of semigroup varieties which have been studied, two turned out to be of major importance: the Burnside property and the finite basis property (see definitions later in the paper). In this paper we will show surprisingly close connections between these properties and other properties.

During the last 20-30 years another source of algorithmic problems appeared, namely theoretical computer science. In particular the theory of finite automata and formal languages has served as a source of many algorithmic problems. Several algorithmic problems about the formal language aspects of finite automata are decidable, while their computational complexity tends to be high. For example, the question as to whether two non-deterministic finite automata recognize the same language is PSPACE complete. Many other problems, especially those concerned with the semigroup theoretic aspects of the theory of finite automata, are undecidable. Further influences included an analysis of the complexity of algorithms in the case where the algorithmic problem is decidable. For example it is known that the word problem in any commutative semigroup may be solved in polynomial time. There is also a very large literature devoted to efficient algorithms for computations in finite groups (see [4]) and other areas of algebra.

The present paper is concerned primarily with a study of algorithmic problems in semigroups, groups and inverse semigroups. The study of such problems for semigroups and groups is a classical part of algebra. A survey of the work in group theory is provided by [72]. Inverse semigroups were introduced into the literature in the papers [125] and [93] in the early 1950's. They may be regarded as semigroups of partial one - one transformations (see [89]). Algorithmic problems for inverse semigroups are natural and have received considerable attention in the literature during the past 15 - 20 years. While algorithmic problems for semigroups and groups are basically problems about words, the corresponding problems for inverse semigroups are essentially problems about finite labelled trees. Indeed the free inverse semigroup may be regarded as a semigroup of finite (birooted) labelled trees (see [76]). There is a developing theory of varieties of inverse semigroups (see [89] for a brief introduction) and of algorithmic problems for finitely presented inverse semigroups. Inverse semigroups form in a sense an intermediate class between groups and semigroups and have influenced both group theory and semigroup theory. For example, congruences on free monoids may be classified via inverse submonoids of the polycyclic monoid [67], and finitely generated subgroups of free groups may be classified via finite inverse semigroups [56], [12]. We will provide some details of these and other connections between inverse semigroups, groups and semigroups later in the paper.

We shall employ geometric, topological, and combinatorial methods to study algorithmic problems for groups, semigroups, and inverse semigroups.

2. Some Geometric Methods

2.1. Groups

Geometric methods have played an important role in combinatorial group theory since the inception of work in that field. We shall not attempt to provide a detailed survey of such methods here, but rather we draw the reader's attention to two standard general ways in which geometric methods have been used to study algorithmic problems for group presentations.

Let $G = gp < X : R >$ be a presentation of a group G. Here X is a non-empty set and we may assume that $R = \{R_i : i \in I\}$ is a (possibly empty) set of non-trivial cyclically reduced words in $(X \cup X^{-1})^*$. Thus $G = FG(X)/N$, where $FG(X)$ denotes the free group on X and N is the normal closure of R in $FG(X)$. Denote by π the canonical projection of $FG(X)$ onto G. We shall adhere to this notation throughout this section.

There are various ways to construct 2-complexes from the presentation. For example we denote by $K = K(X : R)$ the two-dimensional CW-complex with a single 0-cell whose 1-cells are in one-one correspondence with X, and whose 2-cells are in one-one correspondence with R, a given 2-cell being attached by the boundary path determined by reading the corresponding member of R. The underlying 1-skeleton (graph) of K is B_X, the bouquet of X circles. It is well known that $\pi_1(K) = G$ (see [50] or [23]).

It is also standard to consider the universal cover $\tilde{K} = \tilde{K}(X : R)$ of K. This is constructed in the following way. We may take $G = \pi_1(K)$ as the set of 0-cells of $\tilde{K}$. Next, for each $g \in G$ and $x \in X$, we attach a 1-cell (g, x) joining g and $g\pi(x)$: we orient (g, x) from g to $g\pi(x)$ and write $(g\pi(x), x^{-1})$ for the corresponding inverse. Then for each $g \in G$ and $r \in R$ we attach a 2-cell (g, r) via the boundary path $(g_1, x_1)(g_2, x_2)...(g_n, x_n)$ where $r = x_1x_2...x_n, x_i \in X \cup X^{-1}, g_1 = g, g_{i+1} = g_i\pi(x_i), i = 1, \ldots, n-1$. The 1-skeleton (graph) of $\tilde{K}$ is called the Cayley graph of the presentation and is denoted by $\Gamma(X : R)$. One obtains the Cayley complex $C(X : R)$ from $\tilde{K}(X : R)$ by identifying all the faces $(g, r), (g\pi(s), r), \ldots, (g\pi(s^{m-1}), r)$ where $r = s^m, m > 1$ and s is not a proper power in $FG(X)$. The 1-skeleton of $C(X : R)$ is again the Cayley graph $\Gamma(X : R)$ of course.

The 2-complexes described above (and in particular the Cayley complex and its 1-skeleton) play a prominent role in the theory. We refer the reader to the third chapter of [50] for basic information along these lines. In particular, the structure of groups with a presentation whose Cayley complex is planar is studied in [50] and it is shown that the word problem for such a presentation is decidable. Also in [50] the notion of asphericity of group presentations is introduced. In fact there are several somewhat different notions of asphericity that have been considered in the literature. We refer the reader to [19] where five notions of asphericity (all related to the complexes discussed above) are studied in detail.

It is easy to see that the word problem for the presentation $G = gp < X : R >$ is decidable if the ball B_n of radius n in the Cayley graph $\Gamma(X : R)$ is, for all n, finite and effectively constructable. There is a standard procedure (essentially the Todd-Coxeter method of coset enumeration) for inductively constructing $\Gamma(X : R)$ as a limit of a sequence of graphs Γ_i under graph morphisms $\varphi_{i,j} : \Gamma_i \to \Gamma_j$ for all $i < j$. Since this procedure is closely related to much of what follows, we sketch a version of it here and at the same time introduce some notation and terminology that will be used in subsequent parts of the paper. The version that we present is a slight variation on the version discussed in [29]. There seem to be several versions of this algorithm that appear throughout the literature.

We restrict attention to the case that the presentation $G = gp < X : R >$ is finite (i.e X and R are finite). The idea is to build the Cayley graph iteratively from the origin (the vertex representing the identity of G) by using the relators to combine vertices and create loops. At each stage of the procedure we will have constructed a finite graph Γ_i and a graph morphism $\varphi_i : \Gamma_i \to \Gamma_{i+1}$. The graphs Γ_i enjoy the following properties:

(P1) each graph Γ_i is connected with directed edges labelled by the elements of $X \cup X^{-1}$ and such that if there is an edge labelled by x from v_1 to v_2 , then there is also an edge labelled by x^{-1} from v_2 to v_1;

(P2) no two directed edges with the same initial vertex have the same label.

Graphs with these properties are referred to as "unambiguous partial Cayley graphs" in [29] and "inverse word graphs over X" in [119]. There is an evident notion of morphism between inverse word graphs (a graph homomorphism that preserves labels and directions). If Γ is an inverse word graph over X there is an obvious morphism $f : \Gamma \to B_X$ from Γ to the bouquet of X circles : f takes an edge labelled by x in Γ onto the edge labelled by x in B_X. It is clear that this morphism is locally injective at each vertex of Γ ; i.e f is an "immersion" in the sense of [117]. We sometimes abuse notation slightly and refer to Γ itself as an immersion over B_X. Graph immersions have been used by a number of authors ([117], [32], [12], [56]) to study subgroups of free groups and submonoids of free inverse monoids. We shall consider these ideas in more detail in the next section of this paper.

The reason for the name "inverse word graph over X" to describe a graph satisfying properties (P1) and (P2) above is that the transition monoid of such a graph (i.e. the monoid of partial transformations of the vertices of Γ induced by the action of the letters in $X \cup X^{-1}$) is an inverse monoid. An inverse monoid (semigroup) is a monoid (semigroup) M such that for each $x \in M$ there is a unique element (denoted by x^{-1}) in M such that

$$x = xx^{-1}x \text{ and } x^{-1} = x^{-1}xx^{-1}.$$

Such monoids may be viewed as monoids of injective (partial) functions on a set A. Indeed the monoid $SIM(A)$ of all injective partial functions on A (with respect to the usual composition of partial functions) is an inverse monoid (called the "symmetric inverse monoid on A") and the Preston-Wagner theorem asserts that every inverse monoid may be embedded in a suitable symmetric inverse monoid. We refer the reader to [89] for much basic information about inverse monoids. We shall return to a discussion of inverse monoids later in this paper.

We also mention at this stage that an inverse word graph Γ over X may be viewed as an automaton over the alphabet $X \cup X^{-1}$ if we distinguish an initial vertex (state) α and a terminal vertex (state) ω of Γ. Such an automaton is the minimal automaton of the language that it accepts and is referred to in this paper and in the literature as an "inverse automaton over X". The transition monoid of such an automaton is an inverse monoid (which is finite if Γ is finite). The inverse word graphs Γ_i that we are about to construct will have a fixed base point, which may be viewed as the initial and terminal vertex of Γ_i, so we may think of each Γ_i as an inverse automaton over X. Recall that an inverse automaton over an alphabet X is an automaton over $X \cup X^{-1}$ such that each $x \in X \cup X^{-1}$ induces an injective function on the state set and such that the inverse letter induces the inverse injective function.

In order to describe the graphs Γ_i associated with the finite presentation $G = gp < X : R >$ we consider three types of operations that we will perform on a finite labelled graph Γ that satisfies property (P1):

(O1) Loop closings: if in Γ there is a path p from vertex v_1 to vertex $v_2 \neq v_1$ labelled by a relator $r \in R$, then we form a new graph Γ' obtained from Γ by identifying v_1 and v_2. Since R and Γ are finite and Γ' has one less vertex than Γ it follows that, by starting with Γ and applying successive loop closings a finite number of times, we will reach a (unique) graph Γ_l to which no further loop closings are applicable.

(O2) Adding hairs: if in Γ there is a vertex v and a letter $x \in X \cup X^{-1}$ such that Γ has no edge with label x starting at v, then we create a new graph Γ' from Γ by adding to Γ a new vertex v' and a new edge e with label x, initial vertex v and terminal vertex v' (and an inverse edge e' from v' to v with label x^{-1}). Since Γ and X are finite, there are only finitely many hairs that can be added to the vertices of Γ, thereby obtaining a new graph Γ_h.

(O3) Folding: if in Γ there are two (directed) edges with the same initial vertex (or terminal vertex) and the same label $x \in X \cup X^{-1}$, then we create a new graph Γ' from Γ by identifying these edges. One can check that folding is confluent, so that after finitely many foldings one reaches a graph Γ_f to which no further foldings apply: clearly Γ_f is an inverse word graph over X.

We construct a sequence of inverse word graphs $\Gamma_0, \Gamma_1, \dots \Gamma_i$, associated with $G = gp < X : R >$ as follows. Start by taking a base point 1 and attaching to it a loop for each relator in R. Each loop starts and ends at 1 and successive edges describe the word in R. Next apply successive foldings until no more can be applied. Then close the resulting graph under loop closings. This may create more vertices at which foldings may be applied. Continue applying loop closings and foldings until no more loop closings or foldings may be applied. Finiteness of Γ, X and R and confluence of the operations guarantees that after finitely many steps we reach an inverse word graph Γ_0. If we regard the base point as the initial (terminal) vertex, then Γ_0 becomes an inverse automaton.

Assume inductively that we have constructed Γ_i. If, for each vertex v in Γ_i and each letter $x \in X \cup X^{-1}$, there is an edge in Γ_i with label x and initial vertex v, then G must be finite and Γ_i is the Cayley graph $\Gamma(X : R)$. Then define $\Gamma_{i+1} = \Gamma_i$ and φ_i to be the identity map on Γ_i. Otherwise we add all missing hairs to the vertices of Γ_i (obtaining the graph $(\Gamma_i)_h$) and then apply all possible loop closings and foldings until no more can be applied. Call the resulting graph Γ_{i+1}.

Note that each vertex and edge of Γ_i has a natural image in Γ_{i+1} after all the foldings and loop closings are applied, so there is a natural morphism $\varphi_i : \Gamma_i \to \Gamma_{i+1}$. This gives rise to morphisms $\varphi_{i,j} : \Gamma_i \to \Gamma_j$ for $i < j$ and it is not too difficult to see that $\Gamma(X : R)$ is the direct limit of this family $\{\Gamma_i, \varphi_{i,j}\}$ of inverse word graphs and morphisms. If B_n is the ball of radius n in $\Gamma(X : R)$, then there is some integer

i (depending on n, but in general not an effectively computable function of n) such that B_n is contained in Γ_i.

Theorem 2.1 *Let $G = gp < X : R >$ be a finitely presented group and let $\{\Gamma_i, \varphi_{i,j}\}$ be the system of inverse word graphs and graph morphisms constructed according to the Todd-Coxeter algorithm described above. Then $\Gamma(X : R)$ is the direct limit of this system of graphs and graph morphisms.*

The theorem essentially shows that every finitely presented group may be naturally approximated by finite inverse monoids (the transition monoids of the graphs Γ_i constructed above). In fact inverse monoids often arise naturally when given partial information about a group. We shall see other examples of this sort of phenomenon later in the paper.

We turn now to a very brief indication of a second standard way in which geometric methods have played a prominent role in combinatorial group theory. Again let $G = gp < X : R >$ be a finitely presented group, where the notation is as specified above. It is standard to study membership in N (the normal closure of R in $FG(X)$) by associating with each word in N a planar map (diagram) which is referred to in the literature as a van Kampen diagram (or a singular disk diagram). Briefly, the definition of such a diagram may be stated in the following way. We define a map to be a finite, planar, connected and simply connected 2-complex. A diagram Δ over an alphabet of the form $X \cup X^{-1}$ is a map such that every edge e (i.e., a 1-cell) of Δ is provided a label $\varphi(e)$ in $X \cup X^{-1}$ such that $\varphi(e^{-1}) = \varphi(e)^{-1} \in (X \cup X^{-1})^*$. The label of a path $p = e_1 e_2 ... e_n$ in Δ is by definition the word $\varphi(e_1)\varphi(e_2)\ldots\varphi(e_n)$. We call a diagram Δ over $X \cup X^{-1}$ a (van Kampen) diagram over the presentation $G = gp < X : R >$ if the label of a boundary path of every face (i.e., a 2-cell) is a cyclic permutation of some relator r (or r^{-1}) in R. The van Kampen Lemma then says that a word $w \in (X \cup X^{-1})^*$ is in N if and only if there exists a diagram over G such that w is the label of the boundary of the diagram. In fact one may assume that the diagram is reduced. A diagram Δ is said to be reduced if there are no opposite faces in Δ. Two faces Π_1 and Π_2 are called "opposite" if there is an edge e in the intersection of their boundaries such that the labels of Π_1 and Π_2, starting at e, coincide. (That is, Π_1 and Π_2 are "mirror images" of each other). For a proof of this lemma, and for further definitions and results about van Kampen diagrams, we refer the reader to Chapter V of [50]. There are slightly different definitions that appear at various places in the literature. The reader is referred to Ol'shanskii's book [83] for a deep analysis of the use of these methods in combinatorial group theory. The methods of small cancellation theory [50] and much of the recent deep work of Ivanov on the Burnside problem [42] is based on these methods.

We remark that there is a natural graph morphism from any van Kampen diagram over G into the Cayley graph $\Gamma(X : R)$. We define such a map γ as follows. Let Δ be a van Kampen diagram over $G = gp < X : R >$ and suppose that p is a boundary path of Δ with initial (and terminal) vertex O. Define $\gamma(O) = 1$ (the vertex representing 1 in $\Gamma(X : R)$) and for an arbitrary vertex v of Δ let $\gamma(v)$ be the element of G represented by the word $\varphi(q)$, where q is any path in Δ between the vertex O and v. In view of the van Kampen lemma the vertex $\gamma(v)$ of $\Gamma(X : R)$ depends only on v and not on the path q. For an edge e of Δ with initial vertex v let

$\gamma(e)$ be the edge of $\Gamma(X : R)$ with initial vertex $\gamma(v)$ and label $\varphi(e)$. It is clear that γ defines a graph morphism from (the 1-skeleton of) Δ into $\Gamma(X : R)$. Of course γ is not in general an embedding since distinct vertices (edges) of Δ may be identified under γ. The interplay between van Kampen diagrams and Cayley graphs indicated above is used throughout the literature in combinatorial group theory. For a nice recent application of these ideas we refer to Ol'shanskii's recent elementary proof [84] that groups with subquadratic isoperimetric inequality are hyperbolic.

2.2. Semigroups

We turn now to the use of geometric methods for semigroup presentations analogous to the methods discussed above for group presentations. Throughout this section we let $S = sgp < X : R >$ be a semigroup presentation. Here X is a non empty set and we will take R to be a finite set of relations of the form $u_i = v_i$ where u_i and v_i are non empty words in the free semigroup X^+. There is a well developed analogue of the planar van Kampen diagrams of combinatorial group theory in this setting and also an appropriate but somewhat less well developed analogue of the Cayley graph for a group presentation. We provide a brief discussion of both methods below. We begin with a description of the planar semigroup diagrams associated with the semigroup presentation S above.

A semigroup diagram (or s-diagram) over the presentation $S = sgp < X : R >$ for a pair (u, v) of non empty words in X^+ is a diagram Δ over $X \cup X^{-1}$ with the following properties:

(SD1) each face (i.e. bounded 2-cell) of Δ is labelled in the clockwise direction by a word of the form rs^{-1} for some (r, s) such that $r = s$ (or $s = r$) is a relation in R;

(SD2) the boundary of Δ carries a clockwise label uv^{-1};

(SD3) there are no interior sources or sinks (a source is a vertex of indegree 0 - i.e a vertex with no positively labelled edge coming in to it; a sink is the dual concept).

It is clear that if F is any face of such a diagram then there is a unique vertex F_- such that the clockwise boundary cycle of F starting at F_- is of the form $\alpha\beta^{-1}$, where $\varphi(\alpha) = r$ and $\varphi(\beta) = s$. We denote the terminal point of α (= the terminal point of β) by F_+ and refer to α [β] as the left [right] boundary of F. Similarly the whole diagram Δ has a left boundary labelled by u starting at Δ_- and a right boundary labelled by v ending at Δ_+. An important property of such diagrams is that every positively labelled edge of Δ lies on some positively labelled directed path running from Δ_- to Δ_+. (Such a path is sometimes referred to as a "transversal" of Δ.) The analogue of the van Kampen lemma in this setting is that there is a sequence of elementary transitions $u = u_0 \to u_1 \to \ldots \to u_n = v$ with respect to the presentation $S = sgp < X : R >$ if and only if there is a (u, v) diagram over this presentation with exactly n regions (bounded faces).

Semigroup diagrams of this type were introduced into the literature by Remmers [97] and have been used by a number of authors to study semigroup presentations. We refer the reader to the book of Higgins [39] for an exposition of some of the uses that have been made of these diagrams - in particular for some results on "small overlap" semigroups and for a version of Remmer's proof of a result of Adian about the embeddability of a semigroup that is left and right "cycle free" in a group.

We turn now to an exposition of some recent (as yet unpublished) work of Kilibarda concerning the algebra of semigroup diagrams. Semigroup diagrams have in general a much simpler structure than group diagrams (van Kampen diagrams). In particular, the fact that every positively labelled path in a semigroup diagram may be extended to a transversal in the sense described above is very useful and imposes severe restrictions on the common boundary of two faces of such a diagram. For example, no edge can be on the left boundary of two distinct faces of a semigroup diagram. Using ideas of this type, one is able to define a process of reduction on semigroup diagrams and a natural multiplication on the set of semigroup diagrams associated with a given presentation.

A pair of faces F_1 and F_2 of a semigroup diagram Δ is called a "reducible pair" if F_1 and F_2 are mirror images of each other; i.e., F_1 has a boundary cycle of the form $\alpha\beta^{-1}$ and F_2 has a boundary cycle of the form $\beta\gamma^{-1}$ where α, β and γ are positively labelled paths and $\varphi(\alpha) = \varphi(\gamma)$. In this case one can form a new semigroup diagram Δ' by eliminating the common boundary β from Δ and then extracting the new region that is formed from F_1 and F_2 by identifying the paths α and γ. In this case we say that Δ' has been constructed from Δ by one elementary reduction. We define a semigroup diagram Δ to be reduced if it does not have any reducible pair of faces. If Δ and Δ' are two semigroup diagrams over the same presentation we say that they are equivalent if and only if either Δ is identical to Δ' or there is a sequence $\Delta = \Delta_0, \Delta_1, \dots, \Delta_k = \Delta'$ for some k, such that for each $j < k$, one of Δ_{j+1} and Δ_j comes from the other by elementary reduction. It is clear that equivalence of diagrams defined this way is an equivalence relation.

Lemma 2.1 (Kilibarda) *There is exactly one reduced diagram in each equivalence class.*

We denote by $r(\Delta)$ the unique reduced diagram that is equivalent to Δ. The lemma enables us to introduce a natural multiplication on the set of reduced semigroup diagrams associated with a fixed presentation $S = sgp < X : R >$ in the following way. Let Γ be a reduced (u, v) diagram over S and Γ' a reduced (t, w) diagram over S. The product of Γ and Γ' will be defined if and only if $v = t$. In this case we can form a new diagram (a (u, w) diagram over S) by identifying the right boundary of Γ with the left boundary of Γ' and taking the union of the two diagrams under this identification. Denote the new diagram by $\Gamma * \Gamma'$. Then define

$$\Gamma \cdot \Gamma' = r(\Gamma * \Gamma').$$

With respect to this multiplication, the set of reduced semigroup diagrams over S becomes a groupoid. The identities (objects) of this groupoid may be identified with the (u, u) diagrams that have no faces - a reduced (u, v) diagram may be viewed as an arrow (morphism) in this category from the trivial (u, u) diagram to the trivial (v, v) diagram. The maximal subgroups (loop monoids) of this category provide a measure of the essentially different elementary transitions with respect to the presentation that take us from a word u to itself. In particular, the maximal subgroups are all trivial if and only if the presentation is aspherical in the sense of Ivanov [41]. This is a somewhat more restricted notion of asphericity than that of Cho and Pride [21] and may be best thought of as an analogue for semigroups of the notion of "diagrammatic asphericity" in the sense of [19].

The groupoid of semigroup diagrams over the presentation $S = sgp < X : R >$ defined above may be identified with the fundamental groupoid $\pi(K_S)$ of a certain complex K_S. We refer to [40] or [23] or [50] for the notion of the fundamental groupoid of a complex. The complex K_S is defined as follows.

$V(K_S)$: the set of vertices of K_S is X^* (the set of all words over X).

$E(K_S)$: there is an edge e in K_S with initial vertex $\sigma(e) = w_1 \in X^*$ and terminal vertex $\tau(e) = w_2 \in X^*$ whenever there exists a pair of words $x, y \in X^*$ and a relation $r = s$ (or $s = r$) in R such that $w_1 = xry$ and $w_2 = xsy$. The edge e depends not only on r and s but also on the context (x, y) in which they occur. Thus there may be multiple edges between w_1 and w_2 corresponding to application of the same relation but in different contexts. The inverse of the edge e defined above has initial edge w_2 and terminal edge w_1 and is dètermined by the same relation that was used to define e, in the same context.

$C(K_S)$: there is a 2-cell (face) F in K_S with boundary path $e_1e_2e_3e_4$ whenever there are two disjoint appearances of relators in a word $w \in X^*$, such as $w = xr_1yr_2z$ for some $x, y, z \in X^*$ and some relations $r_1 = s_1$ and $r_2 = s_2$ in R. Then e_1, e_2, e_3 and e_4 are edges with initial and terminal vertices given by:

$$\sigma(e_1) = w = \tau(e_4), \sigma(e_2) = xs_1yr_2z = \tau(e_1),$$

$$\sigma(e_3) = xs_1ys_2z = \tau(e_2), \sigma(e_4) = xr_1ys_2z = \tau(e_3).$$

This 2-cell has an obvious inverse with boundary path $(e_1e_2e_3e_4)^{-1}$.

Theorem 2.2 (Kilibarda) *Let $S = sgp < X : R >$ be any semigroup presentation. Then the groupoid of semigroup diagrams over S is isomorphic to the fundamental groupoid $\pi(K_S)$ of the complex K_S defined above. In particular, the maximal subgroups of this groupoid of semigroup diagrams are the fundamental groups of K_S.*

REMARKS. (1) Theorem 2.2 may be used to calculate the maximal subgroups of the groupoid of semigroup diagrams over a presentation. For example if $S = sgp < x, y : xy = yx >$ (the free commutative semigroup on two generators) then one sees that all of the corresponding maximal subgroups of the associated groupoid are trivial, so this presentation is aspherical in the sense described above. This observation also essentially follows from the results of Adian [1] or Ivanov [41]. However the free commutative semigroup on more than 2 generators is not aspherical.

(2) It can be shown from Theorem 2.2 that the maximal subgroups of the groupoid of semigroup diagrams over any finitely presented semigroup have decidable word problem. It can also be shown that any finite direct product of finitely generated free groups arises this way. A complete characterization of the class of groups that arise in such a way is at present unknown.

(3) It is interesting to note that a complex isomorphic to K_S has been introduced independently by S.Pride [94] in connection with his work on low dimensional homotopy of monoid presentations.

We turn now to a discussion of the analogue of the Cayley graph of a group presentation in the setting of monoid presentations. For the remainder of this section, M will denote a monoid presentation of the form $M = mon < X : R >$ where R is a set of monoid relations of the form $u = v$, for some $u, v \in X^*$. Such presentations include the semigroup presentations considered above but we do not now explicitly exclude the case that some of the relations may be of the form $u = 1$ for some $u \in X^+$. In any case we want to view M as a monoid now, even if there are no relations of this form in R. We let π denote the natural map from X^* onto M. One may form the Cayley graph $\Gamma(X : R)$ for such a presentation just as for a group presentation. The vertices of $\Gamma(X : R)$ are the elements of M and there is an edge labelled by x from m to $m\pi(x)$ for each m in M and x in X. We regard $\Gamma(X : R)$ as a word graph over X (i.e., a graph whose edges are labelled over X) in this case, but note that it is not an inverse word graph since there is no "inverse edge" labelled by x^{-1} from $m\pi(x)$ to m.

In order to study the word problem for a monoid presentation M as above, Stephen showed in his thesis [118] that it is convenient to introduce a family of birooted word graphs $B\Gamma(w)$, one for each word w in X^*. The graphs $B\Gamma(w)$ are defined as follows. The underlying graph is the restriction of the Cayley graph $\Gamma(X : R)$ to the set of elements n in M such that $\pi(w)M$ is contained in nM in the monoid M (i.e. n is greater than or equal to $\pi(w)$ in the $\mathcal{R}$-class order on M). Thus the set of vertices of $B\Gamma(w)$ is the set of all such elements defined above and the edges are just the edges of $\Gamma(X : R)$ that connect vertices of this form. There are also two distinguished vertices (roots) namely 1 (the initial vertex) and $\pi(w)$ (the terminal vertex). One may view $B\Gamma(w)$ as an automaton over X with states the vertices of the graph and initial state 1 and terminal state $\pi(w)$. Stephen shows in [118] that the language accepted by this automaton is exactly the set of words in X^* that are equivalent to w under the presentation. Thus one solves the word problem for M if one has an effective procedure for constructing the automata $B\Gamma(w)$ (or the language accepted by these automata) for each w in X^*.

There is an iterative construction of the automaton $B\Gamma(w)$ starting from the "linear automaton" of the word w. If w is the word $w = x_1x_2...x_n$ in X^* then the linear automaton of w is the automaton with $n+1$ states $v_0, v_1, ..., v_n$ and an edge labelled by x_i from v_{i-1} to v_i for $i = 1, \ldots, n$. The initial state is v_0 and the terminal state is v_n. Clearly the language accepted by this automaton is just $\{w\}$. Starting with this automaton, one then builds a sequence of automata obtained by applying operations of the following types:

(M1) Expansions: if in an automaton A over X (i.e. a birooted word graph over X) there are two states v_1 and v_2 and a path p from v_1 to v_2 labelled by one side (say u) of a relation $u = v$ in R, but no path from v_1 to v_2 labelled by the other side, then we expand the automaton A by adding a new path from v_1 to v_2 labelled by v. The resulting automaton A' is said to be obtained from A by an expansion.

(M2) Foldings: if in a birooted word graph (automaton) over X there are two edges with the same initial vertex and the same label, then we identify these edges. The resulting automaton A' is said to be obtained from A by folding edges. Note that this is the same as the operation (O3) that was used in the Todd-Coxeter algorithm outlined above, the difference being that in this case we only fold edges with the same initial vertex and the same label, whereas in (O3) we also fold edges with the same terminal vertex and the same label. If we perform all possible foldings to a finite automaton A the result is a deterministic automaton, but not necessarily an injective automaton. By an injective automaton we mean an automaton where every letter induces a partial one-one function.

Stephen has shown [118] that the operations (M1) and (M2) are confluent - i.e., if we start with an automaton A and apply one of these operations to obtain A' and a possibly different operation to A to obtain A'', then there is a third automaton B such that B can be obtained from A' by applying a finite sequence of operations of the form (M1) or (M2) and also B can be obtained from A'' by applying a finite sequence of these operations. The idea then is that we can build $B\Gamma(w)$ by starting with the linear automaton of w and applying operations of the form (M1) or (M2) successively (in any order). The resulting automata become "better and better approximations" to $B\Gamma(w)$ and "in the limit" we obtain $B\Gamma(w)$ this way. These ideas are made precise in [118]. We include some brief explanation of the details below and refer to [118] for more detail. We also refer to [59] for the relevant definitions and concepts from category theory.

In [118] it is shown that the class of birooted word graphs over X is a cocomplete category (see [59] for the definition of this concept) and so in particular this category has colimits. If (α, Γ, β) and $(\alpha', \Gamma', \beta')$ are birooted word graphs obtained from the linear automaton of w by repeated applications of operations (M1) and (M2) and if $(\alpha', \Gamma', \beta')$ is obtained from (α, Γ, β) by one application of one of these operations, then the language accepted by (α, Γ, β) is contained in the language accepted by $(\alpha', \Gamma', \beta')$ and there is a natural morphism of birooted word graphs from (α, Γ, β) to $(\alpha', \Gamma', \beta')$. So there is a natural diagram (directed system) of birooted word graphs obtained from the linear graph of w this way. Stephen shows that $B\Gamma(w)$ is the colimit of this diagram of birooted word graphs in the category of all birooted word graphs over X.

Theorem 2.3 (Stephen, [118]) *Let $M = mon < X : R >$ be a monoid presentation and let w be any word in X^*. Then $B\Gamma(w)$ is the colimit (direct limit) in the category of birooted word graphs over X of the diagram of birooted word graphs that are obtained from the linear graph of w by repeated applications of operations (M1) and (M2) relative to this presentation.*

REMARKS. (1). In order to apply Theorem 2.3 to construct the automata $B\Gamma(w)$ relative to some given presentation, one uses the confluence of the operations (M1) and (M2) to devise a scheme for iteratively applying these operations so that one "eventually" builds in all applications of all possible expansions and all possible foldings. In practice this can be very difficult, especially if the resulting automaton

$B\Gamma(w)$ is infinite, and there has not been extensive use of this technique in the literature to date.

(2). In his paper [63], McCammond constructs automata somewhat similar to the automata $B\Gamma(w)$ to solve the word problem for the free semigroups in the Burnside variety of semigroups defined by an identity of the form $x^a = x^{a+b}$ for $a > 5$. However the automata that McCammond uses are essentially non - deterministic, unlike the automata $B\Gamma(w)$ of Stephen.

2.3. Inverse Monoids

Inverse monoids arise naturally whenever one is concerned with injective partial functions (i.e., partial one-one functions on a set). The class of inverse monoids forms a variety of algebras of type $< 2, 1, 0 >$ defined by associativity and the identities

$$x = xx^{-1}x,\ (x^{-1})^{-1} = x,\ x \cdot 1 = 1 \cdot x = x,\ xx^{-1}yy^{-1} = yy^{-1}xx^{-1}.$$

This last law expresses the fact that idempotents commute and is one of the crucial properties of inverse monoids. As a consequence of this, the set $E(M)$ of idempotents of an inverse monoid forms a semilattice with respect to the multiplication, which may be extended to the natural partial order on M by defining $a \leq b$ iff $a = eb$ for some e in $E(M)$. Inverse monoids have been considered extensively in the literature. We refer to the book of Petrich [89] for much basic information about inverse monoids and in particular for various ways in which they may be constructed from groups and semilattices.

The study of algorithmic problems in inverse semigroups may be traced back to the work of Gluskin [33] who studied the structure of the free inverse monoid on one generator, but it was not until the work of Munn [76] and Scheiblich [111] that the structure of the free inverse monoid on a set X was known and the word problem for the free inverse monoid was solved. Subsequently, Stephen [118], [119] developed a framework for studying the word problem for presentations of inverse monoids. Stephen's work built on the work of Munn [76] and Margolis and Meakin [54], where graphical methods akin to those commonly used in combinatorial group theory were introduced. This work has led to a substantial amount of subsequent work by Margolis, Meakin, Stephen and others on the word problem for inverse monoids [57], [55], [11], [115], [43] and varieties of inverse monoids [58], [68]. Many other results on varieties of inverse monoids, properties of the free inverse monoid and the elementary theory of the free inverse monoid have appeared in the literature. Closed inverse submonoids of the free inverse monoid have been studied by Margolis and Meakin [56] and applications of these ideas to the study of subgroups of free groups have been developed in [56] and [12]. These ideas will be discussed in the next section of this paper. Much of this work on algorithmic problems in inverse semigroups has been discussed in the survey article of Meakin [66], so we shall include only a brief discussion of part of this theory here.

The structure of the free inverse monoid $FIM(X)$ on a set X can be elegantly described via the Cayley graph (tree) of the free group $FG(X)$ on X. We present this as a special case of a more general construction of Margolis and Meakin [54] here, since we will need the ideas of this more general construction in Section 6.7

of this paper. Let $G = gp < X : R >$ be a group presentation and let $\Gamma(X : R)$ be the corresponding Cayley graph. We build an inverse monoid from $\Gamma(X : R)$ in the following way. Define $M(X : R) = \{(\Gamma, g) : \Gamma$ is a finite connected subgraph of $\Gamma(X : R)$ containing the vertices 1 and g of $\Gamma(X : R)\}$ with multiplication

$$(\Gamma, g) \cdot (\Delta, h) = (\Gamma \cup g \cdot \Delta, gh),$$

where $g \cdot \Delta$ denotes the translate of Δ on the left by g in $\Gamma(X : R)$ and $\Gamma \cup g \cdot \Delta$ simply denotes the union of the corresponding subgraphs of $\Gamma(X : R)$. In [54] it is shown that $M(X : R)$ is an inverse monoid with maximal group homomorphic image G and that the passage from G to $M(X : R)$ defines a functor from the category of X–generated groups to the category of X–generated "E-unitary" inverse monoids that is a left adjoint to the maximal group image functor. We recall that an inverse monoid M is said to be E-unitary if the natural map σ from M onto its maximal group image is "idempotent-pure" (i.e., the inverse image of 1 under σ is just $E(M)$).

The monoids $M(X : R)$ constructed above have many pleasant properties. In particular, if $G = gp < X : \varphi > = FG(X)$, then the corresponding inverse monoid $M(X : \varphi)$ is isomorphic to the free inverse monoid $FIM(X)$ on X. Thus the elements of $FIM(X)$ may be thought of as birooted inverse word trees of the form $(1, \Gamma, g)$ where Γ is a finite subtree of the tree $\Gamma(X : \varphi)$ containing 1 and g, and the multiplication is as specified above. If w is a word in $(X \cup X^{-1})^*$ then the birooted word tree associated with w is $(1, MT(w), r(w))$, where $r(w)$ is the reduced form of w in the usual group theoretic sense and $MT(w)$ is the "Munn tree" of w - i.e., $MT(w)$ is the subtree of $\Gamma(X : \varphi)$ traversed when one reads the word w in $\Gamma(X : \varphi)$, starting at the vertex 1 and ending at the vertex $r(w)$. Munn's solution to the word problem in $FIM(X)$ is: if u and v are two words in $(X \cup X^{-1})^*$, then $u = v$ in $FIM(X)$ if and only if $MT(u) = MT(v)$ and $r(u) = r(v)$.

The Munn tree $MT(w)$ of a word w may be regarded as a finite inverse automaton over $X \cup X^{-1}$ in the usual way (with initial state 1 and terminal state $r(w)$). The vertices of this graph (automaton) are the elements of $FIM(X)$ that are related to w via Green's R-relation on $FIM(X)$ and the language accepted by this automaton is the set of words v in $(X \cup X^{-1})^*$ such that $v \geq w$ in the natural partial order on $FIM(X)$. These ideas have been extended by Stephen [118], [119] to study arbitrary presentations of inverse monoids.

We denote by $inv < X : T >$ the inverse monoid presented by the set X of generators and the set T of relations. Here we may consider T as a set of relations of the form $u_i = v_i$ where u_i and v_i are elements of the free monoid $(X \cup X^{-1})^*$ and we interpret $inv < X : T >$ as the image of $(X \cup X^{-1})^*$ obtained by imposing the relations in T together with all of the identities that define the variety of inverse monoids: alternatively, one may view the u_i and v_i as elements of $FIM(X)$ and $inv < X : T >$ as the image of $FIM < X >$ obtained by imposing these relations. It follows immediately by universal considerations that $gp < X : T >$ is the maximal group homomorphic image of $inv < X : T >$.

There is a well developed and useful analogue of the Cayley graph for studying presentations of inverse monoids, but no analogue at the present time of the methods of van Kampen diagrams or semigroup diagrams discussed above. We briefly indicate the situation with respect to the Cayley graph. One may define the Cayley graph $\Gamma(X : T)$ of an inverse monoid presentation in the obvious way, just as for semigroup presentations or group presentations. It turns out to be useful in this setting, however, to restrict this Cayley graph to the set of vertices that are R–related to the word under consideration.

Let $M = inv < X : T >$ and let w be any word in $(X \cup X^{-1})^*$. Denote the natural map from $(X \cup X^{-1})^*$ onto M by π. We define a word graph $S\Gamma(w)$ over $X \cup X^{-1}$ as follows. The vertices of $S\Gamma(w)$ consist of the elements m of M that are related via Green's R–relation to $\pi(w)$. There is an edge labelled by x from m to $m\pi(x)$ for each x in $X \cup X^{-1}$ provided both m and $m\pi(x)$ are R–related to $\pi(w)$ in M. This implies that there is also an edge labelled by x^{-1} from $m\pi(x)$ to m in $S\Gamma(w)$, which we take as the inverse of the original edge. Thus $S\Gamma(w)$ is in fact an inverse word graph over X in the sense discussed above. The birooted inverse word graph $A(w) = (\pi(ww^{-1}), S\Gamma(w), \pi(w))$ may be regarded as an inverse automaton in the usual way (with initial state $\pi(ww^{-1})$ and terminal state $\pi(w)$). This automaton has been referred to as the "Schutzenberger automaton of w" in the literature and the underlying graph $S\Gamma(w)$ as the "Schutzenberger graph of w" since $S\Gamma(w)$ is the graph of the Schutzenberger representation of M relative to the R–class in consideration. Note the distinction between $S\Gamma(w)$ and the graph $B\Gamma(w)$ constructed above in connection with a monoid presentation. Note also that if $M = FIM(X) = inv < X : \varphi >$ then for each word w in $(X \cup X^{-1})^*$, there is an isomorphism between $A(w)$ and the birooted inverse word graph $(1, MT(w), r(w))$.

In his work [119], Stephen shows that the language accepted by $A(w)$ is in fact the set of words v in $(X \cup X^{-1})^*$ such that $v \geq w$ in the natural order on M and that two words u and v in $(X \cup X^{-1})^*$ are equal in M if and only if $A(u)$ and $A(v)$ accept the same language, i.e., if and only if $A(u)$ and $A(v)$ are isomorphic as birooted word graphs over X. Thus we have a solution to the word problem for M if we have an effective procedure for constructing each automaton $A(w)$ for each word w in $(X \cup X^{-1})^*$. Stephen goes on to provide an iterative procedure for constructing the automata $A(w)$. We briefly describe this procedure below.

We again start with the linear automaton of the word w, as described in the previous section. However we now regard this automaton as an inverse automaton over X (i.e., we associate with each edge labelled by x in $X \cup X^{-1}$ an inverse edge labelled by x^{-1} in the usual way). Let us denote the resulting linear automaton of w by $LinA(w)$. Note that w is in the language accepted by $LinA(w)$, but that this language is infinite if $w \neq 1$ (for example, $(ww^{-1})^n w$ is in this language for each $n \geq 1$). It is not difficult to see that the language accepted by $LinA(w)$ is contained in the language accepted by $A(w)$, as described above. Starting with $LinA(w)$, one builds a sequence of automata obtained by applying operations of the following types:

(I1) Expansions: this is exactly the same as the expansion operation (M1) discussed in the previous section. However we build in the additional requirement that each edge with label x from $X \cup X^{-1}$ in the new automaton must come equipped with its natural inverse edge labelled by x^{-1}.The resulting automaton satisfies condition (P1) of section 2.1.

(I2) Foldings: this is exactly the same as the folding operation (O3) discussed in section 2.1. Note that if we apply all possible foldings to a finite automaton that satisfies condition (P1) we obtain an inverse automaton. Note the distinction between this operation and the folding operation (M2) of section 2.2.

Once again the effect of applying an expansion or a folding to an automaton that satisfies (P1) is to obtain another such automaton which accepts all words in the language of the original automaton and there is a morphism from the first birooted word graph to the second. Thus one may again consider the diagram (directed system) of all birooted word graphs obtained from the linear automaton $LinA(w)$ by successive applications of the expansion and folding operations (I1) and (I2), and A(w) is the colimit (direct limit) of this diagram of automata.

Theorem 2.4 (Stephen, [119]) *Let $M = inv < X : T >$ be an inverse monoid presentation and let w be any word in $(X \cup X^{-1})^*$. Then $A(w)$ is the colimit (direct limit) in the category of birooted word graphs over $X \cup X^{-1}$ of the diagram (directed system) of birooted word graphs that are obtained from the linear graph $LinA(w)$ of w by repeated applications of operations (I1) and (I2) relative to this presentation.*

Note that the operations (I1) and (I2) are again confluent in this setting, so that one may proceed to construct $A(w)$ from $LinA(w)$ by applying a sequence of expansions and foldings in any desired order, so long as one eventually builds in all possible expansions and foldings. In practice, one usually arranges that all intermediate automata in such a sequence are inverse automata, by applying all possible foldings to the automaton under consideration. Note also that the Todd-Coxeter algorithm for constructing the Cayley graph of a finitely presented group is in fact a special case of this general construction of Stephen, corresponding to the presentation $M = inv < X : R, xx^{-1} = x^{-1}x = 1$ for all $x \in X >$.

The iterative construction of the automata $A(w)$ associated with an inverse monoid presentation $M = inv < X : T >$ has proved to be a powerful tool in analysing such monoids. In his original paper and thesis [118], [119], Stephen examines several examples where these methods enable him to solve the word problem for classes of inverse monoid presentations that could not be handled effectively using the usual "linear" arguments on words that are often used in semigroup theory. We shall not cite a list of all other instances in the literature where these methods have been used, but we refer the reader to the following papers where very significant use of these methods has been made:

(a) the paper [55] where the word problem is solved for inverse monoids presented by finitely many relations of the form $e_i = f_i$, where e_i and f_i are idempotents of the free inverse monoid;

(b) the paper [58] where the word problem for the free semigroups in the Burnside variety of inverse semigroups defined by the identity $[x^a = x^{a+b}]$ for $b \leq a$ is solved. This problem is surprisingly easy to solve, quite in contrast to the situation for Burnside varieties of semigroups or groups (see [45] for references and discussions);

(c) the paper [95] where free combinatorial strict inverse semigroups are studied.

(d) the paper [43] where these methods are used to study the structure of free products of inverse semigroups;

(e) the paper [68] where undecidability of the word problem for the Mal'cev product of the variety of semilattices and the variety of abelian groups is proved: this result will be discussed in more detail in Section 6;

(f) the paper [24] where wreath products of varieties of inverse semigroups are studied;

(g) the paper [11] where it is shown that the word problem for one relator inverse monoids of the form $M = inv < X : e = 1 >$ (for e an idempotent of $FIM(X)$) can be solved in polynomial time: this is in contrast to the general situation for the inverse monoids considered in (a) above, where the complexity of the solution to the word problem appears very high.

It is perhaps worth mentioning here that the word problem for one relator inverse monoids of the form $M = inv < X : w = 1 >$ remains unsolved. Even in the case where w is a reduced word, it can be proved that a positive solution to this problem would imply a positive solution to the one relator semigroup problem for semigroups of the form $S = sgp < X : u = v >$. The situation here is in sharp contrast to the situation for one relator groups, where the word problem was solved by Magnus in the early 1930's and one relator monoids of the form $M = mon < X : w = 1 >$ where the word problem was solved by Adian [1]. Some recent (unpublished) work of Ivanov, Margolis and Meakin shows that there may be a reasonable chance of settling the word problem for inverse monoids of the form $M = inv < X : w = 1 >$ in the case where w is *cyclically* reduced. In this case it is possible to use the van Kampen diagrams of combinatorial group theory to show that M is E-unitary. We refer the reader to the paper [57] for some early ideas about this problem.

3. Immersions, Inverse Automata and Algorithmic Problems for Subgroups of Free Groups

3.1. Preliminaries

The notion of an *immersion*, that is a locally injective graph morphism, has been used to prove a number of results about free groups [32], [117]. We show here that inverse monoids play the same role in the theory of immersions that groups play in the theory of coverings. Just as a cover over the bouquet of X circles is "essentially" a representation of the free group $FG(X)$ by permutations, we will see that an immersion over the bouquet of X circles is "essentially" a representation of the free inverse monoid $FIM(X)$ by partial one to one maps. We make this precise by showing that the category of immersions over the bouquet of X circles is naturally equivalent to the category of representations of $FIM(X)$.

By picking two distinguished vertices to be regarded as a start and terminal state we can think of immersions as inverse automata. If the terminal state coincides with the start state then the subset of $FIM(X)$ accepted by this inverse automaton is a closed inverse submonoid. This means that this submonoid is a filter in the natural partial order on $FIM(X)$ ($x \leq y \leftrightarrow x = xx^{-1}y$). Conversely, every closed inverse submonoid is accepted by such an inverse automaton. Under this identification, finitely generated closed inverse submonoids correspond to finite inverse automata. Furthermore, we will see that finitely generated closed inverse submonoids are precisely the closed inverse submonoids of finite index in the sense of inverse semigroup theory [89] and that they are also exactly the closed submonoids that are rational subsets of $FIM(X)$ in the sense of formal language theory.

We can also associate an inverse automaton with every subgroup H of the free group $FG(X)$. This construction, to be explained below, is usually more "compact" than the corresponding permutation automaton (that is, the cover corresponding to H). In this way, just as subgroups of finite index correspond to finite covers, that is finite permutation automata, finitely generated subgroups correspond exactly to finite inverse automata. This observation then gives a quick proof of many finiteness results for subgroups of free groups. Furthermore, this allows us to use algebraic properties of the transition (inverse) monoid of an inverse automaton to give a "syntactic" classification of finitely generated subgroups of free groups. We exploit this in two ways. First of all, we develop an analogue of Eilenberg's Theorem that gives a bijection between pseudo-varieties of finite monoids and semigroups and classes of regular languages called varieties of languages [27], [49], [91].

Secondly, we show that certain algorithmic questions about subgroups of free groups can be shown to be polynomial equivalent to corresponding properties of inverse automata and their transition monoids. This allows us to prove for example, that the problem of testing whether a finitely generated subgroup is pure or p-pure for a given prime p is PSPACE-complete.

Finally we return to inverse semigroup theory by showing how to construct all closed inverse submonoids of a free inverse monoid by looking at actions of groups on trees. For detailed proofs of the material presented here, see [56], [12].

3.2. Covers are Permutation Automata

In this section we recall some basic definitions and establish the connection between covers of graphs and permutation automata. All our graphs will be in the sense of [114]. Let Γ' and Γ be graphs. A morphism $\eta : \Gamma' \to \Gamma$ is a *cover* if for all vertices $v' \in \text{Vert}(\Gamma')$, η induces a bijection:

$$\eta_{v'} : Star_{v'} \to Star_{v'\eta}$$

where $Star_v = \{e \in \text{Edge}(\Gamma) \mid e \text{ starts at } v\}$. A permutation automaton over an alphabet X is an automaton over $X \cup X^{-1}$ such that each $x \in X \cup X^{-1}$ induces a permutation on the state set and such that the inverse letter induces the inverse permutation. For this and the next section our automata do not have any initial state or terminal states specified. There is an evident notion of morphism of cover, morphism of automata and morphism of representation of FG(X).

Theorem 3.1 *The following categories are naturally equivalent:*
a) The category of (connected) covers over the bouquet of X circles.
b) The category of (transitive) representations of FG(X) by permutations.
c) The category of (connected) permutation automata over X.

It is well known that a transitive representation of $FG(X)$ is equivalent (in the category of representations of $FG(X)$) to the coset representation of $FG(X)$ modulo the stabilizer of any vertex. In this way, finite connected covers correspond exactly to conjugacy classes of subgroups of $FG(X)$ of finite index. However, finitely generated subgroups need not correspond to finite covers. We shall see later that they do correspond to finite immersions and can be classified by finite inverse monoids.

3.3. Immersions are Inverse Automata

In this section we show that immersions and inverse automata play the same role as covers and permutation automata in the previous section. Let Γ' and Γ be graphs. A morphism $\eta : \Gamma' \to \Gamma$ is a *immersion* if for all vertices $v' \in \text{Vert}(\Gamma')$, η induces an injection:

$$\eta_{v'} : Star_{v'} \to Star_{v'\eta}.$$

There is an evident notion of morphism of immersion, morphism of inverse automata and morphism of representation of FIM(X).

Theorem 3.2 **([56])** *The following categories are naturally equivalent:*
(a) The category of (connected) immersions over the bouquet of X circles.
(b) The category of (transitive) representations of FIM(X) by injective functions.
(c) The category of (connected) inverse automata over X.

Just as transitive representations of $FG(X)$ are determined up to equivalence by subgroups of $FG(X)$ a theorem of Schein [112] shows that transitive representations of $FIM(X)$ are determined up to equivalence by closed inverse submonoids of $FIM(X)$. That is, the stabilizer of any vertex in a transitive representation of $FIM(X)$ is a closed inverse submonoid of $FIM(X)$. Conversely, any closed inverse submonoid N of $FIM(X)$ determines a transitive representation of $FIM(X)$ on the so called ω-cosets of N. Two closed inverse submonoids of $FIM(X)$ determine equivalent representations (in the category of representations) if and only if they are conjugate in an appropriate sense. See [89] for more details. Thus up to conjugacy, closed inverse submonoids and immersions are equivalent notions. We shall see that unlike the case of groups, finite immersions correspond to both finite index and finitely generated closed inverse submonoids.

3.4. Some Classical Applications of Immersions

In this section we review a number of classical applications of immersions. Many of these have appeared in various places in the literature and in many guises. We claim no originality here, but just gather these together for purposes of illustration.

3.4.1. *The Free Group is Residually Finite*

Let $w \neq 1$ be an element of $FG(X)$. We wish to construct a finite group H and a morphism $\varphi : FG(X) \to H$ such that $w\varphi \neq 1$. There are many well known constructions of such an H. Here is a simple construction based on inverse automata.

Step 1 Let $w \neq 1 \in FG(X)$. Construct the linear inverse automaton $LinA(w)$ that reads the word w (see Section 2.3). $LinA(w)$ has a state set Q_w with length(w) + 1 states.

Step 2 Since $LinA(w)$ has a finite number of states, we can complete the automaton to a permutation automaton P_w on the same set Q_w. Since the word w does not fix the initial state, the transition induced by w in the transition group H of P_w is not equal to 1 in H.

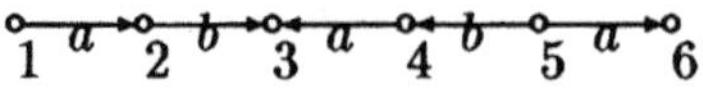

Figure 3.1. $LinA(w)$

EXAMPLE

Let $X = \{a, b\}$ and $w = aba^{-1}b^{-1}a$. Then the automaton $LinA(w)$ is shown in Figure 3.1 below where 1 is the initial state, 6 is the terminal state.

Now we can define P_w to be **any** permutation automaton that extends the action of $LinA(w)$ on the state set $\{1, \ldots, 6\}$. Then since the image of w in the transition monoid H of P_w maps 1 to 6, w does not induce the identity transformation and thus the image of w in H is not the identity.

3.4.2. *The Generalized Word Problem for* $FG(X)$
The generalized word problem for a finitely presented group G is to decide given an element $g \in G$ and a finite subset $Y \subseteq G$ whether or not $g \in < Y >$ where $< Y >$ is the subgroup generated by Y. It is well known that this problem is solvable for the free group $FG(X)$ on a finite set X. One method is to use linear methods related to Neilsen reduction to try to represent g as a product of free generators for $< Y >$. Here we show how to associate a finite inverse automaton $\mathcal{A}(< Y >)$ with $< Y >$ that accepts the unique reduced word representing g if and only if $g \in < Y >$. In [56] it is shown that this automaton just depends on the group $< Y >$ and not the particular generating set Y assuming each element of Y is given by its reduced representative.

ALGORITHM

Input: A finite set $Y \subset FG(X)$ and an element $w \in FG(X)$.

Output: "Yes" if and only if $w \in < Y >$.

Method

Step 1 Construct the flower automaton $\mathcal{F}(Y)$ of Y. $\mathcal{F}(Y)$ has a state i that is the unique initial and terminal state and one "petal" for each $y \in Y$ that spells out y in a loop from i to i.

Step 2 $\mathcal{F}(Y)$ may not be an inverse automaton, so we fold edges (in the sense of Section 2.1) until an inverse automaton is obtained. It can be shown that this process leads to a unique inverse automaton $\mathcal{A}(< Y >)$ that depends only on the subgroup generated by Y and not the set Y itself.

Step 3 Answer "Yes" if and only if the reduced word representing w labels a path from the initial state to itself. It can be shown that a reduced word labels such a path if and only if it is in $< Y >$, so that this algorithm is correct.

It is not difficult to see that $\mathcal{A}(< Y >)$ can be constructed in polynomial time from the input. Furthermore, it is easy to see that testing whether the reduced word representing w is accepted in $\mathcal{A}(< Y >)$ is also done in polynomial time. Thus this algorithm is polynomial. Avenhaus and Madlener [2] studied the complexity of the generalized word problem and proved that this problem is P-complete with respect to logspace reductions.

3.4.3. *Marshall Hall's Theorem*

In [36], Hall proved that every finitely generated subgroup of a free group is closed in the profinite topology of the free group. Since this conference proceedings contains a number of articles on the profinite topology and its relationship to automata theory and semigroup theory, we will not repeat definitions here. We will just call a subgroup H of a group G *closed* if H is the intersection of some collection of subgroups of G each of which has finite index.

We can use the automaton $\mathcal{A}(< Y >)$ constructed in the last section to prove Hall's Theorem. Let Y be a finite set contained in $FG(X)$ and let $w \in FG(X)$ be an element that is not in $< Y >$. Consider $\mathcal{A}(< Y >)$. Since w is not in $< Y >$, either w does not label a loop in $\mathcal{A}(< Y >)$ at the start state or w does not label any path in $\mathcal{A}(< Y >)$. In the first case, let $\mathcal{A}_w(< Y >) = \mathcal{A}(< Y >)$. In the second case, let $\mathcal{A}_w(< Y >)$ be the inverse automaton obtained from $\mathcal{A}(< Y >)$ by sewing on a path reading w from the initial state (adding new states when the path "falls off" $\mathcal{A}(< Y >)$).

In either case, $\mathcal{A}_w(< Y >)$ is a finite inverse automaton with the property that w labels a path starting at the initial state and ending elsewhere. Once again, we complete $\mathcal{A}_w(< Y >)$ to a permutation automaton on its state set in any possible way. We obtain a finite permutation automaton $\mathcal{P}_w(< Y >)$ with the same property. Now let H_w be the stabilizer of the initial state in $\mathcal{P}_w(< Y >)$. Then H_w has the following properties:

- H_w has finite index in $FG(X)$.
- w is not in H_w.
- $< Y > \subseteq H_w$.

It follows that $< Y >$ is the intersection of the collection $\{H_w \mid w$ is not in $< Y >\}$. Since each of these has finite index we have proved that $< Y >$ is closed.

3.4.4. *The Finite Index Problem*

The finite index problem asks whether a finitely generated subgroup of a group has finite index. Again the inverse automaton can be used to solve this problem. The equivalence of the first two conditions of the next theorem appears in [117], but has been known for some time. Let $\mathcal{I}(< Y >)$ be the transition monoid of $\mathcal{A}(< Y >)$. $\mathcal{I}(< Y >)$ is a finite inverse monoid that once again is independent of Y.

Theorem 3.3 *Let Y be a finite subset of $FG(X)$ and $H =< Y >$. Then the following conditions are equivalent:*

a) H has finite index.

b) $\mathcal{A}(H)$ is a permutation automaton.

c) $\mathcal{I}(H)$ is a finite group.

It is easy to see that one can test if a finite inverse automaton $\mathcal{A}$ is a permutation automaton in polynomial time in terms of the number of edges of $\mathcal{A}$. It follows that the finite index problem can be solved in polynomial time.

Theorem 3.3 forms a basis for our ideas expounded below. We wish to translate properties of finitely generated subgroups into properties of the corresponding **finite** inverse automaton and **finite** inverse monoid. If the corresponding property can be effectively checked in the finite automaton or finite inverse monoid, then we have essentially described an algorithmic solution to the original problem. If the problem just depends on the inverse automaton, then since we can construct the automaton in polynomial time from a generating set for the subgroup, we have efficiently reduced the problem in group theory to the corresponding problem in automaton theory. This is the case in Theorem 3.3 above. If on the other hand, the corresponding problem depends on the structure of the inverse monoid, then the complexity of the problem may increase significantly. This is because the size of the inverse monoid may be exponential in the size of the input. We will see this phenomenon in the next section.

3.5. Purity of Subgroups: A PSPACE-complete problem

In this section we use the ideas outlined above to show that the problem of detecting whether a finitely generated subgroup of $FG(X)$ is pure or p-pure is PSPACE-complete. This is the first "natural" problem on free groups that we know of that is PSPACE-complete.

Recall that a subgroup H of a group G is *pure* (*p-pure* for a given prime p) if for all $g \in G$, $g^n \in H$ for some $n > 0$ (n relatively prime to p), implies that $g \in H$.

Theorem 3.4 (**[12]**) *Let H be a finitely generated subgroup of $FG(X)$. Then the following conditions are equivalent:*

a) H is pure (p-pure, for given prime p).

b) Every subgroup of $\mathcal{I}(H)$ is trivial (a p-group).

It is straightforward to see that given a finite inverse automaton $\mathcal{A}$, then checking whether every subgroup in the transition monoid of $\mathcal{A}$ is trivial or a p-group can be done in polynomial space. Thus Theorem 3.4 gives a PSPACE algorithm for detecting whether a finitely generated subgroup of $FG(X)$ specified by a finite set of generators is pure or p-pure. We note that the conditions in Theorem 3.4 depend on the structure of the transition monoid of $\mathcal{A}(H)$. Thus this condition is not necessarily detectable in polynomial time from the input. Note that in the finite index problem of the preceding section, the condition that the transition monoid of $\mathcal{A}(H)$ be a group is equivalent to the easily checkable property that $\mathcal{A}(H)$ itself be a permutation automaton. We can ask whether the property of testing whether every subgroup of the transition monoid of a finite inverse automaton $\mathcal{A}$ is trivial or a p-group is reducible to some easily checkable property of the automaton itself.

The main theorem of [12] shows that these problems are in fact PSPACE-complete so it is unlikely that any of these problems has a polynomial time solution. This is done by showing that these problems are polynomial equivalent to certain problems about inverse automata. A number of interesting problems about inverse automata are needed and are also shown to be PSPACE-complete.

In particular, the following problems are considered in [12] :

The intersection emptiness problem Given a collection of finite injective or inverse automata, $\mathcal{A}_i$, $1 \leq i \leq n$, is there a word accepted by all of the $\mathcal{A}_i$, $1 \leq i \leq n$?

The aperiodicity problem Given an injective or inverse automaton $\mathcal{A}$, is every subgroup of the transition monoid of $\mathcal{A}$ trivial?

The generation problem Given an inverse automaton with state set Q and a test injective function $f : Q \to Q$, is f a member of the transition monoid of the automaton?

The p-periodicity problem Given an injective or inverse automaton $\mathcal{A}$ and a prime number p, is every subgroup of the transition monoid of $\mathcal{A}$ a p-group?

Theorem 3.5 **([12])** *The problems:*

a) the intersection emptiness problem,

b) the aperiodicity problem,

c) the generation problem,

d) the p-periodicity problem,

e) the purity of subgroup problem,

f) the p-purity of subgroup problem.

are all polynomial time reducible to each other. All of these problems are PSPACE-complete.

To do this we make use of injective Turing machines and a theorem of Bennett [9]. These problems were known to be PSPACE-complete for finite automata in general [48], [20]. On the other hand, for group automata, these problems have very fast parallel algorithms and are in the class NC [6]. Thus, with respect to complexity theory, inverse semigroups behave more like arbitrary semigroups than groups.

3.6. Pseudo-varieties of Finite Inverse Monoids and Closed Inverse Submonoids of Free Inverse Monoids

We have seen that we have the following equivalences for a finitely generated subgroup H of a free group $FG(X)$.

- H is finitely generated if and only if $\mathcal{I}(H)$ is a finite inverse monoid.
- H has finite index if and only if $\mathcal{I}(H)$ is a finite group.
- H is pure if and only if $\mathcal{I}(H)$ is a finite inverse monoid with trivial subgroups.

- H is p-pure if and only if $\mathcal{I}(H)$ is a finite inverse monoid and all subgroups in $\mathcal{I}(H)$ are p-groups.

These correspondences suggest that the properties of subgroups of free groups that can be detected by these methods are related to varietal properties of the corresponding inverse monoid. This is in total analogy with the relationship between pseudo-varieties of finite semigroups and monoids and varieties of languages in the sense of Eilenberg. See [27], [49], [91] for an extensive introduction to these ideas.

Ruyle [100] has developed a similar correspondence between pseudo-varieties of finite inverse monoids and pseudo-varieties of finitely generated closed inverse submonoids of free inverse monoids. He also has developed a number of correspondences between classes of finitely generated subgroups of free groups and pseudo-varieties of finite inverse monoids, but the connection is somewhat more subtle, due to the fact that there are many immersions that represent a subgroup of the free group. That is, the inverse automaton $\mathcal{A}(H)$ that we have associated with a subgroup H of $FG(X)$ has the minimal number of states among all immersions representing H as a stabilizer of a vertex. We can obtain all other such immersions by sewing on trees at points on $\mathcal{A}(H)$ that have no edge reading some letter in $X \cup X^{-1}$. The resulting transition monoid of such an inverse automaton can have varietal properties that can be surprisingly different form the transition monoid of $\mathcal{A}(H)$ itself. Depending on the application, it is sometimes useful to consider not only $\mathcal{A}(H)$ but one or all of these larger immersions corresponding to H.

In the cases listed above the corresponding properties are independent of which inverse automaton we consider. For example, one inverse automaton representing H has transition monoid with trivial subgroups if and only if all inverse automata representing H have this property. Details and more examples can be found in [100].

3.7. Structure of Closed Inverse Submonoids of Free Inverse Monoids

The ideas and applications of the preceding sections were geared towards applications to the classification of subgroups of free groups. In this section we return internally to the free inverse monoid itself and show how immersions can be used to give a structure theorem for closed inverse submonoids of $FIM(X)$. We will see that every such object is constructed from the free action of a (free) group on a tree.

Let T be a tree and let G be a group acting freely on the left of T. This means that the stabilizer of each vertex of T is trivial. It is well known that G is a free group and that every free group has such an action on a tree[114].

Fix a vertex $v \in V(T)$. Let $M(T,G,v) = \{(t,g) | t$ is a finite subtree of $T, g \in G$ and $v, gv \in V(t)\}$ and define a multiplication by

$$(t_1, g_1)(t_2, g_2) = (t_1 \cup g_1 t_2, g_1 g_2).$$

Here $g_1 t_2$ denotes the translate of the tree t_2 and $t_1 \cup g_1 t_2$ is the subtree whose set of vertices (edges) is the union of those of the trees t_1 and $g_1 t_2$.

Example Let $G = FG(X)$ and let $T = \Gamma(X)$ be the Cayley graph of G relative to the usual presentation. Then T is a tree and G acts freely on T by left multiplication. By Munn's theorem, [76], $M(T,G,1)$ is isomorphic to $FIM(X)$, the free inverse monoid on X. Clearly in this case we obtain the construction of $FIM(X)$ given in Section 2.3.

One of the main theorems of [56] states that monoids of the form $M(T, G, v)$ are precisely the class of closed inverse monoids of free inverse monoids. We review the main constructions here.

Let N be a closed inverse submonoid of $FIM(X)$. Let Γ_N be the immersion corresponding to N. That is Γ_N is the automaton of right ω-cosets of N and N is represented as the stabilizer of (the right ω-coset) N.

Let T_N be the universal covering graph of Γ_N. It is well known that T_N is a tree and that the fundamental group $G = \pi_1(\Gamma_N)$ acts freely on the left of T_N (by deck transformations). Furthermore, T_N embeds into the Cayley graph of the free group. We can arrange that 1 is a vertex in this embedded image and that 1 covers N in the covering map from T_N to Γ_N. Finally it is known that Γ_N is the quotient of T_N under the action of G which can be represented as a subgroup of $FG(X)$.

Lemma 3.1 **([56])** *Let N be a closed inverse submonoid of $FIM(X)$. Then the maximal group image of N is isomorphic to $G = \pi_1(\Gamma_N)$. Furthermore, N is isomorphic to $M(T_N, G, 1)$.*

Thus, every closed inverse submonoid N of $FIM(X)$ can be naturally constructed form the topological invariants T_N and $G = \pi_1(N)$. Conversely, every monoid constructed this way is isomorphic to a closed inverse submonoid of an appropriate free inverse monoid.

Lemma 3.2 **([56])** *Let G be a group acting freely on a tree T with root v. Let $\Gamma = G\backslash T$ and let X be an orientation of Γ. Then $M(T, G, v)$ is isomorphic to a closed inverse submonoid of $FIM(X)$.*

It is not true that every closed inverse submonoid of $FIM(X)$ is a free inverse monoid. For example, the semilattice of idempotents of $FIM(X)$ is closed, but not a free inverse monoid. In [56] it is proved that every closed inverse submonoid has a retraction onto a free inverse monoid. This is proved by translating the topological notion of contracting a spanning tree into the algebraic setting of inverse semigroups. This and many other facts concerning these monoids can be found in [56]. It is also shown in [56] how to use the free inverse category on a graph Γ to classify immersions over Γ.

3.8. Finiteness Conditions

In this section, we show that the closed inverse submonoids of free inverse monoids satisfy finiteness properties not shared by subgroups of free groups. This allows one to lift properties of finitely generated subgroups of free groups to the closed inverse submonoid generated by the same set. Results such as Howson's Theorem follow easily.

Recall that the set of rational subsets of a monoid M is the smallest collection of subsets of M containing the singletons and closed under finite union, product of subsets and submonoid generation (i.e. Kleene star). A subset S of M is recognizable if there is a finite monoid N and a morphism $f : M \to N$ and a subset P of N such that $S = Pf^{-1}$. See [10] for details. Let $Rat(M)$ be the set of rational subsets of M and let $Rec(M)$ be the set of recognizable subsets of M. We have the following important theorems.

Theorem 3.6 (Kleene) *If M is a finitely generated free monoid, then $Rec(M) = Rat(M)$.*

Theorem 3.7 (Anissimov and Seifert) *Let G be a finitely generated group and let H be a subgroup of G. Then $H \in Rec(G)$ if and only if H has finite index in G. $H \in Rat(G)$ if and only if H is finitely generated.*

It follows from Theorem 3.7 that if G is any infinite group, then the trivial subgroup is rational, but not recognizable. We also list the following consequence of Kleene's theorem due to McKnight.

Theorem 3.8 *Let M be a finitely generated monoid. Then $Rec(M)$ is contained in $Rat(M)$.*

We say that a closed inverse submonoid N of $FIM(X)$ is *finitely generated* if N is the smallest closed inverse submonoid containing a finite set Y. That is $N = \{n \in FIM(X) | n \geq y_1^{\epsilon_1} \ldots y_m^{\epsilon_m}, y_i \in Y, \epsilon_i \in \{1, -1\}\}$. N has *finite index* if N has a finite number of ω-cosets. That is, the immersion Γ_N corresponding to N is finite or equivalently by Theorem 3.2, N is recognized by a finite inverse automaton. Putting this altogether we have the following theorem proved in [56].

Theorem 3.9 ([56]) *Let $M = FIM(X)$ and let N be a closed inverse submonoid of M. Then the following conditions are equivalent:*

a) N is recognized by a finite inverse automaton.

b) N has finite index in M.

c) N corresponds to a finite immersion over the bouquet of circles.

d) N is a recognizable subset of M.

e) N is a rational subset of M.

f) N is finitely generated.

As an application, we can obtain a quick proof of Howson's Theorem: the intersection of two finitely generated subgroups of a free group is also finitely generated. Indeed, if $N_i \subset FG(X)$ is generated by a finite set Y_i, $i = 1, 2$, let $\hat{N}_i$ be the closed inverse submonoid of $FIM(X)$ generated by Y_i. Then $\hat{N}_i$ is recognized by a finite inverse automaton $\mathcal{A}_i$ by Theorem 3.9. A standard construction of the theory of automata allows us to construct a finite automaton recognizing $\hat{N}_1 \cap \hat{N}_2$. Thus, $\hat{N}_1 \cap \hat{N}_2$ is a finitely generated closed inverse submonoid of $FIM(X)$ and so is its image $N_1 \cap N_2$ in $FG(X)$.

4. Burnside-type Properties, Symbolic Dynamics, Identities and Quasi-Identities of Finite Semigroups

We now turn to algorithmic problems in varieties.

First of all we need some basic definitions.

- A semigroup S is *periodic* if all its one-generated subsemigroups are finite, equivalently if for every element $x \in S$ there exist two different numbers m_x and n_x such that $x^{m_x} = x^{n_x}$.
- A nil-semigroup is a semigroup in which a power of every element is equal to zero.
- A nilpotent semigroup of degree n is a semigroup where any product of n elements is zero. Every finite nil-semigroup is nilpotent [22].
- A variety is called periodic if it consists of periodic semigroups. A variety is called non-periodic if it contains a non-periodic semigroup, or, equivalently, if it contains the additive semigroup of natural numbers.

4.1. Burnside Problems in Varieties of Semigroups

As we mentioned in the introduction, the description of varieties where periodic semigroups are locally finite (see Theorems 4.4 and 4.5 below) plays an exceptional role in the study of algorithmic problems in semigroup varieties. Most of the results about algorithmic problems in varieties would be impossible to obtain without it.

The first result about Burnside-type problems in semigroup varieties was published by Morse and Hedlund [75]. The result was the following:

Theorem 4.1 *There exist an infinite semigroup with three generators that satisfies the identity*[1] $x^2 = 0$ *and an infinite semigroup with two generators that satisfies the identity* $x^3 = 0$.

Morse and Hedlund used certain infinite words W_1 and W_2 over a 2-letter alphabet and a 3-letter alphabet, which *avoid* the words x^3 and x^2 respectively. In general if u is a word and φ is an endomorphism of a free semigroup then $\varphi(u)$ is called a *value* of u. A word u is called *avoidable* by a word W if W does not contain any values of u. A word u is called *avoidable* if it is avoided by an infinite word over a finite alphabet.

We present the Thue construction of the word W_1 in Section 6.5 below.

There is a natural correspondence between infinite words over a finite alphabet and finitely generated semigroups (see Section 4.6). This correspondence implies the following connection between the avoidability of words and Burnside-type properties (see [8]).

Theorem 4.2 *A word u is avoidable if and only if the variety given by the identity $u = 0$ is not locally finite.*

The next step was made by D.B.Bean, A.Ehrenfeucht, and G.McNulty [8], and independently by Zimin [126]. They found algorithms for checking if a word is avoidable.

---

[1] We use the short expression $u = 0$ for the pair of identities $ux = u, xu = u$ where x does not occur in u.

We will need the definition of the words Z_n, which we will call Zimin words:

$$Z_1 = x_1, \dots, Z_{n+1} = Z_n x_{n+1} Z_n.$$

The following theorem is a translation of results from [8] and [126] into the language of varieties.

Theorem 4.3 *Let $\mathcal{V}$ be a variety of semigroups given by a (possibly infinite) set of identities $\{u = 0 \mid u \in \Sigma\}$. Assume that the number of variables occurring in words of Σ is n. Then the following conditions are equivalent.*
(1) $\mathcal{V}$ is locally finite.
(2) Z_n contains a value of some word u in Σ.

This theorem gives an algorithmic description of locally finite varieties defined by identities of the form $u = 0$. Notice that all these varieties consist of nil-semigroups.

The next result by Sapir [103] gives an algorithmic description of arbitrary varieties where nil-semigroups are locally finite.

Theorem 4.4 ([103]) *Let $\mathcal{V}$ be a variety of semigroups given by a (possibly infinite) set of identities Σ. Assume that the number of variables occurring in words of Σ is n. Then the following conditions are equivalent.*
(1) All nil-semigroups from $\mathcal{V}$ are locally finite.
(2) All semigroups from $\mathcal{V}$ satisfying the identity $x^2 = 0$ are locally finite.
(3) There exists an identity $u = v \in \Sigma$ such that Z_{n+1} contains a value $\varphi(u)$ for some endomorphism φ but $\varphi(u) \neq \varphi(v)$.

The next theorem, also from [103], describes varieties where all periodic semigroups are locally finite in the class of varieties with "good" groups and in the class of non-periodic varieties.

Theorem 4.5 ([103]) *Let $\mathcal{V}$ be a variety of semigroups given by a (possibly infinite) set of identities Σ. Assume that the number of variables occurring in words of Σ is n. Assume also that the variety $\mathcal{V}$ either is non-periodic or contains no non-locally finite groups of finite exponent. Then the following conditions are equivalent.*
(1) All periodic semigroups from $\mathcal{V}$ are locally finite.
(2) All nil-semigroups from $\mathcal{V}$ are locally finite.
(3) There exists an identity $u = v \in \Sigma$ such that Z_{n+1} contains a value $\varphi(u)$ for some endomorphism φ but $\varphi(u) \neq \varphi(v)$.

These two theorems have many interesting corollaries and applications (see [103], [105]). Let us present just one of them immediately. Others will be discussed later.

Theorem 4.6 ([103]) *A finitely based periodic semigroup variety is locally finite if and only if all its groups and all its nil-semigroups are locally finite.*

4.2. Burnside Properties and Identities of Semigroups

One of the central problems in universal algebra is the following problem connecting the syntactic and semantic ways of describing varieties.

Tarski's Finite Basis Problem Is the set of finite algebras that generate finitely based varieties recursive?

For groups the answer is given by the celebrated theorem of Oates and Powell.

Theorem 4.7 (Oates-Powell) *Every finite group G generates a finitely based variety.*

The same result is not true for semigroups. Let B_2^1 denote the semigroup consisting of the 2×2 matrix units together with the 0 matrix and the identity matrix.

Theorem 4.8 (Perkins [87]) *The variety of semigroups generated by B_2^1 is not finitely based.*

Theorem 4.4 allows one to get a much stronger result. Indeed, one can reformulate Theorem 4.4 in the following way:

Theorem 4.9 (Sapir, [103]) *Let $\mathcal{V}$ be a variety where every nil-semigroup is locally finite. Suppose $\mathcal{V}$ does not satisfy any non-trivial identity of the form $Z_n = W$. Then $\mathcal{V}$ cannot be defined by a finite number of identities.*

This theorem turned out to be a very powerful tool in studying the finite basis property in semigroup varieties.

For example one can easily establish that the Brandt monoid B_2^1 does not satisfy any non-trivial identity of the form $Z_n = W$. This immediately implies the following theorem.

Theorem 4.10 ([103]) *A locally finite variety of semigroups is not finitely based provided it contains the Brandt monoid B_2^1.*

A finite algebra which cannot belong to a locally finite finitely based variety is called *inherently non-finitely based* [88]. Theorem 4.10 actually states that B_2^1 is an inherently non-finitely based semigroup. Theorem 4.10 answered a question by G.McNulty and C.Shallon [65].

It is clear that if an inherently non-finitely based algebra A divides a finite algebra B (that is A is a homomorphic image of a subalgebra of B) then B itself is inherently non-finitely based . Thus if B_2^1 divides a finite semigroup S, then S is not finitely based. Using a result of Volkov [116], it turns out that the converse is also true for inverse semigroups.

Theorem 4.11 ([103]) *Let S be a finite inverse semigroup. Then considered as a semigroup the following conditions are equivalent.*

a) S is finitely based.

b) S is weakly finitely based.

c) B_2^1 does not divide S.

Since it is decidable whether B_2^1 divides a finite inverse semigroup S, given the multiplication table of S, it is decidable whether a given inverse semigroup is finitely based as a semigroup. The Tarski problem for arbitrary finite semigroups remains open.

In [104] Sapir gave the following algorithmic description of all finite inherently non-finitely based semigroups.

Theorem 4.12 ([104]) *A finite semigroup S of order n is inherently non-finitely based if and only if it does not satisfy any non-trivial identity of the form $Z_{n^2} = W$.*

We say that a finite algebra A is *weakly finitely based* if A belongs to some locally finite finitely based variety (that is if A is not inherently non-finitely based). Of course, every finitely based finite algebra is weakly finitely based, but we will see that the opposite implication does not hold.

4.3. Identities of Inverse Semigroups

B_2^1 is also an inverse semigroup. It is easy to see that if $s \in B_2^1$, then s^{-1} is the transpose matrix of s. As we mentioned in Section 2, the class of inverse semigroups is a variety of algebras of type $< 2, 1 >$ consisting of multiplication and inversion.

We can thus study the variety of inverse semigroups generated by B_2^1. Notice that B_2^1 plays a very important role in the theory of varieties of inverse semigroups. In particular by a result of E.Kleiman [47] every inverse semigroup variety that does not contain B_2^1 and that is generated by a finite semigroup, is finitely based. He also proved the following result.

Theorem 4.13 (E. Kleiman [47]) *The variety of inverse semigroups generated by B_2^1 is not finitely based.*

Given Theorem 4.11 and Kleiman's Theorem 4.13, one is led to expect that B_2^1 is inherently non-finitely based even when considered as an inverse semigroup. Thus the following result is very surprising.

Theorem 4.14 ([110]) *Let S be a finite inverse semigroup. Then, considered as an inverse semigroup, S is weakly finitely based.*

That is, every finite inverse semigroup belongs to a locally finite finitely based variety of inverse semigroups. It is a major open problem to decide which inverse semigroups are finitely based when considered as inverse semigroups. It is still possible that every inverse semigroup having B_2^1 as a divisor is not finitely based as an inverse semigroup.

Theorem 4.14 is a consequence of the following two theorems. Let Z_n' be the prefix of Z_n consisting of all but the last letter. We will say that a semigroup S has height $\leq h$ if and only if every descending chain of principal right ideals has length $\leq h$.

Theorem 4.15 *Every inverse semigroup of height $n-2$ satisfies the identity $Z'_n = Z'_n x_1 x_1^{-1}$.*

Theorem 4.16 *Let S be a finitely generated inverse semigroup satisfying the identity $Z'_n = Z'_n x_1 x_1^{-1}$. Let all subgroups of S be locally finite. Then S is finite.*

For example, by Theorem 4.15, B_2^1 belongs to the variety defined by the identities $x^2 = x^3$ and $xyxzxyxtxyxzxy = xyxzxyxtxyxzxyxx^{-1}$. From Theorem 4.16, it follows that the variety defined by these two identities is locally finite.

4.4. Quasi-varieties of Semigroups

Quasi-varieties of algebras are classes defined by quasi-identities, that is formulas of the following form:

$$(1) \qquad (\forall x_1, \ldots, x_n)\ u_1 = v_1 \& \ldots \& u_m = v_m \rightarrow u = v$$

where u_i, v_i, u, v are terms in variables $x_1, \ldots, x_n$.

After varieties, quasi-varieties are the most widely studied classes of algebras. Again, they are studied from both a syntactic and a semantic point of view.

Syntactic A quasi-variety is a class of algebras satisfying a given set of implications.

Semantic A quasi-variety is a class of algebras closed under taking subalgebras, products and "ultraproducts". (See [18] for a definition of ultraproducts).

For example let $Q = \{S | (xz = yz) \rightarrow (x = y)\}$. Q is the quasi-variety of right cancellative semigroups.

There are the evident notions of finitely generated, finitely based, weakly finitely based and inherently non-finitely based quasi-varieties. Ol'shanskii [82] described the collection of finitely generated quasi-varieties of groups that are finitely based.

Theorem 4.17 (Ol'shanskii [82]) *Let G be a finite group. Then the quasi-variety generated by G has a finite basis of implications if and only if every Sylow subgroup of G is Abelian.*

Sapir [101] has studied the question of which finite semigroups have a finite basis of quasi-identities. In particular, he described finite semigroups without two-sided ideals that have this property. But the general problem remains open. The following theorem is the appropriate analogue to Theorem 4.14.

Theorem 4.18 (Margolis and Sapir [60]) *Let S be a finite semigroup. Then S belongs to a locally finite quasi-variety, defined by a finite set of implications. That is, every finite semigroup is weakly finitely based with respect to quasi-identities.*

Theorem 4.18 is a consequence of the following two theorems.

Theorem 4.19 *Every finite semigroup of height $n-2$ satisfies the quasi-identity $(tZ_n = wZ_n) \to (tZ'_n = wZ'_n)$.*

Theorem 4.20 *Let S be a finitely generated periodic semigroup satisfying the quasi-identity $(tZ_n = wZ_n) \to (tZ'_n = wZ'_n)$. Let all subgroups of S be locally finite. Then S is finite.*

For example, by Theorem 4.19, B_2^1 (as a semigroup) belongs to the quasi-variety defined by the identity $x^2 = x^3$ and the implication

$$txyxzxyxuxyxzxyx = wxyxzxyxuxyxzxyx \to$$
$$txyxzxyxuxyxzxy = wxyxzxyxuxyxzxy.$$

From Theorem 4.20, it follows that this quasi-variety is locally finite.

4.5. Semigroups and Inverse Semigroups

It is hard not to notice a similarity between Theorems 4.15 and 4.19. The reason for this similarity is that these theorems express the same structural property of semigroups in two different cases: the case of inverse semigroups and the case of semigroups. In the first case this property may be expressed by an identity in the second case it can be expressed by a quasi-identity.

Let $u, v \in X^+$. We say that S satisfies the condition $u\mathcal{R}^*v$ if for every homomorphism $\varphi : X^+ \to S$, $\varphi(u)$ and $\varphi(v)$ generate the same principal right ideal in some semigroup T containing S. It is known that in the case of inverse semigroups T may be always taken equal to S. The relation $\mathcal{R}^*$ has been studied extensively by J.Fountain [30].

The following lemmas show us how to express the condition "S satisfies $u\mathcal{R}^*v$" in terms of identities for inverse semigroups and quasi-identities for semigroups. Lemma 4.2 appeared in [85].

Lemma 4.1 *Let S be an inverse semigroup. Then S satisfies the condition $u\mathcal{R}^*v$ if and only if S satisfies the identity $uu^{-1} = vv^{-1}$.*

Lemma 4.2 *Let S be a semigroup. S satisfies the condition $u\mathcal{R}^*v$ if and only if S satisfies the bi-implicaton $tu = wu \leftrightarrow tv = wv$, where t and w are variables not appearing in u and v.*

From these lemmas, it is easy to derive the following corollaries.

Corollary 4.21 *Let S be an inverse semigroup. Then S satisfies $Z_n\mathcal{R}^*Z'_n$ if and only if S satisfies the identity $Z'_n x_1 x_1^{-1} = Z'_n$.*

Corollary 4.22 *Let S be a semigroup. Then S satisfies $Z_n\mathcal{R}^*Z'_n$ if and only if S satisfies the quasi-identity $(tZ_n = wZ_n) \to (tZ'_n = wZ'_n)$.*

Now the following theorem implies both Theorems 4.15 and 4.19.

Theorem 4.23 *Let S be a semigroup of height $n-2$. Then it satisfies*

$$Z_n\mathcal{R}^*Z'_n$$

4.6. Semigroups and Symbolic Dynamics

The proofs of Theorems 4.4, 4.16 and 4.20 have one important detail in common. All these proofs employ the following connection between semigroups and symbolic dynamical systems. This connection appeared first in the paper of Sapir [103].

A *symbolic dynamical system* is a closed subset of the Tikhonov product X^Z, where X is a finite set with the discrete topology, which is stable under the shift homeomorphism T (this homeomorphism shifts every sequence from X^Z one position to the right).

The correspondence between semigroups and symbolic dynamics is the following. Let $S =< X >$ be an infinite finitely generated semigroup (the same argument may be applied for any universal algebra). Then there is an infinite set T of words over X such that every element of S represented by a word of T cannot be represented by words over X of shorter length. Such words will be called *geodesic words.* These words label geodesics in the Cayley graph of the semigroup. It is clear that every subword of a geodesic word is also a geodesic word. Now, in every word of T, mark a letter which is closest to the center of this word. There must be an infinite subset T_1 of T of words which have the same marked letters, an infinite subset T_2 of T_1 of words which have the same subwords of length 2 containing the marked letters, ..., an infinite subset T_n of T_{n-1} of words which have the same subwords of length n containing the marked letters, and so on. Therefore there is an infinite word W such that every subword of W is a subword of a word from T. Thus every subword of W is geodesic. Infinite words with this property will also be called *geodesic.* The set $D(S)$ of all infinite geodesic words is a symbolic dynamical system because it is stable under the shift (obviously) and is closed in the Tikhonov topology (this can be easily proved). Conversely, with every symbolic dynamical system D one can associate a semigroup $S(D)$ as follows: $S(D)$ consists of all finite subwords of infinite words from D, and 0. If u and v belong to $S(D)$ then $u \cdot v$ is equal to uv if uv belongs to $S(D)$, or 0 otherwise. It is easy to show that $S(D)$ is a semigroup. The following theorem shows that every symbolic dynamical system is equal to $D(S)$ for some semigroup S.

Theorem 4.24 (Sapir, [45]) *For every symbolic dynamical system D we have*

$$D(S(D)) = D.$$

The correspondence between semigroups and symbolic dynamical systems allows one to show that some important properties of the theory of semigroups and important properties of the theory of symbolic dynamical systems are in fact equivalent. For example, *if a finitely presented semigroup S is periodic then the symbolic dynamical system $D(S)$ does not have cyclic trajectories. The semigroup S is infinite if and only if $D(S)$ is not empty, etc.*

One of the important concepts of the theory of symbolic dynamical systems is the concept of a uniformly recurrent word. An infinite word U is called *uniformly recurrent* if for every finite subword u of U there exists a number $N_U(u)$ such that every subword of U of length $N_U(u)$ contains u as a subword. It is an easy corollary from [31] (see [103] for details) that for every infinite word U there exists a uniformly recurrent word U' such that every subword of U' is a subword of U. It is easy to see that if U belongs to $D(S)$ then U' also belongs to $D(S)$.

Therefore for every infinite finitely generated semigroup $S = < X >$ there exists a uniformly recurrent geodesic word over X (see [103]). Uniformly recurrent words are much more convenient than arbitrary infinite words.

4.7. Uniformly Recurrent Words and Burnside Properties

The following theorem was first proved in [103].

Theorem 4.25 *Let $\alpha \in A^Z$ be uniformly recurrent . Suppose that $a \in A$ and that $\alpha = \alpha_1 a \alpha_2$ is an occurrence of a in α where α_1 is infinite to the left and α_2 is infinite to the right. Then there is a substitution $\varphi : \{x_1, x_2, \dots\} \to A^+$ and a sequence of natural numbers $A(n, \alpha)$ such that:*

a) $\varphi(x_1) = a$

b) $\alpha_1 a = \alpha_3 \varphi(Z_n)$ for some α_3 infinite to the left.

c) $|\varphi(Z_n)| \leq A(n, \alpha)$.

This theorem shows that uniformly recurrent words have strong regularity properties. In order to prove that a finitely generated semigroup S is finite, we must be able to translate these regularities into properties about the structure of S.

For example, let us sketch the proof of Theorems 4.16 and 4.19. Let us take a periodic semigroup S with all subgroups locally finite which have the property

$$Z'_n \mathcal{R}^* Z_n. \tag{2}$$

Suppose that S is infinite. Then there exists a geodesic uniformly recurrent word α. Let $A = A(n, \alpha)$ be the number from Theorem 4.25.

Lemma 4.3 *Let u, v, and w be consecutive subwords of α such that u and v have length $\geq A$, w may be empty. Then $Z'_n \mathcal{R}^* Z_n$ implies $uvwu\mathcal{R}^* uvwuv$.*

Proof. Suppose that p is the longest prefix of v such that the relation

$$uvwu\mathcal{R}^* uvwup$$

follows from (2). If $v = p$ then we are done so suppose $v \neq p$. Let $v = paq$ for some letter a and word q. We have $uvwup = upaqwup$. Since upa is a subword of α and $|upa| \geq A$ by Theorem 4.25 there exists an endomorphism φ of the free semigroup such that $upa = u_1 \varphi(Z_n)$ and $\varphi(x_1) = a$. Therefore $up = u_1 \varphi(Z'_n)$. Thus $Z'_n \mathcal{R}^* Z_n$ implies $up\mathcal{R}^* upa$. Therefore we have $uvwu\mathcal{R}^* uvwup\mathcal{R}^* uvwupa$ which contradicts the choice of the prefix p. The lemma is proved.

Let us consider the (finite) set of all subwords of α of length $2A$. Since α is uniformly recurrent there exists a number B such that every subword of length B of α contains every subword of this set. Let u and v be any two subwords of α of length $\geq 2B$, let a and b be the corresponding elements in S. An induction similar to that used in Lemma 4.3 gives us the following lemma.

Lemma 4.4 *a and b are $\mathcal{J}$-related, i.e $a = c_1bd_1, b = c_2ad_2$ for some elements c_1, c_2, d_1, d_2 from S.*

Notice that so far we did not use the fact that all subgroups of S are locally finite and that U is geodesic. Thus we actually proved the following fact:

Lemma 4.5 *Let $S =< X >$ be a semigroup satisfying $Z_n'\mathcal{R}^*Z_n$. Let U be any uniformly recurrent word over X. Then there exist only finitely many $\mathcal{J}-$classes of S which contain elements represented by subwords of U.*

Now if we take a uniformly recurrent geodesic α, we can find a natural number N such that all subwords of α of length $\geq N$ will represent elements from the same $\mathcal{J}$-class of S. Readers familiar with the basic structure theory of semigroups [3] will see that the condition $Z_n'\mathcal{R}^*Z_n$ makes S semisimple. They will then know that if all subgroups of S are locally finite then these subwords will represent only a finite number of distinct elements of S. It follows that α is not irreducible and thus S is finite.

Thus we see that if a semigroup satisfies the condition (2) then it satisfies the descending chain condition for ideals generated by factors of a uniformly recurrent sequence. The ordinary descending condition for ideals may not hold in such a semigroup (see [110] for a counterexample).

Applications of uniformly recurrent words are very effective, but are not constructive. Indeed, there is no algorithm to find the number $N_U(u)$. The proofs of Theorems 4.4 and 4.5 have been made constructive in [108] where the analogue of the restricted Burnside problem for semigroup varieties is discussed.

5. Varieties with Decidable Word Problem

5.1. Commutative Semigroups

We say that the word problem is decidable (solvable) in a variety $\mathcal{V}$ of semigroups if it is decidable (solvable) in every semigroup which is finitely presented in $\mathcal{V}$.

The first non-trivial (that is non-locally finite) variety of semigroups with solvable word problem was found independently by A.I.Mal'cev [52] and by Ceitin and Emelichev [28]. It was the variety of all commutative semigroups.

Theorem 5.1 *The variety of all commutative semigroups has a solvable word problem.*

Mal'cev showed that every finitely generated commutative semigroup is faithfully representable by matrices over a suitable field, and every finitely generated ring of matrices is residually finite. Therefore every finitely generated commutative semigroup is residually finite and the McKinsey algorithm (see Theorem 5.6 below) gives the solution to the word problem.

As was pointed out by Emelichev in [28] the proof may be easily deduced from an old paper by Hermann [38] devoted to the membership problem for ideals in the ring of polynomials. Indeed, let $S =< X | u_1 = v_1, \dots, u_r = v_r >$ be any finitely presented commutative semigroup. Emelichev [28] proved that a relation $u = v$ holds in S if and only if the polynomial $p = u - v$ belongs to the ideal of the ring of polynomials $Q[X]$ generated by polynomials $p_i = u_i - v_i$. This, in turn, is equivalent to the solvability of the following equation over $Q[X]$:

$$p_1 f_1 + p_2 f_2 + \ldots + p_r f_r = p \tag{3}$$

with unknowns $f_1, \ldots, f_r$. Now we can use the following result from [38]:

Theorem 5.2 *Let $d = max\{deg(p_1), deg(p_2), \ldots, deg(p_r)\}$. If the equation (3) has a solution then there is a solution with $deg(f_i) \leq deg(p) + (rd)^{2^{|X|}}$.*

It is clear that if there is a bound for the degrees of the unknown polynomials in (3) then the number of coefficients of these polynomials is also bounded. Then (3) is equivalent to a finite system of linear equations over the field of rational numbers, which can be solved by, say, the Gauss elimination algorithm.

Ballantyne and Lankford used this connection between commutative semigroups and ideals in the ring of polynomials to apply the Gröbner basis method to the word problem in commutative semigroups [7].

A proof of Theorem 5.1 based on other ideas is presented below in Section 5.3.1.

Taiclin [120] proved the following result which is much stronger than Theorem 5.1.

Theorem 5.3 *The elementary theory of every finitely presented commutative semigroup is decidable.*

5.2. The Description of Varieties of Semigroups with Solvable Word Problem

The first example of a finitely based proper variety of semigroups with an unsolvable word problem is the variety of Murskii [77]. This was the only known example until 1983 when a deep study of varieties with decidable word problem was initiated by I.Mel'nichuk. In particular, she proved [70] that the word problem is undecidable in any variety which contains a non-locally finite variety of semigroups given by identities of the form $u = 0$. Such varieties were described by Bean, Ehrenfeucht, McNulty and Zimin (see Theorem 4.3 above). She also proved that the word problem is decidable in any finitely based variety which satisfies the permutation identity $x_1 x_2 \ldots x_n = x_{\sigma(1)} x_{\sigma(2)} \cdots x_{\sigma(n)}$ where σ is a permutation of the symbols $1, 2, \ldots n$.

Then Mel'nichuk, Sapir, and Kharlampovich [71] found a minimal variety with an undecidable word problem, a boundary between decidability and undecidability. This was the variety generated by the semigroup S_2 from Section 6.4 below. The semigroup S_1 and its dual semigroup $\overleftarrow{S_1}$ from Section 6.4 also appeared in [71].

Finally, Sapir [102], [106] proved that S_1 and $\overleftarrow{S_1}$ also generate boundaries between decidability and undecidability, proved that there are no more boundaries among non-periodic varieties of semigroups, that every periodic semigroup in a non-periodic variety with decidable word problem must be locally finite, and that every non-periodic variety with an undecidable word problem contains one of these three varieties. He also showed that a periodic variety with a solvable word problem and locally finite groups must be locally finite itself. This is everything that one can hope to get, because the problem of describing non-locally finite periodic varieties of groups with solvable word problem is hopeless. It also turned out that many other conditions for finitely presented semigroups in varieties are equivalent to the solvability of the word problem in this variety.

Some of the results from [102] and [106] of Sapir are summarized in the following two theorems.

We will need the following two finite subsemigroups of B_2:

$$P = \left\{ \begin{pmatrix} 1 & 0 \\ 0 & 0 \end{pmatrix}, \begin{pmatrix} 0 & 1 \\ 0 & 0 \end{pmatrix}, \begin{pmatrix} 0 & 0 \\ 0 & 0 \end{pmatrix} \right\},$$

$$T = \left\{ \begin{pmatrix} 1 & 0 \\ 0 & 0 \end{pmatrix}, \begin{pmatrix} 0 & 1 \\ 0 & 0 \end{pmatrix}, \begin{pmatrix} 0 & 0 \\ 0 & 1 \end{pmatrix}, \begin{pmatrix} 0 & 0 \\ 0 & 0 \end{pmatrix} \right\}.$$

For every semigroup S the dual (anti-isomorphic) semigroup is denoted by $\overleftarrow{S}$, the semigroup S with an identity element adjoined is denoted by S^1. For every variety $\mathcal{V}$ let $FP(\mathcal{V})$ be the class of all semigroups which are finitely presented in $\mathcal{V}$.

Theorem 5.4 ([106]) *Let $\mathcal{V}$ be a finitely based non-periodic variety. Then the following conditions are equivalent.*

(1) The word problem is decidable in any semigroup from $FP(\mathcal{V})$.
(2) The elementary theory is decidable for any semigroup from $FP(\mathcal{V})$.
(3) Every semigroup from $FP(\mathcal{V})$ is residually finite.
(4) Every semigroup from $FP(\mathcal{V})$ is representable by matrices over a field.
(5) Every semigroup from $FP(\mathcal{V})$ is Hopfian.

A. Every nil-semigroup from $\mathcal{V}$ is locally finite and $\mathcal{V}$ does not contain any of the semigroups $\overleftarrow{P} \times P^1$ or $\overleftarrow{P}^1 \times P$ or T.

B. Every nil-semigroup of $\mathcal{V}$ is locally finite and for some natural numbers k, m, n, p the variety $\mathcal{V}$ satisfies one of the following identities:

$$x^n y (z^k t^k)^p z^m = x^m (t^k x^k)^p y z^n, \quad n = m + kp; \tag{4}$$

$$xy^n z = y^k x y^m z y^p, \quad n > m. \tag{5}$$

Theorem 5.5 ([106]) *Let $\mathcal{V}$ be a finitely based periodic variety of semigroups in which all groups are locally finite. Then each of the conditions 1, 2, 3, 4, from Theorem 5.4 is equivalent to the condition that $\mathcal{V}$ is locally finite.*

Let us analyze these theorems. From Theorems 5.4, 5.5 it follows that in the class of non-periodic semigroup varieties where all nil-semigroups are locally finite every variety which has an undecidable word problem must contain one of the three finite semigroups:

$$\overleftarrow{P} \times P^1,\ \overleftarrow{P}^{1} \times P,\ T. \tag{6}$$

On the other hand, every non-periodic variety, containing one of these three semigroups, has an undecidable word problem. Therefore the following three varieties are the only minimal varieties with undecidable word problem among non-periodic varieties with locally finite nil-semigroups:

$$\mathrm{var}(\overleftarrow{P} \times P^1 \times N),\ \mathrm{var}(\overleftarrow{P}^{1} \times P \times N),\ \mathrm{var}(T \times N)$$

where N is the semigroup of natural numbers with respect to addition. In fact, these three varieties coincide with the varieties generated by the semigroups S_1, $\overleftarrow{S}_1$, and S_2 respectively, presented in Section 6.4 below.

It is easy to see that each of these varieties is a join of a locally finite variety (generated by a finite semigroup) and the variety of all commutative semigroups. Both have solvable word problem. Therefore we have *examples of varieties with unsolvable word problem that are joins of varieties with solvable (and even polynomially solvable) word problem.*

If a finitely based variety does not contain any of the three semigroups (6) but still has an unsolvable word problem then either it contains a non-locally finite nil-semigroup or it is periodic and contains a "bad" group, a non-locally finite group of finite exponent (see Theorem 5.4). Theorems 4.4 and 4.5 show that there are no minimal finitely based non-locally finite varieties of nil-semigroups, that is every non-locally finite finitely based variety of nil-semigroups contains a proper subvariety with the same properties. Thus in the first case the variety does not contain any minimal variety with unsolvable word problem. In the second case the situation is more complex. We do not know any minimal periodic variety of groups with unsolvable word problem. Moreover an analogy between nil-semigroups and periodic groups make the following two conjectures of Sapir seem reasonable:

Conjecture 5.1 *Every finitely based periodic group variety with undecidable word problem contains a proper subvariety with an undecidable word problem.*

Conjecture 5.2 *Every non-locally finite finitely based variety of periodic groups has an undecidable word problem.*

Of course, we do not have any hope of proving these conjectures. So far we have only a few examples of periodic group varieties with undecidable word problem (see Section 6.6).

Thus, modulo groups, there are exactly three minimal varieties of semigroups with an undecidable word problem.

5.3. Methods of Proving Decidability of The Word Problem in a Variety

The first and most universal method of proving the decidability of the word problem uses the connection between residual finiteness and the solvability of the word problem discovered by Mal'cev [52].

Theorem 5.6 *Let $\mathcal{V}$ be a finitely based variety of algebras. Let A be an algebra finitely presented in $\mathcal{V}$. If A is residually finite then the word problem in A is decidable.*

In order to solve the word problem in a residually finite algebra Mal'cev used an algorithm which was essentially due to McKinsey [64].

As we have mentioned above, in many cases (see, for example, the previous section) a variety has decidable word problem only if all algebras finitely presented in this variety are residually finite. But this is not always so. For example the group variety $\mathcal{N}_2\mathcal{A}$, which is the Mal'cev product of the variety of nilpotent groups of class two and the variety of Abelian groups, has a decidable word problem [44] but the following theorem holds.

Theorem 5.7 **([46])** *There exists a non-residually finite group finitely presented in the variety $\mathcal{N}_2\mathcal{A}$.*

Thus one needs other methods of proving the decidability of the word problem. There are, of course, methods that apply when we are solving the word problem in an algebra given by a specific system of defining relations. For example the word problem is decidable in the case when one can find a terminating Church-Rosser presentation of a finitely presented semigroup or group S. Then there exists a "canonical" form for every word over the alphabet of generators and an effective procedure which transforms every word to its canonical form.

It is very unusual, however, that a terminating Church-Rosser presentation is known for an arbitrary algebra finitely presented in a variety. In connection with the role that is played by the Church-Rosser method in computer algebra [17], it would be very interesting to describe varieties where every finitely presented algebra has a Church-Rosser presentation.

An example is the variety of commutative semigroups where terminating Church-Rosser presentations were found by Ballantyne and Lankford [7].

Experience shows that in general, in order to solve the word problem in an arbitrary finitely presented algebra (semigroup, group, etc.) A in a variety, one has to use methods which are in a sense opposite to the "canonical form" methods described above. Instead of finding a "canonical" word in the set of words which are equal to a given word u one has to consider this set as a whole and find some "hidden" structure on this set. To illustrate this idea we will consider two examples: a variety of semigroups and a variety of groups.

5.3.1. *Semigroups*

Here we would like to present some ideas for proving the decidability of the word problem in varieties of semigroups, employed initially in [13], and later in [71], [69], and [106]. To illustrate these ideas we will show how they work in the variety of commutative semigroups. Other methods of proving the decidability of the word problem in this variety have been discussed in Section 5.1.

First of all recall that the free n-generated commutative semigroup A_n with an identity element is simply a direct product of n semigroups of natural numbers. Therefore every element of this semigroup may be represented by a vector of natural numbers[2]. We can consider A_n as a partially ordered set with the natural coordinatewise order. It is clear that $u < v$ in A_n if and only if u divides v. It is easy to prove that every subset M of A_n has minimal elements, and every element of M is greater than or equal to a minimal element. The first part of the following simple statement is attributed to Dickson [26]. The second part is a constructive version of Dickson's result and is also well known.

Lemma 5.1 *Every infinite set T of elements in A_n contains two comparable elements. If no two elements in a set $T \subset A_n$ are comparable and T contains an element with all coordinates $\leq s$ then $|T| \leq n!s^n$.*

Let R be a finite set of defining relations, i.e. a finite subset of $A_n \times A_n$. We want to show that the word problem is decidable in the factor semigroup of A_n modulo the congruence generated by R. In other words, we want an algorithm, which, given a pair of elements (u, v) in $A_n \times A_n$, says if u equals v modulo the relations of R.

Take an element u in A_n. We will describe the set $M(u)$ of all elements of A_n which are equal to u modulo R.

Let $M_{\min}$ be the set of minimal elements in $M(u)$. Every element v in $M(u)$ is greater than or equal to an element w from $M_{\min}$. Hence $v = we$ for some e. Since both v and w are equal to u modulo R we have that $u = ue$ (Mod R). Such an element e is called a *unit* for u. Let $E(u)$ be the set of all units. Then $E(u)$ is a subsemigroup of A_n and is closed under taking quotients, that is if $e_1 f = e_2$ and e_1, e_2 are units then f is also a unit. This implies easily that the semigroup $E(u)$ is generated by the subset $E_{\min}$ of its minimal elements. Notice also that if e is a unit for u and $w \in M(u)$ then, of course, $we \in M(u)$.

By Lemma 5.1 both sets $M_{\min}$ and $E_{\min}$ are finite. Therefore we have the following description of $M(u)$:

$$M(u) = \{(\prod_{e \in E_{\min}} e^{k_e})w \mid w \in M_{\min}, k_e \in N\}.$$

This description would give us a solution of the word problem, if we had a process of finding the sets $M_{\min}$ and $E_{\min}$. This process is almost straightforward. We simply apply the relations from R to u until we find all elements from $M_{\min}$ and $E_{\min}$. Of course, the most tricky thing in such processes is a stop sign: the sign which shows that we have found all elements that we need, and we can stop and relax. This is organized as follows.

Denote the maximal length of words from R by $\ell(R)$. If M is a subset of A_n then let $T(M)$ be the set of all elements of A_n which can be obtained from elements of M by applying relations of R (at most one application for each relation and for each element of M).

Let us construct a sequence of sets M_n. Let $M_0 = \{u\}$. Suppose we have constructed the set M_n. Let E_n be the set of all quotients of elements from M_n.

Let us apply T to M_n many times until we obtain all elements of the form

$$(\prod_{e \in \min(E_n)} e^{k_e})w$$

[2] Zero also is a natural number.

where $w \in \min(M_n)$ and the sum of the k_e does not exceed $\ell(R)$. This set is finite and each element in it is equal to u modulo R, so we will find all of these elements eventually. Then let M_{n+1} stand for the union of these sets.

By Lemma 5.1 there is a number n such that

$$\min(M_n) = \min(M_{n+1}), \min(E_n) = \min(E_{n+1}).$$

We claim that then $M_{\min}$ is equal to $\min(M_n)$, $E_{\min}=\min(E_n)$. Indeed, it is enough to show that if we apply a relation from R to a word

$$v = \Big(\prod_{e \in \min(E_n)} e^{k_e}\Big)w \tag{7}$$

where $w \in \min(M_n)$ then we obtain a word w of the same form. If the sum of the k_e does not exceed $\ell(R)$ then this follows from the definition of M_n and from the equalities $\min(M_n)=\min(M_{n+1})$, $\min(E_n)=\min(E_{n+1})$.

Let this sum be greater than $\ell(R)$. Any application of a relation from R touches at most $\ell(R)$ units e from the right hand part of (7). Therefore $v = (\prod_{e\in\min(E_n)} e^{m_e})v_1$, and $w = (\prod_{e\in\min(E_n)} e^{m_e})w_1$ where v_1 belongs to M_n and w_1 is obtained from v_1 by applying a relation from R. But then w_1 is of the form (7), and so is w. This completes the proof.

5.3.2. *Nilpotent-by-Abelian Groups*

Let us consider the group variety $\mathcal{N}_2\mathcal{A}$ and prove that the word problem is decidable there. The proof we give essentially belongs to Kharlampovich [44]. See also the survey [45]

Let us take any group $G =< X >$, finitely presented in $\mathcal{N}_2\mathcal{A}$. The group G may be represented as a factor of the relatively free group $\bar{F} =< X >$ of $\mathcal{N}_2\mathcal{A}$.

Therefore $G = \bar{F}/R$ for some normal subgroup R which is finitely generated as a normal subgroup. Given a word w over X, we want to decide if $w = 1$ in G. We can consider w as an element of $\bar{F}$. Then $w = 1$ in G if and only if w belongs to R as an element of $\bar{F}$. Thus the word problem in G is decidable if and only if the membership problem for R in $\bar{F}$ is decidable.

We can find a somewhat bigger normal subgroup for which the membership problem *is* decidable. Consider the subgroup $\bar{F}''R$ (here $\bar{F}''$ is the second derived subgroup of $\bar{F}$). The factor group $\bar{F}/\bar{F}''R$ is finitely presented in the variety of metabelian groups. It is known that the word problem in the variety of metabelian groups is decidable [37]. Therefore the membership problem for the subgroup $\bar{F}''R$ is decidable.

So we can check if our word w belongs to $\bar{F}''R$. If $w \notin \bar{F}''R$ then $w \notin R$. So suppose that $w \in \bar{F}''R$. Since $w \in \bar{F}''R$, we can represent w as a product pr with $p \in \bar{F}''$, $r \in R$. This may be done effectively. Since $r \in R$ it is enough to decide if $p \in R$. The word p belongs to $\bar{F}''$. Thus we may suppose that $w \in \bar{F}''$. Now we have to consider the membership problem for the subgroup $R \cap \bar{F}''$.

Consider $\bar{F}''$ first. This is an Abelian subgroup. If $\bar{F}''$ were finitely generated then $R \cap \bar{F}''$ would be a finitely generated Abelian subgroup and the membership problem for $R \cap \bar{F}''$ would be trivially decidable. Unfortunately $\bar{F}''$ is not finitely generated as an Abelian subgroup (if $\bar{F}$ is not a cyclic group).

The idea is to define a module structure on $\bar{F}''$ to make $\bar{F}''$ a finitely generated module over a Noetherian commutative ring, because such modules have almost as nice algorithmic properties as finitely generated Abelian groups. There is a standard way to make $\bar{F}''$ a module.

Any Abelian normal subgroup A in a group H may be viewed as a right module over the group algebra ZH where the action of ZH on A is defined by the following formula:

$$a \circ (\sum \pm g_i) = \Pi a^{g_i^{\pm 1}}.$$

It is clear that this module has a big annihilator. For example ZA and even $ZC_H(A)$ (here $C_H(A)$ is the centralizer of A in H) is in this annihilator. Therefore A may be considered as a module over ZH/N for every normal subgroup $N \leq C_H(A)$.

In our case $\bar{F} \in \mathcal{N}_2\mathcal{A}$. Therefore $\bar{F}''$ is Abelian and $C_{\bar{F}}(\bar{F}'') \geq \bar{F}'$. Hence $\bar{F}''$ may be considered as a right module over the ring $K = Z\bar{F}/\bar{F}'$. This ring is just the ring of polynomials over Z with $X \cup X^{-1}$ as the set of unknowns factored by the ideal generated by elements $xx^{-1} - 1$ for all $x \in X$. Thus this is a commutative, finitely generated Noetherian domain.

Unfortunately $\bar{F}''$ is not finitely generated as a module over K. But we can *make* it a finitely generated module by increasing K! Indeed, by definition, the second derived subgroup $\bar{F}''$ consists of all elements which are products of double commutators $[[a,b],[c,d]]$ where $a,b,c,d \in \bar{F}$. Let us define a new action of generators from X on $\bar{F}''$:

$$[[a,b],[c,d]] \circ x = [[a,b]^x,[c,d]][[a,b],[c,d]^x].$$

This action is well defined and can be extended to another action of $K = Z\bar{F}/\bar{F}'$ on F''. Moreover this action commutes with the first action of K. Thus we can consider $\bar{F}''$ as a module over the tensor product $K \otimes K$. The last ring is, of course, also a finitely generated domain. It can be verified that $\bar{F}''$ is a finitely generated module over $K \otimes K$. Thus *we have found a hidden module structure on* $\bar{F}''$.

But we are mainly interested in R, or more precisely, in $R \cap \bar{F}''$. This is a normal subgroup and thus a K-submodule under the first action of K (normal subgroups are closed under conjugations). But this is not necessarily a $K \otimes K$-submodule. Fortunately we can split $(R \cap \bar{F}'')$ into a a sum of three Abelian groups $A_1 + A_2 + A_3$ where A_1 is finitely generated as a subgroup (Z-module), A_2 is finitely generated as a normal subgroup (K-module under the first action of K) and A_3 is finitely generated as a $K \otimes K$-module. Moreover generators of A_1, A_2, A_3 may be found effectively (see [44] for details). Therefore the membership problem for R has been reduced to the membership problem for the sum of finitely generated modules A_1, A_2, A_3 over different rings.

Now we can apply the following powerful result by Romanovskii.

Lemma 5.2 ([98]) *Let M be a D−module where D is a finitely generated commutative domain. Let $D_1 < D_2 < \ldots < D_m < D$ be a sequence of finitely generated subrings of D. Let N_i ($i = 1, \ldots, m$) be a finitely generated D_i-submodule of M. Then the membership problem for $N = N_1 + N_2 + \ldots + N_m$ is decidable.*

This lemma gives us the decidability of the membership problem for $R \cap \bar{F}''$, which, in turn, implies the decidability of the word problem in $\mathcal{N}_2\mathcal{A}$.

5.3.3. *Other Hidden Structures*

It is interesting that a hidden module structure appears in the case of semigroups also. We will call a semigroup S a *semi-module* over a semigroup A if there exists an action $\circ$ of A on S (i.e. a function $\circ : A \times S \to S$ with $(a_1a_2) \circ s = a_1 \circ (a_2 \circ s)$) which "agrees" with the operation in S: $a \circ (s_1s_2) = (a \circ s_1)s_2 = s_1(a \circ s_2)$ for every $a \in A, s_1, s_2 \in S$. The following result was proved in [105].

Theorem 5.8 ([105]) *Let T be a semigroup finitely presented in a variety $\mathcal{V}$ satisfying one of the following identities*[3]

$$x^{nk}y(z^kt^k)^pz^{(n-p)k} = x^{(n-p)k}(t^kx^k)^pyz^{nk};$$

$$xy^nz = y^kxy^mzy^p, n > m.$$

and let every periodic group in $\mathcal{V}$ be locally finite. Then there exists a semi-module S over a commutative semigroup A such that T is "almost" a subsemigroup of S. More precisely there is a semigroup S_1 such that S is an ideal of S_1 with S_1/S finite and nilpotent and T is a factor-semigroup of S_1 over a congruence with all congruence classes finite.

This theorem is one of the main steps in describing semigroup varieties with decidable word problem.

The module structure is not the only possible hidden structure on the set of words equal to a given word. For example if we are dealing with inverse semigroups, it is useful to consider the Schützenberger automata discussed in Section 2.3. Recall that these automata accept all words which are greater than or equal to the given word. Then these automata may have nice geometric properties as in [68], which helps to solve the word problem (or to prove its undecidability). Another possiblity was discovered by Margolis and Meakin in [55]. They considered inverse semigroups S given by finitely many defining relations of the form $e = f$ where e and f are words which are idempotents in the free inverse semigroup (equivalently which are equal to 1 in the free group). In this case the Schützeberger graphs embed into the Cayley graph of the free group which is, of course, a labeled tree. Then the set of vertices of the Schützenberger graph may be defined by a formula from a decidable fragment of the second order theory of this tree. This implies the decidability of the word problem in S.

---

[3] As it was mentioned earlier in this section, this condition is necessary for the decidability of the word problem.

6. Minsky Machines and the Undecidability of the Word Problem in Varieties

6.1. Minsky Machines

One of the most powerful tools in proving the undecidability of the word problem is the so-called Minsky Machine. It was invented by Marvin Minsky in 1961 (see [73], [74], [53]). Yu. Gurevich was the first to use Minsky machines to prove the undecidability of an algorithmic problem in algebra [35].

Let us give a definition of a Minsky machine. The hardware of a (two-tape) Minsky machine consists of two tapes and a head. The tapes are infinite to the right and are divided into infinitely many cells numbered from the left to the right, starting with 0. The first cells on both tapes always contain 1, all other cells have 0. The head may acquire one of several internal states: $q_0, \ldots, q_N$; q_0 is called *the terminal state.* At every moment the head looks at one cell of the first tape and at one cell of the second tape. So the *configuration* of the Minsky machine may be described by the triple (m, q_k, n) where m (resp. n) is the number of the cell observed by the head on the first (resp. second) tape, q_i is the state of the head.

Every command has the following form:

$$q_i, \epsilon, \delta \longrightarrow q_j, T^{\alpha}, T^{\beta}.$$

where $\epsilon, \delta \in \{0,1\}, \alpha, \beta \in \{-1,0,1\}$. This means that if the head is in the state q_i and it observes a cell containing ϵ on the first tape and a cell containing δ on the second tape then it acquires the state q_j and the first (the second) tape is shifted α (resp. β) cells to the left relative to the head. If, say, $\alpha = -1$ then the first tape is shifted one cell to the right. The machine always starts working at state q_1 and ends at the *stop state* q_0. The program (software) for a Minsky machine is a set of commands of the above form.

One says that a Minsky machine *calculates* a function $f(m)$ if for every m starting at the configuration $(m, q_1, 0)$ it ends at the configuration $(f(m), q_0, 0)$. If m does not belong to the domain of f then the machine works forever and never gets to the terminal state.

The main property of Minsky machines is contained in the following theorem.

Theorem 6.1 ([73]) *For every partially recursive function $f(m)$ there exists a Minsky machine which calculates the partial function $g_f : 2^m \to 2^{f(m)}$.*

Remark. It is interesting that, in this theorem, the function g_f cannot be replaced by the function f. In particular, there is no Minsky machine which calculates the function m^2 (see [53]).

The "canonical" definition of a Minsky machine seems to be very long and complex while in fact it is very simple and can be understood by a high school or even elementary school student. Let us give a "high school" definition of a Minsky machine.

Consider two glasses. We assume that these glasses are of infinite height. Another (more restrictive!) assumption is that we have infinitely many coins. There are four *operations*: "Put a coin in a glass", "Take a coin from a glass if it is not empty". We are able to check if a glass is/is not empty. A *program* is a numbered sequence of instructions.

An *instruction* has one of the following forms:

- Put a coin in the glass # n and go to instruction # j;
- If the glass # n is not empty then take a coin from this glass and go to instruction # j otherwise go to instruction # k.
- Stop.

A program starts working with the command number 1 and ends when it comes to the Stop instruction which will always have number 0.

We say that a program calculates a function $f(m)$ if, starting with m coins in the first glass and empty second glass, we end up with $f(m)$ coins in the first glass and empty second glass.

This "high school" version of a Minsky machine is known also as a Minsky algorithm. One can prove that Minsky machines are equivalent to Minsky algorithms, that is, given a program for Minsky machine (resp. given a Minsky algorithm), it is easy to construct a Minsky algorithm (resp. a program for Minsky machine) which calculates the same function.

A configuration of a Minsky algorithm is a triple (m, k, n), where m is the number of coins in the first glass, n is the number of coins in the second glass, and k is the number of the instruction we are executing. So the number of an instruction in the algorithm plays the role of an inner state.

Notice that although Minsky machines and Minsky algorithms are equivalent, Minsky algorithms are sometimes better tools in proving the undecidability of algorithmic problems. The main reason: there are four possible commands of a Minsky machine which correspond to the same inner state q_k, while there is only one instruction of a Minsky algorithm with a given number k.

6.2. Minsky Machines and the Word Problem for General Algebras

Here we will show how to apply Minsky machines (algorithms) to prove the undecidability of the word problem. All applications of Minsky machines (algorithms) are based on the following idea. We will show this idea for Minsky machines; exactly the same idea works for Minsky algorithms. Take a Minsky machine M calculating a function g_f with a non-recursive domain X.

One can define an "equality" on the set of all possible configurations of the Minsky machine M: two configurations (m, q_k, n) and $(m', q_{k'}, n')$ are "equal" or "equivalent" (we will write $(m, q_k, n) \equiv (m', q_{k'}, n')$) if there exists another configuration $(m'', q_{k''}, n'')$ such that M transforms both configurations (m, q_k, n) and $(m', q_{k'}, n')$ into $(m'', q_{k''}, n'')$. It is easy to see that this "equality" is symmetric, transitive and reflexive. By the choice of M, $(m, q_1, 0) \equiv (1, q_0, 0)$ if and only if $m \in X$. Therefore we have a sequence of configurations γ_m and a special configuration γ_0 such that $\gamma_m \equiv \gamma_0$ iff $m \in X$.

Suppose now that we want to construct a finitely presented universal algebra $A(M)$ with an undecidable word problem. The idea of doing this has its origin in the works of Markov and Post [61], [92]. First, with every configuration ψ one associates a word $w(\psi)$. This word is usually called a *canonical word*. Then with every command κ of the Minsky machine M one associates a finite set of defining relations R_κ. The algebra $A(M)$ will be defined by the relations from the union R of all R_κ (which is finite since we have only a finite number of commands) and usually some other relations Q which are in a sense "independent" of R. We need Q, for example, to make $A(M)$ satisfy a particular identity.

The algebra $A(M)$ will have an undecidable word problem if the following property holds:

$$\psi_1 \equiv \psi_2 \text{ if and only if } w(\psi_1) = w(\psi_2) \text{ in } A. \tag{8}$$

Indeed, in this case one cannot algorithmically decide if $w(m, q_1, 0) = w(1, q_0, 0)$ for the given number m.

We will say that we have an *interpretation* of the Minsky machine M in the algebra $A(M)$ if we have an assignment $\psi \to w(\psi)$ with the property (8).

Usually, in order to prove the property (8) one has to prove two lemmas.

Lemma 6.1 *If we can proceed from configuration ψ_1 to configuration ψ_2 using command κ, then we can proceed from the word $w(\psi_1)$ to the word $w(\psi_2)$ using relations from R_κ.*

Lemma 6.2 *If we can proceed from the word $w(\psi_1)$ to the word $w(\psi_2)$ by using relations from the union of all R_κ, then $\psi_1 \equiv \psi_2$.*

It is easy to see that Lemmas 6.1 and 6.2 imply property (8). Lemma 6.2 in most cases is more difficult to prove than Lemma 6.1. It is worth mentioning also that in order to prove Lemmas 6.1 and 6.2 we usually do not need any information about the function that is calculated by M.

Let us also make a remark about the case where we are constructing an algebra with undecidable word problem which is finitely presented in a variety $\mathcal{V}$. In this case we are allowed to use identities of $\mathcal{V}$ when we deduce relations of $A(M)$. Notice that unlike the relations of R_κ corresponding to commands of M, the relations obtained from identities of $\mathcal{V}$ have no connections with the Minsky machine, and can spoil the canonical words. So we have to make the canonical words resistant to applications of identities of $\mathcal{V}$. In the best case, the identities of $\mathcal{V}$ are not applicable to canonical words.

The procedure for constructing an algebra finitely presented in a variety $\mathcal{V}$ with undecidable word problem is roughly the following. First we temporarily forget about $\mathcal{V}$ and construct an interpretation of a Minsky machine M in an "absolutely" finitely presented algebra $A(M)$. We prove Lemma 6.1 for $A(M)$. Then we consider the factor algebra $\hat{A}(M)$ of $A(M)$ by the verbal congruence corresponding to $\mathcal{V}$, that is we identify all pairs of terms in $A(M)$ which are identically equal in $\mathcal{V}$. The algebra $\hat{A}(M)$ is finitely presented in $\mathcal{V}$. Then we have to prove Lemmas 6.1 and 6.2 for $\hat{A}(M)$. Fortunately we have Lemma 6.1 for free because the statement of this lemma is stable under homomorphic images. To prove Lemma 6.2 we usually need the above mentioned independence of canonical words from the identities of $\mathcal{V}$.

6.3. Why Minsky Machines?

The concept of an interpretation of a machine in an algebra can be applied to any kind of Turing machine. Why did we choose the Minsky machines? The first answer is: because Minsky machines are in some important sense the simplest universal Turing machines possible.

Indeed, the first and the main step in any interpretation of a machine in an algebra is the choice of canonical words. The canonical words must encode configurations of the machine. Therefore the smaller the number of parameters which determine the configurations, the more freedom we have in simulating the parameters. A configuration of a Minsky machine is determined just by three numbers: m, i, n. Here i runs over a finite set. Therefore one can encode each i by a separate letter q_i. It is also important that the commands of Minsky machines change those three numbers in a natural way. They add 0, 1 or -1 to m and n, and change i according to some simple rule. Therefore we can simulate m and n by, say, powers of different letters, say, a and b, and the relations corresponding to the relations of the Minsky machine will increase (decrease) the powers of these letters. Therefore we can encode the configuration (m, i, n) by a canonical word $a^m q_i b^n$ (we suppose for simplicity that we have a binary associative operation).

True, after a little pondering one can conclude that something is missing in this encoding. Indeed, recall that the relations will simulate the commands. The action of a command depends on whether m or n is equal to 0 or not, so there are four different situations ($m \neq 0, n \neq 0$; $m = 0, n \neq 0$; $m \neq 0, n = 0$; $m = 0, n = 0$). Therefore for each one of these four situations, the corresponding canonical word must have a special small subword which tells us that this situation occurs (then the corresponding relation will replace this subword by some other word).

Notice that the canonical words corresponding to these four situations have the following form: $a^n q_i b^m$, $q_i b^m$, $a^n q_i$, q_i where $m, n > 0$. All these words are subwords of the first one. So every subword of the second (the third or the fourth) word is a subword of the first word. Thus we cannot distinguish between these situations.

The solution to this problem is simple: we have to add two more letters A and B, which we will call "locks", and encode the configuration (m, i, n) by the word $Aa^m q_i b^n B$. Then each of the four situations is characterized by a small subword of the canonical word: The canonical word has a subword:

$aq_i b$ iff $m \neq 0, n \neq 0$;
$Aq_i b$ iff $m = 0, n \neq 0$;
$aq_i B$ iff $m \neq 0, n = 0$;
$Aq_i B$ iff $m = 0, n = 0$.

Thus it is very easy to find an interpretation of the Minsky machine. But this is not the only reason why one has to use it.

Recall that we want to simulate a machine in an algebra which satisfies as many identities as possible. And as everybody knows, those identities tend to change words. For example suppose that we have an identity $xy = yx$, and we simulate the configuration (m, i, n) by the word $Aa^m q_i b^n B$. Then the words $Aa^m q_i b^n B$, $a^m A q_i b^n B$, $Aa^m q_i B b^n$, and $a^m A q_i B b^n$ are equal and, again, we cannot find small subwords which distinguish one of the four situations from another.

The important feature of Minsky machines and their interpretations is that the canonical words, which we obtain, are very stable with respect to identities.

For example, in the case of semigroups, Theorem 5.4 means the following. There exist a few basic encodings of configurations of Minsky machines, and if a nonperiodic semigroup variety with locally finite nil-semigroups satisfies identities which can change all these encodings, then the word problem is solvable in this variety, so the interpretation of a Minsky machine (and any other universal Turing machine) in this variety is impossible.

It is much more difficult to find stable interpretations of other kinds of machines. For example, the configuration of the general (one-tape) Turing machine is a triple u, q_i, v where q_i is an inner state of the head, u is the word written on the tape to the left of the head, and v is the word written to the right of the head. So if we want to interpet the general Turing machine, we need to encode somehow arbitrary words u and v. Since any identity can change some words, we cannot encode the word by itself, thus the encoding must be unnatural. This will lead to great difficulties with simulating commands of the Turing machine, and so on.

6.4. The Word Problem for Semigroup Varieties. The Nonperiodic Case

There are two important semigroup interpretations of Minsky machines: the semigroups S_1 and S_2 below. Let M be a Minsky machine with internal states $q_0, \dots, q_N$. Then both S_1 and S_2 are generated by the elements $q_0, \dots, q_N$ and a, b, A, B. The correspondences between commands of M and relations of S_1 and S_2 are given by the following tables. Every command corresponds to one relation in S_1 and one relation in S_2.

(9)

Command	S_1
$q_i, 0, 0 \to q_j, T^\alpha, T^\beta$	$aq_ib = a^{1+\alpha}q_jb^{1+\beta}$
$q_i, 1, 0 \to q_j, T^\alpha, T^\beta$	$Aq_ib = Aa^\alpha q_jb^{1+\beta}$
$q_i, 0, 1 \to q_j, T^\alpha, T^\beta$	$aq_iB = a^{1+\alpha}q_jb^\beta B$
$q_i, 1, 1 \to q_j, T^\alpha, T^\beta$	$Aq_iB = Aa^\alpha q_jb^\beta B$

(10)

Command	S_2
$q_i, 0, 0 \to q_j, T^\alpha, T^\beta$	$q_i ab = q_j a^{1+\alpha} b^{1+\beta}$
$q_i, 1, 0 \to q_j, T^\alpha, T^\beta$	$q_i Ab = q_j a^\alpha Ab^{1+\beta}$
$q_i, 0, 1 \to q_j, T^\alpha, T^\beta$	$q_i aB = q_j a^{1+\alpha} b^\beta B$
$q_i, 1, 1 \to q_j, T^\alpha, T^\beta$	$q_i AB = q_j a^\alpha Ab^\beta B$

The canonical words in S_i are the following:

(11)

Configuration	S_1	S_2
(m, q_k, n)	$Aa^m q_k b^n B$	$q_k a^m Ab^n B$

To make these interpretations work and to make these semigroups satisfy as many identities as possible we need also some additional relations independent of the commands of M.

In the semigroup S_2 we need the following commutativity relations:

(12) $$ab = ba,\ aB = Ba,\ bA = Ab,\ AB = BA.$$

Also we need all relations of the type

$$xy = 0$$

where xy is a two letter word which is not a subword of $w(m, q_k, n)$ for some m, n or of any word obtained from $w(m, q_k, n)$ by the commutativity relations above.

These relations "kill" all "wrong" words. Basically only canonical words and their subwords are distinct from zero in S_1 and S_2.

Thus we have the following additional relations in S_1: all two letter words are equal to 0 except Aa, Aq_i, a^2, aq_i, q_ib, q_iB, b^2, bB. And we have the following additional relations in S_2: all two letter words are equal to 0 except q_ia, q_ib, q_iA, q_iB, a^2, aA, ab, aB, ba, b^2, bB, bA, Ab, AB, Ba, BA.

These semigroups are very convenient for demonstrating the standard proofs of Lemmas 6.1 and 6.2.

To prove Lemma 6.1 one needs to show that if we pass from a configuration (m, q_k, n) to another configuration $(m', q_{k'}, n')$ by a command κ then we pass from the word $w(m, q_k, n)$ to the word $w(m', q_{k'}, n')$ by the relation corresponding to κ.

Let us prove this only for the case of the semigroup S_2 and the command κ : $q_k, 1, 0 \to q_{k'}, T^\alpha, T^\beta$. All other cases are similar.

Since the command κ is applicable to the configuration (m, q_k, n), in this configuration, the head observes the first cell on the first tape and not the first cell on the second tape. Thus $m = 0, n \neq 0$. Then $m' = \alpha, n' = n + \beta$ (in this case α cannot be negative). Now $w(m, q_k, n) = q_k b^n AB$ and the relation corresponding to κ is $q_k bA = q_{k'} a^\alpha b^{1+\beta} A$. Since $AB = BA$ and $aB = Ba$ we have $w(m, q_k, n) = q_k b^n AB = q_k bAb^{n-1}B$. Thus we can apply our relation and replace $q_i bA$ by $q_{k'} a^\alpha b^{1+\beta} A$. As a result we obtain the word $q_{k'} a^\alpha b^{1+\beta} Bb^{n-1} A$ which is equal to $q_{k'} a^\alpha Ab^{n+\beta} B$ since $bA = Ab$. The last word is equal to $w(m', q_{k'}, n')$ as desired.

The proof of Lemma 6.2 is based on the following two standard observations. First, for every canonical word $w(m, q_k, n)$ there exists at most one relation corresponding to a command of M which is applicable to this word from the left to the right (this means that one replaces the left hand side of this relation by the right hand side of it).

Second, any application of a relation from tables (9) or (10) to any canonical word — from the left to the right or from the right to the left — gives us another canonical word (we do not distinguish words in S_2 which are obtained from each other by the commutativity relations[4]).

Now consider two words $w(m, q_k, n)$ and $w(m', q_{k'}, n')$ in S_1 or S_2. Suppose these words are equal in this semigroup. Therefore there exists a sequence of words

$$w(m, q_k, n), w_1, \dots, w_n, w(m', q_{k'}, n')$$

where each word is obtained from the previous one by applying a defining relation corresponding to a command of the machine M. By the second observation each w_i corresponds to a configuration of M.

These relations may be applied from the left to the right and from the right to the left. Now suppose that in the passage $w_{r-1} \to w_r$, a relation was applied from the right to the left and in the passage $w_r \to w_{r+1}$, a relation was applied from the left to the right. Then w_{r-1} and w_{r+1} are obtained from w_r by applying relations from the left to the right. By the first observation these two relations must coincide and the words w_{r-1} and w_{r+1} must coincide also. Therefore in this case we can shorten our sequence of w_i. Thus we can suppose that in our sequence, there is a word $w_r = w(m'', q_{k''}, n'')$ such that all relations before w_r are applied from the left to the right and all relations after w_r are applied from the right to the left. But this means that the machine M passes from both configurations (m, q_k, n) and $(m', q_{k'}, n')$ to the configuration $(m'', q_{k''}, n'')$. Therefore $(m, q_k, n) \equiv (m', q_{k'}, n')$, as desired.

Thus we have proved that if the Minsky machine computes a non-recursive function then S_1 and S_2 have undecidable word problems. These semigroups are important because every (finitely based) semigroup variety with undecidable word problem, whose periodic semigroups are locally finite, contains either S_1 or S_2 or the semigroup anti-isomorphic to S_2 (see [105]). Therefore we do not need any other semigroup with undecidable word problem to treat non-periodic varieties with good

[4] Actually — important! — we assign to each configuration not a single word $w(m, q_k, n)$ but a set of words which may be obtained from each other by the commutativity relations. It is worth mentioning that it is almost always better to consider $w(m, q_k, n)$ as a set of words; relations corresponding to commands of M connect these sets of words; auxiliary relations connect words in the same set.

periodic semigroups. But if the periodic semigroups are not locally finite we need something else, and we will discuss it in the next subsection.

6.5. The Word Problem for Semigroup Varieties. The Periodic Case

Theorem 4.5 above implies that a finitely based variety of semigroups in which not every periodic semigroup is locally finite contains either a periodic group or a nil-semigroup which is not locally finite. The periodic group case will be considered further (see Section 6.6). Now let us consider the nil-case. So suppose that we have a finitely based variety $\mathcal{V}$ containing a non-locally finite nil-semigroup. Then by virtue of Theorem 5.5, $\mathcal{V}$ has an undecidable word problem and now we are going to explain how to prove this using Minsky machines.

To show only the principal details of the proof let us prove that the word problem is undecidable in the variety of semigroups given by the identity $x^3 = 0$. This variety consists of nil-semigroups and it was proved by Morse and Hedlund [75] that it contains an infinite finitely generated semigroup. We will need a slight modification of the Morse and Hedlund construction.

Let us start with the following Thue endomorphism φ [123] of the free semigroup with generators a, b:

$$\varphi(a) = ab, \ \varphi(b) = ba.$$

Now let us iterate φ and consider the words $a, \varphi(a), \varphi^2(a), \varphi^3(a), \ldots$. For every n, $\varphi^n(a)$ is an initial segment of the word $\varphi^{n+1}(a)$, so all these words are initial segments of an infinite sequence. Let us denote this sequence by $T(\varphi)$. Thue [123] proved that these words do not have subwords of the form uuu where u is any non-empty word. Given $T(\varphi)$, Morse and Hedlund construct a semigroup $S(\varphi)$ as follows. Let $S(\varphi)$ be the set consisting of all subwords of the words from this sequence, and a special symbol 0. Define an operation on this set by the following rule:

$$u \cdot v = \begin{cases} uv & \text{if } uv \text{ is a subword of } \varphi^n(a) \text{ for some } n; \\ 0 & \text{if } uv \text{ is not a subword of any } \varphi^n(a). \end{cases}$$

Here uv is the result of concatenation of u and v. [5]

It is easy to prove that $S(\varphi)$ is an infinite semigroup generated by a and b. This semigroup satisfies the identity $x^3 = 0$. Indeed, as we have mentioned above, for every word u the word u^3 is not a subword of any $\varphi^n(a)$ and so it is equal to 0 in $S(\varphi)$.

One can easily see that it is not possible to use the interpretations from the previous subsection to simulate a Minsky machine in a semigroup satisfying $x^3 = 0$. Indeed, we can no longer encode the number of the cell observed by the head of the machine by a power of a letter: there is a shortage of powers (only 3). The idea is to use powers of the Thue endomorphism φ instead of powers of letters.

Thus we want to encode the configuration (m, q_k, n) by a word like $\varphi^m(a)q_k\varphi^n(a)$. Notice that such a word will be cube-free (will not contain subwords of the form uuu) for any m and n, so we won't be able to apply our identity $x^3 = 0$ to this word.

[5] As we mentioned in Section 4 there is a general construction which associates a semigroup $S(D)$ with every symbolic dynamical system D. The semigroup $S(\varphi)$ is equal to $S(D)$ where D is the symbolic dynamical system generated by the limit of words $\varphi^n(a)$.

Now we have to assign a relation to every command of M. This relation must increase (decrease) m and n in $w(m, q_k, n)$ if the command shifts the tapes to the left (to the right). Unfortunately it is impossible to pass from $\varphi^m(a)q_k\varphi^n(a)$ to $\varphi^{m+1}(a)q_{k'}\varphi^n(a)$ by using one relation independent of m and n. Indeed, for "large" m it is impossible to proceed from $\varphi^m(a)$ to $\varphi^{m+1}(a)$ by replacing a "small" subword by another "small" subword.

But we notice that $\varphi^{m+1}(a) = \varphi(\varphi^m(a))$, so we need to find some auxiliary relations which will simulate the application of φ. This can be done by adding one letter say, c_1 and relations $ac_1 = c_1\varphi(a), bc_1 = c_1\varphi(b)$. Indeed, for every m we will then have $\varphi^m(a)c_1 = c_1\varphi^{m+1}(a)$.

A practical realization of this idea is the following (see Section 6.6 for another realization).

Our semigroup, — let us denote it by $S(M, \varphi)$,— will be generated by the set $\{q_0, \dots, q_N, a, b, c_1, c_0, c_{-1}, d_1, d_0, d_{-1}, A, B\}$.

For every configuration (m, q_k, n) let

$$w(m, q_k, n) = A\overline{\varphi^m(a)}c_0 q_k d_0 \varphi^n(a)B.$$

where $\bar{u}$ is the word u written from the right to the left.

The correspondence between commands of the Minsky machine M and relations in $S(M, \varphi)$ is given by the following table.

(13)

Command	$S(M, \varphi)$
$q_i, 0, 0 \to q_j, T^\alpha, T^\beta$	$bac_0q_id_0ab = bac_\alpha q_jd_\beta ab$
$q_i, 1, 0 \to q_j, T^\alpha, T^\beta$	$Aac_0q_id_0ab = Aac_\alpha q_jd_\beta ab$
$q_i, 0, 1 \to q_j, T^\alpha, T^\beta$	$bac_0q_id_0aB = bac_\alpha q_jc_\beta aB$
$q_i, 1, 1 \to q_j, T^\alpha, T^\beta$	$Aac_0q_kd_0aB = Aac_\alpha q_jd_\beta aB$

The auxiliary relations are the following:

(i) $ac_0 = c_0a$, $bc_0 = c_0b$;

(ii) $ac_1 = c_1\varphi(a)$, $bc_1 = c_1\varphi(b)$, $Ac_1 = Ac_0$;

(iii) $\varphi(a)c_{-1} = c_{-1}a$, $\varphi(b)c_{-1} = c_{-1}b$, $Ac_{-1} = Ac_0$;

(iv) $d_0a = ad_0$, $d_0b = bd_0$;

(v) $d_1a = \varphi(a)d_1$, $d_1b = \varphi(b)d_1$, $d_1B = d_0B$;

(vi) $d_{-1}\varphi(a) = ad_{-1}$, $d_{-1}\varphi(b) = bd_{-1}$, $d_{-1}B = d_0B$.

The role of the new generators c_i and d_i is clear from these relations: d_1 and c_1 increase the power of φ, c_{-1} and d_{-1} decrease the power of φ.

It is not very difficult to prove Lemmas 6.1 and 6.2. Therefore we have obtained an interpretation of the machine M. Now, since all words $w(m, q_k, n)$ and all words which can be obtained from these words by the defining relations of $S(M, \varphi)$ are cube-free, the identities of our variety won't change these words. Therefore the statements of Lemmas 6.1 and 6.2 hold for the factor-semigroup $\hat{S}(M, \varphi)$ of the semigroup $S(M, \varphi)$ over the verbal congruence corresponding to the identity $x^3 = 0$. Indeed, the statement of Lemma 6.1 is stable under homomorphic images. The statement of Lemma 6.2 holds because canonical words which are distinct in $S(M, \varphi)$ are also distinct in $\hat{S}(M, \varphi)$. It remains to notice that the semigroup $\hat{S}(M, \varphi)$ is finitely presented in the variety given by the identity $x^3 = 0$. Therefore this variety has an undecidable word problem.

In the general case, when the identities of the variety $\mathcal{V}$ are more (sometimes much more) complicated, one has to use the endomorphisms constructed in [109] instead of φ, and the interpretation is slightly different also (see [106]).

6.6. The Word Problem for Varieties of Groups. The Periodic Case

Here we present the method of Sapir from [107] of constructing a group with undecidable word problem which is finitely presented in the variety $\mathcal{A}_r\mathcal{B}_p$ for every odd $p \geq 665$ and every prime $r \neq p$.

The free group in the variety $\mathcal{A}_r\mathcal{B}_p$ is a semidirect product of an Abelian group A of exponent r and the free Burnside group F of exponent p. This Abelian group may be considered as a module over the group ring Z_rF. Standard group theoretic observations show that in order to construct a group which is finitely presented in this variety and has an undecidable word problem, it is enough to construct a finitely generated right ideal in the group ring Z_rB with undecidable membership problem, where B is a finitely presented group from $\mathcal{B}_p$.

In order to find this ideal we do the following. First of all we find a semigroup interpretation of a Minsky algorithm, similar to the interpretation given in the previous section, which satisfies the following properties.

(1) Every canonical word is cube-free.
(2) Every relation corresponding to a command of the Minsky algorithm applies to a prefix of the canonical word.
(3) The auxiliary relations are some commutativity relations like $ab = ba$, $Ab = bA$, etc. and do not contain letters q_i which correspond to the numbers of commands of the algorithm.

The interpretation used in [107] is the following.

The alphabet consists of letters

$$\{q_i, q_{i1}, a_1, a_2, b_1, b_2, c_1, c_2, d_1, d_2, A, B, C, D \mid i = 0, 1, \dots, N\}.$$

Recall that a Minsky algorithm deals with coins in two glasses and can add (take off) a coin to (from) one of these glasses (see Section 6.1 for the definition of Minsky algorithms). A configuration of the algorithm is a triple (m, i, n) where i is the number of the command which is being executed, m is the number of coins in the first glass, n is the number of coins in the second glass.

Let ϕ be the Thue-like substitution (see Section 6.5):

$$\phi(x_1) = x_1x_2,\ \phi(x_2) = x_2x_1,\ x \in \{a,b,c,d\}.$$

Then the canonical word $w(m,i,n)$ is equal to $q_i\phi^m(a_1)\phi^n(b_1)ABCD$.

Auxiliary relations are the following:

$$x_iy_j = y_ix_j,\quad x_iY = Yx_i \tag{14}$$

where $x,y \in \{a,b,c,d\}, x \neq y, Y$ is the capital y for $y \in \{a,\ b,\ c,\ d\}$.

These relations make letters a, b, c, d "independent": they can move inside the canonical words without disturbing each other.

The relations corresponding to the commands of the Minsky algorithm are in the following table.

(15)

Command	Semigroup relations
[add a coin to the first glass | j]	$q_ia_t = q_i\phi(c_t), t=1,2$ $q_iA = q_{i1}A$ $q_{i1}c_t = q_{i1}a_t, t=1,2$ $q_{i1}C = q_jC$
[add a coin to the second glass | j]	$q_ib_t = q_i\phi(d_t), t=1,2$ $q_iB = q_{i1}B$ $q_{i1}d_t = q_{i1}b_t, t=1,2$ $q_{i1}D = q_jD$
[take a coin from the first glass | $j\|k$]	$q_ia_1A = q_ka_1A$ $q_i\phi(a_t) = q_ic_t, i=1,2$ $q_iA = q_{i1}A$ $q_{i1}c_t = q_{i1}a_t$ $q_{i1}C = q_jC$
[take a coin from the second glass |$j\|k$]	$q_ib_1B = q_kb_1B$ $q_i\phi(b_t) = q_id_t, i=1,2$ $q_iB = q_{i1}B$ $q_{i1}d_t = q_{i1}b_t$ $q_{i1}D = q_jD$

Then we take the alphabet A of this interpretation and produce a new alphabet A_1 which consists of of four copies of letters from A. If x is a letter of A then A_1

contains letters x_{ij} where $i, j \in \{1, 2\}$. We assume that new letters x_{ij} satisfy the same auxiliary commutativity relations as x. For example, since we had a relation $ab = ba$, we now have relations $a_{ij}b_{k\ell} = b_{k\ell}a_{ij}$ for all i, j, k, ℓ.

We replace every letter x in our interpretation of the Minsky algorithm (that is in the canonical words and in the relations) by the sum $x_{11} + x_{12} + x_{21} + x_{22}$ which we consider as an element of the group algebra ZB where B is the Burnside group generated by A_1 subject to the new auxiliary relations.

The polynomials, obtained from a word u over A by this substitution will be denoted by $\bar{u}$.

Now let $R(M) = \{u_i = v_i \mid i = 1, 2, \ldots\}$ be the set of relations which simulates our Minsky algorithm. Consider the right ideal $I(M)$ in ZB generated by the polynomials $\bar{u}_i - \bar{v}_i$, $i = 1, 2, \ldots$.

It is not difficult to prove that if two configurations (m, i, n) and (m', i', n') of the Minsky algorithm are equivalent then the polynomial $\bar{w}(m, i, n) - \bar{w}(m', i', n')$ belongs to $I(M)$. One has to use the properties 2 and 3 of our interpretation (this is why the interpretation presented in the previous section does not work in the present case).

Thus in order to prove that $I(M)$ has an undecidable membership problem, it is enough to show that if $\bar{w}(m, i, n) - \bar{w}(m', i', n')$ belongs to $I(M)$ then the configurations (m, i, n) and (m', i', n') of the Minsky algorithm are equivalent.

Suppose that $\bar{w}(m, i, n) - \bar{w}(m', i', n')$ belongs to $I(M)$. This means that

$$\bar{w}(m, i, n) - \bar{w}(m', i', n') = \sum(\bar{u}_i - \bar{v}_i)f_i \tag{16}$$

where $u_i = v_i \in R$, $f_i \in B$.

Both sides become sums of monomials. Now take a monomial U in the sum $\bar{w}(m, i, n)$. From the theory of Burnside groups (see Adian [2]) we know that cube-free words are not equal to each other, and are not equal to smaller subwords in the free Burnside group. Therefore this monomial cannot coincide with other monomials from $\bar{w}(m, i, n)$ and with monomials from $\bar{w}(m', i', n')$. Therefore this monomial must coincide with a monomial from the right side of (16) (we are in a group ring!).

Hence we have $U = u_i' f_i$ or $U = v_i' f_i$. Here w' (for w equal to u_i or v_i) is obtained from w by replacing some letters by their "brothers" — the same letters with different indices). Suppose that the first equality holds. Suppose that u_i' is an initial part of the word U, i.e. the equality $U = u_i' f_i$ is an equality in the free semigroup. Then we can take a monomial v_i' in the sum $\bar{v}_i$ and replace U by $U_1 = v_i' f_i$. The monomial U_1 must cancel with some other monomial in a sum $(\bar{u}_s - \bar{v}_s)f_s$. Again suppose that $U_1 = u_s' f_s$ in the free semigroup, and replace U_1 by $U_2 = v_j' f_s$. Finally we will hit all sums $(\bar{u}_s - \bar{v}_s)f_s$ in the right side of the equality (16) and so one of U_g will coincide with a monomial from the sum $\bar{w}(m', i', n')$. Now look at the sequence of monomials $U, U_1, U_2, \ldots, U_k$. If we identify letters and their "brothers" then U will coincide with $w(m, i, n)$, U_k will coincide with $w(m', i', n')$, and every step in this sequence of words is obtained by replacing one side of relations from R by another side of this relation. Therefore we have a semigroup deduction of the relation $w(m, i, n) = w(m', i', n')$. But we already know that if such a relation holds then corresponding configurations of the Minsky machine are equivalent.

This was only an idea of the proof. In order to make this proof work we have to understand what to do if U, the monomial we are dealing with, is equal to $u_i'f_i$ but u_i' is not an initial part of U. Of course, we still can take a monomial v_i' and replace U by $U_1 = v_i'f_i$ but first it may eventually lead to big powers of letters which is bad since we are working with a periodic group B. Second, the transformations $U \to U_1 \to U_2 \dots$ will not simulate semigroup transformations.

In order to avoid these difficulties Sapir defines a set of words E over the alphabet of generators A_1 with the following properties.

E1. Every word in E is cube-free.

E2. If w is a canonical word then one and only one monomial in $\bar{w}$ belongs to E.

E3. If the word f_i is contained in E and the word $u_i'f_i$ is also contained in E, then there exists a unique v_i' such that $v_i'f_i$ belongs to E.[6]

E4. If the word f_i does not belong to E but $u_i'f_i$ belongs to E then there exists a unique monomial u_i'' in $\hat{u}_i$ which is distinct from u_i' and is such that $u_i''f_i$ belongs to E.

E5. If w is a positive word from E, $f \in E$, and $w = u_i'f$, then u_i' is an initial part of w.

Given this set E, we can describe our transformations $U \to U_1 \to U_2 \to \dots$ precisely. Indeed, suppose again that we have the equality (16). We know that there exists (unique) monomial U in $\bar{w}(m, i, n)$ which belongs to E. Since all monomials in the left part of (16) are cube-free, U is not equal to any monomial in the left side, and so it must be equal to some $u_i'f_i$ from the right side.

Now if f_i belongs to E, we choose a (unique by E3) word v_i' in $\bar{v}$, and replace U by $U_1 = v_i'f_i$. Since U is a positive word, u_i' is an initial part of U, and so U_1 is also a positive word. In this case we use all monomials from the sum $(\bar{u}_i - \bar{v}_i)f_i$ which belong to E. Let us call this transformation $U \to U_1$ an $R1$-transformation.

If f_i does not belong to E we choose a (unique by E4) monomial u_i'' in $\bar{u}_i$ which is distinct from u_i' and such that $U_1 = u_i''f_i$ belongs to E. In this case we will call our transformation $U \to U_1$ an $R2$-transformation.

If U_1 belongs to the sum $\bar{w}(m', i', n')$, our process ends. If not, we can find another sum on the right side of (16) where it belongs. Then we can proceed from U_1 to U_2 by an $R1$- or $R2$-transformation.

Why does this process simulate the process of semigroup deductions? Notice that if all transformations in the sequence $U \to U_1 \to U_2 \to \dots$ are $R1$–transformations then all these words are positive and the sequence indeed simulates a deduction of the relation $w(m, i, n) = w(m', i', n')$.

In the case when some of these transformations are $R2$-transformations Sapir proves that there exists another (perhaps even shorter) sequence of $R1$-transformations only, which connects the same words.

---

[6] Here $u_i = v_i$ is a relation from (10).

6.7. The Word Problem for Varieties of Inverse Semigroups

6.7.1. *Inverse Semigroups with Abelian Covers*

Recall that inverse semigroups form a variety of algebras of type $< 2, 1 >$. While inverse semigroups appear at first sight to form a class of algebras intermediate between groups and semigroups, there are many aspects of the theory that are drastically different from their counterparts in group theory or semigroup theory. Even the theory of the free inverse semigroup [76], [113], [55], [99] offers many surprises and interesting unsolved problems. In particular, the free inverse semigroup is not finitely presented as a semigroup [112]. The structure of the free inverse semigroup $FIS(X)$ on X was described in Section 2 above.

Our concern in this section is with the variety of those inverse semigroups which possess an E-unitary cover over some Abelian group. The concept of E-unitary cover is one of the fundamental concepts in the theory of inverse semigroups. It has many equivalent definitions (see [89]). One of the definitions is the following. Let S be an inverse semigroup of partial injective transformations of a set X. We say that S has an E-unitary cover over a group G if G has a faithful representation by permutations of a set Y containing X such that every transformation of S is a restriction of some permutation of G.

For every variety of groups $\mathcal{V}$ the class of inverse semigroups

$$\hat{\mathcal{V}} = \{S|S \text{ has an E-unitary cover over some group in } \mathcal{V}\}$$

is a variety of inverse semigroups. The following result summarizes some of the work of Petrich and Reilly [90] and Pastijn [86].

Theorem 6.2 *Let $\mathcal{V}$ be a variety of groups. Then $\hat{\mathcal{V}}$ is defined by the identities $\{u^2 = u \mid u = 1$ is an identity in $\mathcal{V}\}$.*

The variety $\hat{\mathcal{V}}$ may also be characterized in a different way. Recall that if $\mathcal{U}$ and $\mathcal{V}$ are varieties of universal algebras then the *Mal'cev product* $\mathcal{UV}$ is the variety generated by all algebras A for which there exists a congruence ρ on A such that all ρ-classes which are subalgebras are in $\mathcal{U}$ and $A/\rho \in \mathcal{V}$.

For varieties of inverse semigroups $\mathcal{U}$ and $\mathcal{V}$ we also denote by $\mathcal{U} * \mathcal{V}$ the variety generated by all semidirect products $S * T$ where $S \in \mathcal{U}$, $T \in \mathcal{V}$. It follows from the work of Tilson [122] and O'Carroll [81] and the fact that the variety of semilattices $\mathcal{S}$ is "local", that for every group variety $\mathcal{V}$ we have $\mathcal{SV} = \mathcal{S} * \mathcal{V}$. From Petrich [89] (Theorem 12.9.11) we also have:

Theorem 6.3 *Let $\mathcal{V}$ be any variety of groups. Then*

$$\hat{\mathcal{V}} = \mathcal{S} * \mathcal{V} = \mathcal{SV}.$$

The free semigroups in the variety $\hat{\mathcal{V}}$ have been described by Margolis and Meakin [54]. Let $\mathcal{V}$ be a variety of groups, X a nonempty set and $F_X(\mathcal{V})$ the relatively free X-generated group in $\mathcal{V}$. Let $\Gamma_X(\mathcal{V})$ denote the Cayley graph of $F_X(\mathcal{V})$ relative to the set of generators X. Define

$$M_X(\mathcal{V}) = \{(\Gamma, g)|\Gamma \text{ is a finite connected subgraph of } \Gamma_X(\mathcal{V}) \text{ containing 1 and } g \text{ as vertices}, \Gamma \neq \{1\}\}$$

with multiplication

$$(\Gamma_1, g_1)(\Gamma_2, g_2) = (\Gamma_1 \cup g_1 \cdot \Gamma_2, g_1 g_2).$$

(Here $g_1 \cdot \Gamma_2$ denotes the natural action (left translation) of g_1 on Γ_2.)

The inverse monoid $M_X(\mathcal{V})^1 = M_X(\mathcal{V}) \cup \{1\}$ is just the inverse monoid $M(X : R)$ constructed in Section 2, corresponding to the natural presentation $gp < X : R >$ of $F_X(\mathcal{V})$.

From Margolis and Meakin ([54], Corollary 2.10) we have the following description of the relatively free semigroups in $\hat{\mathcal{V}}$

Theorem 6.4 *If $\mathcal{V}$ is any variety of groups and X is any non-empty set then $M_X(\mathcal{V})$ is the relatively free X-generated inverse semigroup in the variety $\hat{\mathcal{V}}$.*

In particular, if $\mathcal{V} = \mathcal{G}$, the variety of all groups, then $\hat{\mathcal{G}}$ is the variety of all inverse semigroups, and $M_X(\mathcal{G})$ is the free inverse semigroup on X.

Notice that from Theorem 6.4 it follows that the word problem in $M_X(\mathcal{V})$ is decidable if and only if the group $F_X(\mathcal{V})$ has decidable word problem.

If $\mathcal{A}$ is the variety of Abelian groups and $|X| = n$, then the Cayley graph $\Gamma_X(\mathcal{A})$ is the lattice Z^n. The multiplication in $M_X(\mathcal{A})$ may be easily visualized. The word problem in $M_X(\mathcal{A})$ may be solved in the following manner. Let w be a word in $(X \cup X^{-1})^+$ and let Γ_w be the finite subgraph of the lattice $\Gamma_X(\mathcal{A})$ traversed when the word w is read as the label of a path in $\Gamma_X(\mathcal{A})$, starting at the vertex 1. Then $w_1 = w_2$ in $M_X(\mathcal{A})$ if and only if $\Gamma_{w_1} = \Gamma_{w_2}$ and the paths labelled by w_1 and w_2 in $\Gamma_X(\mathcal{A})$ have the same end (that is these words are equal in the free Abelian group). For example, if $X = \{a, b\}$, $w_1 = abaa^{-1}bb^{-1}a^{-1}b^{-1}b^2aa^{-2}b^{-1}a^2$, $w_2 = ba^{-1}ba^2a^{-1}b^{-1}abb^{-2}a^{-1}ba^2a^{-1}$ then $w_1 = w_2$ in $F_X(\mathcal{A})$ and $\Gamma_{w_1} = \Gamma_{w_2}$ is the graph shown in Figure 6.1 below.

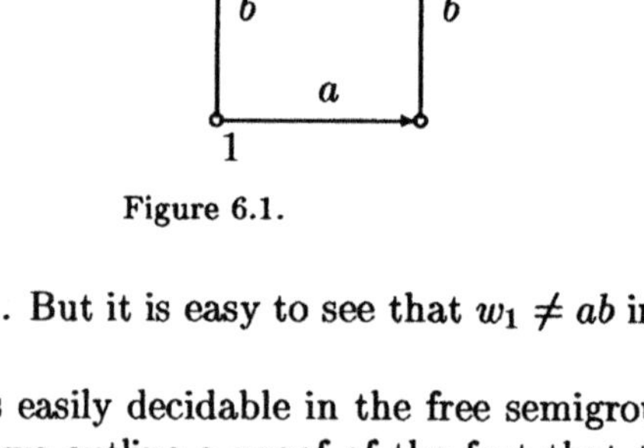

Figure 6.1.

Hence $w_1 = w_2$ in $M_X(\mathcal{A})$. But it is easy to see that $w_1 \neq ab$ in $F_X(\mathcal{A})$, because $\Gamma_{ab} \neq \Gamma_{w_1}$.

Thus the word problem is easily decidable in the free semigroups of the variety $\hat{\mathcal{A}}$. In this section, however, we outline a proof of the fact that the word problem is undecidable in the variety $\hat{\mathcal{A}}$; that is we can add finitely many relations to the defining relations of $M_X(\mathcal{A})$ and obtain a semigroup with undecidable word problem. Full details of the proof of this result may be found in the paper [68].

More precisely we have the following

Theorem 6.5 *If $\mathcal{B}$ is any variety of inverse semigroups containing the variety $\hat{A}$ of inverse semigroups possessing an E-unitary cover over an Abelian group, then $\mathcal{B}$ has undecidable word problem.*

The proof of this theorem uses the Schützenberger automata introduced in Section 2 associated with an inverse semigroup presentation and also an interpretation of a Minsky algorithm.

It is well-known that the variety $\hat{A}$ contains all Brandt semigroups B_n and the bicyclic semigroup $B = Inv < a : aa^{-1} = 1 >$. The inverse semigroup $Inv < a, b : ab = ba >$ is in this variety but the inverse semigroup $Inv < a, b, c : ab = ba,\ ac = ca,\ bc = cb >$ is not because the Schützenberger automaton of the word $ab^{-1}ca^{-1}bc^{-1}$ is linear, so this word is not an idempotent in this semigroup, but it is a commutator and so it must be an idempotent in every semigroup of $\hat{A}$ (see theorem 6.2).

6.7.2. *Minsky Algorithms and the Word Problem in Inverse Semigroups*

Let us take a Minsky algorithm M which calculates the characteristic function $f(n)$ of a recursively enumerable but non-recursive set X (i.e. $f(x) = 0$ if $x \in X$ and is not defined if $x \notin X$). We assume that M has $N+1$ commands, $0, 1, \dots, N$ (where 0 is the Stop command and 1 is the start command).

Recall that by the definition of an interpretation of a Minsky algorithm, to interpret a Minsky algorithm in a semigroup means that one must do the following:

- Associate with each configuration (m, i, n) of a Minsky algorithm, a canonical word $w(m, i, n)$;
- Associate with each command κ of the Minsky algorithm a set of defining relations R_κ.

These canonical words and sets of relations must have the following property:

$$(m, i, n) \equiv (m', i', n') \text{ iff } w(m, i, n) = w(m', i', n) \text{ modulo the union of all } R_\kappa.$$

Notice that the choice of the relations R_κ is in some sense determined by the choice of canonical words because these relations must simulate the execution of the commands of the Minsky algorithm.

The canonical word must encode the information contained in the configuration. The simplest words which encode the triple (m, i, n) are Aa^mqb^nB and qa^mAb^nB (see Section 6.4)

In the case of inverse semigroups, these choices of canonical words do not suffice. Consider, for example the first choice. The Minsky algorithm M may certainly have equivalent configurations (m, i, n) and (m', i', n') with $i = i', n = n'$ but $m \neq m'$. In such a case the Schützenberger automaton of $w(m, i, n)$ must accept $w(m', i, n)$ and vice versa. For simplicity let $m = 2, m' = 4, n = 1$ (the general case is similar).

Thus this automaton must have paths as shown in Figure 6.2.

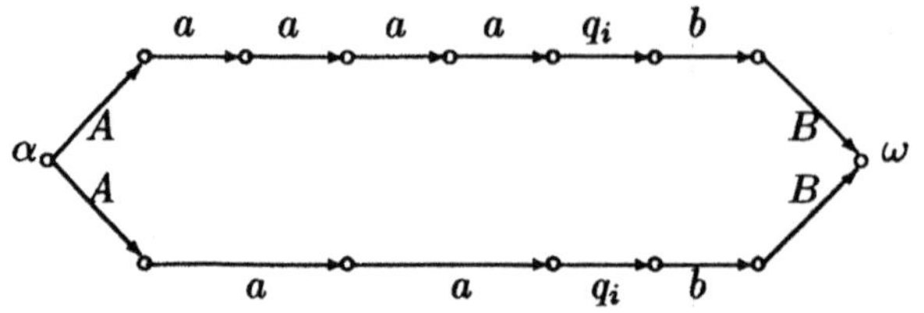

Figure 6.2.

The requirement that the Schützenberger automaton of $w(m,i,n)$ be an inverse automaton (i.e. it must be both deterministic and injective), and the fact that $m \neq m'$ would then give rise to a subgraph of the Schützenberger automaton of the form shown in Figure 6.3.

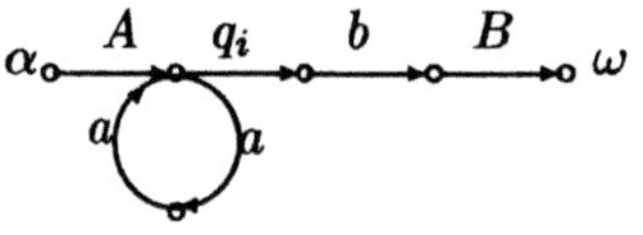

Figure 6.3.

It follows that the Schützenberger automaton of $w(m,i,n)$ would accept words of the form $Aa^{2r}q_ib^nB$ for all r. This would force $(2r,i,n) \equiv (2,i,n)$ for every r.

In general it is easy to see that if $(m,i,n) \equiv (m+t,i,n)$ for some $m,n,t > 0$ then $(m+tr,i,n) \equiv (m,i,n)$ for every r. From this, it is easy to deduce that our Minsky algorithm cannot compute the characteristic function of a non-recursive set X. Indeed, let Y be the set of all numbers m such that $(m,1,0) \equiv (1,0,0)$. We proved that with every two numbers $m, m+t$ the set Y contains all numbers of the form $m+rt$, $r = 0,1,\ldots$. Let us take $m \in Y$ and $t > 0$ such that $m+t \in Y$ and t is the smallest difference between elements of Y. Then it is clear that all elements from Y which are greater than m belong to the arithmetic progression $m, m+t, m+2t\ldots$. Hence Y contains only finitely many numbers outside this progression, so it is recursive. This contradicts the fact that by Minsky's theorem, 2^k belongs to Y if and only if k belongs to X. It is only slightly more difficult to prove that our algorithm cannot compute any non-recursive function: we leave this to the reader as an exercise. Difficulties of this kind lead to the consideration of longer and more symmetric canonical words, and hence to a longer list of defining relations than has been used for similar problems in varieties of semigroups and groups.

In [68] the following canonical word corresponding to the configuration (m,i,n) was chosen:

$$w(m,i,n) = A(i)a(i)^mB(i)b(i)^nQ(i)\bar{b}(i)^n\bar{B}(i)\bar{a}(i)^m\bar{A}(i).$$

The task then is to construct defining relations corresponding to each type of command of the Minsky algorithm M so that the relations simulate the execution of the algorithm. For example, in [68] the following set of defining relations corresponds to a command number i of the type

(17) [take a coin from the first glass $\mid j \mid k$].

(This means that if the first glass is not empty, then take a coin from it and pass to the command number j, and if the first glass is empty then pass to the command number k.)

1. $Q(i) = f(i)P(i)\bar{f}(i)$,
2. $b(i)f(i) = f(i)b_1(i), \quad \bar{f}(i)\bar{b}(i) = \bar{b}_1(i)\bar{f}(i)$,
3. $a(i)B(i)f(i) = a(i)e(i)B_1(i), \quad \bar{f}(i)\bar{B}(i)\bar{a}(i) = \bar{B}_1(i)\bar{e}(i)\bar{a}(i)$,
4. $a(i)e(i) = e(i)a_1(i), \quad \bar{e}(i)\bar{a}(i) = \bar{a}_1(i)\bar{e}(i)$,
5. $A(i)e(i) = A(j)e_{3,1}(i), \quad \bar{e}(i)\bar{A}(i) = \bar{e}_{3,1}(i)\bar{A}(j)$,
6. $e_{3,1}(i)a_1(i) = a(j)e_{3,1}(i), \quad \bar{a}_1(i)\bar{e}_{3,1}(i) = \bar{e}_{3,1}(i)\bar{a}(j)$,
7. $e_{3,1}(i)a_1(i)B_1(i) = B(j)f_{3,1}(i), \quad \bar{B}_1(i)\bar{a}_1(i)\bar{e}_{3,1}(i) = \bar{f}_{3,1}(i)\bar{B}(j)$,
8. $f_{3,1}(i)b_1(i) = b(j)f_{3,1}(i), \quad \bar{b}_1(i)\bar{f}_{3,1}(i) = \bar{f}_{3,1}(i)\bar{b}(j)$,
9. $f_{3,1}(i)P(i)\bar{f}_{3,1}(i) = f_{3,1}(i)f_{3,1}^{-1}(i)Q(j)$
10. $A(i)B(i)f(i) = A(k)B(k)f_{3,2}(i), \quad \bar{f}(i)\bar{B}(i)\bar{A}(i) = \bar{f}_{3,2}(i)\bar{B}(k)\bar{A}(k)$,
11. $f_{3,2}(i)b_1(i) = b(k)f_{3,2}(i), \quad \bar{b}_1(i)\bar{f}_{3,2}(i) = \bar{f}_{3,2}(i)\bar{b}(k)$,
12. $f_{3,2}(i)P(i)\bar{f}_{3,2}(i) = f_{3,2}(i)f_{3,2}^{-1}(i)Q(k)$

Similar relations were introduced to simulate the execution of the other types of commands of the Minsky algorithm. The Stop command is simulated by the single defining relation $w(1,0,0) = 0$. This leads to a long list of defining relations and to an inverse semigroup $S(M)$ generated by the set $G(M)$ of all symbols involving in the defining relations and subject to the union of all of these relations.

Now let $\mathcal{B}$ be any variety of inverse semigroups containing $\hat{A}$ and let $S_{\mathcal{B}}(M)$ be the inverse semigroup in $\mathcal{B}$ obtained by adding to the defining relations of $S(M)$ the identities of $\mathcal{B}$ (i.e. $S_{\mathcal{B}}(M)$ is defined in $\mathcal{B}$ by the same relations which define $S(M)$ in the class of all inverse semigroups).

We want to show that $w(2^m, 1, 0)$ is equal to 0 ($=w(1,0,0)$) in $S_{\mathcal{B}}(M)$ if and only if $m \in X$. This is accomplished by proving the following two lemmas.

Lemma 6.3 *If we can pass from configuration (m,i,n) to configuration (m',j,n') using command i of M then $w(m,i,n) \leq w(m',j,n')$ in $S(M)$ where $\leq$ is the natural order on the inverse semigroup $S(M)$.*

Lemma 6.4 *If $m \notin X$ then $w(2^m,1,0)$ is not equal to 0 in $S_{\mathcal{B}}(M)$.*

Indeed, suppose that we have proved these two lemmas. If $m \in X$ then our algorithm M takes configuration $(2^m, 1, 0)$ to configuration $(1, 0, 0)$ in a finite number of steps. By Lemma 6.3 we can deduce that $w(2^m, 1, 0) \leq w(1, 0, 0) = 0$. But zero is the smallest element in the natural order of any inverse semigroup. Therefore $w(2^m, 1, 0) = 0$ in $S(M)$. Since $S_{\mathcal{B}}(M)$ is a homomorphic image of $S(M)$ we have that $w(2^m, 1, 0) = 0$ in $S_{\mathcal{B}}(M)$. On the other hand if $m \notin X$ then by Lemma 6.4 the element $w(2^m, 1, 0)$ is not equal to 0 in $S_{\mathcal{B}}(M)$. This will complete the proof of Theorem 6.5.

In order to prove lemma 6.3 it suffices to show that if we can proceed from configuration (m, i, n) to configuration (m', j, n') using command i of M then $w(m', j, n')$ must be accepted by the Schützenberger automaton $A(w(m, i, n))$ corresponding to this presentation. For example, suppose that command i is of the type 17. Start with the linear automaton of the word

$$w(m, i, n) = A(i)a(i)^m B(i)b(i)^n Q(i)\bar{b}(i)^n \bar{B}(i)\bar{a}(i)^m \bar{A}(i).$$

If $m \neq 0$ then application of command i to $w(m, i, n)$ must produce $w(m-1, j, n)$ while if $m = 0$, it must produce $w(m, k, n)$.

Notice that the relation 1 may be used to apply an expansion to the automaton (add a new path labelled by $f(i)P(i)\bar{f}(i)$ to the automaton starting at the initial vertex of the edge labelled by $Q(i)$ and ending at the terminal vertex of this edge). This then introduces a path labelled by $b(i)f(i)$ and a path labelled by $\bar{f}(i)\bar{b}(i)$, to which we can apply expansions by use of the relations 2 and so on. Continue in this manner: one easily checks that if one applies only the relations 1-12 corresponding to command i, then the automaton obtained from the linear automaton of $w(m, i, n)$ and closed with respect to application of expansions corresponding to these relations is in fact finite.

Sketches of these automata in the cases $m = 3$, $n = 3$ and $m = 0$, $n = 3$ are provided in figures 6.4 and 6.5.

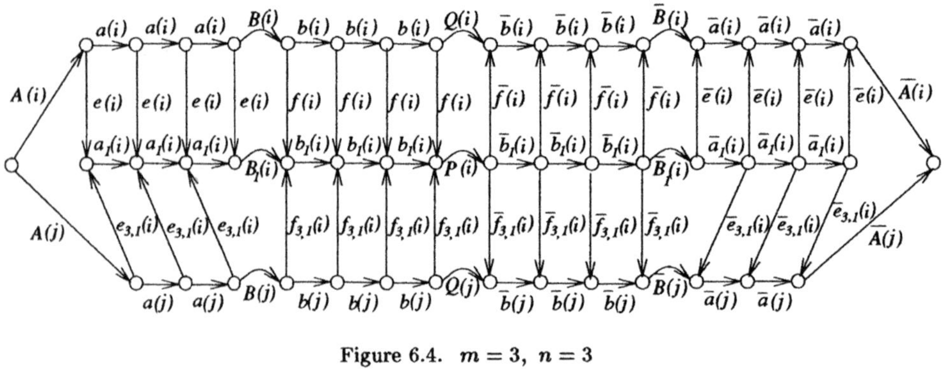

Figure 6.4. $m = 3$, $n = 3$

Notice that in the first case there is a path in this automaton from the initial state to the terminal state labelled by $w(2, j, 3)$ and in the second case there is such a path labelled by $w(0, k, 3)$ (the lower boundaries of the graphs). Hence in each

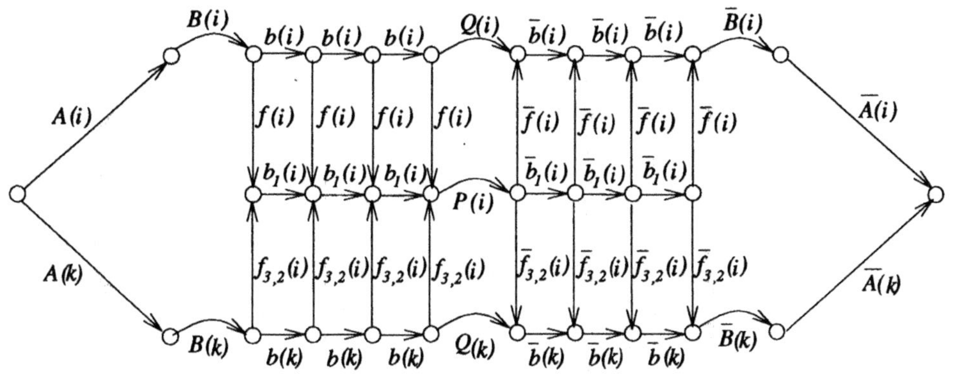

Figure 6.5. $m = 0,\ n = 3$

case $A(w(m,i,n))$ accepts the required word. Similar analysis applies in the other cases with corresponding sets of defining relations.

The basic idea involved in proving lemma 6.4 is to construct an inverse automaton $I(M)$ associated with the Minsky algorithm M and a configuration $(2^{m_0}, 1, 0)$ where $m_0 \notin X$. This inverse automaton is such that the transition semigroup $T(M)$ of $I(M)$ is a homomorphic image of the semigroup $S(M)$ constructed above and is in the variety $\hat{\mathcal{A}}$. In addition one shows that the word $w(2^{m_0}, 1, 0)$ labels a path which is not a loop in $I(M)$, and so it is not equal to 0 in $T(M)$. It follows immediately of course that $w(2^{m_0}, 1, 0) \neq 0$ in $S_{\mathcal{B}}(M)$ for every variety $\mathcal{B}$ containing $\hat{\mathcal{A}}$ and hence that Lemma 6.4 holds.

The idea is that the underlying graph of $I(M)$ should contain the Schützenberger automaton of the canonical word $w(2^{m_0}, 1, 0)$ and have the following properties.

a) The automaton $I(M)$ is closed under application of all of the defining relations of the semigroup $S(M)$. That is, if in $I(M)$ there are two vertices p and q and a path starting at p and ending at q labelled by one side of one of the relations then there is also a path in $I(M)$ from p to q labelled by the other side of that relation. From this it follows that the transition semigroup $T(M)$ of $I(M)$ is a homomorphic image of $S(M)$.

b) The transition semigroup $T(M)$ of $I(M)$ is in the variety $\hat{\mathcal{A}}$.

We refer to the paper [68] for full details.

We note that the semigroup $S(M)$ itself is not in the variety $\hat{\mathcal{A}}$. For example, the word $A(i)a(i)A(i)^{-1}a(i)^{-1}$ is not an idempotent in $S(M)$ since its Schützenberger automaton is linear (no relations apply), but the word is 1 in the free Abelian group. Hence by Theorem 6.2, $S(M) \notin \hat{\mathcal{A}}$. We can add the relations like

$$A(i)a(i)A(i)^{-1}a(i)^{-1} = (A(i)a(i)A(i)^{-1}a(i)^{-1})^2$$

to $S(M)$, but still there will be infinitely many (more sophisticated) commutators which will not be idempotents in the resulting semigroup. Thus we are unable to push $S(M)$ into $\hat{\mathcal{A}}$ by adding only finitely many relations. This raises the following question.

Problem 6.1 Is there a finitely presented inverse semigroup with undecidable word problem, which belongs to $\hat{A}$?

Finally we raise the question as to whether or not the variety $\hat{A}$ is a "bound" between decidability and undecidability of the word problem in the class of varieties of inverse semigroups.

Problem 6.2 Is it true that every variety of inverse semigroups that is strictly contained in $\hat{A}$ has decidable word problem? In particular, does the variety of inverse semigroups generated by the bicyclic semigroup have decidable word problem?

In [66] we conjectured that the variety generated by the bicyclic semigroup has undecidable word problem. At this stage we are unable to either prove or disprove this conjecture. It is clear that this variety is properly contained in $\hat{A}$.

References

1. S. I. Adian, *Defining relations and algorithmic problems for groups and semigroups*, Tr. Mat. Inst. Steklov 85 (Russian); Am. Math. Soc. (translation) 152, 1967.
2. S. I. Adian, *Burnside problem and identities in groups*, Moscow, Nauka, 332 p. (1975), translated by John Lennox and J. Wiegold, Ergebnisse der Math and ihrer Grenzgebiete, 95, Springer-Verlag, 1979.
3. S. I. Adian, G. S. Makanin, *Studies in algorithmic questions of algebra*, Algebra, Mat. Logic, Number theory, Topology, Proc. Steklov Institute of Mathematics, Vol. 168, (1984), English translation (1986), issue 3 of 4, 207–226.
4. M. D. Atkinson, *Computational group theory*, Academic Press, 1984.
5. J. Avenhaus and K. Madlener, *The Nielsen reduction and P-complete problems in free groups*, Theor. Comp. Science, 32(1984), 61–76.
6. L. Babai, E. Luks and A. Seress, *Permutation groups in NC*, Proc. 19th ACM STOC (1987), 409–420
7. A. M. Ballantyne and D. S. Lankford, *New decision algorithms for finitely presented commutative semigroups*, Computers and Math. with Appl., 7 (1981), 159–165.
8. D. B. Bean, A. Ehrenfeucht, and G. McNulty, *Avoidable patterns in strings of symbols*, Pacific J. Math., 85 (1979), no. 2, 261–294.
9. C. Bennett, *Time/Space tradeoffs for reversible computations*, SIAM J. of Computing, 18 (1989), 766–776.
10. J. Berstel, *Transductions and Context-Free Languages*, Teubner Studienbucher, 1979.
11. J-C. Birget, S. Margolis, J. Meakin, *The word problem for inverse monoids presented by one idempotent relator*, Theor. Comp. Science, 1993 (to appear).
12. J-C. Birget, S. W. Margolis, J. C. Meakin and P. Weil, *PSPACE - complete problems for subgroups of free groups and inverse automata*, preprint, 1993.
13. A. P. Biryukov, *Some algorithmic problems for finitely defined commutative semigroups*, Siberian Math. J., 8 (1967), 384–391.
14. L. A. Bokut', *The algorithmic problems and the embedding theorems: some open questions for rings, groups and semigroups*, Izv. Vyssh. Uchebn. Zaved. Mat., (1982), No. 11, 3–11.
15. L. A. Bokut', G. P. Kukin, *Undecidable algorithmic problems for semigroups, groups and rings*, Algebra, Topology, Geometry, Vol. 25. Itogi Nauki i Techniki, 3–66, (1987).
16. W. W. Boone, *Certain simple unsolvable problems in group theory I-VI*, Proc. Kon. ned. akad. Wetensch. A, 57 (1954) 231–237, 492–497, 58 (1955), 252–256, 571–577, 60 (1957), 22–27, 227–232.
17. B. Buchberger and R. Loos, *Algebraic simplification*, in "Computer Algebra. Symbolic and algebraic computation." B. Buchberger, G. E. Collins, and R. Loos (ed.), Springer-Verlag, 1982, 23–65.
18. S. Burris, H. P. Sankappanavar, *A Course in Universal Algebra*, Grad. Texts in Math 78, Springer-Verlag, 1981.

19. I. M. Chiswell, D. J. Collins and J. Huebschmann, *Aspherical group presentations,* Math. Z. 261 (1981), 1–36.
20. S. Cho, D. Huynh, *Finite-automaton aperiodicity is PSPACE-complete*, Theor. Comp. Science 88 (1991), 99-116.
21. J. R. Cho and S. J. Pride, *Embedding semigroups into groups, and the asphericity of semigroups,* Int. J. Algebra and Computation, Vol. 3, No. 1, (1993), 1–15.
22. A. H. Clifford and G. B. Preston, *The algebraic theory of semigroups, v. 1, 2*, Amer. Math. Soc., Providence, 1961.
23. D. E. Cohen, *Combinatorial Group Theory: a Topological approach,* Lond. Math. Soc. Student Texts 14 (1989).
24. D. Cowan, *A class of varieties of inverse semigroups*, J. Algebra 141 (1991), 115–142.
25. M. Dehn, *Über unendliche diskontinuerliche Gruppen,* Math. Ann., 1912, Bd. 71, S. 116–144.
26. L. E. Dickson, *Finiteness of the odd perfect and primitive abundant numbers with n distinct prime factors,* Amer. J. Math. 35 (1913), 413–422.
27. S. Eilenberg, *Automata, languages and machines,* Pure and Applied Mathematics, Vol. B, Academic Press, 1976.
28. V. A. Emelichev, *Commutative semigroups with one defining relation,* Uchenye zap. Shuiskogo ped. in-ta, 6 (1958), 227–242.
29. D. B. A. Epstein , J. W. Cannon, D. F. Holt, S. V. F. Levy, M. S. Patterson and W. P. Thurston, *Word Processing in Groups,* Jones and Bartlett, Boston, 1992.
30. J. Fountain, *Abundant semigroups,* Proc. London Math. Soc. 44 (1982), 103–129.
31. H. Furstenberg, *Poincaré recurrence and number theory,* Bull. Amer. Math. Soc. 5 (1981), 211–234.
32. S. M. Gersten, *Intersections of finitely generated subgroups of free groups and resolutions of graphs,* Invent. Math. 71 (1983), 567–591.
33. L. M. Gluskin, *Elementary generalized groups,* Mat. Sbornik 41 (1957), 23–37 (in Russian).
34. M. Gromov, *Hyperbolic groups,* IHES, France, Preprint. P. 187.
35. Yu. Sh Gurevich, *The problem of equality of words for certain classes of semigroups,* Algebra i Logika, 25–35.
36. M. Hall Jr., *A topology for free groups and related groups,* Ann. of Math. 52 (1950), 127–139.
37. P. Hall, *On the finiteness of certain soluble groups,* Proc. London Math. Soc. Vol. 3, No. 9, 595–622 (1959).
38. G. Hermann, *Die Frage der enlich vielen Schritte in der Theorie der Polynomideale,* Math. Ann. 95 (1926), 736–788.
39. Peter Higgins, *Techniques of semigroup theory,* Oxford Univ. Press (1992).
40. Philip Higgins, *Categories and Groupoids,* Van Nostrand - Rheinhold, New York (1971).
41. Sergei Ivanov, *Relation modules and relation bimodules of groups, semigroups and associative algebras,* Int. J. Algebra and Computation, Vol. 1, No. 1, (1991) 89–114.
42. Sergei Ivanov, *The Burnside problem for all sufficiently large exponents,* Int. J. Algebra and Computation, to appear.
43. P. Jones, S. Margolis, J. Meakin, J. Stephen, *Free products of inverse semigroups II,* Glasgow Math J. 33 (1991) 373–387.
44. O. G. Kharlampovich, *The word problem for subvarieties of the variety* $\mathcal{N}_2\mathcal{A}$, Alg. i Logika, Vol. 26, 481–501 (1987); translation in Algebra and Logic, Vol. 26 (1987).
45. O. Kharlampovich and M. Sapir, *Algorithmic problems in varieties of groups, semigroups, associative and Lie algebras,* in preparation.
46. O. G. Kharlampovich, M. Sapir, *Non-residually finite relatively finitely presented groups in the variety* $\mathcal{N}_2\mathcal{A}$., preprint.
47. E. I. Kleiman, *On bases of identities of Brandt semigroups,* Semigroup Forum, 13, 3 (1977), 209–218.
48. D. Kozen, *Lower bounds for natural proof systems,* Proc. 18th Annual Symosium on Foundations of Computing (FOCS) (1977), 254–266.
49. G. Lallement, *Semigroups and combinatorial applications,* Wiley & Sons, New York, 1979.
50. Roger C. Lyndon and Paul E. Schupp, *Combinatorial group theory,* Springer-Verlag, 1977.
51. I. G. Lysenok, *The infinitude of Burnside groups of period* 2^k *for* $k \geq 13$, Russian Math. Surveys, 47 (1992), 2, 229–230.
52. A. I. Mal'cev, *On the homomorphisms onto finite groups,* Uchebn. Zapiski Ivanovsk. ped.instituta, 1958, V. 18, No. 5, 49–60; also in "Selected papers", Vol. 1, Algebra, (1976),

450–461.
53. A.I.Mal'cev, *Algorithms and recursive functions*, Nauka, Moscow (1965), 391 pp.
54. S. Margolis and J. Meakin, *E-unitary inverse monoids and the Cayley graph of a group presentation,* J. Pure Appl. Algebra 58 (1989), 45–76.
55. S. W. Margolis, J. C. Meakin, *Inverse monoids, trees and context-free languages*, Transactions of the AMS, V. 335, No. 1, 1993, 259–276.
56. S. W. Margolis, J. C. Meakin, *Free inverse monoids and graph immersions* Int. J. Algebra and Computation, Vol. 3, No. 1, (1993) 79–99.
57. S. Margolis, J. Meakin, J. Stephen, *Some decision problems for inverse monoid presentations,* in "Semigroups and their applications", D. Reidel, 1987, 99–110.
58. S. Margolis, J. Meakin, J. Stephen, *Free objects in certain varieties of inverse semigroups,* Canad. J. Math. XLII, 6 (1990), 1084–1097.
59. S. MacLane, *Categories for the working mathematician,* Grad. Texts Math. No. 5, Springer, New York (1971).
60. S. Margolis, M. V. Sapir, *Quasi-identities of finite semigroups and symbolic dynamics,* to appear.
61. A. Markov, *On the impossibility of certain algorithms in the theory of associative systems,* Dokl. Acad. Sci. URSS 55 (1947), 583-586.
62. Yu. V. Matiyasevich, *Studies in certain algorithmic problems in algebra and number theory,* Algebra, Mat. Logic, Number theory, Topology, Proc. Steklov Institute of Mathematics, Vol. 168, (1984), English translation (1986), issue 3 of 4, 227–252.
63. J. McCammond, *The solution of the word problem for the relatively free semigroups satisfying $x^a = x^{a+b}$ with $a \geq 6$*, Int. J. Algebra and Computation, 1 (1991), no. 1, 1–32.
64. J. C. C. McKinsey, *The decision problem for some classes of sentences without quantifiers,* J. Symbolic Logic, 8 (1943), 61–76.
65. G. McNulty, C. Shallon, *Inherently non-finitely based finite algebras*, Lecture Notes Math. 1004 (1982), 206–231.
66. J. Meakin, *An invitation to inverse semigroups,* Proc. Conf. on Ordered Structures and Algebra of Computer Languages, Hong Kong, June 1991, K. P. Shum and P. C. Yuen (Eds.), World Scientific 1993, 91–115.
67. J. Meakin, M. V. Sapir, *Congruences on free monoids and submonoids of polycyclic monoids,* J. Austr. Math. Soc. (Series A), 54 (1993), 236–253.
68. J. Meakin, M. V. Sapir, *On the word problem in varieties of inverse semigroups*, to appear.
69. I. L. Mel'nichuk, *Solvability of problems of equality and divisibility in certain finitely based varieties of semigroups,* in "Properties of semigroups," Leningrad, 1984, 91–103.
70. I. L. Mel'nichuk, *Undecidability of problems of equality and divisibility in certain varieties of semigroups,* Algebra i Logika, 23 (1984), no. 4, 430–438.
71. I. L. Mel'nichuk, M. V. Sapir and O. G. Kharlampovich, *The word problem in varieties of semigroups, rings and Lie algebras*, Sibirsk. Mat. Zh. 1986, 27, 6, 144–156 (translated from Russian into English by Plenum Corp. in Sib. Math. J.).
72. C. F. Miller III, *Decision problems for groups—survey and reflections*, in "Algorithms and classification in combinatorial group theory, G. Baumslag and C. F. Miller III (eds.)," Springer-Verlag, 1992, 1–59.
73. M. L. Minsky, *Recursive unsolvability of Post's problem of "tag" and other topics in the theory of Turing Machnes* , Ann. of Math. (2) , Vol. 74 (1961), 437–455.
74. M.L.Minsky, *Computation: finite and infinite machines*, Prentice-Hall, Englewood Cliffs, N. J., 1967.
75. M. Morse, G. Hedlund, *Unending chess, symbolic dynamics and a problem in semigroups,* Duke Math. J., **11**, (1944), 1–7.
76. W. D. Munn, *Free inverse semigroups,* Proc. Lond. Math. Soc. 30, 1974, 385–404.
77. V. L. Murskii, *Some examples of varieties of semigroups*, Mat. Zametki, 3 (1968), 663–670.
78. H. Neumann, *Varieties of groups*, Ergebnisse der mathematik und ihrer grengebiete, Band 37, New Series, Vol. 37 (1967), 192 pp.
79. P. S. Novikov, *On the algorithmical undecidability of the problem of identity,* Doklady AN SSSR, 4, 85 (1952), 709–719.
80. P. S. Novikov, S. I. Adian, *On infinite periodic groups, I, II, III,* Izv. Akad. Nauk SSSR, Ser. Mat., 1, 32 (1968), 212–244, 2, 32 (1968), 251–524, 3, 32 (1968), 709–731.
81. L.O'Carroll, *Embedding theorems for proper inverse semigroups*, J. Algebra 42 (1976), 26–40.

82. A. Yu.'Ol'shanskii, *Conditional identities of finite groups*, Sib. Matem. J., 15 (1974), 6, 1409–1413 (Russian, English translation see in Sib. Matem. J., 1975, 1000–1003).
83. A. Yu. Ol'shanski, *Geometry of defining relations in groups*, Moscow, "Nauka", 1989 , 448 pp. (Russian).
84. A. Yu. Ol'shanskii, *Hyperbolicity of groups with subquadratic isoperimetric inequality*, Int. J. Algebra and Computation, Vol. 1, No. 3, (1991), 281–290.
85. F. Pastijn, *A representation of a semigroup by a semigroup of matrices over a group with zero*, Semigroup Forum, 10 (1975), 238–249.
86. F. Pastijn, *Inverse semigroup varieties generated by E-unitary inverse semigroups,* Semigroup Forum, 24 (1982), 87–88.
87. Peter Perkins, *Decision problems for equational theories of semigroups*, Ph.D. Thesis, Univ. of California, Berkeley, 1966.
88. P. Perkins, *Basic questions for general algebras,* Algebra Universalis, 19 (1984), 16–23.
89. M. Petrich, *Inverse semigroups*, Wiley, New York, 1984.
90. M.Petrich, N.Reilly, *E-unitary covers and varieties of inverse semigroups,* Acta Sci. Math. Szeged, 46 (1983), 59-72.
91. J-E. Pin, *Varieties of formal languages,* Plenum, 1986.
92. E. L. Post, *Recursive unsolvability of a problem of Thue,* J. Symbolic Logic, 1, 12 (1947), 1–11.
93. G. B. Preston, *Inverse semigroups*, J. London Math. Soc. 29 (1954), 396–403.
94. S. Pride, *Low-Dimensional homotopy theory for monoids*, preprint, 1993.
95. N. Reilly, *Free combinatorial strict inverse semigroups*, J. London Math. Soc. 39 (1989), 102–120.
96. V. N. Remeslennikov, V. A. Roman'kov, *Model-theoretical and algorithmic questions in group theory,* Algebra, Topology, Geometry, Vol. 21, Akad. Nauk SSSR, Vsesoyuzn. Itogi Nauki i Techniki 3–79.
97. J. H. Remmers, *On the geometry of semigroup presentations,* Adv. Math. 36 (1980), 283–296.
98. N. S. Romanovskii, *The embedding problem for abelian-by-nilpotent groups*, Siberian Math. Journal, Vol. 21, No. 2, 170–174 (1980).
99. B. Rosenblat, *Diophantine theories of free inverse semigroups,* Transl. Siberian Math. J. 26 (1986), 101–107 (translation in pp. 860–865).
100. R. Ruyle, Ph. D. Thesis, University of Nebraska-Lincoln, to appear, 1994.
101. M. V. Sapir, *On quasivarieties generated by finite semigroups*, Semigroup Forum, 20 (1980), 1, 73–88.
102. M. V. Sapir, *Algorithmic problems in semigroup varieties*, Izv. Vyssh. Uchebn. Zaved. Mat. Matematica, 1985, 12; 71–74 (Translated from Russian into English by Allerton press).
103. M. V. Sapir, *Problems of Burnside type and the finite basis property in varieties of semigroups* , Izv. Akad. Nauk SSSR, ser. Mat., 1987, 51, 2, 319–340, translation in Math USSR-Izv, Vol. 30 (1988), No. 2, 295–314.
104. M. V. Sapir, *Inherently nonfinitely based finite semigroups*, Mat. Sb., 1987, 133, 2, 154–166 (translated from Russian into English by American MS).
105. M. V. Sapir, *Sur la propriete de base finie pour les pseudovarietes de semigroupes finis*, C. R. Acad. Sci, Paris, 1988, 306, Serie I, 705–708 (in English and French).
106. M. V. Sapir, *Algorithmic problems in varieties of semigroups*, Algebra i Logika, (1988), 27,4, 440–463; translation in Algebra and Logic, 27 (1988), No. 4, 275–292.
107. M. V. Sapir, *On the word problem in periodic group varieties*, Int. J. Algebra and Compution, Vol. 1, No. 1, 115–126 (1991).
108. M. V. Sapir, *Finite semigroups in varieties*, Izv. Akad. Nauk SSSR, ser. Mat., 1991, 55, 5, 245–260.
109. M. V. Sapir, *Weak word problem for finite semigroups*, in "Monoids and semigroups with applications. Proceedings of the Berkeley Workshop in Monoids. Berkeley, 1989. J. Rhodes (ed.)", World Scientific, 1991, 206–219.
110. M. V. Sapir, *Identities of finite inverse semigroups,* Int. J. Algebra and Computation, 3 (1993), no. 1, 115–124.
111. H. E. Scheiblich, *Free inverse semigroups*, Proc. Amer. Math. Soc. 38 (1973), 1–7.
112. B. M. Schein, *Representations of generalized groups,* IZV., Vyss. Uchebn. Zav. Matem., issue 3 (1962), 164–176 (Russian).
113. B. M. Schein, *Free inverse semigroups are not finitely presentable*, Acta. Math. Acad. Sci.

Hung. 26 (1975), 41–52.

114. J. P. Serre, *Trees*, Springer-Verlag, 1980.

115. P. V. Silva, *Rational languages and inverse monoid presentations*, Int. J. Algebra and Computation, 2 (1992), 187–208.

116. L. N. Shevrin, M. V. Volkov, *Identities of semigroups*, Izv. vuzov. Matematika, 11 (1985), 3–47.

117. J. Stallings, *Topology of finite graphs*, Invent. Math. 71 (1983), 551–565.

118. J. B. Stephen, *Application of automata theory to presentations of monoids and inverse monoids*, Ph. D. Thesis, Univ. of Nebraska, 1987.

119. J. B. Stephen, *Presentations of inverse monoids*, J. Pure Appl. Algebra 63 (1990), 81–112.

120. M. A. Taiclin, *Algorithmic problems for commutative semigroups*, Soviet Math. Dokl. 9 (1968), 201–204.

121. H. Tietze, *Uber die topologischen Invarianten mehrdimensionaler Mannigfaltigkeiten* Monatsh. Math. Phys., 19 (1908), 1–118.

122. B. Tilson, *Categories as algebras: an essential ingredient in the theory of finite monoids*, J. Pure Appl. Algebra, 48 (1987), 83–198.

123. Axel Thue, *Über die gegenseitigen Lage gleicher Terlegewisser Zeichenreilren*, Norske Vid. Selsk. Skr., I Mat. Nat. Kl., Christiania, 7 (1906), 1–22.

124. Axel Thue, *Probleme über Veränderlungen von Zeichenreihen nach gegebenen Regeln, Kristiania Videnskapsselskapets Skr.* I: Mat.-Naturvid. Kl. 1914, no. 10; reprinted in his Selected mathematical papers, Universitetsforlaget, Oslo, 493–524 (1977).

125. V. V. Wagner, *Generalized groups*, Dokl. Akad. Nauk. SSSR 84 (1952), 1119–1122 (in Russian).

126. A. I. Zimin, *Blocking sets of terms*, Matem. Sbornik, 119, 3 (1982), 363–375.

GEOMETRIC METHODS IN COMBINATORIAL SEMIGROUP THEORY

STEPHEN J. PRIDE
Department of Mathematics
University of Glasgow
Glasgow G12 8QW
Scotland

The aim of this document is to discuss various geometric ideas, which have been fruitful in the study of groups given by presentations, and which should prove equally beneficial in the study of monoids given by presentations. Three "research programs" are highlighted: the study of isoperimetric functions for monoids (§2); the study of the monoid analogue of the second homotopy module for group presentations (§3.3); the relationship between the monoid and group defined by a given monoid presentation (§4).

As far as I am aware, the theorem in §2 on isoperimetric functions for monoids is new, though the analogous result for groups is known. A fair amount of §3 is an exposition of new material which is given in more detail in [28]. The remainder of the paper is discursive.

I thank Calum Kilgour, Rick Thomas and Xiaofeng Wang for some useful comments. I also thank John Fountain and Victoria Gould for their hospitality in York.

1. Elementary concepts

1.1. Preliminaries

Let

$$\mathcal{P} = [\mathbf{x}; \mathbf{r}] \tag{1}$$

be a monoid presentation. Here $\mathbf{x}$ is a set (the generating symbols), and each $R \in \mathbf{r}$ is an ordered pair (R^{+1}, R^{-1}) where R^{+1}, R^{-1} are words on $\mathbf{x}$. (We remark that we allow the empty word.) We assume that for all $R_1, R_2 \in \mathbf{r}$, (R_1^{+1}, R_1^{-1}) is not equal to the ordered pair (R_2^{-1}, R_2^{+1}). We will write $R^{+1} = R^{-1}$ instead of (R^{+1}, R^{-1}).

It will be convenient (although not always necessary) to assume throughout that $\mathcal{P}$ is *finite*, that is, $\mathbf{x}$ and $\mathbf{r}$ are finite sets.

Let F be the free monoid on $\mathbf{x}$. If we have an element W of F which is of the form $UR^{\epsilon}V$ $(U, V \in F, R \in \mathbf{r}, \epsilon = \pm 1)$ then we may perform an *elementary transformation* replacing R^{ϵ} by $R^{-\epsilon}$ giving the word $W' = UR^{-\epsilon}V$. This can be represented geometrically by means of an *atomic (monoid) picture* $\mathbb{A} = (U, R, \epsilon, V)$.

J. Fountain (ed.), Semigroups, Formal Languages and Groups, 215–232.

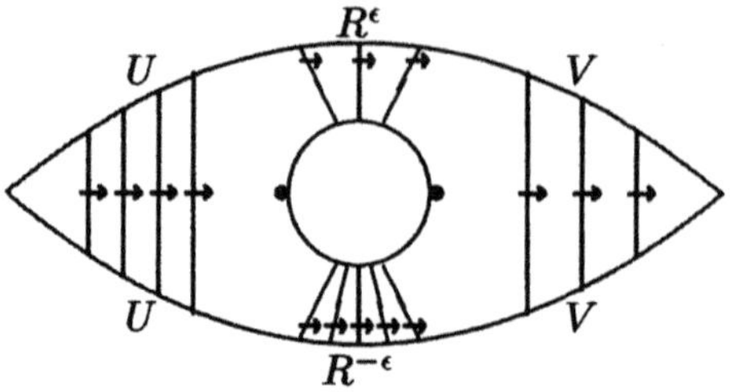

(2)

Here the arcs are transversely orientated and labelled by generating symbols. The word we read off by travelling along the top of the picture from left to right is W and the word we read off by travelling along the bottom is W'.

The atomic picture $\mathbb{A}$ is said to be *positive* if $\epsilon = 1$.

Two words are said to be *equivalent* (relative to $\mathcal{P}$) if one can be obtained from the other by a finite sequence of elementary transformations. The *monoid* $S(= S(\mathcal{P}))$ *defined by* $\mathcal{P}$ consists of the equivalence classes $[W]$ $(W \in F)$, with multiplication given by $[W_1][W_2] = [W_1W_2]$ $(W_1, W_2 \in F)$. Sometimes for emphasis we will write $[W]_\mathcal{P}$ or $[W]_S$ instead of $[W]$.

1.2. Graphs and 2-complexes

We will use the term "graph" in the sense of Serre [31]. Thus a graph consists of two disjoint sets $\mathbf{v}$ (vertices), $\mathbf{e}$ (edges), and three functions

$$\iota\colon \mathbf{e} \to \mathbf{v}, \qquad \tau\colon \mathbf{e} \to \mathbf{v}, \qquad {}^{-1}\colon \mathbf{e} \to \mathbf{e}$$

satisfying $\iota(e^{-1}) = \tau(e)$, $(e^{-1})^{-1} = e$, $e^{-1} \neq e$ for all $e \in \mathbf{e}$. An *orientation* $\mathbf{e}^+$ of $\mathbf{e}$ is a choice of one element from each edge pair $\{e, e^{-1}\}$. If α is a path in the graph starting at the vertex u and ending at the vertex v, then we define $\iota(\alpha)$, $\tau(\alpha)$ to be u, v respectively.

A *(combinatorial) 2-complex* $\mathcal{K}$ is a pair $\langle \Lambda; \mathbf{s} \rangle$ where Λ is a graph and $\mathbf{s}$ is a set of closed paths in Λ (called *defining paths*). We define elementary operations on paths as follows.

(a) Insertion/deletion of an inverse pair ee^{-1} (e an edge of Λ).

(b) Insertion/deletion of an element of $\mathbf{s} \cup \mathbf{s}^{-1}$.

Two paths are said to be *equivalent* (or *homotopic*) (relative to $\mathcal{K}$) if one can be obtained from the other by a finite number of the above operations.

Let $\mathbf{e}^+$ be an orientation of $\mathbf{e}$. We then obtain the *chain complex*

$$C(\mathcal{K})\colon 0 \to C_2(\mathcal{K}) \xrightarrow{\partial_2} C_1(\mathcal{K}) \xrightarrow{\partial_1} C_0(\mathcal{K}) \to 0,$$

where $C_0(\mathcal{K})$ is the free abelian group on $\mathbf{v}$, $C_1(\mathcal{K})$ is the free abelian group on $\mathbf{e}^+$, and $C_2(\mathcal{K})$ is the free abelian group on $\mathbf{s}$. The boundary maps are defined as follows. Let $\sigma \in \mathbf{s}$ and write $\sigma = e_1^{\epsilon_1}e_2^{\epsilon_2}\cdots e_n^{\epsilon_n}$ $(e_i \in \mathbf{e}^+, \epsilon_i = \pm 1$ for $i = 1, \dots, n)$. Then

$$\partial_2(\sigma) = \sum_{i=1}^{n} \epsilon_i e_i.$$

Also for $e \in \mathbf{e}^+$,

$$\partial_1(e) = \iota(e) - \tau(e).$$

The homology of $\mathcal{K}$ is the homology of this chain complex. In particular, $H_0(\mathcal{K}) \cong \mathbb{Z}^d$, where d is the number of components of $\mathcal{K}$. Also

$$H_1(\mathcal{K}) = \frac{\operatorname{Ker} \partial_1}{\operatorname{Im} \partial_2}.$$

(The isomorphism type of $H_1(\mathcal{K})$ is independent of the choice of orientation.)

1.3. The graph and 2-complex associated with a monoid presentation

Let $\mathcal{P}$ be a monoid presentation as in (1). We associate a graph $\Gamma(= \Gamma(\mathcal{P}))$ with $\mathcal{P}$ as follows. The vertex set is F, and the edge set is the collection of atomic monoid pictures. For an atomic picture $\mathbb{A} = (U, R, \epsilon, V)$ as in (2) we define

$$\iota(\mathbb{A}) = UR^{\epsilon}V, \qquad \tau(\mathbb{A}) = UR^{-\epsilon}V,$$

and we define $\mathbb{A}^{-1}$ to be the *mirror image* $(U, R, -\epsilon, V)$ of $\mathbb{A}$.

The edge set of Γ has a natural orientation, namely the collection of all ***positive*** atomic pictures.

There is a left action of F on Γ defined as follows. Let $Z \in F$. If W is a vertex of Γ then we define $Z.W$ to be ZW (product in F). If $\mathbb{A} = (U, R, \epsilon, V)$ is an edge of Γ then $Z.\mathbb{A}$ is the picture obtained from $\mathbb{A}$ by inserting a succession of arcs with total label Z to the left of $\mathbb{A}$, that is, $Z.\mathbb{A} = \{ZU, R, \epsilon, V)$.

There is a similar right action of F on Γ. The left and right actions are clearly compatible with each other, in the sense that for $\gamma \in \Gamma$, $W_1, W_2 \in F$ we have $(W_1.\gamma).W_2 = W_1.(\gamma.W_2)$.

A path in Γ will be called a (monoid) ***picture***; a closed path will be called a ***spherical*** monoid picture. We can obviously talk about ***subpictures*** (= subpaths) of pictures. The left and right actions of F on Γ extends to actions on pictures: if $\mathbb{P} = \mathbb{A}_1 \mathbb{A}_2 \ldots \mathbb{A}_n$ is a picture ($\mathbb{A}_i$ atomic pictures for $i = 1, \ldots, n$) and if $W, V \in F$ then

$$W.\mathbb{P}.V = (W.\mathbb{A}_1.V)(W.\mathbb{A}_2.V) \cdots (W.\mathbb{A}_n.V).$$

Example. $\mathcal{P} = [a, b; \quad a^5 = 1, \quad ab = ba^2]$

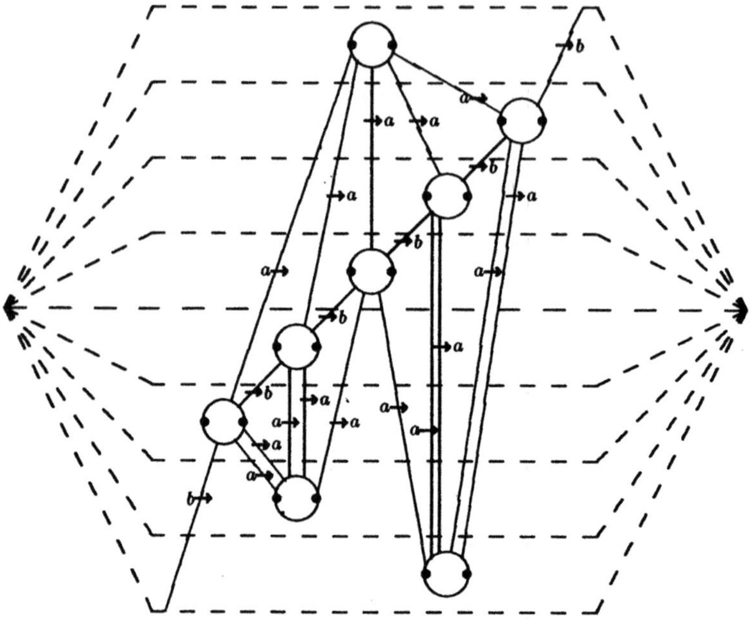

Remarks.

(i) The dual notion of a picture is a *semigroup diagram* as discussed in [29, 17].

(ii) There is a notion of picture over *group* presentations—see [26] or [5] for an exposition (see also §4 below). Pictures (and also their duals, Dehn [or van Kampen] diagrams) are a useful tool in combinatorial group theory.

We wish to associate with $\mathbb{P}$ a certain 2-complex $\mathcal{D}(= \mathcal{D}(\mathbb{P}))$ where the underlying graph is the graph Γ. By way of motivation we consider the above example. Notice that the picture given there could be modified to obtain a new picture by "pulling" the disc second from the bottom down to the bottom level, and "pushing" the bottom disc up to the level above. Or we could "pull" the three discs at levels 5, 6, 7 down to levels 6, 7, 8 respectively, and "push" the disc at level 8 up to level 5. We want to be able to regard two pictures obtained from each other by "switching levels of disjoint discs" in this way as essentially the same. With this in mind, we now single out certain closed paths in Γ which we will want to designate as defining paths for our 2-complex $\mathcal{D}$.

Let $\mathbb{A}, \mathbb{B}$ be atomic monoid pictures. We let $[\mathbb{A}, \mathbb{B}]$ denote the closed path

$$\mathbb{A}.\iota(\mathbb{B}), \tau(\mathbb{A}).\mathbb{B}, \mathbb{A}^{-1}.\tau(\mathbb{B}), \iota(\mathbb{A}).\mathbb{B}^{-1}$$

in Γ. We let $\mathcal{D}$ denote the 2-complex with underlying graph Γ and where the defining paths are all the paths $[\mathbb{A}, \mathbb{B}]$. Notice that in $\mathcal{D}$ the two pictures

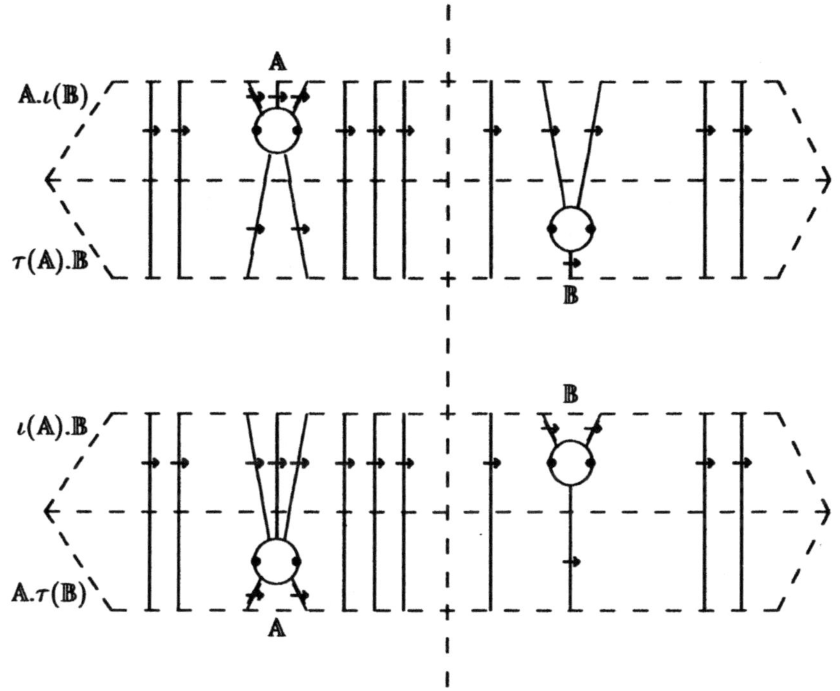

are equivalent.

Note that the set of defining paths of $\mathcal{D}$ is fixed under the left and right action of F on Γ: for $Z \in F$ and $[\mathbb{A}, \mathbb{B}]$ a defining path of $\mathcal{D}$ we have

$$Z.[\mathbb{A}, \mathbb{B}] = [Z.\mathbb{A}, \mathbb{B}],$$
$$[\mathbb{A}, \mathbb{B}].Z = [\mathbb{A}, \mathbb{B}.Z]$$

Remark. We could define $[\mathbb{P}_1, \mathbb{P}_2]$ for any monoid pictures $\mathbb{P}_1$, $\mathbb{P}_2$ (not just for atomic pictures). However, $[\mathbb{P}_1, \mathbb{P}_2]$ can be expressed in terms of the $[\mathbb{A}, \mathbb{B}]$'s. In fact there is a "commutator calculus" analogous to the standard calculus for groups ([25, Chapter 5]), which the reader might find amusing to work out.

2. Isoperimetric functions, Tietze transformations

Notice that two elements U, V of F represent the same element of S if and only if they lie in the same component of Γ. For such a pair we define $d(U, V)$ $(= d_{\mathcal{P}}(U, V))$ to be the distance between U and V in Γ, that is $d(U, V)$ is the length of a shortest path joining U and V.

For $n \in \mathbb{N}_0 = \{0, 1, 2, \ldots\}$ we define

$$\delta_{\mathcal{P}}(n) = \max\{d(U, V) \colon [U] = [V], \quad L(U) + L(V) \leq n\}.$$

(Here $L(\ )$ denotes the length of a word.) This is the *monoid isoperimetric* (or *Dehn*) *function* for $\mathcal{P}$.

We wish to show that the function $\delta_{\mathcal{P}}$ is essentially independent of the choice of finite presentation of a given monoid S. For any two functions $f, g \colon \mathbb{N}_0 \to \mathbb{N}_0$ we write $f \preceq g$ if there are positive constants α, β, γ such that $f(n) \leq \alpha g(\beta n) + \gamma n$ for all $n \in \mathbb{N}_0$, and we write $f \sim g$ if $f \preceq g$ and $g \preceq f$.

Theorem. *Let $\mathcal{P}$ and $\mathcal{Q}$ be two finite presentations for a given monoid S (that is, $S(\mathcal{P}) \cong S \cong S(\mathcal{Q})$). Then $\delta_{\mathcal{P}} \sim \delta_{\mathcal{Q}}$.*

Remark. The analogue for groups of the above theorem is due to Alonso [3] and Short [32]. See also [6] and [14].

To prove the theorem, we must introduce the notion of a *Tietze transformation.* Let $\mathcal{K} = [\mathbf{a}; \mathbf{t}]$ be a monoid presentation. We consider various operations on $\mathcal{K}$.

($T1$) Let b be a symbol not in $\mathbf{a}$, and let Z be a word on $\mathbf{a}$. Then replace $\mathcal{K}$ by $\mathcal{K}_1 = [\mathbf{a}, b; \mathbf{t}, b = Z]$.

($T2$) Let $[U]_{\mathcal{K}} = [V]_{\mathcal{K}}$. Then replace $\mathcal{K}$ by $\mathcal{K}_2 = [\mathbf{a}; \mathbf{t}, U = V]$.

The mapping

$$S(\mathcal{K}) \to S(\mathcal{K}_1) \qquad [a]_{\mathcal{K}} \mapsto [a]_{\mathcal{K}_1} \qquad (a \in \mathbf{a})$$

is an isomorphism, with inverse given by

$$S(\mathcal{K}_1) \to S(\mathcal{K}) \qquad [a]_{\mathcal{K}_1} \mapsto [a]_{\mathcal{K}} \qquad (a \in \mathbf{a}), \qquad [b]_{\mathcal{K}_1} \mapsto [Z]_{\mathcal{K}}.$$

Also the mapping

$$S(\mathcal{K}) \to S(\mathcal{K}_2) \qquad [a]_{\mathcal{K}} \mapsto [a]_{\mathcal{K}_2} \qquad (a \in \mathbf{a})$$

is an isomorphism.

The operations ($T1$), ($T2$) and their inverses are called *Tietze transformations.* Two finite presentations are said to be *Tietze equivalent* if one can be obtained from the other by a finite number of Tietze transformations. We see from the above that if $\mathcal{L}$ and $\mathcal{L}'$ are Tietze equivalent then $S(\mathcal{L}) \cong S(\mathcal{L}')$. Conversely it can be shown that if $S(\mathcal{L}) \cong S(\mathcal{L}')$ then $\mathcal{L}$ and $\mathcal{L}'$ are Tietze equivalent. This is a well known basic result for group presentations (see [25]) and the proof can be easily adapted to monoids.

We now see that in order to prove the theorem, it suffices to show that if $\mathcal{Q}$ is obtained from $\mathcal{P}$ by an operation ($T1$) or ($T2$) then $\delta_{\mathcal{P}} \sim \delta_{\mathcal{Q}}$.

Let $\mathcal{P} = [\mathbf{x}; \mathbf{r}]$.

Case 1. $\mathcal{Q} = [\mathbf{x}, y; \mathbf{r}, y = Z]$ (Z a word on $\mathbf{x}$).
Let us first note that if U_1, U_2 are words on $\mathbf{x}$ then $[U_1]_{\mathcal{P}} = [U_2]_{\mathcal{P}}$ if and only if $[U_1]_{\mathcal{Q}} = [U_2]_{\mathcal{Q}}$. Clearly $d_{\mathcal{P}}(U_1, U_2) \geq d_{\mathcal{Q}}(U_1, U_2)$. In fact we have equality. For consider a picture $\mathbb{P}$ over $\mathcal{Q}$ with $\iota(\mathbb{P}) = U_1$, $\tau(\mathbb{P}) = U_2$. Then any disc labelled by $y = Z$ will be connected to another disc with opposite label by a single arc labelled y. Each of these pairs may be eliminated to give a picture $\mathbb{P}$ over $\mathcal{P}$. This shows that $d_{\mathcal{P}}(U_1, U_2) \leq d_{\mathcal{Q}}(U_1, U_2)$.

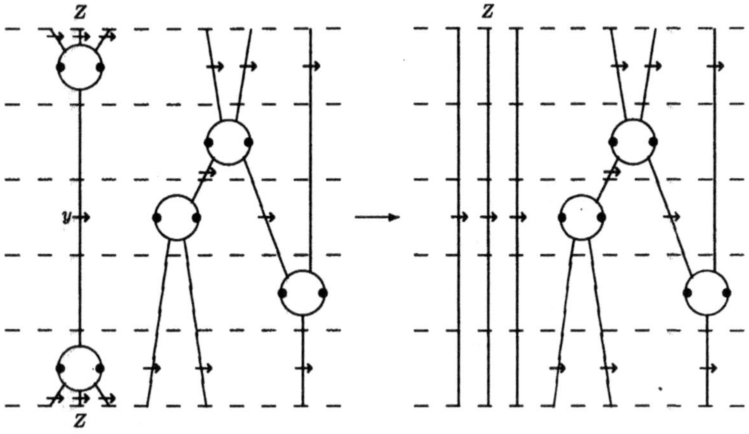

We deduce immediately from the above that $\delta_{\mathcal{P}}(n) \leq \delta_{\mathcal{Q}}(n)$ for all n.

Now let V_1, V_2 be two words on $\mathbf{x} \cup \{y\}$ with $L(V_1)+L(V_2) \leq n$ and $[V_1]_{\mathcal{Q}} = [V_2]_{\mathcal{Q}}$. Let V_1^*, V_2^* be obtained from V_1, V_2 respectively by replacing each occurence of y by the word Z. Then certainly $d_{\mathcal{Q}}(V_i, V_i^*) \leq L(V_i)$ $(i = 1, 2)$. Moreover, $d_{\mathcal{P}}(V_1^*, V_2^*) \leq \delta_{\mathcal{P}}(kn)$, where $k = \max\{L(Z), 1\}$. Thus

$$d_{\mathcal{Q}}(V_1, V_2) \leq \delta_{\mathcal{P}}(kn) + n,$$

and so

$$\delta_{\mathcal{Q}}(n) \leq \delta_{\mathcal{P}}(kn) + n.$$

Case 2. $\mathcal{Q} = [\mathbf{x}; \mathbf{r}, U^{+1} = U^{-1}]$ $([U^{+1}]_{\mathcal{P}} = [U^{-1}]_{\mathcal{P}})$.

Let V_1, V_2 be two words on $\mathbf{x}$ with $L(V_1) + L(V_2) \leq n$ and $[V_1]_{\mathcal{P}} = [V_2]_{\mathcal{P}}$ (or equivalently $[V_1]_{\mathcal{Q}} = [V_2]_{\mathcal{Q}}$). Clearly $d_{\mathcal{Q}}(V_1, V_2) \leq d_{\mathcal{P}}(V_1, V_2)$.

Now let $\mathbb{B}$ be a picture over $\mathcal{P}$ with $\iota(\mathbb{B}) = U^{+1}$, $\tau(\mathbb{B}) = U^{-1}$. Suppose $\mathbb{P}$ is a picture over $\mathcal{Q}$ with $\iota(\mathbb{P}) = V_1$, $\tau(\mathbb{P}) = V_2$ and having $d_{\mathcal{Q}}(V_1, V_2)$ discs. Thus we can convert it to a picture $\mathbb{P}'$ over $\mathcal{P}$ by replacing each term (Z_1, U, ϵ, Z_2) by $Z_1.\mathbb{B}^{\epsilon}.Z_2$. Thus if ℓ is the number of discs in $\mathbb{B}$, then $\mathbb{P}'$ has at most $\ell d_{\mathcal{Q}}(V_1, V_2)$ discs. Hence

$$d_{\mathcal{P}}(V_1, V_2) \leq \ell d_{\mathcal{Q}}(V_1, V_2).$$

Our discussion shows that

$$\delta_{\mathcal{Q}}(n) \leq \delta_{\mathcal{P}}(n) \leq \ell \delta_{\mathcal{Q}}(n) \qquad \text{for all } n \in \mathbb{N}_0.$$

As a consequence of the above theorem, if we agree to identify functions up to $\sim$-equivalence, then we may speak about the isoperimetric function δ_S for a given finitely presented monoid S. In particular, we may speak about the *growth* of δ_S.

Research program 1. Study the growth of Dehn functions of finitely presented monoids.

The study of Dehn functions of *groups* is a very active area of research at present. See [15] for a recent survey.

3. Modules associated with $\mathcal{P}$

3.1. Relation modules

The natural monoid epimorphism $F \to S$ induces a ring epimorphism

$$\theta\colon \mathbb{Z}F \to \mathbb{Z}S \tag{3}$$

of the corresponding integral monoid rings, with kernel J say. Then J/J^2 is a $(\mathbb{Z}S, \mathbb{Z}S)$-bimodule. The left and right actions of S are given by

$$[W].(\zeta + J^2) = W\zeta + J^2 \qquad (\zeta + J^2).[W] = \zeta W + J^2$$

$(W \in F, \zeta \in J)$. Following Ivanov [21], we will call this module the *relation bimodule* of $\mathcal{P}$. We will denote it by $M^{(b)}(\mathcal{P})$ (or simply $M^{(b)}$).

There are also the *left* and *right* relation modules $M^{(l)}$ $(= M^{(l)}(\mathcal{P}))$, $M^{(r)}$ $(= M^{(r)}(\mathcal{P}))$, of $\mathcal{P}$, also considered by Ivanov in [21]. Let $I(F)$ denote the *augmentation ideal* of F, that is $I(F)$ is the kernel of the augmentation map

$$\text{aug}\colon \mathbb{Z}F \to \mathbb{Z} \qquad W \mapsto 1 \qquad (W \in F).$$

Then

$$M^{(l)} = J/JI(F), \qquad M^{(r)} = J/I(F)J,$$

with left/right S-action given by

$$[W].(\zeta + JI(F)) = W\zeta + JI(F), \qquad (\zeta + I(F)J).[W] = \zeta W + I(F)J$$

$(W \in F, \zeta \in J)$. It is not hard to show (see [28, Lemma 3.1]) that

$$M^{(l)} \cong M^{(b)} \otimes_{\mathbb{Z}S} \mathbb{Z}, \qquad M^{(r)} \cong \mathbb{Z} \otimes_{\mathbb{Z}S} M^{(b)}$$

(where $\mathbb{Z}S$ acts on $\mathbb{Z}$ with trivial S-action).

3.2. The first homology of $\mathcal{D}$

Consider the chain complex

$$0 \to C_2(\mathcal{D}) \xrightarrow{\partial_2} C_1(\mathcal{D}) \xrightarrow{\partial_1} C_0(\mathcal{D}) \to 0$$

of $\mathcal{D}$. Here $C_0(\mathcal{D})$ is the free abelian group with basis F, $C_1(\mathcal{D})$ is the free abelian group with basis the positive atomic pictures, and $C_2(\mathcal{D})$ is the free abelian group on the defining paths $[\mathbb{A}, \mathbb{B}]$ of $\mathcal{D}$. These chain groups are $(\mathbb{Z}F, \mathbb{Z}F)$-bimodules with F-action given by the left and right action of F on the bases. Moreover, the boundary maps are bimodule homomorphisms. Consequently $H_1(\mathcal{D})$ $(= \operatorname{Ker}\partial_1/\operatorname{Im}\partial_2)$ is a $(\mathbb{Z}F, \mathbb{Z}F)$-bimodule. Now it can be shown (see [28, Lemma 4.1]) that J annihilates $H_1(\mathcal{D})$ on the left and the right, so $H_1(\mathcal{D})$ has an induced $(\mathbb{Z}S, \mathbb{Z}S)$-bimodule structure.

Let $P^{(b)}$ be the free $(\mathbb{Z}S, \mathbb{Z}S)$-bimodule with basis $\{t_R^{(b)} : R \in \mathbf{r}\}$ in one-to-one correspondence with $\mathbf{r}$. Then there is an exact sequence [28, Theorem 4.2]

$$H_1(\mathcal{D}) \xrightarrow{\rho} P^{(b)} \xrightarrow{\mu} M^{(b)} \to 0. \tag{4}$$

Here

$$\mu(t_R^{(b)}) = (R^{+1} - R^{-1}) + J^2 \qquad (R \in \mathbf{r}),$$

and ρ is defined as follows. Let

$$\alpha = n_1\mathbb{B}_1 + n_2\mathbb{B}_2 + \cdots + n_s\mathbb{B}_s \tag{5}$$

be an element of $\operatorname{Ker}\partial_1$, where $\mathbb{B}_i = (U_i, R_i, +1, V_i)$ say, and $n_i \in \mathbb{Z}$ $(i = 1, \ldots, s)$. Then

$$\rho(\alpha + \operatorname{Im}\partial_2) = \sum_{i=1}^{s} n_i[U_i]t_{R_i}^{(b)}[V_i].$$

Now let $P_2^{(l)}$ be the free left $\mathbb{Z}S$-module with basis $\{t_R^{(l)} : R \in \mathbf{r}\}$ in one-to-one correspondence with $\mathbf{r}$. Applying the right exact functor $- \otimes_{\mathbb{Z}S} \mathbb{Z}$ to (4) then leads to an exact sequence

$$H_1(\mathcal{D}) \otimes_{\mathbb{Z}S} \mathbb{Z} \xrightarrow{\phi} P_2^{(l)} \to M^{(l)} \to 0 \tag{6}$$

where

$$t_R^{(l)} \mapsto (R^{+1} - R^{-1}) + JI(F) \qquad (R \in \mathbf{r}),$$

and

$$(\alpha + \operatorname{Im}\partial_2) \otimes 1 \overset{\phi}{\mapsto} \sum_{i=1}^{s} n_i[U_i]t_{R_i}^{(l)}$$

(α as in (5)). There is of course a sequence analogous to (6) for right modules.

3.3. Resolutions

Consider the left $\mathbb{Z}S$-module $\mathbb{Z}$ where S acts trivially. A *free resolution* of $\mathbb{Z}$ is an exact sequence

$$\cdots \to B_i \xrightarrow{\delta_i} B_{i-1} \to \cdots \to B_2 \xrightarrow{\delta_2} B_1 \xrightarrow{\delta_1} B_0 \xrightarrow{\delta_0} \mathbb{Z} \to 0$$

where the B_i's are free left $\mathbb{Z}S$-modules. Such a resolution always exists.

The monoid S is said to have the *left* FP_n *property* $(n \geq \infty)$ if there is a free resolution as above with B_i finitely generated for $i \leq n$.

A presentation $\mathcal{P}$ of S as in (1) gives rise to the start of a free resolution as follows. Let $P_1^{(l)}$ be the left free $\mathbb{Z}S$-module with basis $\{t_x^{(l)} : x \in \mathbf{x}\}$ in one-to-one correspondence with $\mathbf{x}$. Then we have an exact sequence (see [28])

$$H_1(\mathcal{D}) \otimes_{\mathbb{Z}S} \mathbb{Z} \xrightarrow{\phi} P_2^{(l)} \xrightarrow{\delta_2} P_1^{(l)} \xrightarrow{\delta_1} \mathbb{Z}S \xrightarrow{\text{aug}} \mathbb{Z} \to 0 \tag{7}$$

where ϕ is as in (6),

$$\delta_2(t_R^{(l)}) = \sum \theta \frac{\partial(R^{+} - R^{-1})}{\partial x} t_x^{(l)} \qquad (R \in \mathbf{r})$$

$$\delta_1(t_x^{(l)}) = 1 - [x] \qquad (x \in \mathbf{x}).$$

(Here $\partial/\partial x\colon \mathbb{Z}F \to \mathbb{Z}F$ is left Fox derivation [13] and θ is as in (3).) Consequently if we know a set T of module generators of $H_1(\mathcal{D}) \otimes_{\mathbb{Z}S} \mathbb{Z}$ then we may take a free left $\mathbb{Z}S$-module $P_3^{(l)}$ with basis in one-to-one correspondence with T and obtain a partial resolution

$$(8) \qquad P_3^{(l)} \xrightarrow{\delta_3} P_2^{(l)} \xrightarrow{\delta_2} P_1^{(l)} \xrightarrow{\delta_1} \mathbb{Z}S \xrightarrow{\text{aug}} \mathbb{Z} \to 0.$$

Note that from (8) we could compute the second (co)homology of S:

$$H_2(S,A) = \frac{\operatorname{Ker}(\delta_2 \otimes id_A)}{\operatorname{Im}(\delta_3 \otimes id_A)} \qquad (A \text{ a right } \mathbb{Z}S\text{-module}),$$
$$H^2(S,B) = \frac{\operatorname{Ker}(\operatorname{Hom}_{\mathbb{Z}S}(\delta_3, B))}{\operatorname{Im}(\operatorname{Hom}_{\mathbb{Z}S}(\delta_2, B))} \qquad (B \text{ a left } \mathbb{Z}S\text{-module}).$$

We can of course consider right modules rather than left modules and hence define the *right FP_n property*. We have an exact sequence analogous to (7) for right modules, and hence a partial resolution of length 3 analogous to (8), assuming we know a set of module generators of $\mathbb{Z} \otimes_{\mathbb{Z}S} H_1(\mathcal{D})$.

An example is given in [11] of a monoid which is right FP_∞ but is not even left FP_1! (By considering the opposite monoid with multiplication given by $a * b = ba$ we then also have an example of a monoid which is left FP_∞ but not right FP_1.)

It is known that if S has a finite presentation giving rise to a complete rewriting system then S is both left and right FP_∞. See [12] for an excellent exposition of this result.

Observe from the above that if S has a finite presentation $\mathcal{P}$ such that $H_1(\mathcal{D})$ is finitely generated as a bimodule, then S is both left and right FP_3.

For some other recent results on resolutions for monoids see [24].

3.4. Generators of $H_1(\mathcal{D})$

Let $\mathbb{P} = \mathbb{A}_1^{\epsilon_1} \mathbb{A}_2^{\epsilon_2} \ldots \mathbb{A}_n^{\epsilon_n}$ be a spherical picture. Here $\mathbb{A}_1, \mathbb{A}_2, \ldots, \mathbb{A}_n$ are positive atomic pictures and $\epsilon_i = \pm 1$ $(i = 1, \ldots, n)$. We let

$$\alpha_{\mathbb{P}} = \epsilon_1 \mathbb{A}_1 + \epsilon_2 \mathbb{A}_2 + \cdots + \epsilon_n \mathbb{A}_n.$$

Then $\alpha_{\mathbb{P}} \in \operatorname{Ker} \partial_1$.

Definition. A set X of spherical pictures will be said to *generate* $H_1(\mathcal{D})$ if the elements $\alpha_{\mathbb{P}} + \operatorname{Im} \partial_2$ $(\mathbb{P} \in X)$ generate $H_1(\mathcal{D})$ as a $(\mathbb{Z}S, \mathbb{Z}S)$-bimodule.

We note that from §3.2 that there is some interest in computing a (hopefully "nice") set of generators of $H_1(\mathcal{D})$. We note also from (4) that having such a set X enables us to write down a module presentation

$$\left[t_R^{(b)} \quad (R \in \mathbf{r}); \quad \rho(\alpha_{\mathbb{P}} + \operatorname{Im} \partial_2) = 0 \qquad (\mathbb{P} \in X)\right]$$

for $M^{(b)}$ (and similarly for $M^{(l)}$). Since

$$H^{k+2}(S,-) = \operatorname{Ext}^k_{\mathbb{Z}S}(M^{(l)},-)$$
$$H_{k+2}(S,-) = \operatorname{Tor}_k^{\mathbb{Z}S}(-,M^{(l)}),$$

($k \geq 1$) it is conceivable that if the presentation of $M^{(l)}$ is nice enough then we could get a handle on the higher (co)homology of S. (In the case of groups it has been possible in a number of interesting situations to obtain useful information about (co)homology by an analysis of spherical pictures — see [4, 18, 19, 20, 27].)

In order to give a geometric method for finding generators of $H_1(\mathcal{D})$ we consider certain operations on pictures.

(I) Insert/delete an inverse pair $\mathbb{A}\mathbb{A}^{-1}$ ($\mathbb{A}$ an atomic picture).

(II) Replace a subpicture $[\mathbb{A} \cdot \iota(\mathbb{B})][\tau(\mathbb{A}) \cdot \mathbb{B}]$ by $[\iota(\mathbb{A}) \cdot \mathbb{B}][\mathbb{A} \cdot \tau(\mathbb{B})]$ or vice versa.

Now suppose that X is a set of spherical pictures.

(III) Insert/delete subpictures of the form $W \cdot \mathbb{P}^{\pm 1} \cdot V$ ($\mathbb{P} \in X$, $W, V \in F$).

We will say that X *collapses* $\mathcal{D}$ if every spherical picture can be reduced to the empty picture by a finite sequence of operations (I), (II), (III). It is not hard to show that *if X collapses $\mathcal{D}$ then X generates $H_1(\mathcal{D})$*. The usefulness of this remark is as follows. Given a presentation $\mathcal{P}$ one looks for a set X of "obvious" spherical pictures over $\mathcal{P}$. One then tries to show geometrically that every spherical picture can be reduced to the empty picture by a finite sequence of operations (I), (II), (III).

Research program 2. Develop a calculus of spherical pictures for monoids.

In developing such a calculus, a point of view which has been very successful in the case of groups (see [5]) is the following.

Let $\mathcal{P}_i$ ($i \in I$) be a family of subpresentations of $\mathcal{P}$, and let X_i denote the collection of all spherical pictures over $\mathcal{P}_i$ ($i \in I$). We will say that a set Y of spherical pictures over $\mathcal{P}$ *collapses* $\mathcal{D}$ relative to $\{\mathcal{P}_i : i \in I\}$ if $Y \cup (\cup_{i \in I} X_i\}$ collapses $\mathcal{D}$. Note that we may equivalently take X_i to be a set which collapses $\mathcal{D}_i$ ($= \mathcal{D}(\mathcal{P}_i)$) ($i \in I$).

This concept is of interest in connection with constructions. Suppose a monoid S is constructed in some way from monoids $S_1, \ldots S_k$, and suppose that this can be modelled at the presentation level by putting together presentations $\mathcal{P}_1, \ldots, \mathcal{P}_k$ of $S_1, \ldots, S_k$ in some "canonical" way to obtain a presentation $\mathcal{P}$ of S. (For example, if $S = S_1 \times S_2$ and $\mathcal{P}_i = [\mathbf{x}_i; \mathbf{r}_i]$ is a presentation of S_i ($i = 1, 2$) then

$$\mathcal{P} = [\mathbf{x}_1, \mathbf{x}_2; \mathbf{r}_1, \mathbf{r}_2, xy = yx \quad (x \in \mathbf{x}_1,\, y \in \mathbf{x}_2)]$$

is a "canonical" presentation of S.) Now since the monoids $S_1, \ldots, S_k$ are assumed arbitrary, we cannot assume we know anything about the spherical pictures over the $\mathcal{P}_i$'s. However, it could still be possible to determine a set of "canonical" spherical pictures which together with the pictures over the $\mathcal{P}_i$'s collapse $\mathcal{D}$.

For groups, a calculus of spherical pictures for various standard constructions has been successfully developed (see [5] for a survey). A similar program for monoids would be useful.

4. Relationship between the monoid and group associated with a presentation

Given a presentation $\mathcal{P} = [\mathbf{x}; \mathbf{r}]$ as in (1), we not only have the *monoid* S associated with $\mathcal{P}$, but also the *group* G ($= G(\mathcal{P})$) associated with $\mathcal{P}$. This is defined as follows. Let $\hat{F}$ be the free group on $\mathbf{x}$. Thus $\hat{F}$ consists of free equivalence classes $< W >$ of words on $\mathbf{x} \cup \mathbf{x}^{-1}$ (where $\mathbf{x}^{-1}$ is a set in one-to-one correspondence $x \leftrightarrow x^{-1}$ with $\mathbf{x}$). Let N be the smallest normal subgroup of $\hat{F}$ containing the elements $< R^{+1} >< R^{-1} >^{-1}$ $(R \in \mathbf{r})$. Then $G = \hat{F}/N$. We write $[W]_G$ instead of $< W > N$.

We have a natural homomorphism (of monoids)

$$\psi\colon S \to G \qquad [W]_S \mapsto [W]_G \qquad (W \in F). \tag{9}$$

Research program 3. Study the relationship between S and G.

A fair amount of work has been done in this area already.

It is easy to see that S is embeddable in a group if and only if ψ is injective. Consequently, considerable attention has been directed at obtaining sufficient conditions on $\mathcal{P}$ which guarantee that ψ is injective. The injectivity of ψ is open to geometric analysis. In order to discuss this we must introduce the notion of a general picture over $\mathcal{P}$. (This will be a slight variation of the definition given in [26] and [5].)

A *picture* $\mathbb{P}$ over $\mathcal{P}$ consists of the following.

a) A disc D^2 with basepoints $0, 0'$ on ∂D^2.

b) Disjoint discs $\Delta_1, \ldots, \Delta_n$ in the interior of D^2. The disc Δ_λ has two basepoints $0_\lambda, 0'_\lambda$ on $\partial\Delta_\lambda$ $(\lambda = 1, \ldots, n)$.

c) A finite number of disjoint arcs $\alpha_1, \ldots, \alpha_m$. Each arc lies in the closure of $D^2 - \cup_{\lambda=1}^{n}\Delta_\lambda$ and is either a simple closed curve having empty intersection with $\partial D^2 \cup \partial\Delta_1 \cup \cdots \cup \partial\Delta_n$, or is a simple non-closed curve which joins two points of $\partial D^2 \cup \partial\Delta_1 \cup \cdots \cup \partial\Delta_n$, neither point being a basepoint. Each arc has a normal orientation, indicated by a short arrow meeting the arc transversely, and is labelled by an element of $\mathbf{x}$.

d) For each disc Δ_λ $(\lambda = 1, \ldots, n)$, if we travel along $\partial\Delta_\lambda$ in the clockwise (respectively, anticlockwise) direction from 0_λ to $0'_\lambda$ then the arcs we encounter are crossed in the direction of their orientation, and if we successively read off the labels on these arcs we obtain a word $R_\lambda^{\epsilon_\lambda}$ (respectively $R_\lambda^{-\epsilon_\lambda}$ for some $R_\lambda \in \mathbf{r}$, $\epsilon_\lambda \in \{-1, 1\}$.

Example. $\mathcal{P} = [a, b, c; ab = ba, (bc)^2 = 1, ca = ac, c^2 = 1]$

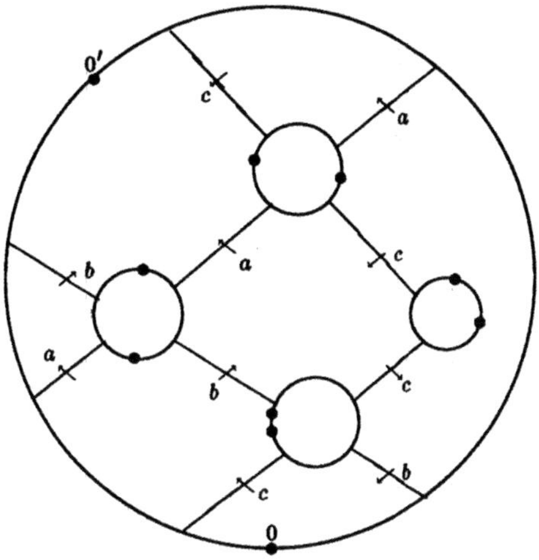

We define $\iota(\mathbb{P}), \tau(\mathbb{P})$ to be the words (on $\mathbf{x} \cup \mathbf{x}^{-1}$) we obtain by successively reading off the labels on the arcs encountered by travelling from 0 to 0′ along ∂D^2 in the clockwise, anticlockwise direction respectively. When reading off labels we use the convention that if we cross an arc labelled x in the direction of its orientation then we read x, whereas if we cross an arc labelled x against its orientation then we read x^{-1}.

By the *discs* of $\mathbb{P}$ we mean the discs $\Delta_1, \ldots, \Delta_n$ and not the ambient disc D^2.

We say that $\mathbb{P}$ is *minimal* if there is no other picture $\mathbb{P}'$ with $\iota(\mathbb{P}') = \iota(\mathbb{P})$, $\tau(\mathbb{P}) = \tau(\mathbb{P}')$ and where $\mathbb{P}'$ has fewer discs than $\mathbb{P}$.

A *transverse path* in $\mathbb{P}$ is a path in the closure of $D^2 - \cup_{\lambda=1}^{n} \Delta_\lambda$ which intersects the arcs only finitely many times (moreover, if the path intersects an arc then it crosses it, and does not just touch it). If we travel along a transverse path from its initial point to its terminal point we will cross several arcs and we can read off the labels on these arcs (adopting the convention mentioned in last-but-one paragraph) giving a word on $\mathbf{x} \cup \mathbf{x}^{-1}$, which we call the *label on the path.*

We say that $\mathbb{P}$ is *stratifiable* if there are transverse paths $\gamma_0, \gamma_1, \ldots, \gamma_n$ each starting at 0 and ending at 0′, where γ_0 is the part of ∂D^2 from 0 to 0′ in the clockwise direction, γ_n is the part of ∂D^2 from 0 to 0′ in the anticlockwise direction, and where the part of $\mathbb{P}$ between γ_i and γ_{i+1} is an atomic monoid picture for $i = 0, \ldots, n-1$.

If $\mathbb{P}$ is stratifiable then it can be regarded as a *monoid* picture over $\mathcal{P}$. (Obviously, in order for $\mathbb{P}$ to be stratifiable we must have that $\iota(\mathbb{P}), \tau(\mathbb{P})$ are words on $\mathbf{x}$.)

It can be shown that for any two words W_1, W_2 on $\mathbf{x} \cup \mathbf{x}^{-1}$, $[W_1]_G = [W_2]_G$ if and only if there is a picture $\mathbb{B}$ over $\mathcal{P}$ with $\iota(\mathbb{B}) = W_1$, $\tau(\mathbb{B}) = W_2$. We then define $\hat{d}(W_1, W_2)$ $(= (\hat{d}_{\mathcal{P}}(W_1, W_2))$ to be the number of discs in a minimal such picture. Notice that if there exists a minimal picture which is stratifiable then $[W_1]_S = [W_2]_S$ and $\hat{d}(W_1, W_2) = d(W_1, W_2)$.

We define

$$\hat{\delta}_{\mathcal{P}} : \mathbb{N}_0 \to \mathbb{N}_0$$

by

$$\hat{\delta}_{\mathcal{P}}(k) = \max\{\hat{d}(W_1, W_2) : [W_1]_G = [W_2]_G, \qquad L(W_1) + L(W_2) \le k\}.$$

This is the *group isoperimetric function* associated with $\mathcal{P}$.

The next result follows from the above discussion

Lemma. *Suppose the following condition is satisfied.*

(*) *Every minimal picture* $\mathbb{P}$ *over* $\mathcal{P}$ *with* $\iota(\mathbb{P}), \tau(\mathbb{P})$ *words on* $\mathbf{x}$ *is stratifiable.*

Then

a) $\psi: S \to G$ *is injective;*

b) $\delta_{\mathcal{P}}(k) \leq \hat{\delta}_{\mathcal{P}}(k)$ *for all* $k \in N_0$.

Remark. In relation to (*) some caution is necessary. A picture may contain a *floating circle*, that is, a closed arc which does not encircle any other discs or arcs of the picture.

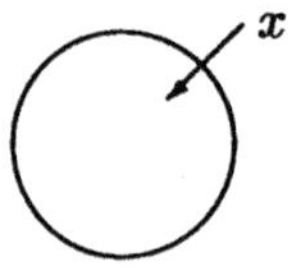

Obviously a picture containing floating circles could not be stratifiable, so we would need to allow ourselves to remove floating circles if necessary.

Also it may be possible to apply a *bridge move*

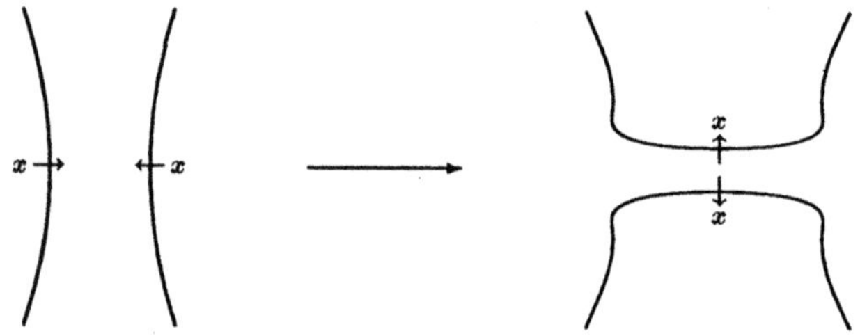

to a picture, which could affect its stratifiability. For example the two pictures below differ by a bridge move; the first is not stratifiable, whereas then second is. *In interpreting (*) we must therefore allow ourselves to alter minimal pictures by elimination of floating circles and by application of bridge moves if necessary.*

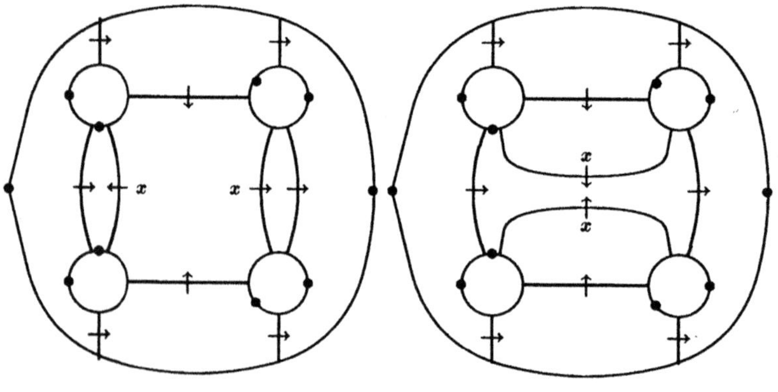

Part (b) of the above lemma is of interest because it shows that the growth of the Dehn function for S is bounded by the growth for G. As previously mentioned, the growth of Dehn functions for groups has received a lot of attention.

There are several classes of presentations for which (*) holds.

(i) **Cycle free presentations.**

Suppose R^{+1} and R^{-1} are non-empty for each $R \in \mathbf{r}$. We define the *left graph* $LG(\mathcal{P})$ of $\mathcal{P}$ as follows. The vertex set is $\mathbf{x}$ and for each $R \in \mathbf{r}$ there is an edge pair e_R, e_R^{-1}, where $\iota(e_R), \tau(e_R)$ are defined to be the first letters of R^{+1}, R^{-1} respectively. The *right graph* $RG(\mathcal{P})$ is defined analogously, except that now $\iota(e_R), \tau(e_R)$ are taken to be the last letters of R^{+1}, R^{-1} respectively.

Example. $\mathcal{P} = [a, b, c;\ a^2 = ab,\ bc = a^2b^2,\ b^2ca = ac]$

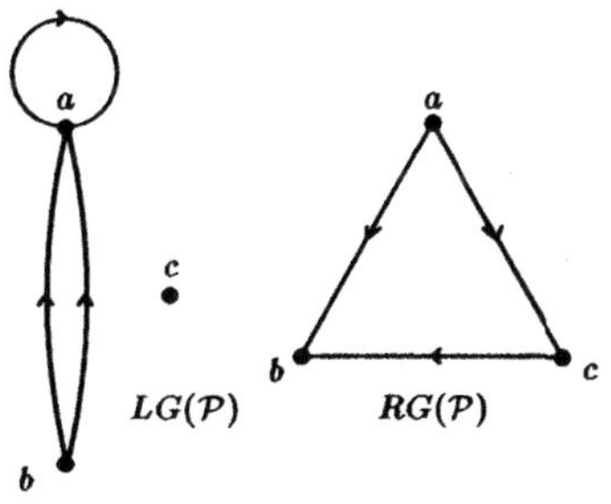

The presentation $\mathcal{P}$ is said to be *cycle-free* if $LG(\mathcal{P})$ and $RG(\mathcal{P})$ are forests. For such presentations (*) holds. This can be proved by translating the geometric arguments in [29] into the language of pictures (see also [17, 33]). The fact that $S \to G$ is injective for monoids given by cycle-free presentations was originally proved algebraically in [1, 2]. For further recent work on monoids given by cycle-free presentations see [16].

(ii) **Small cancellation presentations.**

As in (i) we assume that R^{+1} and R^{-1} are non-empty for each $R \in \mathbf{r}$.

We say that $D(q)$ holds for $\mathcal{P}$ if there is no non-empty reduced closed path in $LG(\mathcal{P})$ or $RG(\mathcal{P})$ of length less than q.

Also, let $\mathcal{Y}$ be the set of words $\{R^{+1}, R^{-1} \colon R \in \mathbf{r}\}$. A non-empty word W on $\mathbf{x}$ is said to be a *piece* (relative to $\mathcal{Y}$) if there are elements U_1WU_2, V_1WV_2 of $\mathcal{Y}$ with either U_1 distinct from V_1 or U_2 distinct from V_2. We say that $\mathcal{P}$ satisfies $C_0(p)$ if no element of $\mathcal{Y}$ is a product of less than p pieces.

If $\mathcal{P}$ satisfies $C_0(4)$ and $D(2)$, or $C_0(3)$ and $D(3)$, then (*) holds [22]. (The work in [22] is done in terms of van Kampen diagrams, but can be easily translated into the language of pictures. The condition of the Basic Lemma on p. 415 of [22] guarantees that (*) holds.)

(iii) **New presentations from old.**

A method of putting together a collection of certain presentations satisfying (*) to obtain a new presentation satisfying (*) can be deduced from [10].

Let Λ be a graph with vertex set $\mathbf{x}$ and edge set $\mathbf{e}$, and let $\mathbf{e}^+$ be an orientation of $\mathbf{e}$. We assume that Λ is *simple*, that is, no edge starts and ends at the same vertex, and if e, f are edges with $e \neq f^{\pm 1}$ then the sets $\{\iota(e), \tau(e)\}$ $\{\iota(f), \tau(f)\}$ are distinct. We can then identify $\mathbf{e}^+$ with a collection of ordered pairs of elements of $\mathbf{x}$. For each edge $e = (x, y)$ in $\mathbf{e}^+$ let $\mathbf{r}_e$ be a set of relations involving the letters x and y, and put $\mathcal{P}_e = [x, y; \mathbf{r}_e]$. Then let $\mathcal{P}$ be the monoid presentation $[\mathbf{x}; \mathbf{r}]$ where $\mathbf{r} = \bigcup_{e \in \mathbf{e}} \mathbf{r}_e$.

We consider the following conditions.

($C1$) For each edge $e = (x, y)$, if $R \in \mathbf{r}_e$ then R^{+1} and R^{-1} both involve x and y, and the first (resp. last) letter of R^{+1} is different from the first (resp. last) letter of R^{-1}.

($C2$) For each edge e, the natural map $S(\mathcal{P}_e) \to G(\mathcal{P}_e)$ is injective.

($C3$) For each edge e, the condition $(*)$ holds for $\mathcal{P}_e$.

It is shown in [10, Theorems 3.1, 3.2] that if ($C1$) and ($C2$) hold then the natural map $S(\mathcal{P}) \to G(\mathcal{P})$ is injective, provided that Λ is triangle-free, or if Λ is not triangle-free then certain other compensating conditions ((A) and (B) of [10, Theorem 3.2]) hold. Now an analysis of the proof of Theorems 3.1, 3.2 of [10], translated into the language of pictures, shows that if in the statements of these theorems we replace ($C2$) by ($C3$), then we can obtain the conclusion that $(*)$ holds for $\mathcal{P}$.

(iv) **Relative presentations.**

A *relative monoid presentation* $\mathcal{P}$ is a triple $[C, \mathbf{x}; \mathbf{r}]$ where C is a monoid, $\mathbf{x}$ is a set, and a typical element $R \in \mathbf{r}$ has the form $R^{+1} = R^{-1}$, where R^{+1}, R^{-1} are words on $C \cup \mathbf{x}$ each containing some letter from $\mathbf{x}$. The monoid S defined by $\mathcal{P}$ is the quotient of the free product $C * F$ (where F is the free monoid on $\mathbf{x}$) by the congruence generated by the relations.

Now if S is embeddable in a group then so is C, since the natural homomorphism $C \to S$ is injective. Consequently, if we are concerned with questions of embeddability, then there is no great loss of generality in assuming at the outset that the coefficient monoid of C is actually a group. We will assume this from now on.

The group G defined by $\mathcal{P}$ is the quotient of the free product $C * \hat{F}$, (where $\hat{F}$ is the free group on $\mathbf{x}$) by the normal subgroup generated by the relations. There is then the natural monoid homomorphism $S \to G$. The injectivity of this homomorphism (as well as other questions about S) can be investigated geometrically by considering ***pictures over relative presentations***, as discussed in [4]. There is a notion of stratification of such pictures, and a relative version of the above lemma holds.

For results on monoids defined by relative presentations see [23].

Finally, we mention some interesting recent work [7] concerning a situation where the map ψ in (9) is not injective.

For $n \geq 1$, $r \geq 2$ consider the presentation

$$\mathcal{P}(r, n) = [a_1, a_2, \ldots, a_n; a_1 a_2, \ldots, a_r = a_{r+1}, a_2 a_3 \ldots a_{r+1} = a_{r+2}$$
$$a_{n-1} a_n a_1 \ldots a_{r-2} = a_{r-1}, a_n a_1 a_2 \ldots a_{r-1} = a_r],$$

where all subscripts are reduced modulo n. The group defined by this presentation is the *Fibonacci group* $G(r,n)$. Such groups have been well studied—see [34] for a survey. We also have of course the *Fibonacci monoid* $S(r,n)$.

Notice that $RG(\mathcal{P}(r,n))$ is connected, and that $LG(\mathcal{P}(r,n))$ has $\ell = (r,n)$ components.

More generally, consider a presentation

$$\mathcal{P} = [a_1, a_2, \ldots, a_n; Z_1 = a_1, Z_2 = a_2, \ldots, Z_n = a_n]$$

where $Z_1, \ldots, Z_n$ are words on $\{a_1, a_2, \ldots, a_n\}$ of length at least 2, and assume that each a_i occurs as the first, second and last letter respectively of three of the Z_j's. Then the following result holds [7]: *If $RG(\mathcal{P})$ is connected and if $LG(\mathcal{P})$ has ℓ components, then the monoid S defined by $\mathcal{P}$ is the disjoint union of subsets $\{1\}$, $G_1, \ldots, G_\ell$, where each G_i is a group isomorphic to G. The restriction of $\psi\colon S \to G$ to G_i is an isomorphism. Each G_i is a right ideal of S.*

Although the above result may seem rather specialised, it nevertheless points in a new and interesting direction. For related work see [8, 9, 30].

References

1. S. I. Adyan, *On the embeddability of semigroups in groups*, DAN SSSR(2) **133** (1960), 255–257.
2. ———, *Defining relations and algorithmic problems for groups and semigroups*, Trudy Mat. Inst. AN SSSR **85** (1966).
3. J. Alonso, *Inégalités isopérimetriques et quasi-isométries*, C. R. Acad. Sci. Paris Série 1, **331** (1991), 761–764.
4. W. A. Bogley and S. J. Pride, *Aspherical relative presentations*, Proc. Edin. Math. Soc. **35** (1992), 1–39.
5. ———, *Calculating generators of π_2*, Two-Dimensional Homotopy and Combinatorial Group Theory (C. Hog-Angeloni, W. Metzler and A. Sieradski, eds.), London Mathematical Society Lecture Notes Series, no. 197, Cambridge University Press, 1993, pp. 157–188.
6. S. G. Brick, *Dehn functions and products of groups*, Trans. Amer. Math. Soc. **335** (1993), 369–384.
7. C. M. Campbell, E. F. Robertson, N. Ruškuc, and R. M. Thomas, *Fibonacci semigroups*, J. Pure Appl. Alg., to appear.
8. ———, *Semigroup and group presentations*, Bull. London Math. Soc., to appear.
9. C. M. Campbell, E. F. Robertson, and R. M. Thomas, *Symmetric semigroup presentations*, Semigroup Forum **46** (1993), 286–306.
10. P. R. Cho and S. J. Pride, *Embedding semigroups into groups, and the asphericity of semigroups*, Internat. J. Alg. Comp. **3** (1993), 1–13.
11. D. E. Cohen, *A monoid which is right FP_∞ but not left FP_1*, Bull. London Math. Soc. **24** (1992), 340–342.
12. ———, *String rewrinting-a survey for group theorists*, Geometric Group Theory (G. A. Niblo and M. A. Roller, eds.), vol. 1, London Mathematical Society Lecture Series, no. 181, Cambridge University Press, 1993, pp. 79–96.
13. R. H. Crowell and R. H. Fox, *Introduction to Knot Theory*, Graduate Text in Mathematics, no. 57, Springer-Verlag, Berlin-Heidelberg-New York, 1977.
14. S. M. Gersten, *Dehn functions and ℓ_1-norms of finite presentations*, Algorithms and Classification in Combinatorial Group Theory (G. Baumslag and C. F. Miller III, eds.), MSRI Publications, no. 23, Springer-Verlag, 1992, pp. 195–224.
15. ———, *Isoperimetric and isodiametric functions of finite presentations*, Geometric Group Theory (G. A. Niblo and M. A. Roller, eds.), vol. 1, London Mathematical Society Lecture Note Series, no. 181, Cambridge University Press, 1993, pp. 79–96.
16. R. Goldstein and J. Teymouri, *Adjan's theorem and conjugacy in semigroups*, Semigroup Forum **47** (1993), 299–304.

17. P. M. Higgins, *Techniques of Semigroup Theory*, Oxford Science Publications, 1992.
18. J. Howie, *The quotient of a free product of groups by a single high-powered relator. I. Pictures. fifth and higher powers*, Proc. London Math. Soc. **59** (1989), 507–540.
19. ———, *The quotient of a free product of groups by a single high-powered relator. II. Fourth powers*, Proc. London Math. Soc. **61** (1990), 33–62.
20. J. Howie and R. M. Thomas, *The groups $(2,3,p;q)$; asphericity and a conjecture of Coxeter*, J. Algebra **154** (1993), 289–309.
21. S. V. Ivanov, *Relation modules and relation bimodules of groups, semigroups and associative algebras*, Internat. J. Alg. Comp. **1** (1991), 89–114.
22. E. S. Kashintsev, *Small cancellation conditions and embeddability of semigroups in groups*, Internat. J. Alg. Comp. **2** (1992), 433–441.
23. C. W. Kilgour, *Relative presentations of monoids*, in preparation, University of Glasgow.
24. L. A. Lambe, *Resolutions which split off the bar construction*, J. Pure Appl. Alg. **84** (1992), 311–329.
25. W. Magnus, A. Karrass, and D. Solitar, *Combinatorial Group Theory*, Intersciences, New York, 1966.
26. S. J. Pride, *Identities among relations of group presentations*, Group Theory from a Geometric Viewpoint, Trieste 1990 (E. Ghys, A. Haefliger, and A. Verjovsky, eds.), World Scientific Publishing, Singapore, 1991, pp. 687–717.
27. ———, *The (co)homology of groups given by presentations in which each defining relator involves at most two types of generators*, J. Austral. Math. Soc. (Series A) **52** (1992), 205–218.
28. ———, *Low-dimensional homotopy theory for monoids*, preprint, University of Glasgow, 1993.
29. J. H. Remmers, *On the geometry of semigroup presentations*, Advances in Math. **36** (1980), 283–296.
30. E. F. Robertson and Y. Ünlü, *On semigroup presentations*, Proc. Edin. Math. Soc. **36** (1992), 55–68.
31. J.-P. Serre, *Trees*, Springer-Verlag, Berlin, Heidelberg, New York 1980.
32. H. Short, *Groups and combings*, preprint, ENS Lyon, 1990.
33. J. R. Stallings, *A graph-theoretic lemma and group embeddings*, Combinatorial Group Theory and Topology (S. Gersten and J. Stallings, eds.), Princeton University Press, 1987, pp. 145–155.
34. R. M. Thomas, *The Fibonacci groups revisited*, Groups-St. Andrews 1989 (C. M. Campbell and E. F. Robertson, eds.), vol. 2, London Mathematical Society Lecture Notes Series, no. 160, Cambridge University Press, 1991, pp. 445–454.

AN INTRODUCTION TO AUTOMATIC GROUPS

HAMISH SHORT*
U.F.R. Mathématiques,
Université de Provence,
Place Victor Hugo,
Marseille cedex 3,
France.

1. Introduction

The first five sections of this article form the text of two talks given at the conference. The last three sections are an exploration of some possible extensions of the theory of automatic groups, using more general kinds of machine, and more general classes of normal forms. Some of the ideas in this part were previously presented at the conference in honour of Wilhelm Magnus in Brooklyn Polytechnic, 1992.

The study of automatic groups, initiated by Epstein, Cannon, Holt, Levy, Paterson and Thurston in their book [3], has opened up new links between group theory and the theory of machines and languages. While basic problems in combinatorial group theory like the word and conjugacy problems have clear links with logic, connections between finite state automata and the elementary geometry of (the Cayley graph of) a group which Epstein *et al* have uncovered, form a new and interesting area.

We shall present most of the basic ideas of the theory of *automatic groups* introduced in [3]. For more comprehensive introductions, see for instance [1], [6], [4]. An automatic group is a finitely generated group which is characterized by a finite set of finite state automata. One of these defines a regular language of normal forms as a subset of the free semigroup on a set of semigroup generators for the group. Another determines when two words in normal form represent the same element of the group, the others determine whether two words in normal form represent elements which differ by multiplication on the right by a generator (these are the so-called *comparator* automata). In this way the operation of multiplication in an automatic group can be defined in terms of the simplest type of machine. Finite groups, finitely generated abelian groups, finitely generated free groups, word hyperbolic groups, and most fundamental groups of three manifolds are automatic. A larger class of groups can be defined by allowing the comparator automata for multiplication to be *asynchronous* two tape automata.

The existence of such a structure has group theoretic consequences, as is born

* I would like to thank NATO and the other conference sponsors for financial support during the conference. I would also like to thank the United States National Science Foundation for its support via grant #DMS-9103098 while most of this work was carried out.

J. Fountain (ed.), Semigroups, Formal Languages and Groups, 233–253.

out for instance by the result of Epstein *et al* that an automatic group which is nilpotent is a finite extension of an abelian group.

We shall begin with the formal definitions concerning regular languages of representatives for the elements of a group and then in section 2 give a formal definition of automatic and biautomatic groups. To illustrate the power of the theory, I shall sketch the proof of two results obtained in collaboration with S.M. Gersten [6], concerning the subgroup strucure of *biautomatic groups* — a group is said to be biautomatic if, in addition to the collection of automata described above, there are also finite state automata for multiplication on the left by the generators. It is not known whether or not there exist automatic groups which are not biautomatic.

Theorem 1.1 ([6]) *A nilpotent subgroup of a biautomatic group is virtually abelian.*
A biautomatic group has no subgroup of the form $< x, y | y^{-1}x^p y = x^q >$ *with* $|p| \neq |q|$.

(To establish the first part, what is actually shown is that there is no commutator $[a,b] = a^{-1}b^{-1}ab$ of infinite order in a biautomatic group which commutes with both a and b. It is not really necessary for the reader to know that a nilpotent group is one in which the lower central series terminates.)

In Section 5 we show how the conditions for the existence of an automatic structure can be reformulated in terms of the geometry of the Cayley graph (the so–called *fellow traveller* property). (We shall not discuss isoperimetric inequalities in this article.) The geometric idea of quasiconvexity leads to a class of subgroups which are also automatic. A rather surprising link between the geometry and the language theory can then be seen: a subgroup is quasiconvex if and only if it is regular, i.e. it corresponds to a regular sublanguage of the set of normal forms. In these sections proofs, when given, are only sketched. Details can be found in [3], [1] or [6].

Some ideas for extending and furthering the theory into the realm of transducers and pushdown automata are presented in section 6. This part is basically independent of the rest of the article. Here the main point of view is to think of the generators of the group not as elements, but as machines which operate on the elements of the group. From the point of view of actual computing, it does not seem essential for the representatives of the elements of the group to be words in the generators, but rather we can allow them to be finite strings over some finite alphabet. Rather, we take the point of view that it suffices to be able to obtain a representative from a word in the generators, and be able to multiply a representative by a generator to obtain a new representative. In this way many more kinds of group can be considered (including infinitely presented ones), though the connection with the geometry described in section 5 is apparently lost.

For example, let D be the additive group of dyadic fractions, i.e., the set of all rational numbers that can be written in the form ℓ/m, where m is an integral power of 2. Each element of this group can be naturally represented in the usual way, in base 2, e.g.,

$$13 + 1/2 + 1/8 + 1/32 = 1101.10101 .$$

Despite the fact that the group is not finitely generated, these representatives of the elements of D form a regular language. If we take

$$1/2, 1/4, 1/8, \ldots$$

to be an infinite set of generators of D, then for each generator $\frac{1}{2^i}$ there is a comparator automaton M_i as in the definition of an automatic group. So, in some sense, D is as tractable as any automatic group.

We shall suppose that the reader is familiar with regular languages, finite state automata, two–tape automata, free semigroups and free monoids, and finite presentations of groups, though we shall try to avoid any advanced results from any of these areas.

I should like to thank G. Baumslag, C.F.Miller III, and M. Shapiro for many useful discussions on some of the topics presented here. I would also like to thank the NATO for their support, and the conference organisers for an enjoyable and stimulating time.

2. Rational Structures

R. Gilman discussed several ideas linking automata and group theory in the seventies.

Definition 2.1 Let A be a finite set of semigroup generators for a group G, and let $\mu : A^* \to G$ denote the natural map. Let $L \subset A^*$ be a regular language.

We say that (L, A) is a *rational structure* for G if $\mu(L) = G$, and that (L, A) is a *rational cross–section* for G if μ maps L bijectively onto G.

Regularity means that there is a finite state automaton which, when a word in A^* is input, decides whether or not it is in the set of representatives L. For instance let x, y, X, Y be a set of standard generators for $\mathbb{Z} \times \mathbb{Z}$, where $x^{-1} = X, y^{-1} = Y$. Then the set

$$\{x^n y^m | n, m \geq 0\} \cup \{X^n y^m | m \geq 0, n > 0\} \cup \{x^n Y^m | n \geq 0, m > 0\} \cup \{X^n Y^m | n, m > 0\}$$

is a rational cross–section for $\mathbb{Z} \times \mathbb{Z}$.

Gilman [7] studied various properties of groups with rational cross–sections. Recall that a group is said to have a solvable word problem if there is an algorithm which, when given a word in the generators of the group, decides whether or not the word represents the trivial element of the group. (The solvability of the problem is independent of the finite set of generators chosen.)

Theorem 2.1 *Let G be a group with a rational cross–section.*
(1) If G is finitely presented, then G has solvable word problem.
(2) If G is infinite, then G contains an element of infinite order.

Proof. Let $L \subset A^*$ be a rational cross-section for G, where A is a finite set of semigroup generators. There is a finite set R of relators for G, i.e. such that $\langle A | R \rangle$ is a finite presentation for G.

The set $\mathfrak{R}$ of all products of conjugates of the relators and their inverses is a recursively enumerable set. As L contains at least one representative for each element of G, one of these elements appears in the list L. It can be found in the usual way, by listing the first N elements of both sets, and looking for an intersection. If no element is found common to both, then another element is generated from each set. Eventually an element will be found. This is the word ε of L which is the representative of the identity element of G.

Let w be a word to be tested for triviality. Premultiply each element of $\mathfrak{R}$ by w, to give the recusively enumerable set $w\mathfrak{R}$. This generates all of the words in $F(X)$ which represent the same element of the group as w does. As before, L contains a unique representative for each element of G, and the unique element of $L \cap w\mathfrak{R}$ can be eventually found. If this element is ε, then w is trivial, else w is not trivial.

The second part of the theorem is a simple application of the pumping lemma. For an infinite regular language L over a finite alphabet there is a constant $k > 0$ such that if w is a word in L, and $\ell(w) > k$, then w can be rewritten in the form $w = w_1w_2w_3$ with $\ell(w_2) > 0$, such that $w_1(w_2)^n w_3 \in L$ for all $n \geq 0$.

It now follows that the element $\mu(w_2)$ of G has infinite order in G, else each element $\mu(w_1(w_2)^k w_3)$ has infinitely many representatives. (Notice however that w_2 is not in general an element of L.) □

Notice that in part (1), the regularity of the set L was not really used, just the fact that the set is recursive. Also, in part (2), the uniqueness of the representatives in L is not necessary, just the fact that there are only finitely many representatives for each element of G.

Gilman also shows that the class of groups with rational cross–sections includes finite groups, free groups, finitely generated abelian groups, and is closed under free products, direct products and extensions.

The most interesting result of Gilman's is a generalization of a result of Anissimov and Seifert. First we require a definition.

Definition 2.2 Let (L, A) be a rational structure for the group G, and let H be a subgroup of G. We say that H is an *L–rational subgroup* of G if the preimage of H in L, $\mu^{-1}(H) \cap L$, is a regular language.

For instance a finite index subgroup of a group is an L–rational subgroup, for any rational structure. This can be seen easily, as follows. Let (L, A) be a rational structure for G, and let H be a subgroup of finite index in G. Build a finite state automaton $M(H)$ with one state for each coset, and a change of state function corresponding to multiplication by the generators in A, i.e. there is an edge in the state graph from Hg to Hg' labelled $a \in A$ when $g\mu(a) \in Hg'$. If the initial state is the unique accept state, then the language accepted by this automaton is exactly the set of words in A which represent elements of H. Intersecting this with L gives a rational language representing the elements of H. (See [6] for a geometric version of this construction.)

Anissimov and Seifert proved the following result about free groups. Let $F(X)$ be the free group on the finite set X. Let L be the regular language of all reduced words over the alphabet $A = X \cup X^{-1}$. Then a subgroup H of $F(X)$ is L-rational if and only if H is finitely generated.

Theorem 2.2 ([7]) *Let (L, A) be a rational structure for the group G. If H is an L–rational subgroup of G, then H is finitely generated.*

Proof. Let $L' = L \cap \mu^{-1}(H)$, and let M' be a finite state automaton accepting L'. The problem is to finite a finite set of generators B for H, and a regular language L'' in B^*.

During acceptance of a word in L' by M', from each state visited there is a shortest path to an accept state. The length of such a shortest path is bounded by the number of states in the machine. Let $c_1, \dots, c_k$ be the words labelling such paths. Let word $w = \alpha_1 \dots \alpha_n$ be a word in L' ($\alpha_i \in A$). For each j, there is a c_{i_j} such that $\alpha_1 \dots \alpha_j c_{i_j}$ is an element in L', and $(c_{i_j})^{-1}\alpha_{j+1}c_{i_{j+1}}$ represents an element of H. The set of elements $c_{i_j}{}^{-1}a_{j+1}c_{i_j}$ generates H, and translation of the elements of L' into this form provides a regular language. □

3. Automatic groups

We now add more structure. We have associated an automaton to the group G as a set. Now attach automata to the operation of multiplication in G. At the moment we consider two main ways to do this:

Firstly we consider finite state automata which read two tapes at the same rate (at each step a pair of letters is read, one from each tape). In general the two words u, v will have different lengths, so we pad out the end of the shorter of the two with \$ symbols until they have the same length. Write $(u, v)^\$$ to denote this padded pair.

Definition 3.1 Let (L, A) be a rational structure for G, $\mu : A^* \to G$ the natural map. If the sets

$$L_= = \{(u,v)^\$ | \mu(u) = \mu(v),\ u,v \in L\}$$

and for all $a \in A$

$$L_a = \{(u,v)^\$ | \mu(u) = \mu(va),\ u,v \in L\}$$

are regular languages over the alphabet $A^\$ = \{(a,b)\ | a,b \in A \cup \{\$\}\}$, then we say that (L, A) is a ***synchronous automatic*** structure for G (the word "synchronous" is often omitted)

If in addition the languages

$${}_aL = \{(u,v)^\$ | \mu(u) = \mu(av),\ u,v \in L\}$$

are regular over $A^\$$ for each $a \in A$, then we say that (L, A) is a ***biautomatic structure***.

A larger class of groups is obtained if we take the comparator automata for the multiplication process to be ***two–tape automata***. These read from two tapes, but may read an arbitrary number of symbols from one tape before reading from the other. If the languages $L_=$ and L_a (without the padding \$'s) are the languages of two–tape automata, then we say that (L, A) is an ***asynchronous automatic*** structure.

We say that a group is synchronously automatic (resp. asynchronously automatic) if it has a synchronously automatic (resp. asynchronously automatic) structure, with respect to some set of generators.

Examples: Finite groups, finitely generated free groups, finitely generated abelian groups are easily seen to have automatic structures. Surface groups, braid groups and many knot groups and 3–manifold groups also have automatic structures. The so–called Baumslag–Solitar group $< x, y | y^{-1}xy = x^2 >$ is asynchronously automatic but is not synchronously automatic [3]. The classes are closed under free and direct products, and amalgamated products of abelian groups are automatic.

Another interesting class of examples is provided by the class of *word hyperbolic* groups introduced by Gromov [8]. A group G is *word hyperbolic* if, with respect to any finite set of group generators X, the set of geodesic words, i.e. shortest representatives in $(X \cup X^{-1})^*$, forms an automatic structure for G. That this definition is equivalent to the other definitions follows from work of P. Papasoglu [10]. An interesting equivalent definition is that a group is word hyperbolic if and only if it has a finite presentation $< X | R >$ where each relator can be written in the form $u(r) = v(r)$, with $\ell(u(r)) > \ell(v(r))$, such that if a word in $F(X)$ represents the trivial element of G, then there is a sequence of length reducing operations $w_1 u(r) w_2 \to w_1 v(r) w_2$ which convert w into the empty word. We shall give a more geometric characterization of hyperbolicity in section 4, and we shall show geometrically that $\mathbb{Z} \times \mathbb{Z}$ is never a subgroup of a word hyperbolic group.

Some of the basic results about automatic structures:

Lemma 3.1 *Let (L, A) be an automatic structure for G.*

(1) There is a rational cross–section (L', A) with $L' \subset L$ which is also an automatic structure for G.

(2) If (L, A) is also a rational cross-section, then there is a constant $k > 0$ such that if $(u, v)^\$ \in L_a$ then $|\ell(u) - \ell(v)| < k$.

(3) For all $w \in A^*$, the language

$$L_w = \{(u, v)^\$ | \mu(u) = \mu(vw), \ u, v \in L\}$$

is regular.

(4) If B is another finite set of semigroup generators then G has an automatic structure (L', B).

An automatic structure which is also a rational cross–section, as in (1), is often called an automatic structure with uniqueness. It follows from (1) and (4), that if a group is automatic, then for any finite set of generators there is an automatic structure with uniqueness whose language is in the free semigroup on that finite set.

The extra structure of biautomatic groups allows one to easily prove that the conjugacy problem for biautomatic groups is solvable. Recall that the conjugacy problem for G is solvable (with respect to a finite set of generators X) if there is an algorithm which, when given two words u and v in $F(X)$ will decide whether or not they are conjugate in G.

Proposition 3.1 *The conjugacy problem is solvable for biautomatic groups.*

Proof. Let G be a biautomatic group, and let X be a finite set of group generators. Take as semigroup generators $A = X \cup X^{-1}$, and let (L, A) be a biautomatic structure. Let $u, v \in A^*$; if these words represent conjugate elements of G, then there is some word $w \in L$ such that $\mu(w^{-1}uw) = \mu(v)$ or $\mu(uw) = \mu(wv)$. But this means that there is some word $z \in L$ such that $(z, w)^\$ \in L_v$ and $(z, w)^\$ \in {}_uL$. Whether or not the intersection of two regular languages is empty is a decidable property, so the proof is complete. □

Basically the same method produces a slightly stronger result, which has many interesting consequences:

Theorem 3.2 *Let (L, A) be a rational structure which is also a biautomatic structure for the group G.*

The centralizer of any element $g \in G$ is an L–rational subgroup.

Proof. Let h be an element of the centralizer of g, and let $w(h), w(g), w(hg)$ denote the representatives of the elements g, h, hg in the language L. Then $h^{-1}gh = g$ is equivalent to $gh = hg$, and $(w(hg), w(h))^\$ \in L_g$ and $(w(gh), w(h))^\$ \in {}_gL$. Thus the centralizer of g consists of the projection of the intersection $L_g \cap {}_gL$ onto the second factor. It follows from the elementary properties of regular languages that the centralizer is thus an L–rational subgroup. □

The usefulness of this result comes from the following:

Theorem 3.3 ([6]) *If (L, A) is an automatic (resp. biautomatic, word hyperbolic, asynchronously automatic) structure for a group G, and H is an L–rational subgroup, then H too is an automatic (resp. biautomatic, word hyperbolic, asynchronously automatic) group.*

The proof of this proposition follows the proof of Gilman's original result, i.e., Theorem 2.2 paying a little more attention to details. The word hyperbolic case is a little trickier and requires a somewhat more subtle property of word hyperbolic groups, namely that quasigeodesics lie close to geodesics.

It is not too hard to show that word hyperbolic groups are biautomatic. Let X be a set of generators, $A = X \cup X^{-1}$ and let (L, A) be the word hyperbolic structure of all geodesic words in A^*. The formal inverse of a geodesic word is also geodesic, so $\mu(u) = \mu(av)$ if and only if $\mu(u^{-1}) = \mu(v^{-1}a^{-1})$. Thus the language

$$_aL = \{(u, v)^\$ | (u^{-1}, v^{-1})^\$ \in L_{a^{-1}}\} = \{(u, v)^\$ | (v^{-1}, u^{-1})^\$ \in L_a\}$$

This is a regular language.

Already we can deduce an interesting consequence, namely: if a word hyperbolic group were to contain $\mathbb{Z} \times \mathbb{Z}$ as a subgroup, then some free abelian group of rank at least two would also be a word hyperbolic group (which is not possible).

This follows by considering the centralizer of one of the generators of such a subgroup. This must also be a word hyperbolic group, as it is an L–rational subgroup. The centre of this subgroup is in turn a rational subgroup of that. This (abelian) subgroup thus contains a free abelian group of rank at least 2 as a finite index subgroup, which is thus a rational subgroup.

That free abelian groups of rank greater than one are not word hyperbolic will be shown in section 5.

4. Translation Lengths

A useful technique, suggested originally by Gromov, and developed in [6], allows one to measure a sort of asymptotic length for elements in a biautomatic group.

Definition 4.1 Let A be a finite set of semigroup generators for the group G.

The *translation length* of an element $g \in G$ with respect to A is defined to be

$$\tau_{G,A}(g) = \lim_{n>0} \frac{|g^n|_A}{n}$$

Here $|g^n|_A$ denotes the length of the shortest word in A^* representing g^n.

In a free abelian group, using a standard set of generators, it is not hard to see that a non-trivial element has non-zero translation length. In general, this length depends on the set of generators chosen — if we choose a set of generators which includes the element g, then the translation length of g is not more than 1. However whether the translation length of an element is zero or not is an invariant. The basic properties of translation lengths are summed up in:

Lemma 4.1
(1) $\tau_{G,A}(g)$ is a conjugacy class invariant.
(2) $\tau_{G,A}(g) \neq 0$ is independent of the finite generating set chosen.
(3) $\tau_{G,A}(g^k) = k\tau_{G,A}(g)$.
(4) If (L, A) is an automatic structure for G and H is an L–rational subgroup, then for $h \in H$, $\tau_{H,B}(h) \neq 0$ if and only if $\tau_{G,A}(h) \neq 0$, where B is any finite generating set for H.

The last part here follows from a careful study of Gilman's original proof of Theorem 2.2.

From this lemma it is easy to see that:

Proposition 4.1 ([6]) *Let G be a biautomatic group.*
(1) An element has finite order if and only if it has zero translation length.
(2) If x is an element of infinite order and there is an element y such that $y^{-1}x^py = x^q$ with $p, q > 0$, then $p = q$.
(3) Let $x, y \in G$ be elements of infinite order and suppose that $[x, y] = x^{-1}y^{-1}xy$ commutes with x and with y. Then $[x, y]$ has finite order in G.

Proof. For part (1), let $C_G(g)$ denote the centralizer of g in G. This is a biautomatic group, and so too is the centre of this, $Z(C_G(g))$. But this is an abelian group, and so if g has infinite order then it has non-zero translation length.

Part (2) now follows from parts 1,2 and 3 of the lemma. The third part follows by observing that the fact that the commutator commutes with the elements x and y implies that $[x,y]^{n^2} = [x^n, y^n]$. This in turn implies that the commutator $[x,y]$ has zero translation length. □

In a nilpotent group which is not a finite extension of an abelian group there exist commutators with the properties in (3) above, and so a nilpotent subgroup of a biautomatic group is virtually abelian.

5. Geometry

Recall that the Cayley graph of a group G with respect to a finite set of group generators X is a graph $\Gamma_X(G)$ whose vertices are the elements of G, and for each element x of X, and each $g \in G$, there is a directed edge labelled x from the vertex g to the vertex $g\mu(x)$.

We shall require here that X be a set of group generators, because we shall have to consider non–directed paths in this graph. Traversing the edge labelled x in the opposite direction to its given orientation corresponds to the label x^{-1}. So here we shall always assume that the set of semigroup generators under consideration is $A = X \cup X^{-1}$.

Given a word w in the free semigroup A^* and an element $g \in G$, there is a path labelled w beginning at g and terminating at $g\mu(w)$. We shall usually consider a word w of length $\ell(w) = n$ to correspond to a map of the interval $[0, n]$ to $\Gamma_X(G)$ beginning at the identity vertex and terminating at the element of the group $\mu(w)$ represented by this word. At integral time t, the value $w(t)$ is the vertex of G corresponding to the inital segment of w of length t, and when $t \geq n$, $w(t)$ is $\mu(w)$.

Because X is a set of generators for G, the graph is connected. We make it into a metric space by assigning length 1 to each edge, and defining the distance between vertices to be the length of the shortest path joining them. This distance is invariant under the natural left action of G on its Cayley graph.

Definition 5.1 For a constant $K \geq 0$, two words $u, v \in A^*$ are called *K–fellow travellers* if, for all t, the distance between the vertices $u(t)$ and $v(t)$ is at most K. This means that the paths labelled u and v beginning at the identity vertex are uniformly close in the Cayley graph.

The somewhat surprising fact about automatic structures is

Proposition 5.1 *A rational structure (L, A) for G is an automatic structure if and only if there is a $K \geq 0$ such that for all $u, v \in L$, and all $a \in A$, if $\mu(u) = \mu(va)$, then u and v are K-fellow travellers.*

We can restate this as: words which begin at the identity vertex and end unit distance apart in the Cayley graph are uniformly K close. The analagous condition for biautomatic groups is that words which begin unit distance apart, and end unit distance apart are uniformly close.

For the proof of this (which is not hard to follow) we refer the reader to [3]. In some way the automata for the automatic structure seem to "roll up" the Cayley graph, and the homogeneity of graph gives the finiteness of the automata.

Notice that immediately this means that a free abelian group of rank at least two is not word hyperbolic. If we take x, y to be two of the standard generators for $\mathbb{Z}^k$, then the geodesic words $x^n y^n$ and $y^n x^n$ (based at the identity vertex) end at the same vertex, but at time n are distance $2n$ apart (remember distance is measured in the 1 dimensional graph, not in the plane). If the group were word hyperbolic then the set of all geodesic words would form the language of an automatic structure, contradicting the proposition.

Another consequence of the fellow traveller property is:

Proposition 5.2 *Automatic groups are finitely presented.*

Proof. Let (L, A) be an automatic structure for the group G, where $A = X \cup X^{-1}$, for some finite set of group generators X. Suppose that the word $w \in F(X)$ represents the trivial element of G. The path labelled w beginning at the identity vertex of $\Gamma_X(G)$ is then a loop. Let $w = a_1 \dots a_n$. For each vertex $v_i = \mu(a_1 \dots a_i)$ on this loop, let w_i be a shortest element of L such that $\mu(w_i) = v_i$. By the fellow traveller property this means that, for some K, the words w_i and w_{i+1} are uniformly K–close. It follows that the loop labelled w can be expressed as a sum of small loops, each of length at most $2K+2$. (See [3] for the standard picture here.) This is equivalent to saying that there is a finite set of relators of length at most $2K + 2$ for the group (as X is finite). □

For asynchronous automatic groups, there is a similar equivalent condition, but it is somewhat more complicated. It does however imply a simply stated necessary condition:

Proposition 5.3 *If a rational structure (L, A) for G is an asynchronous automatic structure then there is a $K \geq 0$ such that for all $u, v \in L$, and all $a \in A$, if $\mu(u) = \mu(va)$, then the paths u and v in $\Gamma_X(G)$ lie in K-neighbourhoods of each other.*

In other words, in an asynchronously automatic group representative words which finish close together, are always close together, though perhaps not uniformly so. It can also be deduced that asynchronously automatic groups are finitely presented.

There is also a geometric version of the rational subgroups discussed in section 1.

Definition 5.2 Let (L, A) be a rational structure for G, and C be a positive number. A subgroup is *C-quasiconvex* with respect to L if for all $w \in L$ such that $\mu(w) \in H$, the path w in $\Gamma_X(G)$ from the identity vertex to $\mu(w)$ lies in a C–neighbourhood of H.

The rather surprising result concerning this definition is:

Proposition 5.4 ([6]) *A subgroup is L–rational if and only if it is L–quasiconvex.*

Again this is proved by interpreting the proof of Theorem 2.2 in this geometrical manner.

We finish this section with the more usual definition of a hyperbolic group: Given three vertices a, b, c in a Cayley graph, a ***geodesic triangle*** is a set of three geodesic (i.e. minimal length) paths from a to b, b to c, and from c to a. Such a triangle is said to be δ–thin if each side is in a δ–neighbourhood of the union of the other two.

A group is word hyperbolic with respect to finite set of generators X if and only if there is a δ such that all geodesic triangles in $\Gamma_X(G)$ are δ–thin.

In fact this definition is independent of the set of generators chosen, though of course the value of δ depends on the particular finite set of generators used. Using this definition it is easy to see that $\mathbb{Z} \times \mathbb{Z}$ is not word hyperbolic — there is no K such that all geodesic triangles in the plane (even in the 1 dimensional complex consisting of points with at least one integral coordinate), are K–thin.

6. Transducers

One of the most attractive features of an automatic group is the existence of an efficient algorithm for constructing its Cayley graph. If we restrict our viewpoint to the comparator automata described in the definition, we get the following inefficient algorithm.

Let (A, L) be an automatic structure for the group G. The Cayley graph of G with respect to the set X of generators can be constructed as follows.

Consider first the free group $F = F(X)$. Let $X(0) = \{\varepsilon\}$, where ε denotes the empty word, and let $X(i)$ be the set of all words in F of length at most i. The set $X(i+1)$ is obtained by adjoining to these words each of the generators and their inverses. The set $X(i)$ is easily generated. Those elements of $X(i)$ which lie in L can be found by simply testing each element of $X(i)$ in the finite state automaton which accepts the language L. This works as a kind of filter, generating labels (in L) for the vertices of the Cayley graph. To see whether two vertices are to be joined by an edge labelled x, the elements of $X(i) \cap L$ are fed in pairs (suitably padded) into the finite state automaton M_x which accepts the language L_x.

Because one can build the Cayley graph of an automatic group to any desired finite size, the word problem is solvable.

However there is a more efficient algorithm which solves the word problem in quadratic time.

What this means is that there exists a Turing machine, which, when given a word in the generators and their inverses, decides whether of not the word represents the identity element of the group. The construction of the Cayley graph described above provides an inefficient such algorithm. There is another algorithm, to be described later, which is more efficient, in that the number of machine operations, i.e., the number of state changes, the number of letters read, the number of letters deleted, and the number of letters written, is proportional to the square of the length of the word given. Thus, if each kind of machine operation takes unit time, the total time taken is a quadratic function of the length of the input word.

We shall see that in an automatic group, right multiplication by the generator x can be described by a finite state transducer.

Here is a formal definition of a finite state transducer (see e.g., [2], sect. III.6):

Definition 6.1 (A finite state transducer) A finite state transducer is a quintuple $(A, B, S, s_0, Y, \Delta)$ where:
(1) A is a finite input alphabet;
(2) B is a finite output alphabet;
(3) S is a finite set of states;
(4) $s_0 \in S$ is the initial or start state;
(5) Y is a subset of S, the set of final or accept states;
(6) $\Delta \subset (A \cup \varepsilon) \times S \times S \times B^*$ is a finite set of transitions (or edges).

We think of a finite state transducer as a finite, labelled, directed graph, with a distinguished vertex, the initial vertex s_0. An element $(a, s, s', w) \in \Delta$ we think of as an edge from the vertex s to the vertex s' labelled by a pair a/w, where $a \in A \cup \{\varepsilon\}$, $w \in B^*$, where ε denotes the empty word. On an edge labelled a/w, the generator a (or ε) is regarded as being the input letter, and the word w is regarded as being the word output as that edge is traversed. A successful computation by this transducer corresponds to a directed path from the initial vertex to one of the accept vertices. The sequence u of input letters is thought of as the *input word* and the corresponding sequence v of output words is thought of as the result, or output, of the computation. We say $v \in T(u)$, and that the path is labelled u/v.

The transducer defined here is non-deterministic; that is, at any vertex there may be several edges leaving with the same input generator a, but with possibly different output words w. Thus a transducer defines a map from a subset of A^* to the set of subsets of B^*. It is slightly misleading to describe a finite state transducer as an input/output machine. An initial segment of a given input word may lead to more than one state of the transducer, and have several different corresponding output words (this will not happen when L is closed under initial segments). Thus when attempting to find $T(u)$ for some $u \in L$, we have to remember several possible partial solutions until we reach an accept vertex having gotten to the end of the word u.

Suppose now that $A = B$, and in addition that the finite state transducer T defines a *function* from $L \subset A^*$ to itself. Then it follows from the work of Eilenberg (see for example [2] chap. IV, 4.3.) that there is a transducer T' with the following property. For each word $u \in L$ there is exactly one $v \in L$ such that $v \in T(u) = T'(u)$, and there is exactly one sequence of edges from the initial state to an accept state labelled u/v, and so we can write $T'(u) = v$. A transducer with this property is said to be *unambiguous*. Notice that in an unambiguous transducer there is a bound on the length of a sequence of consecutive edges with input symbol ε (and on the length of a sequence of consecutive edges with output symbol ε) which can occur in a path to an accept vertex.

A finite state transducer is *deterministic* if (a) each edge with input label ε leads to a final state, i.e., one which is not the initial state of any edge., and (b) for each state s and generator a, there is at most one edge leaving the state s with input label a. We shall say that a transducer is *restricted* if there is a bound on the number of edges with empty input label which can occur in a path to an accept vertex, and all edges whose output label is of length greater than one terminate in a final state.

Notice that in the case of a restricted finite state transducer, there is a constant K such that if $v = T(u)$ then $\ell(v)$ is at most $\ell(u) + K$. On the other hand, when T is a general unambiguous finite state transducer, the most we can say is that there is a constant C such that if $v = T(u)$, then $\ell(v)$ is at most $C\ell(u)$.

We introduce now the notion of a ***transducer group***. To this end, let G be a group, and let X be set of generators of G, closed under inverses, and let A be a finite alphabet (*not* necessarily X).

Definition 6.2 (Transducer group) The group G is a transducer group if
(1) there is a language L over A and a bijection ν from L to G;
(2) L contains the empty word denoted ε and $\nu(\varepsilon) = 1$, the identity element of G;
(3) for each element x in the set X, there is an unambiguous finite state transducer T_x such that for each $u, v \in L$,

$$T_x(u) = v \text{ if and only if } \nu(u)x =_G \nu(v).$$

We say that (A, L, X) is a ***transducer structure*** for G. Notice that X need not be finite, and L need not be regular. If all the transducers in (3) are deterministic (restricted) we say that G is a ***deterministic (restricted) transducer group***.

Notice that the set L is recursive, as it can be built up from the representative ε of the identity element by successive multiplications by the transducers T_{x_i}. Moreover, whether or not a word in A^* is in L, can be decided (using the length estimates given above).

Notice also that, in the definition, we are concerned only with the effect of the transducers on the words in L. The effect of the transducers upon the elements of $A^* - L$ is not determined.

Let (A, L, X) be a transducer structure for the group G. The Cayley graph of G with respect to the set X of generators can be constructed as follows.

Let $L(0) = \{\varepsilon\}$, and let $L(i)$ denote the set of words in L representing elements of G which can be expressed as a product of at most i of the generators in X (which is closed under inverses). The set $L(i+1)$ can be obtained from the set $L(i)$ by applying the transducers T_x to each of the elements of $L(i)$. That is

$$L(i+1) = L(i) \cup \Big(\bigcup_{x \in X} T_x(L(i)) \Big).$$

This generates labels (in L) for the vertices of the Cayley graph. In addition, the Cayley graph has an edge labelled x from the vertex labelled u (in L) to the vertex labelled v (and an edge labelled x^{-1} from v to u) if and only if $T_x(u) = v$. In this way, the ball of radius $i+1$ in the Cayley graph is constructed from the ball of radius i.

Examples:
(1) The additive group of dyadic numbers (as described in the introduction), is an example of an infinitely generated transducer group. Notice that the transducers for addition are non–deterministic.
(2) The so–called Baumslag–Solitar group (Actually due to Graham Higman.)

Let $G = \langle a, t \mid t^{-1}at = a^2 \rangle$. This group can be thought of as an HNN extension of the dyadic rationals, in such a way that the elements $t^{-i}at^i$ and $t^{-j}at^j$ commute.

A unique description for each element of this group can be obtained by pushing t-symbols as far to the left as possible. In this case t^{+1} can be moved leftwards past an $a^{\pm 1}$ leaving $a^{\pm 2}$ behind, while t^{-1} can be moved leftwards past an $a^{\pm 2}$ leaving an $a^{\pm 1}$ behind.

This gives rise to two normal forms:

$$t^n a^{\epsilon_0} t^{-n_0} a^{\epsilon_1} t^{-n_1} a^{\epsilon_2} t^{-n_2} \dots a^{\epsilon_k} t^{-n_k} a^N, \quad n_i > 0\,, \; N, n \in \mathbb{Z}, \; \epsilon_i = \pm 1$$

and

$$t^m a_{m_0}^{\epsilon_0} a_{m_1}^{\epsilon_1} \dots a_{m_k}^{\epsilon_k} a^N, \text{ where } a_j = t^{-j} a t^j, \; m_j < m_{j+1} < 0.$$

It is convenient to use the fact that the a_i commute, and write

$$t^m a^N a_{m_k}^{\epsilon_k} a_{m_{k-1}}^{\epsilon_{k-1}} \dots a_{m_1}^{\epsilon_1} a_{m_0}^{\epsilon_0}\,, \quad 0 > m_i > m_j \text{ if } i > j.$$

We can re-express the element above as a sequence of m $1's$ (preceded by a – symbol if $m < 0$) (corresponding to the initial power of t), followed by a marker of some kind (we shall use a dot '.' here), followed by a sequence of 1's and 0's (preceded by a – - symbol if N is negative), representing the power N of a in base two. This is followed by a "decimal point", followed by a sequence of ±1's and 0's, with a ±1 in the mth position if the power of $a_{-m} = t^m a t^{-m}$ is ±1 (not zero). Thus for example the element 1111.1101.10101 represents the element

$$t^4 a^{13} a_{-1} a_{-3} a_{-5}$$

and −111. –10010.0011 represents

$$t^{-3} a^{-18} a_{-3} a_{-4}.$$

Now consider multiplication by generators. Multiplication on the right by a is easily described; multiplication by t has the effect

$$w \to t^n a^{\epsilon_0} t^{-n_0} a^{\epsilon_1} t^{-n_1} a^{\epsilon_2} t^{-n_2} \dots a^{\epsilon_k} t^{-n_k+1} a^{2N}.$$

The effect of multiplication by t^{-1} depends upon whether N is odd or even (for simplicity we describe the case when N is positive).

$$M_{t^{-1}} : w \to t^n a^{\epsilon_0} t^{-n_0} a^{\epsilon_1} t^{-n_1} a^{\epsilon_2} t^{-n_2} \dots a^{\epsilon_k} t^{-n_k-1} a^{N/2} \quad \text{when } N \text{ is even}$$
$$w \to t^n a^{\epsilon_0} t^{-n_0} a^{\epsilon_1} t^{-n_1} a^{\epsilon_2} t^{-n_2} \dots a^{\epsilon_k} t^{-n_k} a t^{-1} a^{(N-1)/2} \quad \text{when } N \text{ is odd.}$$

If we use the final set of normal forms in terms of –, 1's and 0's, multiplication on the right by t corresponds to increasing the initial sequence of 1's by one (or decreasing the number if the first symbol read is a '–' symbol). Furthermore, the binary number represented by the subsequent sequence of 1's and 0's must be multiplied by 2. This means that the "decimal point" must be shifted one place to the right. Thus multiplying 1111.1101.10101 by t gives 11111.11011.0101. This can be realised by a deterministic finite state transducer. Multiplication by t^{-1} decreases the initial power of t (if it is positive) and shifts the decimal point one place to the left: multiplying 1111.1101.10101 by t^{-1} gives 111.110.110101. It seems that this cannot be realised by a deterministic finite state transducer: the machine must be able to "look ahead" on the input tape to see if the next symbol is to be the decimal point symbol. The operation can be realised by a non–deterministic finite state automaton (or by a machine with "look ahead"; see for example [9], chap. 10.8).

Notice that G is an asynchronously automatic group, but it is not (synchronously) automatic [3].

(3) An infinitely presented group.

Consider the group

$$G = \langle a, b, t \mid t^{-1}at = a^2,\ t^{-1}b^2t = b,\ [a, b^{t^n}] = 1\ (n \in \mathbb{Z}) \rangle\ .$$

G is a semidirect product of the direct product of two copies of the group of dyadic fractions by an infinite cyclic group. The generator of this infinite cyclic group acts on the direct product by squaring the elements of the first factor and taking the square root of the elements in the other factor. The group G is not finitely presented. We can construct normal forms for the elements of G by mimicking the approach used in the previous example, and transducers for multiplication by the generators similarly.

This is a finitely generated non–finitely presented transducer group.

The following theorem sums up many properties of transducer groups.

Theorem 6.1

(1) If a group G has a transducer structure (A, L, X) and X' is another set of generators for G which is also closed under inverses, then (A, L, X') is a transducer structure for G.

(2) Direct products and free products of transducer groups are transducer groups.

(3) If H is a subgroup of a transducer group G, then H is a transducer group.

(4) If a group is automatic, then it is a restricted transducer group. If a group is asynchronously automatic, then it is a transducer group.

(5) If G is a finitely generated deterministic restricted transducer group, then the word problem is solvable in quadratic time.

(6) If G is a finitely generated deterministic transducer group, then the word problem is solvable in exponential time.

Proof.

(1) We must construct a transducer for multiplication by each element of X'. We shall in fact show that for any element g of G, there is a finite state transducer T_g such that for each $u \in L$, $T_g(u) = v$ if and only if $\nu(v) = \nu(u)g$. We show how to construct the transducer for a word w in $F(X)$ of length 2. A finite state transducer for multiplication on the right by a word of any length can be built by repeating this construction.

Let $T_x = (A, A, S, s_0, Y, \Delta)$ and $T_y = (A, A, S', s_0', Y', \Delta')$ be finite state transducers for multiplication by the generators x and y in X.

Let $T_{xy} = (A, A, S \times S', (s_0, s_0'), Y \times Y', \Delta'')$ be a finite state transducer, with the set of transitions Δ'' defined as follows. There is a transition $(a, (s_1, s_1'), (s_2, s_2'), c)$ in Δ'' if there is a word $c' = a_1 \ldots a_k \in A^*$ such that

(a) $(a, s_1, s_2, c') \in \Delta$,

(b) there is a sequence of states $s'_1 = t_0, t_1, \ldots, t_k = s'_2$ in S', and

(c) a sequence of edges $(a_1, t_0, t_1, b_1), (a_2, t_1, t_2, b_2), \ldots, (a_k, t_{k-1}, t_k, b_k)$ in Δ', such that $b_1 b_2 \ldots b_k = c$.

Notice that the set of such c' is finite, and that if the transducers T_x and T_y are both restricted, then so is T_{xy}.
(2) Let (A, L, X) and (B, L', X') be transducer structures for the groups G and H, where A and B are assumed to be disjoint. Let T_x (respectively $T_{x'}$) be a transducer for multiplication by $x \in X$ (resp. $x' \in X'$) in the transducer structure for G (resp. H).

We show that there is a transducer structure $(A \cup B, LL', X \cup X')$ for the direct product $G \times H$. Here LL' denotes the language $\{ww' \mid w \in L, w' \in L'\}$. A transducer for multiplication by a generator x in the product structure is obtained from the transducer T_x by adjoining at each accept state of T_x an edge labelled b/b for each $b \in B$. Notice that in general this gives a non–deterministic transducer, which has some edges labelled ε/x which end in a non–final state (the b/b edges are attached there). Similarly a transducer for multiplication by x' in the product structure can be obtained by adjoining an edge labelled a/a for each $a \in A$ at the initial vertex s'_0 of the transducer $T_{x'}$.

For the free product $G*H$, it is not hard to see that there is a transducer structure

$$(A \cup B, L(L' - \{\varepsilon\})(L - \{\varepsilon\}))^* L', X \cup X').$$

(3) This follows immediately from the proof of (1).
(4) Let G be an automatic group, and let (A, L) be an automatic structure which is a rational cross–section. Let M_x be the comparator automata for multiplication on the right by the generator $x \in X \cup X^{-1}$.

Each edge is labelled by a pair (a, b) with $a, b \in X \cup X^{-1} \cup \{\$\}$. To create a transducer, replace each such label by the label a/b. We thus consider the lefthand letter a to be the input letter, the righthand one b to be the output word. A sequence of \$ symbols on one side or the other cannot be followed (on that side) by any other symbol. Each such sequence can then be replaced by a single edge appropriately labelled ε/c or c/ε, for some $c \in A^*$. This defines a restricted non–deterministic transducer, which is unambiguous. It is non–deterministic as there may well be edges labelled (a, b) and (a, c) starting from the same vertex of M_x.

For asynchronously automatic groups a similar construction converts a two tape automaton into a transducer. Edges labelled a which leave a state in S_L (respectively S_R) are relabelled a/ε (resp. ε/a).
(5) One way of solving the word problem for a transducer group is to construct the Cayley graph. That is, to see whether a word w of length n is trivial in the group, build the ball of radius n in the Cayley graph, using the algorithm described earlier.

A more efficient solution is to proceed as follows. Let $w = a_1 a_2 \ldots a_n$ be a word in the generators (and their inverses). Input the representative ε for the identity element of G into the transducer T_{a_1}; let $w_1 \in A^*$ be the result. Now input this word into the transducer T_{a_2}, and let w_2 be the result. Continuing in this way, the word w_n obtained from the last transducer T_{a_n} is either ε, in which case w is equal to the identity element of G, or it is not ε, in which case w does not represent the identity element of G.

If all of the transducers are restricted, there is a constant K such that $\ell(w_i) < \ell(w_{i-1})+K$. If in addition each transducer is deterministic, the number of operations performed by the transducer T_{a_i} while calculating that $T_{a_i}(w_{i-1}) = w_i$ is bounded by $\max(\ell(w_{i-1}), \ell(w_i))$. It follows that the number of calculations required to see whether w is equal to the identity element of G is at most

$$\sum_{i=1}^{n} \max(\ell(w_{i-1}), \ell(w_i)) \leq \sum_{i=1}^{n} Ki \leq K'n^2.$$

If a transducer is not deterministic, then at each stage while tracing out an input word u, one must record each of the states that arise (after having read the first j letters of u). There are at most a bounded number C of states, (and non–ambiguity means that there is just one accept path to each of them for the word being read). The number of operations is thus bounded above by a constant multiple of the length of the input word (plus a constant). It follows that while doing the calculation $T_{a_i}(w_{i-1}) = w_i$, the number of operations is at most the $\max\{\ell(w_{i-1}), \ell(w_i)\}$–th power of C. So while doing the complete calculation for the word w, the number of operations is bounded by a function which is exponential in the length of w.
(6) Using the notation of (5), if T_{a_i} is a general finite state transducer, and $T_i(w_{i-1}) = w_i$, then there is a constant C such that $\ell(w_i) \leq C\ell(w_{i-1}) + C$. The number of operations is bounded by

$$\sum_{i=1}^{n} \max(\ell(w_{i-1}), \ell(w_i)) \leq 2\sum_{i=1}^{n} C^i \leq C'e^n.$$

□

The previous proposition does not apply in the case of the Baumslag–Solitar group with the language described in example (2) above. The transducers are non–deterministic, when "moving the decimal point to the left". If the machine is allowed to "look ahead" ([9] chap. 10.8) to see if the next character after the one being read is the point or not, the machine becomes deterministic. This introduces another operation for each letter read, so that the number of operations performed while reading a word is bounded above by a linear function of the length of the word input. As in part (5) of the above theorem, it follows that:

Proposition 6.2 *The word problem for the Baumslag–Solitar group is solvable in quadratic time.*

In fact as this is a matrix group, there is a solution to the word problem which depends on how fast rational matrices can be multiplied. During such a multiplication the entries in the matrices may have entries which grow exponentially in the length of the word. It is not clear to the author whether this gives a faster algorithm (asymptotically).

Let (A, L, X) be a transducer structure for the group G. We say that it is a *bi–transducer* structure for G if in addition there are finite state transducers for left multiplication by the generators. That is, for each $x \in X \cup X^{-1}$ there is an additional transducer T'_x such that for each $u \in L$, $T'_x(u) = v$ if and only if $x\nu(u) = \nu(v)$.

As in the proof of (1) in the theorem, for each $g \in G$ there is then a transducer T'_g such that for $u \in L$, $T'_g(u) = v$ if and only if $g\nu(u) = \nu(v)$.

Proposition 6.3 *Suppose that G is a finitely generated group with a restricted bi-transducer structure (A, L, X) where X is finite. If in addition L is a regular language, then the conjugacy problem is solvable for G.*

Proof. Let M be a finite state automaton accepting the language L. This automaton can be combined with each transducer T_x and T'_x so that only pairs u/v with $u, v \in L$ lead to accept states. To see this, let $M = (A, S, \delta, s_0, Y)$ be a finite state automaton which accepts the language L. Let $T_x = (A, A, S', s'_0, Y', \Delta)$ be a finite state transducer for multiplication by x. Define a new finite state transducer T $(A, A, S \times S', (s_0, s'_0), Y \times Y', \Delta')$ by defining Δ' to contain those edges $(a, (s_1, s_2), (s_2, s'_2), b)$ satisfying

(i) $\delta(s_1, a) = s_2$; and

(ii) (a, s'_1, s'_2, b) is an edge in Δ.

It is not hard to see that $v = T(u)$ if and only if $u, v \in L$ and $v = T_x(u)$. Notice that the new transducer T is restricted when T_x is.

Assume now that the transducers T_x all have this property. Let g, h be two elements of G given as words w, w' in $F = F(X)$. As above, construct restricted finite state transducers T_g and T'_h for right multiplication by g and for left multiplication by h respectively. If there is some element $c \in G$ such that $cgc^{-1} = h$, then $cg = hc$, and there is some element γ of L such that $T_g(\gamma) = T'_h(\gamma) = \alpha$ and $\nu(\gamma) = c$. Let $\gamma = a_1 a_2 \ldots a_m$.

Let N be the maximum of the number of states in the transducers T_g and T'_h. We show that if g and h are conjugate then there is some conjugating element c as above, whose representative has length at most $N^N + N$.

Suppose that $m > N^N + N$. In the accept path through the graph of the transducer T_g with input label γ, the same vertex s is encountered more than N times. That is, for at least N values of j, say $j_1, \ldots, j_N$, the initial segment $a_1 \ldots a_j$ of γ is the input label on a path to s. As T_g is restricted, the word output while reading the initial segment $a_1 \ldots a_j$, call it α_j, has length j. As T'_h is restricted, during the calculation that $T'_h(\gamma) = \alpha$, i.e. while following the accept path with input label γ, the word α_j is output while the initial segment $a_1 \ldots a_j$ is input.

Meanwhile, in the transducer T'_h, for at least two values of j_k, say j_p and j_q $(p < q)$ the same vertex t of the graph of T'_h is encountered on reading the initial segment $a_1 \ldots a_{j_k}$ of the input word γ. It follows that the input word $a_1 \ldots a_{j_p} a_{j_q+1} \cdots a_n$ is the label of a input word in both the transducer T_g and the transducer T'_h. This word therefore lies in L, and the accept path for it in the two transducers is a subpath of the accept path for γ. What is more, as the two transducers are restricted, the output words corresponding to this word are the same.

Thus if g and h are conjugate, a representative in L can be found of length at most $N^N + N$. This solves the conjugacy problem. □

7. Other directions

It is possible to define more complicated types of transducer. The next degree of complication would be a single stack transducer. This is like the well known single–stack pushdown automaton but with an output function where the letter output depends on the input letter, the state, and the letter on top of the stack. Consider for example the Heisenberg group

$$H = \langle a, b, c \mid [a, b] = c, [a, c] = 1 = [b, c] \rangle.$$

This group is nilpotent and not automatic (it is not even asynchronously automatic) [3]. The group H may be thought of as the group of upper triangular integer matrices, of the form

$$\begin{pmatrix} 1 & p & r \\ 0 & 1 & q \\ 0 & 0 & 1 \end{pmatrix}.$$

The generators then correspond to the following matrices:

$$a = \begin{pmatrix} 1 & 1 & 0 \\ 0 & 1 & 0 \\ 0 & 0 & 1 \end{pmatrix} b = \begin{pmatrix} 1 & 0 & 0 \\ 0 & 1 & 1 \\ 0 & 0 & 1 \end{pmatrix} c = \begin{pmatrix} 1 & 0 & 1 \\ 0 & 1 & 0 \\ 0 & 0 & 1 \end{pmatrix}.$$

Thus a set of normal forms can be taken to be the set of triples of integers (p, q, r), or the set of words which begin with a string of a's (or A's if $p < 0$), followed by a string of b's (or B's if $q < 0$), followed by a string of c's (or C's if $r < 0$).

Using the first set of normal forms, the transducers for multiplication realise the functions

$$M_a : (p, q, r) \to (p+1, q, r)$$

$$M_b : (p, q, r) \to (p, q+1, r+p)$$

$$M_c : (p, q, r) \to (p, q, r+1)$$

The second set requires a machine which can output the sum of the numbers $r + p$, while at the same time being able to recall the number p for output in its turn (or vice versa). A finite state transducer cannot do this, though a "push down transducer" can. It is not hard to see how to define such transducers.

Also it is possible to make a deterministic single stack transducer from a finite state comparator automaton like those defined for multiplication in an automatic group (or from the two tape automata defined for an asynchronously automatic group). The only problem is that the output word is 'back to front', that is, if the representative for $\nu(u)x$ is $v = a_1 \dots a_n$, then the word which is output is $a_n a_{n-1} \dots a_1$. (This construction is essentially the one described in [3] sect. 2.3.10.)

A problem with this definition arises if one attempts to define a machine for multiplying by a word $w \in F(X)$ from the machines for multiplying by the generators. In order to obtain a definition which is independent of the set of generators used, it appears that one must allow finite sequences of single stack automata. Thus if $x, y \in X$, and we have machines M_x, M_y for multiplication by x and by y, then we can connect them in series. In this way the output of M_x becomes the input of M_y, and the result is a machine M_{xy} for multiplication by the product xy. This would also allow a 'reversing' machine which would read a word from the tape, putting the letters one after the other onto the stack. When the last letter is read, the letters are 'popped' off the stack one after the other.

Another more powerful type of machine to consider is an automaton which can read an input tape in both directions – a so–called two–way automaton. These can accept non–context free languages, for instance $\{a^n b^n c^n\}$. If one allows this type of machine one does not have the 'reversing' problem described above.

8. Questions

We end with some questions. A larger list of questions about automatic groups was compiled by S.M. Gersten [5], and most of those problems are still open. I repeat some of them here.

(1) Are all automatic groups biautomatic? Is the conjugacy problem for automatic groups solvable? The second part would follow immediately from the first.

(2) Is a direct factor of an automatic group also automatic? It is true for free factors.

(3) Is the isomorphism problem solvable in the class of automatic groups?

(4) What groups have all of their finitely generated subgroups L–rational for some rational cross–section L?

(5) Which groups have *deterministic* transducer structures? Clearly finitely generated abelian groups and free groups have this property. Some examples of small cancellation groups show that it is not possible for the usual regular language to provide this sort of structure.

(6) Is there an algorithm which decides whether a finitely presented group is a transducer group? There is such an algorithm for automatic groups [3, Chapter 5].

(7) Is there a transducer structure for the Heisenberg group, or for any nilpotent group (which is not abelian–by–finite)?

(8) Is the conjugacy problem solvable for a general finitely generated bi–transducer group? For a general finitely generated transducer group?

(9) What is the class of groups which have a linear (quadratic, etc.) time solution to the word problem?

References

1. G. Baumslag, S. M. Gersten, M. Shapiro and H. Short, *Automatic groups and amalgams*, J. Pure Appl. Algebra, **76**, (1991), 229–316.
2. J. Berstel, *Transductions and Context–free Languages*, B. G. Teubner, Stuttgart, 1979.
3. D. B. A. Epstein, with J. W. Cannon, D. F. Holt, S. V. F. Levy, M. S. Paterson and W. P. Thurston, *Word processing in Groups*, Jones and Barlett, Boston–London, 1992.
4. B. Farb, *Automatic groups; a guided tour*, L'enseignement mathématique, **38**, (1992), 291–313.

5. S. M. Gersten, *Problems on Automatic Groups*, in "Algorithms and Classification in Combinatorial Group Theory", MSRI Publications No. 23, edited by G. Baumslag and C. F. Miller III, Springer-Verlag, 1991.
6. M. Gersten and H. Short *Rational subgroups of biautomatic groups*, Annals of Math, **134**, (1992), 125–158.
7. R. H. Gilman *Groups with rational cross–sections*, in "Essays in group theory", edited by S. M. Gersten and J. R. Stallings P. U. P. Annals of Math series, # 111, 1987.
8. M. Gromov, *Hyperbolic groups*, in "Essays on group theory", MSRI Publications No. 8, edited by S. M. Gersten, Springer-Verlag, 1987
9. J. E. Hopcroft and J. D. Ullman *Introduction to Automata Theory, Languages and Computation*, Addison Wesley, 1979
10. P. Papasoglu, *Geometric properties of groups*, thesis, Columbia University, New York, 1993.

INTRODUCTION TO Λ-TREES

I. M. CHISWELL
School of Mathematical Sciences
Queen Mary and Westfield College
Mile End Road
London E1 4NS
England

Introduction

This article is not intended to be a comprehensive treatment of the basic theory of Λ-trees. There are good accounts of this in [1] and [14], and we shall often refer to them for arguments which have been omitted. Equally, it is not a comprehensive survey of results in this area; for this, we recommend the articles by Morgan and by Shalen [33], [45], [46]. What we have mainly tried to do is to look in more detail than usual at some aspects of the basic theory, in some cases (in §3) suggesting new arguments, while giving only a summary of other parts of the theory. However, in the final section (§5), there is a survey of recent results on free actions of groups on Λ-trees. I should like to thank the organisers, John Fountain and Victoria Gould, for a very enjoyable and interesting conference.

1. Simplicial Trees

A Λ-tree is a generalisation of a tree in the usual sense, that is, a connected circuit-free graph. It is a special kind of metric space, where the metric takes values in Λ, which is an ordered abelian group. Any graph induces an integer-valued metric on its vertex set, and to arrive at a definition of Λ-tree we have to look at the special properties of this metric in the case that the graph is a tree. We shall do this later in this introductory section, but we begin by recalling the relevant basic facts about graphs and trees. This is followed by an account of the results about ordered abelian groups which are needed.

There are several different definitions of a graph in the literature, and we start by surveying those we shall use. The definition of a directed graph is standard. It consists of a set V (whose elements are called vertices), a set E (whose elements are called edges), and two mappings $o : E \to V$ and $t : E \to V$. The vertex $o(e)$ is the *origin*, and $t(e)$ is the *terminus* of the edge e. It is required that $V \cap E = \emptyset$.

For us a graph Γ will mean a directed graph with involution, that is, a directed graph with, additionally a mapping $E \to E$, denoted by $e \mapsto \bar{e}$, such that, for all $e \in E$, $\bar{\bar{e}} = e$, $\bar{e} \neq e$, $o(\bar{e}) = t(e)$ and (consequently) $t(\bar{e}) = o(e)$. We write $V(\Gamma)$, $E(\Gamma)$ for V, E. A pair $\{e, \bar{e}\}$ is called an unoriented edge. We call $o(e)$ and $t(e)$

J. Fountain (ed.), Semigroups, Formal Languages and Groups, 255–293.

the endpoints of e, and of the corresponding unoriented edge. Vertices are depicted by small circles, and an unoriented edge is depicted by a line segment joining its endpoints.

A morphism of graphs (or graph mapping) from Γ to Δ is a mapping $f : V(\Gamma) \cup E(\Gamma) \to V(\Delta) \cup E(\Delta)$ which maps vertices to vertices and edges to edges, such that, for all edges $e \in V(\Gamma)$, $f(o(e)) = o(f(e))$, $f(t(e)) = t(f(e))$ and $f(\bar{e}) = \overline{f(e)}$. A graph Γ is a subgraph of a graph Δ if $V(\Gamma) \subseteq V(\Delta)$, $E(\Gamma) \subseteq E(\Delta)$ and the inclusion map $V(\Gamma) \cup E(\Gamma) \hookrightarrow V(\Delta) \cup E(\Delta)$ is a graph mapping, that is, if $e \in E(\Gamma)$ then $o(e)$, $t(e)$ and $\bar{e}$ have the same meaning in Δ as they do in Γ. The graph mapping f is called an isomorphism if it is bijective, and an isomorphism from Γ to Γ is called an automorphism of Γ.

Note that there can be more than one unoriented edge with the same endpoints in a graph Γ. If this does not happen we say that Γ has no multiple edges. Also, there is nothing to prevent the occurrence of an edge e with $o(e) = t(e)$. An edge with this property is called a loop.

Graphs in this sense may be viewed as combinatorial descriptions of the class of topological spaces known as 1-dimensional CW-complexes (see [31; Ch.6]). A special case is the class of 1-dimensional polyhedra, that is, geometric realisations of abstract 1-dimensional simplicial complexes. Such a simplicial complex can be described as a set V (the set of vertices) together with a set S of two-element subsets of V (the set of 1-simplices). Given such a simplicial complex, we can construct a graph Γ by defining $V(\Gamma) = V$, $E(\Gamma) = \{(u, v) \in V \times V \mid \{u, v\} \in S\}$ and $o(u, v) = u$, $t(u, v) = v$, $\overline{(u, v)} = (v, u)$. This graph has no loops or multiple edges. Conversely, given a graph Γ with no loops or multiple edges, we may define a 1-dimensional simplicial complex by putting $V = V(\Gamma)$ and

$$S = \{\ \{u, v\} \mid \text{there is an edge in } \Gamma \text{ with endpoints } u,\ v\}.$$

For this reason, we shall call a graph with no loops or multiple edges a simplicial graph.

We are interested mainly in connected graphs, which are the combinatorial analogues of path-connected topological spaces. Let L_n be the simplicial graph corresponding to the simplicial complex with vertex set $\{0, 1, \ldots, n\}$ and 1-simplices $\{0, 1\}, \{1, 2\}, \ldots, \{n-1, n\}$. Here n is a non-negative integer.

L_n:

We define a path of length n in a graph Γ to be a graph mapping $p : L_n \to \Gamma$. We denote $p(0)$ by $o(p)$ and $p(n)$ by $t(p)$ and call these vertices the endpoints of p (they may coincide). We say that p *joins* $o(p)$ to $t(p)$, and that Γ is connected if, given u, $v \in V(\Gamma)$, there exists a path in Γ joining u and v.

This is perhaps the most aesthetic definition of path since it mimics the definition for topological spaces. However, for each vertex v of Γ there is a unique path of length 0 with v as its only endpoint, while a path p of length $n > 0$ can be thought of as a sequence of edges $e_1, \dots, e_n$ of Γ with $t(e_i) = o(e_{i+1})$ for $1 \leq i \leq n-1$, and $o(p) = o(e_1)$, $t(p) = t(e_n)$. This will be the most convenient way to think of paths.

A path $e_1, \dots, e_n$ is called *reduced* if $e_i \neq \bar{e}_{i+1}$ for $1 \leq i \leq n-1$ and a path is called *closed* if its endpoints coincide (paths of length 0 are regarded as reduced). A *circuit* is a closed reduced path of positive length, $e_1, \dots, e_n$ such that $o(e_i) \neq o(e_j)$ for $1 \leq i, j \leq n$ with $i \neq j$. The image of such a path is the combinatorial analogue of a homeomorphic image of a circle. We can now define a tree.

Definition. A *tree* is a connected graph having no circuits.

Note that a graph is simplicial if and only if it has no circuits of length ≤ 2. Thus trees are simplicial graphs, and we shall sometimes refer to them as simplicial trees to distinguish them from the Λ-trees which will be considered later. We use the term subtree to mean a subgraph which is also a tree. The only fact about trees which we shall use is the following characterisation.

Lemma 1.1 *A graph Γ is a tree if and only if, given vertices u, v of Γ, there is a unique reduced path which joins u to v.*

Proof. See [12; Ch.5, Lemma 3] or [44; Ch.I §2.2, Prop. 8]. □

The degree of a vertex v in a graph Γ is defined to be the number of edges e of Γ such that $o(e) = v$. A vertex of degree 1 is called a terminal vertex of Γ, and an edge having an endpoint which is a terminal vertex is called a terminal edge of Γ.

Before discussing the metric induced by a connected graph on its vertex set, we shall consider metrics with values in an ordered abelian group. First, we make some basic observations about ordered abelian groups.

By an ordered abelian group we mean a totally ordered abelian group, that is, an abelian group Λ, which will be written additively, together with a linear (total) ordering $\leq$ on Λ, such that for all a, b and $c \in \Lambda$, $a \leq b$ implies $a + c \leq b + c$. Given an ordered abelian group Λ, let $P = \{a \in \Lambda | \ a > 0\}$. Then
(1) Λ is the disjoint union of P, $-P$ and $\{0\}$
(2) $P + P \subseteq P$.
Conversely, given an ordered abelian group Λ and a subset P satisfying (1) and (2), Λ can be made into an ordered abelian group by defining $a \leq b$ if and only if $b - a \in P \cup \{0\}$.

Examples.
(1) Any subgroup of the additive group of $\mathbb{R}$, in particular, $\mathbb{R}$ and $\mathbb{Z}$.
(2) If Λ_1 and Λ_2 are ordered abelian groups, the direct sum $\Lambda_1 \oplus \Lambda_2$ can be made into an ordered abelian group by defining $(a, b) \leq (c, d)$ if and only if either $a < c$, or $a = c$ and $b \leq d$ (this is the *lexicographic ordering*). In particular, $\mathbb{Z} \oplus \mathbb{Z}$ is an ordered abelian group in this way.

These will, in fact, be sufficient examples to illustrate the general theory, although ordered abelian groups can be much more complicated.

By a morphism of ordered abelian groups we mean a group homomorphism $h : \Lambda \to \Lambda'$, where Λ and Λ' are ordered abelian groups, such that $a \leq b$ implies $h(a) \leq h(b)$. For example, if $\mathbb{Z} \oplus \mathbb{Z}$ has the lexicographic ordering, then projection onto the first factor is a morphism of ordered abelian groups. If the morphism h is one-to-one, we shall call it an embedding of ordered abelian groups.

A subset X of an ordered abelian group Λ is called *convex* if $a,\ b \in X$ and $a \leq c \leq b$, where $c \in \Lambda$, implies $c \in X$. In particular, it makes sense to describe a subgroup of Λ as convex. (for example, the trivial subgroup and the whole group are convex). It is an easy exercise to show that the set of all convex subgroups of Λ is linearly ordered by inclusion.

If $h : \Lambda \to \Lambda'$ is a morphism of ordered abelian groups, then $\mathrm{Ker}(h)$ is a convex subgroup of Λ. If Λ' is a convex subgroup of Λ, then the quotient Λ/Λ' can be made into an ordered abelian group by defining $a+\Lambda' > 0$ if and only if $a > b$ for all $b \in \Lambda'$. (This defines the set of positive elements P, and P satisfies conditions (1) and (2) above, so this specifies the ordering). Further, the canonical projection $\pi : \Lambda \to \Lambda/\Lambda'$ is a morphism of ordered abelian groups. There is also a correspondence theorem: $M \to \pi^{-1}(M)$ gives a $1-1$ correspondence between the set of convex subgroups of Λ/Λ' and the set of convex subgroups of Λ containing Λ'.

The rank of an ordered abelian group is defined to be the order type of its set of non-trivial convex subgroups. If this set is finite, with n elements, we may take the rank to be n. Thus Λ is of rank 0 if and only if $\Lambda = 0$. The group $\mathbb{Z} \oplus \mathbb{Z}$ with the lexicographic ordering is of rank 2, the proper non-trivial convex subgroup being $0 \oplus \mathbb{Z}$. Rank 1 ordered abelian groups are characterised by the next theorem, but before stating it we need a definition.

Definition. An ordered abelian group Λ is called *archimedean* if, given $a,\ b \in \Lambda$ with $b \neq 0$, there exists $n \in \mathbb{Z}$ such that $a < nb$.

Theorem 1.1 *For an ordered abelian group Λ, the following are equivalent.*
(1) Λ has rank less than or equal to 1
(2) Λ is archimedean
(3) there exists an embedding of ordered abelian groups $\Lambda \to \mathbb{R}$.

Proof. The proof of (1)$\Longleftrightarrow$(2) is left as an exercise, (3)$\Rightarrow$(2) follows because $\mathbb{R}$ is archimedean, and for the proof that (2)$\Rightarrow$(3) see §10, Ch. VI in [49]. □

Let Λ be an ordered abelian group. Viewing abelian groups as $\mathbb{Z}$-modules, we can form the abelian group $\mathbb{Q} \otimes_{\mathbb{Z}} \Lambda$. It is well-known from commutative algebra that this group can be described as follows. It is the set of equivalence classes of the equivalence relation $\sim$ defined on the set $\Lambda \times (\mathbb{Z} \setminus \{0\})$ by: $(\lambda, m) \sim (\mu, n)$ if and only if $n\lambda = m\mu$. Denoting the equivalence class of (λ, m) by $\frac{\lambda}{m}$, addition is given by

$$\frac{\lambda}{m} + \frac{\mu}{n} = \frac{(n\lambda + m\mu)}{mn}.$$

We can make $\mathbb{Q} \otimes_{\mathbb{Z}} \Lambda$ into an ordered abelian group by defining $\frac{\lambda}{m} > 0$ if and only if $m\lambda > 0$ (again it suffices to specify the positive elements). The map $\Lambda \to \mathbb{Q} \otimes_{\mathbb{Z}} \Lambda$ given by $\lambda \mapsto \frac{\lambda}{1}$ is an embedding of ordered abelian groups. In particular, Λ embeds in the subgroup $\frac{1}{2}\Lambda = \{\frac{\lambda}{2} | \lambda \in \Lambda\}$, which is an ordered abelian group isomorphic to Λ. This group $\frac{1}{2}\Lambda$ will play a role in what follows.

Now we can consider metric spaces. This is a standard notion, the only novelty being that the metric can take values in an arbitrary ordered abelian group.

Definition. Let X be a non-empty set, Λ an ordered abelian group. A Λ-*metric* on X is a mapping $d : X \times X \to \Lambda$ such that:

(i) $d(x,y) \geq 0$ for all x, $y \in X$

(ii) $d(x,y) = 0$ if and only if $x = y$, for all x, $y \in X$

(iii) $d(x,y) = d(y,x)$ for all x, $y \in X$

(iv) $d(x,y) \leq d(x,z) + d(z,y)$ for all x, y, $z \in X$.

We call the inequality in (iv) the triangle inequality, as usual. A Λ-metric space is a pair (X,d), where X is a non-empty set and d is a Λ-metric on X. If (X,d) and (X',d') are Λ-metric spaces, an isometry from (X,d) to (X',d') is a mapping $f : X \to X'$ such that $d(x,y) = d'(f(x), f(y))$ for all x, $y \in X$. Such a mapping is necessarily one-to-one, but need not be onto. If it is onto, we call it a metric isomorphism, and (X,d), (X',d') are said to be metrically isomorphic if there exists a metric isomorphism $f : X \to X'$.

One example of a Λ-metric space is (Λ, d), where $d(a,b) = |a-b|$ and $|x| = \max\{x, -x\}$ for $x \in \Lambda$. We shall need to know about the metric automorphisms of Λ, that is, metric isomorphisms from Λ to Λ. Examples are the translations $t_a(x) = a + x$ and the reflections $r_a(x) = a - x$, for $a \in \Lambda$. These are in fact the only examples.

Lemma 1.2 *(1) The group G of metric automorphisms of Λ consists of the reflections and translations. The set T of translations is a normal subgroup of G, and G is the semidirect product of T and the cyclic subgroup of order 2 generated by r_0 $(x \mapsto -x)$.*

(2) If A is a subset of Λ with at least two elements, then any metric automorphism of A is the restriction to A of a metric automorphism of Λ.

Proof. See [1] 2.5 and 2.6. □

Given a, $b \in \Lambda$ with $a \leq b$, where Λ is an ordered abelian group, we define $[a,b]_\Lambda = \{x \in \Lambda | a \leq x \leq b\}$, and $[b,a]_\Lambda = [a,b]_\Lambda$.

Definition. A *segment* in a Λ-metric space (X,d) is the image of an isometry $\alpha : [a,b]_\Lambda \to X$ for some a, $b \in \Lambda$ with $a \leq b$. Note that $a = b$ is allowed. The endpoints of the segment are $\alpha(a)$, $\alpha(b)$. We shall say that the segment joins $\alpha(a)$, $\alpha(b)$.

The endpoints are characterised by the fact that $d(\alpha(a), \alpha(b))$ is the maximum distance apart of any pair of points in the segment. In fact, replacing α by $\alpha \circ t_a$, we can assume $a = 0$. If $\beta : [0,b]_\Lambda \to X$ is another isometry with the same image as α, then $\alpha^{-1} \circ \beta$ is a metric automorphism of $[0,b]_\Lambda$, so by 1.2 is the restriction of a metric automorphism of Λ. This can only be either $t_0 = id$ or r_b, so either $\beta = \alpha$

or $\beta(x) = \alpha(b - x)$. This explains why we usually need only consider the image of α rather than α itself, which is more convenient in the theory which follows.

In case $\Lambda = \mathbb{Z}$, $[0, n]_{\mathbb{Z}} = \{0, 1, \dots, n\}$, and a segment in a $\mathbb{Z}$-metric space (X, d) is the image of an isometry $\alpha : [0, n]_{\mathbb{Z}} \to X$ for some $n \geq 0$. Putting $v_i = \alpha(i)$, the segment is a finite set $\{v_0, \dots, v_n\}$ with $d(v_i, v_j) = |i - j|$.

Definition. A Λ-metric space (X, d) is *geodesic* if for all x, $y \in X$, there is a segment in X with endpoints x, y, and (X, d) is *geodesically convex* (or *geodesically linear*) if, for all x, $y \in X$, there is a unique segment in X whose set of endpoints is $\{x, y\}$.

If σ is a segment with endpoints x, y and $z \in \sigma$, then $d(x, y) = d(x, z) + d(z, y)$, and the segment is the union of two segments, $\sigma_1 = \{u \in \sigma | d(x, u) \leq d(x, z)\}$ with endpoints x, z, and $\sigma_2 = \{u \in \sigma | d(x, u) \geq d(x, z)\}$ with endpoints z, y.

Lemma 1.3 *Let (X, d) be a Λ-metric space.*

(1) Let σ be a segment in X with endpoints x, y, let τ be a segment in X with endpoints y, z, and suppose that, for all $u \in \sigma$ and $v \in \tau$, $d(u, v) = d(u, y) + d(y, v)$. Then $\sigma \cup \tau$ is a segment with endpoints x and z.

(2) Assume (X, d) is geodesically convex. Let x, y and $z \in X$, and let σ be the segment with endpoints x, y. Then $z \in \sigma$ if and only if $d(x, y) = d(x, z) + d(z, y)$.

Proof. (1) By using translations we can find surjective isometries $\alpha : [0, a]_{\Lambda} \to \sigma$ and $\beta : [a, b]_{\Lambda} \to \tau$, where $0 \leq a \leq b$. Then $\alpha \cup \beta$ is an isometry with image $\sigma \cup \tau$.

(2) Assume $d(x, y) = d(x, z) + d(z, y)$, and let τ be the segment with endpoints x, z, ρ the segment with endpoints z, y. Take $u \in \tau$, $v \in \rho$. Then

$$d(u, z) = d(x, z) - d(x, u)$$
$$d(z, v) = d(z, y) - d(y, v)$$

Using this and the triangle inequality,

$$\begin{aligned} d(u, v) &\leq d(u, z) + d(z, v) \\ &= d(x, y) - d(x, u) - d(y, v) \\ &\leq d(u, v) \end{aligned}$$

so there is equality throughout, and by Part (1) $\tau \cup \rho$ is a segment with endpoints x, y. Hence $\sigma = \tau \cup \rho$ and $z \in \sigma$. The converse has already been noted. □

We now return to a connected graph Γ and consider the metric induced on its vertex set. For x, $y \in V(\Gamma)$, we define $d(x, y)$ to be the minimal length of a path joining x and y. This is easily seen to be a $\mathbb{Z}$-metric on $V(\Gamma)$ (to see the triangle inequality, if $e_1, \dots, e_n$ is a path joining x and y, and $f_1, \dots, f_m$ is a path joining y and z, then $e_1, \dots, e_n, f_1, \dots, f_m$ is a path joining x and z). We call d the path metric of Γ. We say that Γ is of finite diameter if there exists an integer n such that, for all x, $y \in V(\Gamma)$, $d(x, y) \leq n$, and the least such n is called the diameter of Γ. Before proceeding, we make a remark that will be useful later.

Remark. If $e_1, \dots, e_n$ is a path in a graph Γ and $e_{i+1} = \bar{e}_i$ for some i, then removing e_i, e_{i+1} gives a path of shorter length joining the same vertices. Thus paths of shortest length joining two given vertices are reduced. Further, if $t(e_i)$ is a terminal vertex for some i with $1 \leq i < n$, then $e_{i+1} = \bar{e}_i$. It follows that, if Γ is connected and Γ' is the graph obtained by removing all terminal vertices and edges from Γ, then Γ' is connected. If Γ is a tree then the shortest paths joining vertices are precisely the reduced paths, by Lemma 1.1. If Γ is a tree of finite diameter ≥ 2, the endpoints of a reduced path of maximal length are terminal vertices. Consequently, Γ' is a subtree of smaller diameter.

Lemma 1.4

(1) Let Γ be a connected graph, let $X = V(\Gamma)$ and let d be the path metric of Γ. Then (X, d) is a geodesic $\mathbb{Z}$-metric space. Further, a mapping $\alpha : [0, n]_{\mathbb{Z}} \to X$, $i \mapsto v_i$ (where $n > 0$) is an isometry if and only if there is a path $e_1, \ldots, e_n$ in Γ with $v_0 = o(e_1)$, $v_i = t(e_i)$ for $1 \leq i \leq n$ and $d(v_0, v_n) = n$.

(2) Let (X, d) be a geodesic $\mathbb{Z}$-metric space. Then there exists a connected simplicial graph Γ such that $X = V(\Gamma)$ and d is the path metric of Γ.

Proof. (1). Suppose $e_1, \ldots, e_n$ is a path of shortest length joining two vertices u and v, and put $v_i = t(e_i)$, $v_0 = o(e_1) = u$. Then clearly $e_i, \ldots, e_j$ is a path of shortest length joining v_{i-1} to v_j for $i < j$, and $\alpha : [0, n]_{\mathbb{Z}} \to X$ given by $\alpha(i) = v_i$ is an isometry. Hence (X, d) is geodesic. Conversely, given the isometry α, we have $d(v_{i-1}, v_i) = 1$ for $1 \leq i \leq n$, so by definition of d there is an edge e_i (a path of length 1) with $o(e_i) = v_{i-1}$, $t(e_i) = v_i$. Then $e_1, \ldots, e_n$ is a path with the required properties.

(2). Define Γ to be the graph associated to the 1-dimensional simplicial complex with vertex set X, where $\{x, y\}$ is a 1-simplex if and only if $d(x, y) = 1$. Let u, $v \in X$, let $n = d(x, y)$ and let $\alpha : [0, n]_{\mathbb{Z}} \to X$ be an isometry with $\alpha(0) = u$, $\alpha(n) = v$. Put $v_i = \alpha(i)$. Then $d(v_{i-1}, v_i) = 1$, so there is an unoriented edge with endpoints v_{i-1}, v_i for $1 \leq i \leq n$, hence a path in Γ joining u to v. In particular Γ is connected. If there is a shorter path joining u, v, say $f_1, \ldots, f_m$, then by definition of Γ, $d(o(f_i), t(f_i)) = 1$, so by the triangle inequality $d(u, v) \leq \sum_{i=1}^{m} d(o(f_i), t(f_i)) = m < n$, a contradiction. Hence d is the path metric. □

If in a graph Γ there are two vertices u, v and more than one unoriented edge with these vertices as endpoints, then replacing these edges by a single edge does not change the path metric. Also, it is not changed by removing loops. Thus we can replace Γ by a simplicial graph without changing the path metric. This explains why the graph Γ in Part (1) of the lemma is not assumed simplicial, but the graph Γ in Part (2) is simplicial. In the next result we consider the path metric in a tree.

Lemma 1.5 *Let Γ be a connected simplicial graph, let $X = V(\Gamma)$ and let d be the path metric of Γ. Then Γ is a tree if and only if the following condition is satisfied.*

If two segments of (X, d) intersect in a single point, which is an endpoint of both, then their union is a segment.

Proof. Suppose that the condition is satisfied, and assume that there is a circuit $e_1, \dots, e_n$ in Γ. Then $n \geq 3$ since Γ is simplicial. Put $v_i = t(e_i)$ for $1 \leq i \leq n$ and $v_0 = o(e_1) = v_n$. We claim that, for $r \leq n-1$, $\{v_0, \dots, v_r\}$ is a segment with endpoints v_0, v_r. The proof is by induction on r, and it is clearly true if $r \leq 1$. Suppose it is true for r with $r < n-1$. Then $\{v_r, v_{r+1}\}$ is a segment, and $\{v_0, \dots, v_r\} \cap \{v_r, v_{r+1}\} = \{v_r\}$, so by assumption $\{v_0, \dots, v_{r+1}\}$ is a segment. Its endpoints are v_0 and v_{r+1}, because v_i is the unique point of this segment at distance i from v_0, for $i \leq r$, hence $d(v_0, v_{r+1}) = r+1$. Thus $\{v_0, \dots, v_{n-1}\}$ is a segment. But so is $\{v_0, v_{n-1}\}$ (because the edge e_n has these vertices as its endpoints), and so $d(v_0, v_{n-1}) = n-1 = 1$, that is, $n = 2$, a contradiction. It follows that Γ is a tree.

Conversely assume Γ is a tree. Recall from the remark preceding 1.5 that the shortest paths joining vertices are precisely the reduced paths. Let $\sigma = \{v_0, \dots, v_n\}$, $\tau = \{w_0, \dots, w_m\}$ be segments, such that $d(v_i, v_j) = |i-j|$ and $d(w_i, w_j) = |i-j|$ for all relevant values of i and j. Assume they intersect only in the single point $v_n = w_0$. We can assume $n > 0$ and $m > 0$. By Lemma 1.5(1) they correspond to reduced paths $e_1, \dots, e_n$ and $f_1, \dots, f_m$ with $v_0 = o(e_1)$ and $v_i = t(e_i)$ for $i \geq 1$, with similar equations for the w_i. By assumption $v_{n-1} \neq w_1$, so $e_n \neq \bar{f}_1$. Hence $e_1, \dots, e_n, f_1, \dots, f_m$ is a reduced path, and the set of endpoints of the edges in this path is a segment by Lemma 1.5(1). This set of endpoints is $\sigma \cup \tau$. □

It follows from Lemma 1.4(1) and Lemma 1.1 that, in the situation of Lemma 1.5, (X, d) is a geodesically convex $\mathbb{Z}$-metric space. This has a generalisation.

Lemma 1.6 *Let Λ be an ordered abelian group and let (X, d) be a geodesic Λ-metric space such that the following condition is satisfied.*

If two segments of (X, d) intersect in a single point, which is an endpoint of both, then their union is a segment.

Then (X, d) is a geodesically convex Λ-metric space.

Proof. Let σ, τ be segments, both with endpoints u, v. Let $w \in \sigma$, and let w' be the point of τ such that $d(u, w) = d(u, w')$ (so that $d(v, w) = d(v, w')$). We have to show $w = w'$. Let ρ be a segment with endpoints w, w'. Now $\sigma = \sigma_1 \cup \sigma_2$, where σ_1 is a segment with endpoints u, w, and σ_2 is a segment with endpoints w, v. We claim that either $\sigma_1 \cap \rho = \{w\}$ or $\sigma_2 \cap \rho = \{w\}$. For if $x \in \sigma_1 \cap \rho$ and $y \in \sigma_2 \cap \rho$, then $d(x, y) = d(x, w) + d(w, y)$, and either $d(w, y) = d(w, x) + d(x, y)$ or $d(w, x) = d(w, y) + d(x, y)$, depending on how x, y are situated in the segment ρ. It follows that either $x = w$ or $y = w$, establishing the claim. If $\sigma_1 \cap \rho = \{w\}$, then by assumption $\sigma_1 \cup \rho$ is a segment, and it is easily seen that its endpoints must be u, w'. Thus $d(u, w') = d(u, w) + d(w, w')$, so $w = w'$. Similarly if $\sigma_2 \cap \rho = \{w\}$ then $w = w'$. □

Lemma 1.7 *Let (X, d) be a $\mathbb{Z}$-metric space. Then there is a tree Γ, such that $X = V(\Gamma)$ and d is the path metric of Γ, if and only if (a) and (b) below hold.*

(a) *(X, d) is geodesic*

(b) *if two segments of (X, d) intersect in a single point, which is an endpoint of both, then their union is a segment.*

Proof. This is immediate from Lemmas 1.4 and 1.5. □

If G is a group acting as automorphisms on a connected graph Γ, and d is the path metric of Γ, then G acts as isometries on the $\mathbb{Z}$-metric space (X, d), where $X = V(\Gamma)$, because G preserves shortest paths. Conversely, if G is a group acting as isometries on a geodesic $\mathbb{Z}$-metric space (X, d), then G acts as automorphisms of the corresponding simplicial graph in Lemma 1.4(2), because vertices are adjacent if and only if they are at distance 1 apart.

The next lemma applies to general Λ-metric spaces, and appears in [34; Lemma II.1.1]

Lemma 1.8 *Let Λ be an ordered abelian group and let (X, d) be a geodesically convex Λ-metric space. Then the intersection of two segments with a common endpoint is the image of a convex subset of Λ with least element 0. If $\Lambda = \mathbb{Z}$ or $\Lambda = \mathbb{R}$ then the intersection is a segment.*

Proof. Let σ and τ be segments with a common endpoint. Using appropriate reflections and translations, we can find surjective isometries $\alpha : [0, a]_\Lambda \to \sigma$, $\beta : [0, b]_\Lambda \to \tau$ with $\alpha(0) = \beta(0)$. Define $I = \{c \in [0, \min\{a, b\}] \mid \alpha(c) = \beta(c)\}$. Clearly $\alpha(I) = \beta(I) = \sigma \cap \tau$. If x, $y \in I$ and $z \in \Lambda$ with $x \leq z \leq y$, then $z \in I$, for otherwise the images of $\alpha|_{[x,y]_\Lambda}$ and $\beta|_{[x,y]_\Lambda}$ are distinct segments in X with the same endpoints, contrary to assumption. Thus I is a convex subset of Λ with least element 0. If $\Lambda = \mathbb{Z}$ then I is finite and equal to $[0, n]_\mathbb{Z}$ for some n, and if $\Lambda = \mathbb{R}$, $\sigma \cap \tau$ is a compact connected subset of X homeomorphic to I, so I is $[0, c]_\mathbb{R}$ for some c. □

We end with a caution. A graph such that, for any vertices u, v there is a unique shortest path joining u, v is not necessarily a tree. A counterexample is provided by (the image of) a circuit of odd length.

2. Λ-trees; the Definition

After investigating $\mathbb{Z}$-metric spaces we are ready to define a Λ-tree. After doing so, we shall consider examples and elementary but important properties of such trees. In what follows, Λ will always denote an ordered abelian group.

Definition. A Λ-*tree* is a Λ-metric space (X, d) such that:

(a) (X, d) is geodesic

(b) if two segments of (X, d) intersect in a single point, which is an endpoint of both, then their union is a segment.

(c) the intersection of two segments with a common endpoint is also a segment.

We can immediately give two examples of Λ-trees.

Example 1. (Λ, d), where d is the usual metric ($d(a,b) = |a-b|$) is a Λ-tree. It is easy to see that a segment in Λ must be of the form $[a,b]_\Lambda$ for some a, $b \in \Lambda$. More generally, any convex subset of Λ is a Λ-tree.

Example 2. A $\mathbb{Z}$-metric space (X,d) is a $\mathbb{Z}$-tree if and only if there is a simplicial tree Γ such that $X = V(\Gamma)$ and d is the path metric of Γ. This follows from Lemmas 1.6, 1.7 and 1.8.

Lemma 2.1 *A Λ-tree is geodesically convex.*

Proof. This follows from Axioms (a) and (b) by Lemma 1.6. $\square$

In the cases $\Lambda = \mathbb{R}$ and $\Lambda = \mathbb{Z}$, Axiom (c) follows from (a) and (b), by 1.8. On the other hand, we leave it as an exercise to show that Lemma 2.1 follows from Axioms (a) and (c) alone. We shall henceforth denote the unique segment in a Λ-tree with endpoints x, y by $[x,y]$. Thus $[x,y] = [y,x]$ and $[x,x] = \{x\}$. Indeed we shall use this notation in any geodesically convex space. Axiom (c) for a Λ-tree implies that, if x, y, z are points in a Λ-tree, then $[x,y] \cap [x,z] = [x,w]$ for some w (that is, the intersection has x as an endpoint, which is an easy exercise). In fact, more can be said. The following is essentially the "Y-proposition" of [1], in a slightly generalised form.

Lemma 2.2 *Let x, y and z be points of a Λ-tree (X,d), and let w be the point such that $[x,y] \cap [x,z] = [x,w]$. Then*
(1) $[y,w] \cap [w,z] = \{w\}$ and $[y,z] = [y,w] \cup [w,z]$
(2) If y_m is the point on $[x,y]$ at distance m from x, and z_n is the point on $[x,z]$ at distance n from x, where $0 \le m \le d(x,y)$ and $0 \le n \le d(x,z)$, then $d(y_m, z_n) = m + n - 2\min\{m, n, d(x,w)\}$.
(3) w depends only on the set $\{x,y,z\}$, not on the order in which the elements are written.

Proof. Apart from (2), which is left as an exercise, see (2.11) in [1]. $\square$

The segments joining x, y and z are thus illustrated by the following picture, which explains the name "Y-proposition".

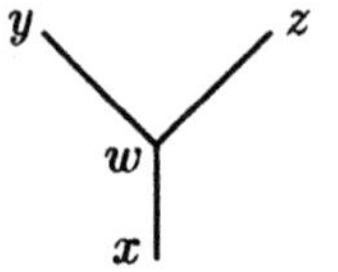

The next result is elementary, but is important in the basic theory of Λ-trees, and is called the "Piecewise Geodesic Proposition" in [1]. The proof of Part (3) generalises part of the argument for Lemma 1.5.

Lemma 2.3 *Let (X,d) be a Λ-tree and let $x_0, \ldots, x_n$ be points of X. Then*
(1) $[x_0, x_n] \subseteq \bigcup_{i=1}^n [x_{i-1}, x_n]$ and $d(x_0, x_n) \le \sum_{i=1}^n d(x_{i-1}, x_i)$;

(2) *if* $d(x_0, x_n) = \sum_{i=1}^n d(x_{i-1}, x_i)$ *then* $[x_{i-1}, x_i] \cap [x_i, x_{i+1}] = \{x_i\}$ *for* $1 \leq i \leq n-1$ *and* $[x_0, x_n] = \bigcup_{i=1}^n [x_{i-1}, x_i]$;

(3) *if* $[x_{i-1}, x_i] \cap [x_i, x_{i+1}] = \{x_i\}$ *for* $1 \leq i \leq n-1$ *and* $x_i \neq x_{i+1}$ *for* $1 \leq i \leq n-2$ *then* $d(x_0, x_n) = \sum_{i=1}^n d(x_{i-1}, x_i)$ *and* $[x_0, x_n] = \bigcup_{i=1}^n [x_{i-1}, x_i]$.

Proof. See [1], 2.14. □

We shall now prove a series of lemmas giving simple ways in which to construct Λ-trees, leading to further examples of such trees.

Lemma 2.4 *Let* (X_1, d_1) *and* (X_2, d_2) *be* Λ*-trees with* $X_1 \cap X_2 = \{x_0\}$, *a single point. Let* $X = X_1 \cup X_2$ *and define* $d : X \times X \to \Lambda$ *by:* $d|_{X_i \times X_i} = d_i$ *for* $i = 1$, 2, *and if* $x_i \in X_i$ *for* $i = 1$, 2 *then* $d(x_1, x_2) = d(x_2, x_1) = d_1(x_1, x_0) + d_2(x_0, x_2)$. *Then* (X, d) *is a* Λ*-tree.*

Proof. It is easy to see that d is a metric. We consider only the triangle inequality in detail. Let x, y, $z \in X$; if x, y, z all belong to the same X_i then clearly $d(x, y) \leq d(x, z) + d(z, y)$. If $x \in X_1$, y, $z \in X_2$, then

$$d(x, y) = d_1(x, x_0) + d_2(x_0, y)$$
$$d(y, z) + d(z, x) = d_2(y, z) + d_1(x, x_0) + d_2(x_0, z)$$

and again $d(x, y) \leq d(x, z) + d(z, y)$. The other cases are left to the reader.

Let x, $y \in X$. If they are in the same X_i, then there is a segment with endpoints x, y, otherwise $[x, x_0] \cup [x_0, y]$ (the union of two segments, one in X_1, the other in X_2) is a segment in X with endpoints x, y, by Lemma 1.3(1). Thus (X, d) is geodesic. We claim it is geodesically convex. For suppose x, $y \in X_1$, and there is a segment in X joining x, y in X containing a point $z \in X_2 \setminus X_1$. Then

$$\begin{aligned} d_1(x, y) &= d(x, y) \\ &= d(x, z) + d(z, y) \\ &= d_1(x, x_0) + d_2(x_0, z) + d_1(y, x_0) + d_2(x_0, z) \\ &\geq d_1(x, y) + 2d_2(x_0, z) \end{aligned}$$

which implies $d_2(x_0, z) = 0$, so $z = x_0 \in X_1$, a contradiction. Since (X_1, d_1) is geodesically convex, there is only one segment joining x, y. This is similarly true if x, y both belong to X_2. Otherwise, we can assume $x \in X_1$, $y \in X_2$. If a segment joining x, y contains x_0 it must be the segment $[x, x_0] \cup [x_0, y]$ as above since (X_1, d_1) and (X_2, d_2) are geodesically convex. If σ is a segment in X with endpoints x, y, let z be the point of σ such that $d(x, z) = d(x, x_0)$, so that $d(z, y) = d(x_0, y)$. If $z \in X_1$, then by definition of d, $d(z, y) = d(z, x_0) + d(x_0, y)$, while if $z \in X_2$, then $d(z, x) = d(z, x_0) + d(x_0, x)$. In any case $d(z, x_0) = 0$, so $x_0 = z \in \sigma$. We now verify the remaining axioms (b) and (c) for a Λ-tree.

(b) Suppose x, y, $z \in X$ and $[x, y] \cap [x, z] = \{x\}$. We show that $[y, z] = [x, y] \cup [x, z]$. This is clear if all three points belong to the same X_i. Otherwise we may assume without loss of generality that $x \in X_1$, and by symmetry there are only two cases we need to consider.

Case 1. $y \in X_1$, $z \in X_2$. Then from the above, $\{x\} = [x,y] \cap ([x,x_0] \cup [x_0,z])$ hence $\{x\} = [x,y] \cap [x,x_0]$. Since (X_1,d_1) is a Λ-tree, $[y,x_0] = [y,x] \cup [x,x_0]$, so

$$[y,z] = [y,x_0] \cup [x_0,z] = [y,x] \cup [x,x_0] \cup [x_0,z] = [y,x] \cup [x,z]$$

Case 2. y, $z \in X_2$. Then

$$\{x\} = ([x,x_0] \cup [x_0,y]) \cap ([x,x_0] \cup [x_0,z]) = [x,x_0] \cup ([x_0,y] \cap [x_0,z])$$

so $x = x_0$ (since the right-hand side contains x_0) and x, y, z are all in X_2.

(c) Suppose x, y, $z \in X$. We have to show that $[x,y] \cap [x,z] = [x,w]$ for some $w \in X$. As before, we can assume $x \in X_1$ and need consider only two cases.

Case 1. $y \in X_1$, $z \in X_2$. Then $[x,z] = [x,x_0] \cup [x_0,z]$ and $[x,z] \cap [x,y] = [x,x_0] \cap [x,y]$ since $[x_0,z] \cap [x,y] \subseteq X_2 \cap X_1 = \{x_0\}$, and $x_0 \in [x,x_0]$. Since (X_1,d_1) satisfies Axiom (c), $[x,x_0] \cap [x,y] = [x,w]$ for some $w \in X_1$.

Case 2. y, $z \in X_2$. Then

$$[x,y] \cap [x,z] = [x,x_0] \cup ([x_0,y] \cap [x_0,z]) = [x,x_0] \cup [x_0,w] = [x,w]$$

for some $w \in X_2$, since (X_2,d_2) satisfies Axiom (c).

This concludes the proof. □

The next lemma gives a generalisation of 2.4.

Lemma 2.5 *Let (X,d) be a Λ-metric space, and let $\{(X_j,d_j) \mid j \in J\}$ be a family of Λ-trees, such that, for all $j \in J$, $X_j \cap X = \{x_j\}$, a single point. Also, suppose that $X_j \cap X_k = \{x_j\}$ if $x_j = x_k$, and $X_j \cap X_k = \emptyset$ otherwise. Put $Y = \left(\bigcup_{j \in J} X_j\right) \cup X$, and define $\bar{d} : Y \times Y \to \Lambda$ as follows: $\bar{d}|_{X \times X} = d$, $\bar{d}|_{X_j \times X_j} = d_j$. If $x \in X_j$, $x' \in X_k$ with $j \neq k$, then $\bar{d}(x,x') = d_j(x,x_j) + d(x_j,x_k) + d_k(x_k,x')$. Finally if $x \in X_j$, $x' \in X$, then $\bar{d}(x,x') = \bar{d}(x',x) = d_j(x,x_j) + d(x_j,x')$. Then $(Y,\bar{d})$ is a Λ-tree.*

Proof. If j, k, l are indices in J, then $X_j \cup X_k \cup X_l \cup X$, with the metric obtained from $\bar{d}$ by restriction, can be built up in stages using Lemma 2.4, so is a Λ-tree. It follows that $(Y,\bar{d})$ is a geodesic Λ-metric space, and the lemma will follow if we can show it is geodesically convex. Suppose x, $y \in Y$. Then x, $y \in X_j \cup X_k \cup X$ for some j, k. Suppose there is a segment in Y joining x, y which contains a point in X_l, with $l \neq j$, k. Then the argument in the previous lemma to show uniqueness of segments (with $X_1 = X_j \cup X_k \cup X$, $X_2 = X_l$) gives a contradiction. Uniqueness of segments in Y follows since $X_j \cup X_k \cup X$ is a Λ-tree. □

The last of this series of lemmas is proved in a similar manner.

Lemma 2.6 *Let (X_i,d_i) $(i \in \mathbb{N})$ be an ascending chain of Λ-trees. (Thus $X_0 \subseteq X_1 \subseteq \cdots$, and for $i < j$, d_i is d_j restricted to $X_i \times X_i$). Put $X = \bigcup_{i \geq 0} X_i$, $d = \cup_{i \geq 0} d_i$. Then (X,d) is a Λ-tree.*

Proof. Any three points of X are contained in some X_i, so (X,d) is a geodesic Λ-metric space and as in the previous lemma, it suffices to show that it is geodesically convex. Let x, $y \in X$, so x, $y \in X_i$ for some i. Suppose there is a segment in X joining x, y containing a point $z \notin X_i$. Then x, y, $z \in X_j$ for some $j > i$. In X_j, let $[z,x] \cap [z,y] = [z,w]$. Then using Lemma 2.2,

$$\begin{aligned} d(x,y) &= d(x,z) + d(z,y) \\ &= d(x,w) + d(w,z) + d(y,w) + d(w,z) \\ &= d(x,y) + 2d(z,w) \end{aligned}$$

hence $z = w$. The segment joining x and y in X_i must also be the segment $[x,y]$ joining them in X_j since X_j is geodesically convex, and $w \in [x,y]$ (again by Lemma 2.2). It follows that $z \in X_i$, a contradiction. Uniqueness of segments in Y now follows. □

We continue with some examples of Λ-trees.

Example 3. Our next example is an $\mathbb{R}$-tree, the geometric realisation real(Γ) of a simplicial tree Γ, in which we view the unoriented edges as part of the tree, each of them being metrically isomorphic to the unit interval $I = [0,1]_{\mathbb{R}}$, with endpoints identified appropriately.

Formally, let Γ be a simplicial tree, put $U(\Gamma) = (E(\Gamma) \times I) \cup V(\Gamma)$ (disjoint union). Define a binary relation $\simeq$ on $U(\Gamma)$ by declaring that

$$\begin{aligned} (e,t) &\simeq (\bar{e}, 1-t) \\ o(e) &\simeq (e,0) \end{aligned}$$

for all $e \in E(\Gamma)$ and $t \in I$. Let $\sim$ be the equivalence relation on $U(\Gamma)$ generated by $\simeq$ (note that $t(e) \sim (e,1)$ for all edges e). Define $\mathrm{real}(\Gamma) = U(\Gamma)/\sim$, and denote the equivalence class of $u \in U(\Gamma)$ by $\langle u \rangle$, abbreviating $\langle (e,t) \rangle$ to $\langle e,t \rangle$. If e and e_1 are edges, $0 < t < 1$ and $t_1 \in I$, then $\langle e,t \rangle \sim \langle e_1, t_1 \rangle$ implies either $e = e_1$ and $t = t_1$ or $e = \bar{e}_1$ and $t = 1 - t_1$. Also, $\langle v \rangle \sim \langle w \rangle$, where v, w are vertices, implies $v = w$.

If $f : \Gamma \to \Delta$ is a graph map, where Δ is also a simplicial tree, there is an obvious induced map $U(f) : U(\Gamma) \to U(\Delta)$, which clearly preserves $\simeq$, and so $\sim$, hence there is an induced mapping $\mathrm{real}(f) : \mathrm{real}(\Gamma) \to \mathrm{real}(\Delta)$. If f is one-to-one, it is easily seen that real(f) is one-to-one. Thus if Γ is a subgraph of Δ, we may view real(Γ) as a subset of real(Δ).

For $e \in E(\Gamma)$, we denote by real(e) the set $\{\langle e,t \rangle | t \in I\}$. The mapping $I \to \mathrm{real}(e)$, $t \mapsto \langle e,t \rangle$ is bijective, and we can use it to define a metric d_e on real(e). Thus $d_e(\langle e,t \rangle, \langle e,s \rangle) = |t-s|$.

We want to define an $\mathbb{R}$-metric d_Γ on real(Γ) so that $(\mathrm{real}(\Gamma), d_\Gamma)$ is an $\mathbb{R}$-tree. First assume Γ has finite diameter n. We use induction on n. If $n = 0$, real(Γ) consists of a single point and the metric is obvious. If $n = 1$, Γ has just one unoriented edge, say $\{e, \bar{e}\}$, and real(Γ) = real(e). We take $d_\Gamma = d_e$, so real(Γ) is isomorphic to I, which is an $\mathbb{R}$-tree.

If $n \geq 2$, then (remark preceding Lemma 1.4) removing all terminal edges and vertices from Γ gives a tree Γ′ of smaller diameter, so we can assume inductively that $d_{\Gamma'}$ is already defined and $(\mathrm{real}(\Gamma'), d_{\Gamma'})$ is an $\mathbb{R}$-tree. For each pair $\{e, \bar{e}\}$ of terminal edges take d_e as metric on real(e) = real$(\bar{e})$. Then Lemma 2.5 can be used to define the metric d_Γ, taking X = real(Γ') and the X_j as the spaces real(e) for every terminal pair $\{e, \bar{e}\}$ (e has one endpoint which is a terminal vertex, and its other endpoint is in $V(\Gamma')$, so $X \cap$ real(e) is a single point). This gives an $\mathbb{R}$-tree (real$(\Gamma), d_\Gamma$) by Lemma 2.5. By induction on n, we see that:

(i) if Δ is a subtree of Γ, then d_Δ is d_Γ restricted to real$(\Delta) \times$ real(Δ):

(ii) d_Γ restricted to real$(e) \times$ real(e) is d_e, for any edge e.

(iii) if u, v are vertices of Γ, $x = \langle u \rangle$ and $y = \langle v \rangle$, then $d_\Gamma(x, y) = d(u, v)$, where d is the path metric of Γ.

Now suppose Γ is an arbitrary simplicial tree. Let x, $y \in$ real(Γ). We can find a finite diameter subtree Δ of Γ such that x, $y \in$ real(Δ) (in fact, we can find a finite Δ with this property). We define $d_\Gamma(x, y) = d_\Delta(x, y)$. By property (i) above, this is independent of the choice of Δ, for if Δ_1 is another finite diameter subtree containing x, y, then it is easy to see that $\Delta \cup \Delta_1$ is also a finite diameter subtree containing x, y (the union is formed by taking the unions of the vertex sets and of the edge sets). Note that (i), (ii) and (iii) continue to hold even when Γ is not of finite diameter.

Choose a vertex $v \in V(\Gamma)$, and let Γ_n be the subgraph such that $V(\Gamma_n) = \{u \in V(\Gamma) | d(v, u) \leq n\}$, with $E(\Gamma_n)$ the set of edges whose endpoints belong to this set. Then clearly Γ_n is a finite diameter subtree of Γ. Also, real$(\Gamma) = \bigcup_{n \geq 0}$ real(Γ_n), and it follows from Lemma 2.6 that (real$(\Gamma), d_\Gamma$) is an $\mathbb{R}$-tree. The idea of constructing the metric in this way is taken from [26], where, however, a different definition of $\mathbb{R}$-tree, which we shall consider in §3, is used.

We shall give one final example to show that not all $\mathbb{R}$-trees arise in this way. To do so, we introduce the notion of degree of a point in a Λ-tree. Let x be a point of a Λ-tree (X, d). Then the relation $\equiv$ defined on the set of segments $\{[x, y] | \ y \in X\}$ by $[x, y] \equiv [x, z]$ if and only if $[x, y] \cap [x, z] \neq \{x\}$ (that is, $[x, y] \cap [x, z] = [x, w]$ for some $w \in X$ with $x \neq w$) is an equivalence relation on this set (exercise, or see Lemma 3.4 below). The equivalence classes are called directions (or direction germs) at x. The degree of x is defined to be the cardinal number of the set of directions at x.

Suppose that Γ is a simplicial tree with more than one vertex and x, $y \in$ real(Γ). Note that, by property (iii) of real(Γ), real(e) is the segment joining $\langle o(e) \rangle$, $\langle t(e) \rangle$, for any edge e. We can find a reduced path $e_1, \ldots, e_n$ in Γ such that $x = \langle e_1, t \rangle$, $y = \langle e_n, s \rangle$ for some s, $t \in I$ (exercise). Put $x_i = \langle t(e_i) \rangle$. Then real$(e_{i+1}) \cap$ real$(e_i) = \{x_i\}$ for $1 \leq i \leq n - 1$. It follows from Lemma 2.3(iii) that, if $n > 1$, the segment joining x, y contains x_1. It now follows that, if $0 < t < 1$, then x has degree 2, the two directions at x being the equivalence classes of the segments $[x, x_1]$ and $[x, x_0]$, where $x_0 = \langle o(e_1) \rangle$. Thus the set of points in real(Γ) of degree greater than 2 is contained in the set $\{\langle v \rangle | v \in V(\Gamma)\}$, so by Property (iii) is discrete. (With a little more work one can show that, if $v \in V(\Gamma)$, then the degree of $\langle v \rangle$ is the degree of v in the graph Γ, as defined in §1).

Example 4. Let $Y = \mathbb{R}^2$ be the plane, but with metric d defined by

$$d((x_1, y_1), (x_2, y_2)) = \begin{cases} |y_1| + |y_2| + |x_1 - x_2| & \text{if } x_1 \neq x_2 \\ |y_1 - y_2| & \text{if } x_1 = x_2. \end{cases}$$

(For a picture illustrating this example see 2.9 of [45]). Then (Y, d) is an $\mathbb{R}$-tree by Lemma 2.5, with X equal to the real axis $\mathbb{R} \times \{0\}$, $J = \mathbb{R}$ and X_j the vertical line $\{j\} \times \mathbb{R}$, for $j \in J$, so that X and all the X_j are isomorphic as $\mathbb{R}$-metric spaces to $\mathbb{R}$. In particular, X is not discrete, and it is easy to see that every point of X has degree 4. Thus Y is not isomorphic to real(Γ) for any simplicial tree Γ.

By a subtree of a Λ-tree (X, d) we mean a convex subset of X, that is, a subset A such that $a, b \in A$ implies $[a, b] \subseteq A$. Then A, with the metric obtained by restriction from d, becomes a Λ-tree. A subtree A is called *closed* if it is convex-closed, that is, if the intersection of A with any segment of X is either empty or a segment of X.

For example, a segment is a closed subtree (exercise, or see Cor. 2.16 in [1]; this is related to Lemma 3.4 below). An example of a subtree which is not closed is the interval $(0, 1)$ in $\mathbb{R}$. The intersection of any collection of subtrees is a subtree, and the intersection of finitely many closed subtrees is a closed subtree. However, the intersection of an arbitrary family of closed subtrees need not be a closed subtree. For example, take $\Lambda = \mathbb{Z} \times \mathbb{Z}$ with the lexicographic ordering, $X = \Lambda$ and let A_i be the segment joining $(0, i)$ and $(1, 0)$. Then $\bigcap_{i=0}^{\infty} A_i$ is not a closed subtree (consider its intersection with A_i). Another example is obtained by taking $\Lambda = \mathbb{Q}$, $X = \Lambda$, and A_r to be the segment joining r and 2, where $0 \leq r < \sqrt{2}$, and considering the intersection of $\bigcap_r A_r$ with the segment $[0, 2]_{\mathbb{Q}}$. In case $\Lambda = \mathbb{Z}$, $X = V(\Gamma)$ for some simplicial tree Γ, the subtrees of X correspond to subtrees of Γ, and they are all closed.

The main result about closed subtrees is the following lemma, which is part of the "Bridge Proposition" in [1].

Lemma 2.7 *Let A, B be non-empty closed subtrees of a Λ-tree (X, d) such that $A \cap B = \emptyset$. Then there exist unique points $a \in A$, $b \in B$ such that $[a, b] \cap A = \{a\}$, $[a, b] \cap B = \{b\}$ and if $a_1 \in A$, $b_1 \in B$, then $[a, b] \subseteq [a_1, b_1]$.*

Proof. See 2.17 in [1]. □

We call $[a, b]$ in 2.7 the bridge between A and B. The lemma applies in particular when $B = \{b\}$ is a single point and A is a closed subtree. In this case we also say that a is the bridge from b to A. Also, if A is a closed subtree and $a \in A$, we call $[a, a] = \{a\}$ the bridge between A and a.

Any Λ-metric space (X, d) has a topology with basis the open balls $B(x, r) = \{y \in X \mid d(x, y) < r\}$, where $x \in X$ and $r \in \Lambda$, $r > 0$, just as in the case $\Lambda = \mathbb{R}$. Any closed subtree of a Λ-tree is closed in this topology (use 2.7 to show the complement is open), but the converse is false in general, because of the examples above, where the intersection of a family of closed subtrees is not a closed subtree. However, the converse is true in the case $\Lambda = \mathbb{R}$, for the intersection of a subtree closed in this topology with a segment is a compact connected subset of the segment, hence a segment.

3. Lyndon Length Functions

Despite the title, we shall not meet Lyndon length functions for some while. Firstly, we shall discuss hyperbolic spaces in the sense of Gromov [23]. The usual examples of such spaces are Riemannian manifolds whose sectional curvature in every direction at every point is negative and bounded away from zero by some constant (for example, real hyperbolic space), and word hyperbolic groups. However, the $\mathbb{R}$-trees are precisely the geodesic 0-hyperbolic $\mathbb{R}$-metric spaces, and we begin by establishing a version of this for arbitrary Λ-trees. We then give an important construction, which essentially says that a 0-hyperbolic Λ-metric space can be isometrically embedded in a Λ-tree in a natural way, subject to a minor technical condition needed when dealing with arbitrary Λ. We use this construction to establish an equivalence between actions of a group on Λ-trees and Λ-valued Lyndon length functions defined on the group. We shall not give the definition yet, but the prototype such length function is given by taking a free group G with basis X, and defining $L : G \to \mathbb{Z}$ by $L(g) =$ the length of the reduced word on $X^{\pm 1}$ representing g. This example will be considered in more detail in §5. The notion of Lyndon length function in monoids has been considered by Rhodes [42], and this may be a fruitful area for further study. We shall also use our construction to define a base-change functor. Given a Λ-tree X and a morphism of ordered abelian groups $h : \Lambda \to \Lambda'$, this produces an induced Λ'-tree which we denote by $\Lambda' \otimes_\Lambda X$.

Fix an ordered abelian group Λ, and suppose (X, d) is a Λ-metric space. Choose a point $v \in X$, and for $x, y \in X$, define

$$(x \cdot y)_v = \frac{1}{2}(d(x,v) + d(y,v) - d(x,y)).$$

If (X, d) is a Λ-tree, then $[v, x] \cap [v, y] = [v, w]$ for some w, and $(x \cdot y)_v = d(v, w)$, by Lemma 2.2(2). There is also a simple geometric interpretation of $(x \cdot y)_v$ in real euclidean and hyperbolic space (see [13], Ch.1, §1). In these examples $(x \cdot y)_v \in \Lambda$, but in general it is an element of $\frac{1}{2}\Lambda$. If t is another point of X, we have the identity

$$(*) \qquad (x \cdot y)_t = d(t,v) + (x \cdot y)_v - (x \cdot t)_v - (y \cdot t)_v$$

for all $x, y \in X$.

Definition. Let $\delta \in \Lambda$ with $\delta \geq 0$. Then (X, d) is *δ-hyperbolic* with respect to v if, for all $x, y, z \in X$,

$$x \cdot y \geq \min\{x \cdot z, y \cdot z\} - \delta$$

where $x \cdot y$ means $(x \cdot y)_v$, etc.

Lemma 3.1 *If (X, d) is δ-hyperbolic with respect to v, and t is any other point of X, then (X, d) is 2δ-hyperbolic with respect to t.*

Proof. We need to show that $(x \cdot y)_t \geq \min\{(x \cdot z)_t, (y \cdot z)_t\} - 2\delta$, for all $x, y, z \in X$. On adding $(x \cdot t)_v + (y \cdot t)_v + (z \cdot t)_v - d(t, v)$ to both sides and using $(*)$ above, it suffices to show that

$$(x \cdot y)_v + (z \cdot t)_v \geq \min\{(x \cdot z)_v + (y \cdot t)_v, (y \cdot z)_v + (x \cdot t)_v\} - 2\delta.$$

for all x, y, z and $t \in X$. Now interchanging x and y if necessary, and also interchanging z and t if appropriate, we can assume that $x \cdot z$ is greater than or equal to $x \cdot t$, $y \cdot z$ and $y \cdot t$, where $x \cdot z$ means $(x \cdot z)_v$, etc. Then by assumption $x \cdot y \geq y \cdot z - \delta$ and $z \cdot t \geq x \cdot t - \delta$. The lemma follows on adding these inequalities. □

Definition. (X, d) is *δ-hyperbolic* if it is δ-hyperbolic with respect to all points of X, and is *hyperbolic* if it is δ-hyperbolic for some $\delta \geq 0$.

Lemma 3.2 *The Λ-metric space (X, d) is δ-hyperbolic if and only if, for all x, y, z and $t \in X$,*

$$d(x,y) + d(z,t) \leq \max\{d(x,z) + d(y,t), d(y,z) + d(x,t)\} + 2\delta.$$

Proof. Put $a = d(x,y) + d(z,t)$, $b = d(x,z) + d(y,t)$, $c = d(y,z) + d(x,t)$ and $e = d(x,t) + d(y,t) + d(z,t)$. Then $a \leq \max\{b,c\} + 2\delta$ if and only if $e - a \leq \min\{e-b, e-c\} - 2\delta$, and on dividing by 2, this is equivalent to saying that X is δ-hyperbolic with respect to t. □

If (X, d) is geodesic, then to say that (X, d) is hyperbolic is equivalent to saying that geodesic triangles in X are, in various senses, "thin". For a full discussion of this see [13; Ch. 1] or [22; Ch. 2]. One interpretation of Lemma 2.2 is that, in a Λ-tree, geodesic triangles are (in all these various senses) "0-thin". We shall now prove directly that Λ-trees are 0-hyperbolic in the sense we have defined. Before doing so, we consider the property of being 0-hyperbolic in a little more detail.

Definition. A triple of elements (a, b, c) of Λ is called an *isosceles triple* if $a \geq \min\{b, c\}$, $b \geq \min\{c, a\}$ and $c \geq \min\{a, b\}$. (This means that at least two of a, b, c are equal, and not greater than the third).

Lemma 3.3

(1) If (a, b, c) is any triple then $(\min\{a, b\}, \min\{b, c\}, \min\{c, a\})$ is an isosceles triple.

(2) If (a, b, c) and (d, e, f) are isosceles triples then so is

$$(\min\{a, d\}, \min\{b, e\}, \min\{c, f\}).$$

Proof. This is left as an easy exercise (there is a dual version of this in [25].) □

Note that, by Lemma 3.1, if (X, d) is 0-hyperbolic with respect to one point, then it is 0-hyperbolic with respect to any other point. Also, (X, d) is 0-hyperbolic with respect to v if and only if, for all x, y and $z \in X$, $((x \cdot y)_v, (y \cdot z)_v, (z \cdot x)_v)$ is an isosceles triple. Further, by Lemma 3.2, (X, d) is 0-hyperbolic if and only if, for all x, y, z and $t \in X$,

$$d(x,y) + d(z,t) \leq \max\{d(x,z) + d(y,t), d(y,z) + d(x,t)\}.$$

This means that of the three numbers $a = d(x,y) + d(z,t)$, $b = d(x,z) + d(y,t)$, $c = d(y,z) + d(x,t)$, at least two are equal and not less than the third, or equivalently

$(-a, -b, -c)$ is an isosceles triple, for all x, y, z and $t \in X$. This is referred to in the literature as the 4-*point condition.*

Remark. If d satisfies the 4-point condition, the triangle inequality follows from this and the other axioms for a metric (put $t = z$). Further, the argument of Lemma 3.1 does not use the triangle inequality, so the triangle inequality follows if we know that for some $v \in X$, $((x \cdot y)_v, (y \cdot z)_v, (z \cdot x)_v)$ is an isosceles triple, for all x, y and $z \in X$.

In the next lemma we shall use the following notation from [1]. We write $[x_0, x_n] = [x_0, x_1, \ldots, x_{n-1}, x_n]$ to mean that the points $x_0, x_1, \ldots, x_{n-1}, x_n$ lie in this order on the segment $[x_0, x_n]$. This is described more precisely in Remark 2.10(2) of [1]. Also, if $[x, y] \cap [x, z] = [x, w]$, we denote w by $Y(x, y, z)$.

Lemma 3.4 *Let (X, d) be a Λ-tree. Then (X, d) is 0-hyperbolic, and for all x, y and $v \in X$, $(x \cdot y)_v \in \Lambda$.*

Proof. Fix $v \in X$. To show $((x \cdot y)_v, (y \cdot z)_v, (z \cdot x)_v)$ is an isosceles triple for all x, y and z, it suffices to show that if x, y, $z \in X$ and $m \in \Lambda$,

$$x \cdot y \geq m \text{ and } y \cdot z \geq m \text{ implies } x \cdot z \geq m.$$

where $x \cdot y = (x \cdot y)_v$, etc. Let $w = \mathrm{Y}(x, v, y)$, $u = \mathrm{Y}(y, v, z)$, $t = \mathrm{Y}(z, v, x)$. As we have already observed, $x \cdot y = d(v, w)$ by Lemma 2.2(2) (which establishes the last part of the theorem), and $y \cdot z = d(v, u)$, $z \cdot x = d(v, t)$. Thus we assume $d(v, w) \geq m$, $d(v, u) \geq m$ and have to show that $d(v, t) \geq m$.

Suppose $d(v, w) \leq d(v, u)$. By their definition, w, $u \in [v, y]$, so $[v, y] = [v, w, u, y]$ and $w \in [v, u] \subseteq [v, z]$ (by definition of u) so $w \in [v, x] \cap [v, z] = [v, t]$, hence $d(v, t) \geq d(v, w) \geq m$. Similarly if $d(v, u) \leq d(v, w)$ then $d(v, t) \geq d(v, u) \geq m$. □

It follows that at least two of the points w, u, t in the proof of Lemma 3.4 coincide. In fact there are three possibilities, and a precise version of this is given by the "H-proposition"; see [1; 2.15], where the pictures illustrating the result give a geometric interpretation of the four-point condition. Before proceeding we need to make a few remarks.

Remarks.

(1) If (X, d) is any Λ-metric space, and v, x, $y \in X$, then it follows by the triangle inequality that $0 \leq (x \cdot y)_v \leq d(x, v)$, $d(y, v)$.

(2) If (X, d) is a Λ-metric space then the following are equivalent:

(i) for some $v \in X$ and all x, $y \in X$, $(x \cdot y)_v \in \Lambda$

(ii) for all v, x and $y \in X$, $(x \cdot y)_v \in \Lambda$.

This follows from equation $(*)$ above.

Our next goal is to prove a converse of Lemma 3.4. If (X, d) is a geodesic Λ-metric space satisfying the conditions of Lemma 3.4, then it is a Λ-tree. We shall do this by first proving two lemmas which show that, in a 0-hyperbolic space, segments behave as they do in the Y-proposition (Lemma 2.2) for Λ-trees.

Lemma 3.5 *Assume (X, d) is a 0-hyperbolic metric space, and let σ, τ be segments in X with endpoints v, x and v, y respectively. Let $x \cdot y$ denote $(x \cdot y)_v$. Then*

(i) *If $x' \in \sigma$, then $x' \in \tau$ if and only if $d(v, x') \leq x \cdot y$.*

(ii) *If $x \cdot y \in \Lambda$, and w is the point of σ at distance $x \cdot y$ from v (which exists by Remark 1 above), then $\sigma \cap \tau$ is a segment with endpoints v and w.*

Proof. If $d(x', v) > d(y, v)$ then $x' \notin \tau$, and $d(x', v) > x \cdot y$ by Remark 1, so we can assume that $d(x', v) \leq d(y, v)$. Let y' be the point in τ such that $d(v, x') = d(v, y')$. Define $\alpha = x \cdot y$, $\beta = x' \cdot y$, $\gamma = x \cdot x'$, $\alpha' = x' \cdot y'$.

Since $x' \in \sigma$ and $y' \in \tau$, we have $\gamma = d(v, x') = d(v, y') = y \cdot y'$, hence (α, β, γ) and (α', β, γ) are isosceles triples. We have to show that $x' \in \tau$ if and only if $\alpha \geq \gamma$. By Remark 1 above, $x' \cdot y \leq d(v, x')$, that is, $\beta \leq \gamma$. Also,

$$\alpha' = d(v, x') - \frac{1}{2}d(x', y') = \gamma - \frac{1}{2}d(x', y') \leq \gamma$$

and

$$x' \in \tau \iff x' = y' \iff d(x', y') = 0 \iff \alpha' = \gamma.$$

Now $\alpha' = \gamma$ if and only if $\beta = \gamma$, because (α', β, γ) is an isosceles triple and α', $\beta \leq \gamma$. Since (α, β, γ) is also an isosceles triple, $\beta = \gamma$ is equivalent to $\alpha \geq \gamma$. This proves (i), and Part (ii) of the lemma follows immediately. □

Lemma 3.6 *Once more assume (X, d) is a 0-hyperbolic metric space, and σ, τ are segments in X with endpoints v, x and v, y respectively. Let $x \cdot y$ denote $(x \cdot y)_v$, and assume that $x \cdot y \in \Lambda$. Let w be the point of σ at distance $x \cdot y$ from v (so w is an endpoint of $\sigma \cap \tau$ by Lemma 3.5).*

Suppose $x' \in \sigma$, $y' \in \tau$ and $d(x', v) \geq x \cdot y$, $d(y', v) \geq x \cdot y$. Then

$$d(x', y') = d(x', w) + d(y', w).$$

Proof. The conclusion is clear if $d(x', v) = x \cdot y$ or $d(y', v) = x \cdot y$, so we assume that $d(x', v) > x \cdot y$ and $d(y', v) > x \cdot y$. We again put

$$\alpha = x \cdot y, \ \beta = x' \cdot y, \ \gamma = x \cdot x', \ \alpha' = x' \cdot y'$$

and additionally $\gamma' = y \cdot y'$, so $\gamma = d(v, x')$, and $\gamma' = d(v, y')$. Thus $\alpha < \gamma$, hence $\alpha = \beta$ since (α, β, γ) is an isosceles triple. Also, $\alpha < \gamma'$, so $\beta < \gamma'$, hence $\alpha = \alpha' = \beta$ since $(\alpha', \beta, \gamma')$ is an isosceles triple. By definition of α',

$$d(x', y') = d(v, x') + d(v, y') - 2\alpha' = d(v, x') + d(v, y') - 2\alpha.$$

Since $w \in \sigma \cap \tau$ and $\alpha = d(v, w)$, we have

$$d(x', w) = d(v, x') - \alpha$$
$$d(y', w) = d(v, y') - \alpha$$

and the lemma follows on adding these equations. □

We can now prove a converse of Lemma 3.4.

Lemma 3.7 *Assume (X,d) is a Λ-metric space and v is a point of X with the following properties.*

(i) *for all x, y, $z \in X$, $(x \cdot y)_v \in \Lambda$.*

(ii) *(X,d) is 0-hyperbolic with respect to v*

(iii) *For every $x \in X$, there is a geodesic with endpoints v, x*

Then (X,d) is a Λ-tree.

Proof. Take x, $y \in X$ and let σ, τ be segments with endpoints v, x and v, y respectively. Let w be the point of $\sigma \cap \tau$ at distance $(x \cdot y)_v$ from v (using Lemma 3.5). Then σ is the union $(\sigma \cap \tau) \cup \sigma_1$, where $\sigma_1 = \{u \in \sigma | \ d(v,u) \geq (x \cdot y)_v\}$ is a segment with endpoints w, x. Similarly τ is the union $(\sigma \cap \tau) \cup \tau_1$, where $\tau_1 = \{u \in \tau | \ d(v,u) \geq (x \cdot y)_v\}$ is a segment with endpoints w, y. By Lemma 3.6 and Lemma 1.3, $\sigma_1 \cup \tau_1$ is a segment with endpoints x, y. Thus (X,d) is geodesic. Note that, by Lemma 3.5 if $\sigma \cap \tau = \{w\}$ then $(x \cdot y)_v = 0$ and $\sigma_1 = \sigma$, $\tau_1 = \tau$, hence $\sigma \cup \tau$ is a segment. Also by Lemma 3.5, $\sigma \cap \tau$ is a segment with endpoints v, w.

Now by Lemma 3.1 and Remark 2 preceding Lemma 3.5, we may replace v in this argument by any other point of X, hence (X,d) satisfies the axioms for a Λ-tree. □

The next result is basic in the theory of Λ-trees.

Theorem 3.1 *Let (X,d) be a Λ-metric space and v a point of X such that*

(i) *for all x, y, $z \in X$, $(x \cdot y)_v \in \Lambda$.*

(ii) *(X,d) is 0-hyperbolic with respect to v*

Then there exist a Λ-tree (X',d') and an isometry $\phi : X \to X'$ such that, if $\psi : X \to Z$ is any isometry of X into a Λ-tree Z, there is a unique isometry $\mu : X' \to Z$ such that $\mu \circ \phi = \psi$, that is, the following diagram is commutative.

$$\begin{array}{ccc} X' & & \\ \phi \uparrow & \searrow^{\mu} & \\ X & \underset{\psi}{\longrightarrow} & Z \end{array}$$

Proof. As usual $x \cdot y$ means $(x \cdot y)_v$ for x, $y \in X$. Let

$$Y = \{(x,m) | \ x \in X, \ m \in \Lambda \text{ and } 0 \leq m \leq d(v,x)\}.$$

Define, for (x,m), $(y,n) \in Y$, $(x,m) \sim (y,n)$ if and only if $x \cdot y \geq m = n$. This is an equivalence relation on Y (it is transitive by assumption (ii)). Let $X' = Y/\sim$, and let $\langle x,m \rangle$ denote the equivalence class of (x,m). We define the metric by $d'(\langle x,m \rangle, \langle y,n \rangle) = m + n - 2\min\{m,n,x \cdot y\}$. It follows from assumption (ii) that

d' is well-defined. Note that $d'(\langle x,m\rangle, \langle x,n\rangle) = |m-n|$ and $\langle x,0\rangle = \langle v,0\rangle$ for all $x \in X$, so $d'(\langle x,m\rangle, \langle v,0\rangle) = m$.

Clearly d' is symmetric, and it is easy to see that $d'(\langle x,m\rangle, \langle y,n\rangle) = 0$ if and only if $\langle x,m\rangle = \langle y,n\rangle$. Also, in X',

$$(\langle x,m\rangle \cdot \langle y,n\rangle)_{\langle v,0\rangle} = \min\{m,n,x\cdot y\}$$

which is an element of Λ. If $\langle x,m\rangle$, $\langle y,n\rangle$ and $\langle z,p\rangle$ are three points of X', then $(\min\{m,n\}, \min\{n,p\}, \min\{p,m\})$ is an isosceles triple by Lemma 3.3(1), hence by Lemma 3.3(2) so is $(\min\{m,n,x\cdot y\}, \min\{n,p,y\cdot z\}, \min\{p,m,z\cdot x\})$. It follows that (X',d') is a 0-hyperbolic Λ-metric space (d' satisfies the triangle inequality by the remark just before Lemma 3.4).

If $\langle x,m\rangle \in X'$, then the mapping $\alpha : [0,m]_\Lambda \to X'$ given by $\alpha(n) = \langle x,n\rangle$ is an isometry, so $\mathrm{Im}(\alpha)$ is a segment with endpoints $\langle v,0\rangle$ and $\langle x,m\rangle$. It now follows from Lemma 3.7 that (X',d') is a Λ-tree. Further, the mapping $\phi : X \to X'$ defined by $\phi(x) = \langle x, d(v,x)\rangle$ is easily seen to be an isometry.

Given an isometry $\psi : X \to Z$ and $x \in X$, denote by x_m the point on the segment in Z with endpoints $\psi(v)$ and $\psi(x)$, at distance m from $\psi(v)$. Then the mapping $\mu : X' \to Z$ defined by $\mu(\langle x,m\rangle) = x_m$ is an isometry by Lemma 2.2(2) and is the only one with the required properties, because isometries of Λ-trees preserve segments. □

Corollary 3.2 *In the situation of Theorem 3.1, suppose $\theta : X \to X$ is an isometry. Then there is a unique isometry $\mu : X' \to X'$ such that $\mu \circ \phi = \phi \circ \theta$.*

Proof. Apply the last part of Theorem 3.1 to $\psi = \phi \circ \theta$. □

We can now introduce the idea of a Lyndon length function. They were originally introduced to carry out Nielsen-style cancellation arguments used in free groups in an abstract setting. Let G be a group, Λ an ordered abelian group. A mapping $L : G \to \Lambda$ is called a Lyndon length function if

(1) $L(1) = 0$

(2) for all $g \in G$, $L(g) = L(g^{-1})$

(3) for all g, h, $k \in G$, $(c(g,h), c(h,k), c(k,g))$ is an isosceles triple, where $c(g,h)$ is defined to be $\frac{1}{2}(L(g) + L(h) - L(g^{-1}h))$.

We leave it as an exercise to verify that these axioms imply

(4) for all $g \in G$, $L(g) \geq 0$

(5) for all g, $h \in G$, $L(gh) \leq L(g) + L(h)$.

We refer to (5) as the triangle inequality. For example, let G be a group acting as isometries on a Λ-tree (X,d) and let $x \in X$. Then defining $L_x(g) = d(x,gx)$ gives a Lyndon length function on G. By applying the isometry g^{-1} to $[x,gx] \cup [x,hx]$, we see that $c(g,h)$ is $d(x,w)$, where $[x,gx] \cap [x,hx] = [x,w]$. It follows by Lemma 3.4 that L_x satisfies Axiom (3) above, and it clearly satisfies (1) and (2). Also, it follows that this example satisfies

(6) For all $g, h \in G$, $c(g,h) \in \Lambda$.

Using Theorem 3.1 we can prove a converse.

Theorem 3.3 *Let G be a group and $L : G \to \Lambda$ a Lyndon length function satisfying (6). Then there are a Λ-tree (X', d'), an action of G on X' and a point $x \in X'$ such that $L = L_x$.*

Proof. Define $g \approx h$ if and only if $L(g^{-1}h) = 0$. This is an equivalence relation on G (use the triangle inequality to see it is transitive), and we denote the set of equivalence classes by X, and the equivalence class of g by $\langle g \rangle$. We obtain a Λ-metric space (X, d) by defining $d(\langle g \rangle, \langle h \rangle) = L(g^{-1}h)$. Also, G acts as isometries on X by $g\langle h \rangle = \langle gh \rangle$. Direct computation shows that

$$(\langle g \rangle \cdot \langle h \rangle)_{\langle 1 \rangle} = c(g,h),$$

so by axioms (3) and (6), (X, d) satisfies the hypotheses of Theorem 3.1 (with $v = \langle 1 \rangle$), and we obtain a corresponding Λ-tree (X', d') and an isometry $\phi : X \to X'$. We denote the point $\langle \langle g \rangle, m \rangle$ of X' simply by $\langle g, m \rangle$, and take $x = \langle 1, 0 \rangle$ as basepoint. Thus $\phi(\langle g \rangle) = \langle g, L(g) \rangle$ for $g \in G$, and $x = \phi(v)$. It follows by Corollary 3.2 that the action of G on X induces an action of G on X', such that $g\phi(u) = \phi(gu)$ for all $u \in X$ and $g \in G$. Thus $gx = \langle g, L(g) \rangle$ for $g \in G$, and by definition of d', $d'(x, gx) = L(g)$, as required. □

A more detailed inspection of the proof of Theorem 3.3 yields the formula originally used to define the action of G in [9]. Our final application of Theorem 3.1 will be to introduce the base-change functor.

Theorem 3.4 *Let $h : \Lambda \to \Lambda'$ be a morphism of ordered abelian groups, and let (X, d) be a Λ-tree. Then there exist a Λ'-tree (X', d') and a mapping $\phi' : X \to X'$ satisfying*

$$d'(\phi'(x), \phi'(y)) = h(d(x,y))$$

for all $x, y \in X$, and such that, if (Z, d'') is a Λ'-tree and $\psi : X \to Z$ is a mapping satisfying

$$d''(\psi(x), \psi(y)) = h(d(x,y))$$

for all $x, y \in X$, then there is a unique Λ'-isometry $\mu : X' \to Z$ such that $\mu \circ \phi' = \psi$.

Proof. We can define an equivalence relation on X by $x \approx y$ if and only if $h(d(x,y)) = 0$. We denote the equivalence class of x by $\langle x \rangle$, and obtain a Λ'-metric space $((X/\approx), \delta)$ by defining $\delta(\langle x \rangle, \langle y \rangle) = h(d(x,y))$. Since (X, d) is a Λ-tree, this Λ'-metric space satisfies the hypotheses of Theorem 3.1, on choosing a basepoint v, by Lemma 3.4. Thus Theorem 3.1 gives a Λ'-tree (X', d') and a Λ'-isometry $\phi : (X/\approx) \to X'$. We define $\phi'(x)$ to be $\phi(\langle x \rangle)$; then $d'(\phi'(x), \phi'(y)) = d'(\phi(\langle x \rangle), \phi(\langle y \rangle)) = \delta(\langle x \rangle, \langle y \rangle) = h(d(x,y))$, as required. If ψ is as in the statement of the theorem, then ψ induces a Λ'-isometry $\bar{\psi} : (X/\approx) \to Z$ by $\bar{\psi}(\langle x \rangle) = \psi(x)$. The theorem now follows easily from Theorem 3.1. □

It follows that the Λ'-tree (X', d') in Theorem 3.4 is determined up to a unique metric isomorphism, by a standard category-theoretic argument (in particular, it does not depend on the choice of basepoint in the proof of Theorem 3.4). We denote (X', d') by $\Lambda' \otimes_\Lambda X$.

In proving both Theorems 3.3 and 3.4 we have made use of the standard constuction of a metric space from a pseudo-metric space, which works for any Λ just as in the case $\Lambda = \mathbb{R}$.

Corollary 3.5 *Let $h : \Lambda \to \Lambda'$ be a morphism of ordered abelian groups, and let (X, d) be a Λ-tree. Let $\phi' : X \to \Lambda' \otimes_\Lambda X$ be the isometry in Theorem 3.4. If $\theta : X \to X$ is a Λ-isometry, then there is a unique Λ'-isometry $\mu : X' \to X'$ such that $\mu \circ \phi' = \phi' \circ \theta$.*

Proof. Define $\psi = \phi' \circ \theta$. Then

$$d'(\psi(x), \psi(y)) = d'(\phi'(\theta(x)), \phi'(\theta(y))) = h(d(\theta(x), \theta(y)))$$

by Theorem 3.4, and this equals $h(d(x, y))$ since θ is an isometry. The result now follows from Theorem 3.4. □

Corollary 3.6 *With the hypotheses of Cor. 3.5, suppose G is a group acting as Λ-isometries on X. Then there is an action of G on $\Lambda' \otimes_\Lambda X$ such that, for all $x \in X$ and $g \in G$, $\phi'(gx) = g\phi'(x)$. Moreover, the Lyndon length functions are related by $L_{\phi'(x)}(g) = h(L_x(g))$.*

Proof. The existence of the action follows at once from Corollary 3.5, and the formula for the length functions follows easily. □

In particular, if (X, d) is a Λ-tree and s is an isometry of X onto X, there is an induced isometry from $\Lambda' \otimes_\Lambda X$ onto itself, which we denote by $\Lambda' \otimes_\Lambda s$.

4. Isometries of Λ-trees

We begin by considering the behaviour of a single isometry of a Λ-tree. There is a classification of such isometries similar to that for hyperbolic space (and indeed there is such a classification for any proper geodesic hyperbolic $\mathbb{R}$-metric space in the sense of Gromov-see [13] or [22]). There are three possible kinds of isometries in a Λ-tree, elliptic, hyperbolic and inversions. Inversions do not occur in the case $\Lambda = \mathbb{R}$, or indeed whenever $\Lambda = 2\Lambda$, and unlike hyperbolic space, there are no parabolic isometries.

We shall mention only very briefly the classification of group actions on Λ-trees, and the important work of Morgan and Shalen on compactifying the space of conjugacy classes of discrete and faithful representations of a finitely generated group on hyperbolic space by means of actions on $\mathbb{R}$-trees. We shall prove some results on actions of finitely generated groups on Λ-trees, and describe an interesting method, due to Serre [44; Ch.II §1] in the case $\Lambda = \mathbb{Z}$ and to Morgan and Shalen [34] in general, of constructing an action of $SL_2(F)$ on a Λ-tree, where F is a field with a valuation.

Let (X, d) be a Λ-tree, where Λ is an arbitrary ordered abelian group, and let s be an isometry of X onto X. We call s *elliptic* if it has a fixed point. The behaviour of such isometries is described in the next lemma, where $\langle s \rangle$ denotes the cyclic subgroup generated by s (in the group of metric automorphisms of X). If $[x, y]$ is a segment, there is a unique point p such that $d(x, p) = d(p, y)$ if and only if $d(x, y) \in 2\Lambda$. This point, when it exists, is called the midpoint of $[x, y]$.

Lemma 4.1 *Let s be an elliptic isometry and let A_s denote the set of fixed points of s. Then*

(i) *A_s is a closed non-empty $\langle s \rangle$-invariant subtree of X which meets every $\langle s \rangle$-invariant subtree of X.*

(ii) *A_s is contained in every subtree of X with the property that it meets every $\langle s \rangle$-invariant subtree of X.*

(iii) *If $x \in X$ and $[x, p]$ is the bridge between x and A_s, then p is the midpoint of $[x, sx]$, and for all $a \in A_s$, $[a, x] \cap [a, sx] = [a, p]$. Hence $d(x, sx) = 2d(x, p)$.*

Proof. See [1] Prop. 6.1. □

Lemma 4.2 *The following are equivalent:*

(i) *There is a segment of X invariant under s, and the restriction of s to this segment has no fixed points.*

(ii) *There is a segment $[x, y]$ in X such that $sx = y$, $sy = x$ and $d(x, y) \notin 2\Lambda$.*

(iii) *s^2 has a fixed point, but s does not.*

(iv) *s^2 has a fixed point, and for all $x \in X$, $d(x, sx) \notin 2\Lambda$.*

Proof. For the implications (i)⇒(ii) and (ii)⇒(iii) see [1], Prop. 6.3.
(iii)⇒(iv): Let p be a fixed point of s^2. Then s^2 fixes the endpoints of $[p, sp]$, so fixes this segment pointwise and this segment is invariant under s. Take $x \in X$, and let q be the bridge from x to $[p, sp]$. Then sq is the bridge from sx to $[p, sp]$, and $sq \neq q$ by assumption. It follows from Lemma 2.3(3) that

$$d(x, sx) = d(x, q) + d(q, sq) + d(sq, sx) = 2d(x, q) + d(q, sq)$$

and s^2 fixes q. Hence $d(q, sq) \notin 2\Lambda$, otherwise s fixes the midpoint of $[q, sq]$, so $d(x, sx) \notin 2\Lambda$.
(iv)⇒(i): Let p be a fixed point of s^2. Then $[p, sp]$ is invariant under s, and s has no fixed points at all, by Lemma 4.1(iii). □

Definition. If s satisfies the equivalent conditions of 4.2, it is called an *inversion.*

Examples of inversions are provided by certain reflections, e.g., take $X = \Lambda = \mathbb{Z}$ and $s : x \mapsto 1 - x$. There is yet another criterion for an isometry to be an inversion in [1], 6.3, to which we refer for the proof.

Lemma 4.3 *The isometry s is an inversion if and only if: s has no fixed point, and whenever Λ' is an ordered abelian group containing Λ such that $d(x, sx) \in 2\Lambda'$ for some $x \in X$, then $d(x, sx) \in 2\Lambda'$ for all $x \in X$, and the automorphism $\Lambda' \otimes_\Lambda s$ of $\Lambda' \otimes_\Lambda X$ has a unique fixed point.*

We call an isometry that is not an inversion and not elliptic a hyperbolic isometry. A subtree of a Λ-tree is called *linear* if it is metrically isomorphic to a subtree of Λ.

Theorem 4.1 *Suppose s is hyperbolic, and define A_s to be the set of all points $p \in X$ such that $[s^{-1}p, p] \cap [p, sp] = \{p\}$. Then*

(i) *A_s is a non-empty closed $\langle s \rangle$-invariant subtree of X which meets every $\langle s \rangle$-invariant subtree of X.*

(ii) *A_s is contained in every subtree of X with the property that it meets every $\langle s \rangle$-invariant subtree of X.*

(iii) *A_s is a linear tree, and s restricted to A_s is equivalent to a translation $a \mapsto a + \ell(s)$ for some $\ell(s) \in \Lambda$ with $\ell(s) > 0$.*

(iv) *If $x \in X$ and $[x, p]$ is the bridge between x and A_s, then $[x, sx] \cap A_s = [p, sp]$ and $d(x, sx) = \ell(s) + 2d(x, p)$.*

Proof. See [1] Theorem 6.6. □

When s is not an inversion, A_s is called the characteristic subset of s; if s is elliptic it is the set of fixed points of s, and if s is hyperbolic it is linear and is called the axis of s (which explains the notation). When s is elliptic, we put $\ell(s) = 0$.

Corollary 4.2 *Assume s is not an inversion. Then $\ell(s) = \min\{d(x, sx) | \ x \in X\}$ and $A_s = \{p \in X | \ d(p, sp) = \ell(s)\}$.*

Proof. In case s is elliptic this is clear and in case s is hyperbolic it follows from Theorem 4.1(iv). □

When s is an inversion we put $A_s = \emptyset$ and $\ell(s) = 0$. This is justified by the fact that the extension of s to an isometry of $\frac{1}{2}\Lambda \otimes_\Lambda X$ has a unique fixed point, by Lemma 4.3. Note that s is hyperbolic if and only if $\ell(s) > 0$.

Lemma 4.4 *Let t be another isometry of X onto X. Then*
(1) $A_{tst^{-1}} = tA_s$ and $\ell(tst^{-1}) = \ell(t)$.
(2) If n is an integer then $\ell(s^n) = |n|\ell(s)$, and $A_s \subseteq A_{s^n}$. If $\ell(s) > 0$ and $n \neq 0$ then $A_s = A_{s^n}$.

Proof. See [1] 6.13. □

Recall that the Lyndon length $L_x(s)$ of s relative to a point $x \in X$ is defined to be $d(x, sx)$. If s is not an inversion, Cor. 4.2 says that $\ell(s) = \min\{L_x(s) | \ x \in X\}$. The next result expresses ℓ in terms of L_x where x is a single point of X.

Lemma 4.5 *For any isometry s from X onto X and any $x \in X$,*

$$\ell(s) = \max\{L_x(s^2) - L_x(s), 0\}.$$

Proof. Suppose s is elliptic. From Lemma 4.1 we have $L_x(s) = 2d(x,p)$ where p is a point stabilised by s, so by s^2. By the triangle inequality, $L_x(s^2) \leq d(x,p) + d(p, s^2x) = d(x,p) + d(s^2p, s^2x) = 2d(x,p)$, so $\max\{L_x(s^2) - L_x(s), 0\} = 0$, as required.

If s is an inversion, then the proof of (iii)$\Rightarrow$(iv) in Lemma 4.2 gives a point q stabilised by s^2 such that $L_x(s) = 2d(x,q)+d(q,sq)$. Again by the triangle inequality $L_x(s^2) \leq 2d(x,q)$ and so $\max\{L_x(s^2) - L_x(s), 0\} = 0 = \ell(s)$.

Finally, if s is hyperbolic, let $[p,x]$ be the bridge between x and A_s. By Theorem 4.1, $L_x(s) = \ell(s) + 2d(p, sp)$. By Lemma 4.4, $A_{s^2} = A_s$, $\ell(s^2) = 2\ell(s)$ and s^2 is hyperbolic, hence $L_x(s^2) = 2\ell(s) + 2d(p, sp)$. Therefore $L_x(s^2) - L_x(s) = \ell(s) > 0$ and the result follows. □

Now suppose G is a group acting as isometries on a Λ-tree (X,d). For $g \in G$, the characteristic set of the isometry induced on X by g is denoted by A_g. We have a function $\ell : G \to \Lambda$, where $\ell(g) = \min\{d(x, gx)| \ x \in X\}$, for $g \in G$. This function is known as the hyperbolic or translation length function of the action, also as the unbased length function (to distinguish it from the Lyndon length functions, which are called based length functions since they depend on a choice of basepoint). Like the Lyndon length functions, which are characterised in Theorem 3.3, the hyperbolic length can be characterised axiomatically (see [38]).

Corollary 4.3 *Let $h : \Lambda \to \Lambda'$ be a morphism of ordered abelian groups, and let (X,d) be a Λ-tree. Suppose G is a group acting as Λ-isometries on X, so that there is an induced action of G on $\Lambda' \otimes_\Lambda X$ (see Cor. 3.6). Let ℓ be the hyperbolic length function for the action on X, and ℓ' the hyperbolic length function for the action on $\Lambda' \otimes_\Lambda X$. Then $\ell' = h \circ \ell$.*

Proof. This follows at once from Cor. 3.6 and Lemma 4.5. □

In §7 of [1] there is a classification of group actions on Λ-trees which we shall describe briefly. Suppose G is a group acting on a Λ-tree (X,d) and let $\ell : G \to \Lambda$ be the hyperbolic length function. Put $N = \{g \in G| \ \ell(g) = 0\}$, the set of elements of G which do not act as hyperbolic isometries. The action is called *abelian* if $\ell(gh) \leq \ell(g)+\ell(h)$ for all $g, h \in G$, *dihedral* if it is non-abelian but $\ell(gh) \leq \ell(g)+\ell(h)$ whenever $g, h \notin N$, and *of general type* otherwise. Put $B = \bigcap_{g \in G} A_g$. An abelian action is of exactly one of the following types.

(1) Some element of G acts as an inversion. In this case $N = G$.
(2) $B \neq \emptyset$ and $N = G$. Then B is the set of fixed points of G.
(3) $B \neq \emptyset$ and $N \neq G$. Then B is a G-invariant linear subtree on which G acts as translations. Also, no point of X is fixed by G.
(4) End type.
(5) Cut type.

Types (2) and (3) are called actions of core type. They are the abelian actions for which $B \neq \emptyset$. In type (4) there is a unique end of X fixed by G, and in type (5) X is the disjoint union of two G-invariant subtrees, and the action of G on both subtrees is of end type. We shall say very little about the proof of this classification, which is difficult, other than to note that it makes use of the important idea of ends in a Λ-tree. For information on this topic see [1] and [3].

If the action of G is abelian then there is a group homomorphism $h : G \to \Lambda$ such that $\ell(g) = |h(g)|$ for all $g \in G$. If the action is without inversions and non-abelian then there is a unique minimal G-invariant subtree of X, denoted by $X_{\min}$, and the action of G on $X_{\min}$ has the same hyperbolic length function as the action on X. Further, the action is dihedral if and only if $X_{\min}$ is a linear subtree. Thus in this case the hyperbolic length function is obtained from an action of G by isometries of Λ, by Lemma 1.2 (and the action must include a reflection since it is non-abelian). For proofs of these facts see §7 in [1].

If $\Lambda = 2\Lambda$ then the action must be without inversions by Lemma 4.2, and this applies in particular when $\Lambda = \mathbb{R}$. Also, it is shown in [1] that cut type cannot occur when $\Lambda = \mathbb{R}$ or $\mathbb{Z}$, so in the case $\Lambda = \mathbb{R}$, abelian actions are of type (2), (3) or (4). In [14], actions on $\mathbb{R}$-trees which are not of general type are called reducible, an action is called semisimple if it is non-abelian or abelian of core type, and actions of type (3) are called shifts.

There is one further useful observation that can be made about the classification of actions. The action is called trivial if $N = G$, that is, there are no hyperbolic isometries. Such an action is necessarily abelian. If G is finitely generated, the next result says that a trivial action must be of type (1) or (2). (In general, a trivial action can be of type (4) or (5); see §3.10 in [14] for an example).

Lemma 4.6 *Let G be a finitely generated group acting trivially and without inversions on a Λ-tree (X, d). Then there is a point of X which is fixed by all elements of G.*

Proof. The hypotheses imply that every element of G fixes some point of X. Let G be generated by $x_1, \dots, x_n$. We use induction on n, and the result is clear if $n = 1$. Let H be the subgroup generated by $x_1, \dots, x_{n-1}$, and assume inductively that there is a point $v \in X$ fixed by all of H. By Lemma 4.1, the midpoint u of $[v, x_n v]$ is fixed by x_n, and indeed by $x_n h$ for all $h \in H$, because $x_n hv = x_n v$. This applies in particular to $h = 1, x_1, \dots, x_{n-1}$, so u is fixed by $x_1, \dots, x_n$, hence by G. □

In fact, the argument of Lemma 4.6 can be easily modified to give a stronger result.

Lemma 4.7 *Let G be generated by $x_1, \dots, x_n$ and assume that G acts on a Λ-tree. Suppose that x_i , for $1 \leq i \leq n$, and $x_i x_j$, for $j < i$ all have a fixed point. Then there is point of the tree fixed by all of G.*

Proof. Left to the reader (see [44; Ch. I §6.5] for further discussion) □

The classification of actions given above involves the following result. Let s, t be isometries of the Λ-tree (X, d), and put $\beta(s,t) = \ell(st) - \ell(s) - \ell(t)$, where ℓ denotes hyperbolic length. Then $d(A_s, A_t) = \max\{\frac{1}{2}\beta(s,t), 0\}$. (Note: if A, B are any closed subtrees of X, we put $d(A, B) = 0$ if A intersects B, and $d(A, B) = d(x, y)$ otherwise, where $[x, y]$ is the bridge between A and B). The proof involves a lengthy analysis of cases, which is carried out in §8 of [1]. We shall need one small part of this in what follows.

Lemma 4.8 *Suppose s, t are isometries of the Λ-tree (X, d) with $A_s \cap A_t = \emptyset$. Let $[S, T]$ be the bridge between A_s and A_t. Then $[S, T] \subseteq A_{st}$, and*

$$\ell(st) = \ell(s) + \ell(t) + 2d(A_s, A_t).$$

Proof. See Proposition 8.1 in [1]. □

An obvious question is: if a group has an action on a Λ-tree, what does this imply in general about the structure of the group? Any group has a trivial action on a Λ-tree (e.g., take the tree to have just a single point). Thus it is reasonable to confine attention to non-trivial actions. We shall prove one result on non-trivial actions when the group is finitely generated. First we need a lemma.

Lemma 4.9 *Let G be a group acting without inversions on a Λ-tree X. Suppose G is generated by $x_1, \ldots, x_n$. Then there exist generators $y_1, \ldots, y_n$ for G, obtained from $x_1, \ldots, x_n$ by Nielsen transformations, such that $A_{y_1} \cap \ldots \cap A_{y_n} \neq \emptyset$.*

Proof. We use induction on n. Since the action is without inversions we have $A_{x_1} \neq \emptyset$, so if $n = 1$ we take $y_1 = x_1$. Suppose $n > 1$ and the lemma is true for $n - 1$, so $\langle x_1, \ldots, x_{n-1}\rangle = \langle y_1, \ldots, y_{n-1}\rangle$, where $A_{y_1} \cap \ldots \cap A_{y_{n-1}} \neq \emptyset$. Put $A = A_{y_1} \cap \ldots \cap A_{y_{n-1}}$, a closed subtree of X, being the intersection of finitely many closed subtrees. If $A \cap A_{x_n} \neq \emptyset$, we can take $y_n = x_n$, so suppose $A \cap A_{x_n} = \emptyset$. Let $[u, v]$ be the bridge between A and A_{x_n} with $u \in A_{x_n}$ and $v \in A$. Since A_{y_i} is a closed subtree and $v \in A_{y_i}$, $A_{y_i} \cap [u, v] = [v_i, v]$ for some $v_i \in X$ (where $1 \leq i \leq n - 1$). Re-order the y_i so that the v_i occur in order in $[u, v]$, that is, in the notation of Remark (2.10)2 in [1], $[u, v] = [u, v_1, \ldots, v_{n-1}, v]$. Then $v_{n-1} \in [v_i, v] \subseteq A_{y_i}$ for $1 \leq i \leq n - 1$, so $v_{n-1} \in A$, hence $[v_{n-1}, v] \subseteq A \cap [u, v] = \{v\}$, that is, $v_{n-1} = v$.

Thus $A_{y_{n-1}} \cap [u, v] = \{v\}$ and $A_{x_n} \cap [u, v] = \{u\}$. It follows that $A_{y_{n-1}} \cap A_{x_n} = \emptyset$. For if $w \in A_{y_{n-1}} \cap A_{x_n}$, then by Lemma 2.2 $[u, w] \cap [v, w] = [w, w']$ for some $w' \in [u, v]$; but $[u, w] \subseteq A_{x_n}$ and $[v, w] \subseteq A_{y_{n-1}}$, so $w' \in A_{y_{n-1}} \cap A_{x_n} \cap [u, v]$, which is impossible as this intersection is empty. It follows from Lemma 2.7 that $[u, v]$ is the bridge between $A_{y_{n-1}}$ and A_{x_n}. Now let $y_n = x_n y_{n-1}$, so $G = \langle y_1, \ldots, y_n\rangle$, and inductively $y_1, \ldots, y_n$ is obtained from $x_1, \ldots, x_n$ by Nielsen transformations. By Lemma 4.8, $[u, v] \subseteq A_{y_n}$; in particular, $v \in A \cap A_{y_n}$, so this intersection is non-empty, as required. □

The following result appeared in [10]. We give a simplified proof which has also been noted in [27].

Theorem 4.4 *If a finitely generated group G has a non-trivial action on a Λ-tree for some ordered abelian group Λ, then it has a non-trivial action on some $\mathbb{R}$-tree.*

Proof. Suppose G acts non-trivially on a Λ-tree (X,d). Recall from §1 that there is an embedding $\Lambda \hookrightarrow \frac{1}{2}\Lambda$. By Lemma 4.3, the induced action of G on $\frac{1}{2}\Lambda \otimes_\Lambda X$ in Corollary 3.6 is without inversions, and is non-trivial by Corollary 4.3. Thus we may assume the action is without inversions. By Lemma 4.9, we can choose generators y_i $(1 \leq i \leq n)$ for G such that $A_{y_1} \cap \ldots \cap A_{y_n} \neq \emptyset$. Choose a point $u \in A_{y_1} \cap \ldots \cap A_{y_n}$, and let L_u be the corresponding Lyndon length function. Reorder the generators so that $L_u(y_1) \leq \ldots \leq L_u(y_n)$. Let H be the intersection of all convex subgroups of Λ containing $L_u(y_n)$. Since L_u satisfies the triangle inequality $L_u(gh) \leq L_u(g) + L_u(h)$ and $L_u(g^{-1}) = L_u(g)$, it follows that $L_u(g) \in H$ for all $g \in G$, by writing g as a product of the generators and their inverses. Also $L_u(y_n) \neq 0$, otherwise L_u would be identically zero and the action would be trivial by Lemma 4.5.

Let K be the union of all convex subgroups of Λ not containing $L_u(y_n)$. Since the convex subgroups are linearly ordered by inclusion, K is a convex subgroup contained in H, $L_u(y_n) \notin K$, and there are no convex subgroups strictly between K and H. Thus H/K is a rank 1 ordered abelian group. Also, since $u \in A_{y_n}$, $L_u(y_n) = \ell(y_n)$, the hyperbolic length of y_n, by Cor. 4.2, so $\ell(y_n) \notin K$.

Using Lemma 2.2, the subset $Y = \bigcup_{g\in G}[u, gu]$ of X, with metric obtained from d by restriction, is easily seen to be an H-tree on which G acts by restriction, since H is a convex subgroup of Λ. Also, L_u is the Lyndon length function corresponding to the point u and the action of G on Y, and passing from X to Y does not change the hyperbolic length by Lemma 4.5. There is an induced action of G on $Z := (H/K) \otimes_H Y$ formed using the canonical projection $H \to H/K$, and by Cor. 4.3 y_n has non-zero hyperbolic length for this action, which is therefore non-trivial.

By Theorem 1.1, there is an embedding of ordered abelian groups $(H/K) \hookrightarrow \mathbb{R}$, and there is an induced action of G on the $\mathbb{R}$-tree $\mathbb{R} \otimes_{H/K} Z$, which again by Cor. 4.3 is non-trivial, as required. □

Thus if we want to investigate the class of finitely generated groups which have a non-trivial action on a Λ-tree, it suffices to consider the case $\Lambda = \mathbb{R}$. This is no longer true, however, if we put further restrictions on the kind of action allowed, as we shall see in the next section.

One reason for the importance of Λ-trees is the work of Morgan and Shalen ([32], [34], [35], [36]). They showed, among other things, the following. Let G be a finitely generated group, let $\mathrm{Isom}^+(\mathbb{H}^n)$ denote the group $\mathrm{SO}_0^+(n,1)$ of orientation preserving isometries of real n-dimensional hyperbolic space $\mathbb{H}^n$. Give G the discrete topology, and $\mathrm{Hom}(G, \mathrm{Isom}^+(\mathbb{H}^n))$ the compact-open topology. Let $\mathcal{F}_n(G)$ be the subspace of this consisting of the discrete, faithful representations, and let $\mathcal{H}_n(G)$ be the quotient space $\mathcal{F}_n(G)/\sim$, where, for α, $\beta \in \mathcal{F}_n(G)$, $\alpha \sim \beta$ if and only if there exists $\gamma \in \mathrm{Isom}^+(\mathbb{H}^n)$ such that $\alpha(g) = \gamma\beta(g)\gamma^{-1}$ for all $g \in G$. Suppose G is not abelian by finite. Then $\mathcal{H}_n(G)$ has a compactification $\mathcal{H}_n(G) \cup \mathcal{T}$, where the points of $\mathcal{T}$ arise from non-trivial actions on $\mathbb{R}$-trees in which the stabiliser of any segment with more than one point is abelian by finite.

In the case that G is the fundamental group of a closed surface with negative Euler characteristic, $\mathcal{H}_n(G)$ is one description of the Teichmüller space of G. The work of Morgan and Shalen gives two ways in which Λ-trees arise in nature. One is the "dual" $\mathbb{R}$-tree of a measured lamination, about which we shall say nothing. The other is a means of constructing a Λ-tree on which the group $SL_2(F)$ acts without inversions, where F is a field equipped with a valuation, and Λ is the value group. We shall describe the construction briefly, and refer to [1] and [34] for details and proofs, and to [32] for generalisations.

Let F be a field, let $F^\times = F \setminus \{0\}$ be the multiplicative group of F, and let Λ be an ordered abelian group. The following is a standard definition.

Definition. A Λ-*valuation* on F is a surjective group homomorphism $v : F^\times \to \Lambda$ such that, for all $x, y \in F$, $v(x+y) \geq \min\{x, y\}$, where we put $v(0) = \infty$ with the convention that $\lambda < \infty$ for all $\lambda \in \Lambda$.

It follows that $(v(x), v(y), v(x+y))$ is an isosceles triple in the sense of §3. We call Λ the value group, and $A := \{a \in F \mid v(a) \geq 0\}$ is a subring of F, called the valuation ring. It is a local ring with maximal ideal $\{a \in F \mid v(a) > 0\}$.

Let V be the vector space F^2. An A-lattice in V is a finitely generated A-module $L \subseteq V$ which spans V as an F-module. Given lattices L, L', we define $L \sim L'$ to mean that $L' = aL$ for some $a \in F$. This is an equivalence relation on the set of A-lattices in V and we denote the set of equivalence classes by X_v. The equivalence class of L is denoted by $\bar{L}$. Two equivalence classes can be expressed as $\bar{L}_0$, $\bar{L}_1$, where $L_0 \subseteq L_1$ and L_1/L_0 is isomorphic to a cyclic A-module A/cA for some $c \in A$. We put $d(\bar{L}_0, \bar{L}_1) = v(c)$. This gives a well-defined function from $X_v \times X_v$ to Λ, and we have:

Theorem 4.5 *The pair* (X_v, d) *is a* Λ*-tree.*

The natural action of $GL(V) = GL_2(F)$ on V induces an action on X_v as isometries. Put $L_0 = A^2$, and if $s = \left(\begin{smallmatrix} a & b \\ c & d \end{smallmatrix}\right) \in GL_2(F)$, define

$$v(s) = \min\{v(a), v(b), v(c), v(d)\}.$$

The next result gives the Lyndon length function based at $\bar{L}_0$ and the hyperbolic length function ℓ for this action.

Lemma 4.10 *For* $s \in GL(V)$,
(1) $d(s\bar{L}_0, \bar{L}_0) = v(\det(s)) - 2v(s)$;
(2) $\ell(s) = \max\{v(\det(s)) - 2v(tr(s)), 0\}$.

For $s \in SL_2(F)$, the lemma gives $d(s\bar{L}_0, \bar{L}_0) = -2v(s) \in 2\Lambda$, so by Lemma 4.2(iv) the action of $SL_2(F)$ on X_v is without inversions. Hence $s \in SL_2(F)$ has a fixed point if and only if $\ell(s) = 0$, which by Lemma 4.10 happens if and only if $v(tr(s)) \geq 0$. Thus s acts as a hyperbolic isometry if and only if $v(tr(s)) < 0$.

5. Free Actions

We return to the question: what group theoretic information can be obtained from an action of a group on a Λ-tree? This has a satisfactory answer in the case $\Lambda = \mathbb{Z}$ by the Bass-Serre theory ([44; Ch.I], [12; Ch.8]), since actions on $\mathbb{Z}$-trees correspond to actions on simplicial trees as graph automorphisms. The simplest case of the Bass-Serre theory is when the action is free (and without inversions), and this suggests that the study of free actions on Λ-trees may be simpler than the study of actions on Λ-trees in general. We call an action of a group G on a Λ-tree free if g acts as a hyperbolic isometry, for all $1 \neq g \in G$. This is the opposite extreme to a trivial action. If N denotes the set of non-hyperbolic elements of G, then trivial means $N = G$, while free means $N = 1$. If Λ is an ordered abelian group, we call a group Λ-free if it has a free action on some Λ-tree, and a group is called tree-free if it is Λ-free for some Λ. We shall survey known results on free actions, mainly by giving a list of examples of tree-free groups. Note that any subgroup of a Λ-free group is Λ-free. Also, if a group G acts freely on a Λ-tree (X, d) and Λ embeds in an ordered abelian group Λ', then G acts freely on the Λ'-tree $\Lambda' \otimes_\Lambda X$ by Cor. 4.3. One other useful observation is that tree-free groups are torsion-free, by Theorem 4.1(iii).

Example 1. A group is $\mathbb{Z}$-free if and only if it is a free group (see [44; Ch.I, §3]). In particular, a free group acts freely on its Cayley graph with respect to a basis, which is a simplicial tree, giving a free action on the corresponding $\mathbb{Z}$-tree.

Example 2. Any ordered abelian group Λ is Λ-free, because Λ acts on itself as translations. It follows that any torsion-free abelian group is tree-free, for such a group can be made into an ordered abelian group (see [19; Lemma 2]). This applies to free abelian groups. Finitely generated free abelian groups are $\mathbb{R}$-free since they are isomorphic to subgroups of $\mathbb{R}$.

Example 3. The fundamental group of a closed surface is $\mathbb{R}$-free, except for the non-orientable surfaces of genus 1, 2 and 3 (the connected sum of 1, 2 or 3 real projective planes). This result is due to Morgan and Shalen (see [37]). A simplified construction in the orientable case is given in [47].

Example 4. If $\{G_i \mid i \in I\}$ is a collection of Λ-free groups then the free product $*_{i \in I} G_i$ is Λ-free. This is essentially a consequence of Prop. 1.2 in [24] and Lemma 4.5 (the argument is given for $\Lambda = \mathbb{R}$ and I finite, but works in general).

One of the best results on tree-free groups which has been proved so far is the following.

Theorem (Rips' Theorem) *A finitely generated group G is $\mathbb{R}$-free if and only if it can be written as a free product $G = G_1 * \cdots * G_n$ for some integer $n \geq 1$, where each G_i is either a finitely generated free abelian group or one of the groups in Example 3 (i.e., surface groups, with three exceptions).*

Rips' proof has not been published, but a proof is given in [16]. See also [43].

Example 5. The structure of Λ-free groups for $\Lambda = \mathbb{Z} \times \Lambda_0$, where Λ_0 is an ordered abelian group and Λ has the lexicographic ordering, has been determined in Theorem 4.7 of [3]. We shall not reproduce the result, but as a special case, if G and H are free groups, the free product with amalgamation $G *_{a=b} H$, amalgamating cyclic subgroups generated by a, b, is $\mathbb{Z} \times \mathbb{Z}$-free provided a is not a proper power in G and b is not a proper power in H. Also, the HNN extension $G = \langle F, t \mid tat^{-1} = b \rangle$, where F is a free group, a, b are not proper powers, a is not conjugate in F to b^{-1} and F acts freely on some Λ_0-tree in such a way that $\ell(a) = \ell(b)$, is $\mathbb{Z} \times \Lambda_0$-free.

This provides examples of finitely presented tree-free groups which are not $\mathbb{R}$-free. For instance, let $G = \langle x, y, z \mid x^2y^2z^2 = 1 \rangle$, the non-orientable surface group of genus 3. As noted in Example 3, G is not $\mathbb{R}$-free, but it can be shown to be $(\mathbb{Z} \times \mathbb{Z})$-free by using the result on HNN-extensions in Example 5 (see the example after Theorem 3 in [21]).

Example 6. Other examples of $\mathbb{R}$-free groups have been given in [2] and [39], which we shall not attempt to describe here. Recently M.J. Dunwoody has shown that the group with presentation $\langle a_1, b_1, a_2, b_2, \ldots \mid b_1 = [a_2, b_2], b_2 = [a_3, b_3], \ldots \rangle$ is $\mathbb{R}$-free.

Our final example is of non-standard models of the free groups, and their subgroups. We shall construct these as ultrapowers of free groups, and begin by describing ultrapowers in general. Let I be a set and let $\mathcal{P}(I)$ be the Boolean algebra of all subsets of I. An ultrafilter in $\mathcal{P}(I)$ is a subset $\mathcal{D}$ of $\mathcal{P}(I)$ such that

(i) $A \in \mathcal{D}$ and $A \subseteq B \subseteq I$ implies $B \in \mathcal{D}$

(ii) A, $B \in \mathcal{D}$ implies $A \cap B \in \mathcal{D}$

(iii) for all $A \in \mathcal{P}(I)$, exactly one of A, $I \setminus A$ belongs to $\mathcal{D}$.

Equivalently, $\mathcal{D}$ is an ultrafilter if and only if $\{A \in \mathcal{P}(I) | I \setminus A \in \mathcal{D}\}$ is a maximal ideal in the Boolean ring $\mathcal{P}(I)$ (where addition is symmetric difference).

Let $\{X_i \mid i \in I\}$ be a family of sets indexed by I, and let $\mathcal{D}$ be an ultrafilter in $\mathcal{P}(I)$. The ultraproduct $\prod_{i \in I} X_i/\mathcal{D}$ is defined to be the quotient set $\prod_{i \in I} X_i/\sim$, where $(x_i)_{i \in I} \sim (y_i)_{i \in I}$ if and only if $\{i \in I \mid x_i = y_i\} \in \mathcal{D}$. This is easily checked to be an equivalence relation. We shall denote the equivalence class of an element $(x_i)_{i \in I} \in \prod_{i \in I} X_i$ by $\langle x_i \rangle_{i \in I}$, and usually abbreviate this to $\langle x_i \rangle$. If all $X_i = X$, a single set, then the ultraproduct is $X^I/\mathcal{D}$, which is called an ultrapower of X. Note that X embeds in $X^I/\mathcal{D}$, by mapping x to $\langle x_i \rangle$, where $x_i = x$ for all $i \in I$.

Denoting the ultrapower $X^I/\mathcal{D}$ by ${}^*X$, given a function $f : X^n \to Y$ of n variables, we obtain an extension to a function ${}^*f : ({}^*X)^n \to {}^*Y$ by defining ${}^*f(x^{(1)}, \ldots, x^{(n)}) = \langle f(x_i^{(1)}, \ldots, x_i^{(n)}) \rangle$, where $x^1 = \langle x_i^1 \rangle$, etc. Similarly any finitary relations on X can be extended to ${}^*X$. Thus if X is a group, then we can define a multiplication in ${}^*X$ by $\langle x_i \rangle \langle y_i \rangle = \langle x_i y_i \rangle$, and this makes ${}^*X$ into a group. If X is abelian, so is $X^I/\mathcal{D}$, and if X is an ordered abelian group, then $X^I/\mathcal{D}$ can be made into an ordered abelian group by defining $\langle x_i \rangle \leq \langle y_i \rangle$ if and only if $\{i \in I \mid x_i \leq y_i\} \in \mathcal{D}$. Similarly if X is a ring or a field, so is $X^I/\mathcal{D}$. These assertions can be proved directly, but all our claims follow from Łoš's Theorem (see Corollary 2.2 or Lemma 2.3 in [6], or for those not familiar with first order logic, [48]).

Example 7. Return to Example 1, where F is a free group acting freely on its Cayley graph, relative to some basis X. Then using the vertex 1 as basepoint, it is easy to see that the corresponding Lyndon length function L is given by: $L(u)$ is the length of the reduced word on $X^{\pm 1}$ representing u, and $c(u, v)$ is the length of the longest common suffix of the reduced words representing u and v. Since the action is free, $L(u^2) > L(u)$ for all $u \in F \setminus \{1\}$ by Lemma 4.5. Let I be an index set, let $\mathcal{D}$ be an ultrafilter in $\mathcal{P}(I)$ and denote the ultrapower $X^I/\mathcal{D}$ by ${}^*X$. Then as indicated above we obtain a function ${}^*L : {}^*F \to {}^*\mathbb{Z}$, given by ${}^*L(\langle u_i \rangle) = \langle L(u_i) \rangle$, and ${}^*L$ is a Lyndon length function. This can be checked directly, and is another manifestation of Łoš's Theorem. Further, ${}^*L(u^2) > {}^*L(u)$ for all $u \in {}^*F \setminus \{1\}$. Therefore there is an action of ${}^*F$ on a ${}^*\mathbb{Z}$-tree by Theorem 3.3, and this action is free by Lemma 4.5. Thus ${}^*F$, and all its subgroups, are tree-free.

This example is due to Gaglione and Spellman [19], [20] (see also Remeslennikov [41]). These authors were led to consider ultrapowers of free groups by model-theoretic considerations. We shall give a version (indeed an easy generalisation) of their results in Theorems 5.1 and 5.2 below. Theorem 5.1 will give a group-theoretic characterisation of the groups which occur in Example 7. A group G is called n-residually free, where n is a positive integer, if given a collection $g_1, \dots, g_n$ of elements of $G \setminus \{1\}$, there is a homomorphism $\phi : G \to F$, where F is a free group, such that $\phi(g_i) \neq 1$ for $1 \leq i \leq n$. A group is called fully residually free if it is n-residually free for all $n > 0$. Also, "residually free", without qualification, means 1-residually free. It is easy to see that residually free groups are torsion-free. For information on these classes see [4] and [5]. We shall show that a group G embeds in some ultrapower of F_2, the free group of rank 2, if and only if it is locally fully residually free.

Lemma 5.1 *Let I be a countably infinite set. Then there is an ultrafilter $\mathcal{D}$ in $\mathcal{P}(I)$ such that, if G is any countable fully residually free group, there is an embedding $G \hookrightarrow {F_2}^I/\mathcal{D}$.*

Proof. Just take $\mathcal{D}$ to be a regular ultrafilter in $\mathcal{P}(I)$ (see Lemma 1.14, Ch. 6 in [6]). Let G be a finitely generated fully residually free group. To show G is embedded in the indicated ultrapower, we may obviously assume G is non-trivial, and since it is torsion-free, there is a bijective map from $G \setminus \{1\}$ to I. For the rest of this proof we shall abuse notation by choosing such a map and identifying $G \setminus \{1\}$ with I under this mapping. Thus, for all $g \in I$, there exists $A_g \in \mathcal{D}$ such that for all $h \in I$, the set $\{g \in I | h \in A_g\}$ is finite.

Since all countable free groups embed in F_2, given $h \in I$, there is a homomorphism $\phi_h : G \to F_2$ such that, for all $g \in I$ with $h \in A_g$, we have $\phi_h(g) \neq 1$. Define $\phi : G \to {F_2}^I/\mathcal{D}$ by $\phi(g) = \langle \phi_h(g) \rangle_{h \in I}$. Clearly ϕ is a group homomorphism. For $g \in I$, define $B_g = \{h \in I | \phi_h(g) \neq 1\}$. Now $h \in A_g$ implies that $\phi_h(g) \neq 1$, so $A_g \subseteq B_g$, hence $B_g \in \mathcal{D}$ since $\mathcal{D}$ is an ultrafilter. Thus $\phi(g) \neq 1$, and ϕ is a monomorphism. □

A set Σ of subgroups of a group G is called a *local system* of G if $G = \bigcup_{S \in \Sigma} S$ and for all $S, T \in \Sigma$, there exists $U \in \Sigma$ such that $S, T \subseteq U$.

Lemma 5.2 *If Σ is a local system of a group G, then for some ultrafilter $\mathcal{D}$ in $\mathcal{P}(\Sigma)$, there is an embedding $G \hookrightarrow \prod_{S \in \Sigma} S/\mathcal{D}$.*

Proof. See [28; 1.L.7]. □

We can now prove one implication in our main result.

Lemma 5.3 *Let G be a locally fully residually free group. Then G embeds in some ultrapower ${F_2}^J/\mathcal{D}$.*

Proof. Let Σ be the local system consisting of all finitely generated subgroups of G. Let I be a countably infinite set. By Lemma 5.1, there is an ultrafilter $\mathcal{D}_1$ in $\mathcal{P}(I)$ such that for all $S \in \Sigma$, S embeds in ${F_2}^I/\mathcal{D}_1$. By Lemma 5.2, there is an embedding $G \hookrightarrow \prod_{S \in \Sigma} S/\mathcal{D}_2$ for some suitable ultrafilter $\mathcal{D}_2$, and so an embedding $G \hookrightarrow ({F_2}^I/\mathcal{D}_1)^\Sigma/\mathcal{D}_2$. But this iterated ultrapower is isomorphic to ${F_2}^J/\mathcal{D}$ where $J = I \times \Sigma$ and $\mathcal{D}$ is the ultrafilter $\mathcal{D}_1 \otimes \mathcal{D}_2$. (See Exercises 2.6 and 2.7 in Ch.6 of [6]). This proves the lemma. □

It remains to show that finitely generated subgroups of ultrapowers of F_2 are fully residually free. The proof of this is due to Remeslennikov [40], and we shall give his argument with only minor modifications. Fix a set I and an ultrafilter $\mathcal{D}$ in $\mathcal{P}(I)$, and again denote the ultrapower $X^I/\mathcal{D}$ by $^*X$. We shall view $^*\mathbb{Z}$ as an extension of the ring $\mathbb{Z}$, and there is an obvious identification of $^*SL_2(\mathbb{Z})$ with $SL_2(^*\mathbb{Z})$. The proof of the next lemma refers to the first order language of rings. By this we mean the first order language with equality as its only predicate symbol, binary function symbols $+$ and $\cdot$, unary function symbol $-$, and constant symbols 0, 1. (The proof could be altered to avoid the use of first-order logic, but the result would not be very elegant).

Lemma 5.4 *Let R be a finitely generated subring of $^*\mathbb{Z}$. Then as a ring, R is fully residually $\mathbb{Z}$. That is, given $n \geq 1$ and $x_1, \ldots, x_n \in R \setminus \{0\}$, there is a ring homomorphism $\phi : R \to \mathbb{Z}$ such that $\phi(x_i) \neq 0$ for $1 \leq i \leq n$.*

Proof. There is a short exact sequence

$$J \rightarrowtail \mathbb{Z}[y_1, \ldots, y_k] \overset{\theta}{\twoheadrightarrow} R$$

where the y_i are commuting indeterminates, and J is a finitely generated ideal. Choose generators $f_1, \ldots, f_m$ for J and $g_j \in \mathbb{Z}[y_1, \ldots, y_k]$ such that $\theta(g_j) = x_j$ for $1 \leq j \leq n$. The sentence

$$\exists z_1 \cdots \exists z_k (f_1(\underline{z}) = 0 \wedge \ldots \wedge f_m(\underline{z}) = 0 \wedge \neg(g_1(\underline{z}) = 0) \wedge \ldots \wedge \neg(g_n(\underline{z}) = 0))$$

in the first order language of rings is then valid in $^*\mathbb{Z}$ (assign $\theta(y_j)$ to the variable z_j), hence is valid in $\mathbb{Z}$ by Łoš's Theorem (see Corollary 2.2 or Lemma 2.3 in [6]). Thus there are integers $u_1, \ldots, u_k$ in $\mathbb{Z}$ such that $f_1(\underline{u}) = \cdots = f_m(\underline{u}) = 0$ and $g_j(\underline{u}) \neq 0$ for $1 \leq j \leq n$. The mapping $\psi : \mathbb{Z}[y_1, \ldots, y_k] \to \mathbb{Z}$ which sends y_j to u_j and is the identity on $\mathbb{Z}$ is zero on J, so there is an induced ring homomorphism $\phi : R \to \mathbb{Z}$ using θ. Then $\phi(x_j) = \psi(g_j) = g_j(\underline{u}) \neq 0$, as required. □

Let p be an odd positive prime in $\mathbb{Z}$ and let $K = Ker(SL_2(\mathbb{Z}) \to SL_2(\mathbb{Z}/p\mathbb{Z}))$. Then $SL_2(\mathbb{Z})$ is a free product of two cyclic groups of orders 4 and 6 amalgamating their subgroups of order 2. This is proved in [44; Ch.I, 4.2]. One can obtain explicit generators for the cyclic free factors from the proof (or see Theorem 3.1 of Ch. 3 in [30]). It is easily checked that K intersects these free factors, and so all their conjugates, trivially (alternatively, one can quote an old result of Minkowski which implies K is torsion-free; see Lemma 2.1, Ch 4 in [8]). Hence K is free by the subgroup theorem for amalgamated free products (see Ch. I, §4.3, Prop. 18 in [44]). It is easy to see that K is non-abelian, so there is an embedding $F_2 \hookrightarrow K$, and this has an extension to an embedding ${}^*F_2 \hookrightarrow {}^*K$. Note that matrices in K have the form

$$\begin{pmatrix} 1+pa & pb \\ pc & 1+pd \end{pmatrix}$$

where a, b, c and d are in $\mathbb{Z}$. It follows that the elements of ${}^*K$ are the matrices in $SL_2({}^*\mathbb{Z})$ of the same form, but where a, b, c and d are now in ${}^*\mathbb{Z}$.

Lemma 5.5 *Let G be a finitely generated subgroup of ${}^*F_2$. Then G is fully residually free.*

Proof. By the above we may assume $G \subseteq {}^*K \subseteq SL_2({}^*\mathbb{Z})$. Let $\{g_1, \ldots, g_n\}$ be a set of generators for G, and for simplicity assume this set is closed under taking inverses. Write

$$g_j = \begin{pmatrix} 1+pa_j & pb_j \\ pc_j & 1+pd_j \end{pmatrix}$$

where a_j, b_j, c_j and d_j are all in ${}^*\mathbb{Z}$. Let R be the subring of ${}^*\mathbb{Z}$ generated by $\{a_j, b_j, c_j, d_j \,|\, 1 \le j \le n\}$. If $g \in G$, an easy induction on the length of a word in the g_j representing g shows that we can write

$$g = \begin{pmatrix} 1+pa & pb \\ pc & 1+pd \end{pmatrix}$$

where $a, b, c, d \in R$. In particular, $G \subseteq SL_2(R)$.

Suppose $h_1, \ldots, h_k$ are non-identity elements of G. If

$$h_j = \begin{pmatrix} a_{11}^j & a_{12}^j \\ a_{21}^j & a_{22}^j \end{pmatrix}$$

then by Lemma 5.4 there is a ring homomorphism $\phi : R \to \mathbb{Z}$ such that $\phi(a_{\alpha\beta}^j) \neq 0$ whenever $a_{\alpha\beta}^j \neq 0$ and $\phi(a_{\alpha\beta}^j) \neq 1$ whenever $a_{\alpha\beta}^j \neq 1$. The ring homomorphism ϕ induces a group homomorphism $\psi : SL_2(R) \to SL_2(\mathbb{Z})$, and $\psi(h_j) \neq 1$ for $1 \le j \le k$. Also, for elements of G, we have

$$\psi \begin{pmatrix} 1+pa & pb \\ pc & 1+pd \end{pmatrix} = \begin{pmatrix} 1+p\phi(a) & p\phi(b) \\ p\phi(c) & 1+p\phi(d) \end{pmatrix} \in K.$$

Thus $\psi|_G$ maps G into K which is free. □

Lemmas 5.3 and 5.5 establish our characterisation of subgroups of ultrapowers of F_2.

Theorem 5.1 *Let G be any group. Then G embeds in some ultrapower ${}^*F_2$ of the free group of rank 2 if and only if G is locally fully residually free.*

If F is any free group, then it follows from 5.6 that F embeds in some ultrapower of F_2. By an argument similar to that in Lemma 5.3, any ultrapower of F embeds in some ultrapower of F_2. Thus we can replace F_2 by any other free group F in the statement of Theorem 5.1.

Theorem 5.1 has an interesting model-theoretic consequence. Let L_0 be the first order language of groups, with equality as the only predicate symbol, a binary function symbol $\cdot$, a unary function symbol $^{-1}$ and a constant symbol 1. The non-abelian free groups have the same universal theory with respect to the language L_0 (see the introduction to [20]). Thus the set Φ of all universal and existential sentences which are true in any non-abelian free group is the same as the set of all such sentences true in F_2, the free group of rank two.

Theorem 5.2 *Let G be any group. Then the following are equivalent.*
(1) G is non-abelian and locally fully residually free.
(2) G is a model of the set of sentences Φ.

Proof. Assume (1). To prove (2) we use an observation in the introduction of [19]. By [5], G contains a subgroup isomorphic to F_2. Thus by Theorem 5.1 $F_2 \subseteq G \subseteq {F_2}^J/\mathcal{D}$ for some ultrapower ${F_2}^J/\mathcal{D}$. It follows from Lemma 3.7 in [6] that an existential sentence true in F_2 is true in G, and any existential sentence true in G is true in ${F_2}^J/\mathcal{D}$. Since negations of existential sentences are equivalent to universal sentences, any universal sentence true in ${F_2}^J/\mathcal{D}$ is true in G. By [6; Ch. 5, Lemma 2.3], ${F_2}^J/\mathcal{D}$ is elementarily equivalent to F_2, in particular ${F_2}^J/\mathcal{D}$ is a model of ϕ. Hence G is a model of Φ.

Conversely if (2) holds, then by [6; Ch. 9, Lemma 3.8], G embeds in some ultrapower of F_2, so by Theorem 5.1 G is locally fully residually free. Since F_2 is non-abelian, the existential sentence $\exists x \exists y (\neg(xy = yx))$ belongs to Φ, so G is non-abelian. □

Theorem 5.2 generalises the results that non-abelian fully residually free groups are models of Φ [18], and that non-abelian locally free groups are models of Φ ([17], after Question 3).

Recently, Remeslennikov [41] has shown that if G is a finitely generated fully residually free group, then G is Λ-free for some finitely generated ordered abelian group Λ. Briefly, the argument proceeds as follows. Let $P = \mathbb{Q}(x_1, x_2, y_1, y_2)$ be a field extension of $\mathbb{Q}$ by four algebraically independent elements. There is a valuation v of P given by $v(f/g) = deg(g) - deg(f)$, where f, g are polynomials in x_1, x_2, y_1, y_2 and deg means total degree. Define

$$A = \begin{pmatrix} x_1 & x_1x_2 - 1 \\ 1 & x_2 \end{pmatrix} \qquad B = \begin{pmatrix} y_1 & y_1y_2 - 1 \\ 1 & y_2 \end{pmatrix}$$

so that A, $B \in SL_2(P)$. Let H be the subgroup generated by A and B. If w is a reduced word on $\{A^{\pm 1}, B^{\pm 1}\}$, let $\|w\|$ denote the length of the cyclically reduced word corresponding to w. Remeslennikov shows that, if h is the element of H represented by w, then $v(tr(h)) = -\|w\|$. It follows that H is freely generated by A and B, and that $v(tr(h)) < 0$ for all $1 \neq h \in H$.

Now let G be a finitely generated fully residually free group, so by Theorem 5.1 G embeds in some ultrapower $^*F_2$, which is isomorphic to $^*H$, a subgroup of $SL_2(^*P)$. By a similar argument to that used in Lemma 5.5, G embeds in $^*H \cap SL_2(L)$, where L is some subfield of $^*P$ which is finitely generated over $\mathbb{Q}$. The valuation v extends to a valuation $^*v$ of $^*P$, and by restriction we obtain a valuation v_1 of L. Now v_1 inherits from v the property that $v_1(tr(g)) < 0$ for all $1 \neq g \in G$, and it follows from Lemma 4.10 and the remarks following it that G acts freely on the Λ-tree (X_{v_1}, d), where Λ is the value group of v_1. Also, v_1 is identically zero on $\mathbb{Q}^\times$, and so Λ is finitely generated (see Ch. VII, §10.3, Corollaire 1 in [7]).

It follows from Example 7 and Theorem 5.1 that locally fully residually free groups are tree-free. However, not all tree-free groups are locally fully residually free, for examples of finitely presented non-abelian tree-free groups which are not models of the set of sentences Φ are given in [15], [21] and [41], and by Theorem 5.2 these are not fully residually free. In fact, it is easy to see directly that the example $G = \langle x, y, z \mid x^2y^2z^2 = 1\rangle$, given in [21], is not even residually free. For by a result of Lyndon [29], if ϕ is any homomorphism from G to a free group, the images of x, y and z lie in a cyclic subgroup. Thus $\phi([x,y]) = 1$, but the commutator $[x, y] \neq 1$, since G is the free product of a free group generated by x and y and the cyclic group generated by z, amalgamating the cyclic subgroups generated by x^2y^2 and z^{-2}.

One can ask what properties of (locally) fully residually free groups hold for all tree-free groups. For example, if G is a finitely generated tree-free group, is G Λ-free for some finitely generated Λ? We finish with two such properties which hold for tree-free groups. We call a group G commutative transitive if given a, b, $c \in G$ such that $[a, b] = 1$, $[b, c] = 1$ and $b \neq 1$, then $[a, c] = 1$. Equivalently, centralisers of non-identity elements of G are abelian. It was shown in [4; Theorem 1] that a residually free group is fully residually free if and only if it is commutative transitive. (We note in passing that it is easy to see that a 2-residually free group is commutative transitive. It follows that a group is fully residually free if and only if it is 2-residually free, as has been noted by Remeslennikov [40]). In [5; §4, Lemma 1] it is observed that, in a residually free group, every two-generator subgroup is either free of rank 2 or abelian.

Theorem 5.3 *Let G be a tree-free group. Then G is commutative transitive, and any subgroup which can be generated by at most two elements is either free of rank 2 or abelian.*

Proof. See [11]. □

In the special case of $\mathbb{R}$-free groups, an equivalent formulation of this result, using Lyndon length functions, was proved in [24].

References

1. R. C. Alperin and H. Bass, *Length functions of group actions on Λ-trees.* In: Combinatorial group theory and topology (ed. S. M. Gersten and J. R. Stallings), Annals of Mathematics Studies **111**, pp 265–378. Princeton: University Press 1987.
2. R. C. Alperin and K. N. Moss, *Complete trees for groups with a real-valued length function,* J. London Math. Soc. **31** (1985), 55–68.

3. H. Bass, *Group actions on non-archimedean trees*. In: Arboreal group theory, (ed. R.C. Alperin), MSRI Publications **19**, pp 69-131. New York: Springer-Verlag 1991.
4. B. Baumslag, *Residually free groups*, Proc. London Math. Soc. (3) **17** (1967), 402–418.
5. G. Baumslag, *On generalised free products*, Math. Z. **78** (1962), 423–438.
6. J. L. Bell and A. B. Slomson, *Models and ultraproducts*. Amsterdam: North-Holland 1971.
7. N. Bourbaki, *Éléments de mathematique XXX, algèbre commutative*. Paris: Hermann 1964.
8. J. W. S. Cassels, *Local fields*, London Mathematical Society Student Texts **3**. Cambridge, University Press 1986.
9. I. M. Chiswell, *Abstract length functions in groups*, Math. Proc. Cambridge Philos. Soc. **80** (1976), 451–463.
10. I. M. Chiswell, *Non-trivial group actions on Λ-trees*, Bull. London Math. Soc. **24** (1992), 277–280.
11. I. M. Chiswell, *Harrison's theorem for Λ-trees*, to appear in Quart. J. Math.
12. D. E. Cohen, *Combinatorial group theory: a topological approach*, London Mathematical Society Student Texts **14**. Cambridge, University Press 1989.
13. M. Coornaert, T. Delzant and A. Papadopoulos, *Géométrie et théorie des groupes*, Lecture Notes in Mathematics **1441**. Berlin, Heidelberg, Springer 1990.
14. M. Culler and J. Morgan, *Group actions on* $\mathbb{R}$*-trees*, Proc. London Math. Soc. (3) **55** (1987), 571–604.
15. B. Fine, A. M. Gaglione, G. Rosenberger and D. Spellman, *n-free groups and questions about universally free groups*, preprint.
16. D. Gaboriau, G. Levitt and F. Paulin, *Pseudogroups of isometries of* $\mathbb{R}$ *and Rips' Theorem on free actions on* $\mathbb{R}$*-trees*, preprint.
17. A. M. Gaglione and D. Spellman, *More model theory of free groups*, preprint.
18. A. M. Gaglione and D. Spellman, *Even more model theory of free groups*, to appear.
19. A. M. Gaglione and D. Spellman, *Does Lyndon's length function imply the universal theory of free groups*, preprint.
20. A. M. Gaglione and D. Spellman, *Every 'universally free' group is tree free*, to appear.
21. A. M. Gaglione and D. Spellman, *Generalisations of some free groups: some questions*, preprint.
22. E. Ghys and P. de la Harpe, *Sur les groupes hyperboliques d'après Mikhael Gromov*. Boston: Birkhäuser 1990.
23. M. Gromov, *Hyperbolic groups*. In: Essays in group theory, (ed. S.M. Gersten), MSRI Publications **8**, pp 75–263. New York: Springer-Verlag 1987.
24. N. Harrison, *Real length functions in groups*, Trans. Amer. Math. Soc. **174** (1972), 77–106.
25. A. H. M. Hoare and D. L. Wilkens, *On groups with unbounded non-archimedean elements*. In: Groups-St. Andrews 1981 (eds. C. M. Campbell and E. Robertson), London Math. Soc. Lecture Notes **71**, pp 228-236. Cambridge: University Press 1982.
26. W. Imrich, *On metric properties of tree-like spaces*. In: Beiträge zur Graphentheorie und deren Anwendungen, (edited by Sektion MARÖK der Technischen Hochschule Ilmenau), pp 129–156. Oberhof, DDR 1977.
27. S. Jackson and L. Q. Zamboni, *A note on a theorem of Chiswell*, preprint.
28. O. H. Kegel and B. A. F. Wehrfritz, *Locally finite groups*. Amsterdam, London: North Holland 1973.
29. R. C. Lyndon, *The equation* $a^2b^2 = c^2$ *in free groups*, Michigan Math. J. **6** (1959), 155–164.
30. W. Magnus, *Noneuclidean tesselations and their groups*. New York and London: Academic Press 1974.
31. W. Massey, *Algebraic topology: an introduction*. New York: Harcourt Brace and World 1967.
32. J. W. Morgan, *Group actions on trees and the compactification of the space of classes of* $SO(n,1)$*-representations*, Topology **25** (1986), 1–33.
33. J. W. Morgan, *Λ-trees and their applications*, Bull. Amer. Math. Soc. **26** (1992), 87–112.
34. J. W. Morgan and P. B. Shalen, *Valuations, trees and degenerations of hyperbolic structures: I*, Annals of Math. (2) **122** (1985), 398–476.
35. J. W. Morgan and P. B. Shalen, *Degenerations of hyperbolic structures: II: measured laminations in 3-manifolds*, Annals of Math. (2) **127** (1988), 403–456.
36. J. W. Morgan and P. B. Shalen, *Degenerations of hyperbolic structures: III: actions of 3-manifold groups on trees and Thurston's compactness theorem*, Annals of Math. (2) **127** (1988), 457–519.

37. J. W. Morgan and P. B. Shalen, *Free actions of surface groups on* $\mathbb{R}$*-trees*, Topology (to appear).
38. W. Parry, *Axioms for translation length functions*. In: Arboreal group theory, (ed. R. C. Alperin), MSRI Publications **19**, pp 295–330. New York: Springer-Verlag 1991.
39. D. Promislow, *Equivalence classes of length functions on groups*, Proc. London Math. Soc. (3) **51** (1985), 449–477.
40. V. N. Remeslennikov, ∃*-free groups*, Siberian Math. J. **30** (1989), 193–197.
41. V. N. Remeslennikov, ∃*-free groups and groups with a length function*, preprint.
42. J. Rhodes, *Monoids acting on trees: elliptic and wreath products and the holonomy theorem for arbitrary monoids with applications to infinite groups*, Int. J. Algebra Comp. **1** (1991), 253–279.
43. F. S. Rimlinger, $\mathbb{R}$*-trees and normalisation of pseudogroups*, preprint.
44. J.-P. Serre, *Trees*. New York: Springer 1980.
45. P. B. Shalen, *Dendrology of groups: an introduction*. In: Essays in group theory, (ed. S.M. Gersten), MSRI Publications **8**, 265–319. New York: Springer-Verlag 1987.
46. P.B. Shalen, *Dendrology and its applications*. In: Group theory from a geometrical viewpoint (eds. E. Ghys, A. Haefliger and A. Verjovsky). Singapore, New Jersey, London, Hong Kong: World Scientific 1991.
47. J. R. Stallings, *Adian groups and pregroups*. In: Essays in group theory, (ed. S. M. Gersten), MSRI Publications **8**, 321–342. New York: Springer-Verlag 1987.
48. L. van den Dries and A. Wilkie, *Gromov's Theorem on groups of polynomial growth and elementary logic*, J. Algebra **89** (1984), 349–374.
49. O. Zariski and P. Samuel, *Commutative algebra*, Vol. II. Princeton, New York: van Nostrand 1960.

AN INTRODUCTION TO REDUCTIVE MONOIDS

LOUIS SOLOMON*
Department of Mathematics
University of Wisconsin
Madison, WI 53706
U. S. A.

For Elsbeth Solomon-Aellig

1. Introduction to the Introduction

The aim of this paper is to give an introduction to the theory of reductive monoids, a well developed branch of semigroup theory with good prospects. It has highly structured problems which, like those in the theory of reductive groups, are rooted in the combinatorics of the Weyl group. This theory is the work of Mohan Putcha and Lex Renner, who originated it independently around 1980. There are now at least 50 papers on the subject - most of them by Putcha and/or Renner, although others, notably Okniński, have been involved. Their work has been more or less ignored by those who might enjoy it and profit from it. These include workers in (the order is alphabetic) algebraic groups and related finite groups, algebraic combinatorics, semigroups, and possibly other parts of mathematics. This subject has a marketing problem. My estimate, based on some minimal evidence, is that those who do algebraic groups are sympathetic but uninterested, those who do algebraic combinatorics do not know that the subject exists, and those who do semigroups are put off by prerequisites which seem formidable.

A reductive monoid M is pieced together from its group of units $G = G(M)$ and its set of idempotents $E = E(M)$. The group G is a reductive group. The idempotents are intimately connected to the group structure. Although one can see this connection in the simplest example, where M is the monoid of all matrices over a field and G is the general linear group, it is remarkable that natural questions about idempotents lead to all the standard constructs in semisimple Lie theory: roots, parabolic subgroups, Tits building, and so forth.

My intent in this paper is to give an introduction to the theory of reductive monoids from scratch, for a reader with general background and interest in algebra but no special knowledge of semigroup theory or reductive algebraic groups. The theory of reductive groups is itself a big subject with various prerequisites. To overcome the prerequisites in minimal space we concentrate on examples in the context of both groups and monoids. Statements of the main theorems are accompanied by references but not by proofs. Comments written as footnotes provide some further

* This work was supported in part by the NSA under Grant MDA904-91-H-0026

J. Fountain (ed.), Semigroups, Formal Languages and Groups, 295–352.

references as well as bits of argument and hints concerning the depth of unsupported statements. The footnotes may be ignored at first reading without loss of continuity. A willing beginner should be able to check most of the details in the examples. In the first steps, and perhaps at any time, the examples are more important than the theorems.

The best course is to begin with an example which is elementary but sufficiently interesting. Example 2.3 in Section 2 should do the job. Once this is done there should be some incentive for you to accept the basic definitions in the theory of reductive groups: weights, roots, Coxeter-Dynkin diagram, parabolic subgroup, BN-pair and so forth. Weights are introduced in Section 2 and the rest of the apparatus is introduced in Section 4. The exposition here has been stripped to a minimum; we do what is necessary for an understanding of the monoid problems and no more. This will allow us to describe in Section 5 by example and statement of theorem, some of the main results of Putcha and Renner, in particular Putcha's cross section lattice and Renner's analogue of the Bruhat decomposition via the Renner monoid.

I would like to thank Mohan Putcha and Lex Renner for their encouragement. I would also like to thank NATO and the organizers of this meeting, in particular J. Fountain, S. Goberstein, and V. A. R. Gould, for the chance to give the talks.

2. Algebraic Monoids

A monoid is a semigroup with 1. What is an algebraic monoid? Let K be an algebraically closed field. Let $\mathbf{M}_n = \mathbf{M}_n(K)$ denote the set of all $n \times n$ matrices over K. We may think of $\mathbf{M}_n$ as an affine space of dimension n^2 with coordinate functions X_{ij} defined by

$$X_{ij}(\begin{bmatrix} c_{11} & \cdots & c_{1n} \\ \vdots & \ddots & \vdots \\ c_{n1} & \cdots & c_{nn} \end{bmatrix}) = c_{ij} \qquad c_{ij} \in K.$$

It is also a monoid with the general linear group $\mathbf{GL}_n = \mathbf{GL}_n(K)$ as its group of units. If V is a vector space over K of dimension n, then choice of a basis allows us to identify $\mathbf{M}_n$ with $\mathbf{End}_K(V)$ and $\mathbf{GL}_n$ with $\mathbf{GL}(V)$. In these notes V is usually the space K^n of column vectors and $\{v_1, \ldots, v_n\}$ is the standard basis for K^n.

A *linear algebraic monoid*, for short an *algebraic monoid*, is a submonoid M of $\mathbf{M}_n$ which is a Zariski closed set. [1] This means that M is the set of common zeros of a family of polynomials in the X_{ij}. For example $\mathbf{M}_n$ itself is algebraic, defined by the empty family of polynomials. The monoid $\mathbf{D}_n = \mathbf{D}_n(K)$ of diagonal matrices is defined by vanishing of the X_{ij} with $i \neq j$ and is thus algebraic. Similarly, the monoid of upper triangular matrices is defined by vanishing of the X_{ij} with $i > j$ and is thus algebraic.

If $K = \overline{\mathbf{F}}_q$ is the algebraic closure of a finite field $\mathbf{F}_q$ with q elements, there are closely related finite monoids. Let $\sigma : \mathbf{M}_n \to \mathbf{M}_n$ be the Frobenius map defined

[1] We use Putcha [31] as a reference on algebraic monoids; we have tried, whenever possible, to give a reference to original source.

by $[c_{ij}] \mapsto [c_{ij}^q]$. The set of fixed points of σ is the monoid $\mathbf{M}_n(\mathbf{F}_q)$ of all matrices over $\mathbf{F}_q$. If $M \subseteq \mathbf{M}_n$ is any algebraic monoid which is the zero set of a family of polynomials with coefficients in $\mathbf{F}_q$ then $\sigma M = M$. Let $M_\sigma = \{a \in M \mid \sigma a = a\}$ be the set of fixed points. Then $M_\sigma \subseteq \mathbf{M}_n(\mathbf{F}_q)$ is a finite monoid. For example if $M = \mathbf{M}_n$ then $M_\sigma = \mathbf{M}_n(\mathbf{F}_q)$; if $M = \mathbf{D}_n$ then M_σ is the monoid of diagonal matrices with coefficients in $\mathbf{F}_q$. [2]

An algebraic monoid M is *connected* if it is connected in the Zariski topology. An algebraic monoid is *irreducible* if it is irreducible as affine algebraic set - it cannot be written as a union of proper Zariski closed subsets. Irreducible implies connected but not conversely. For example the monoid $M = \{\text{diag}(c_1, c_2) \in \mathbf{D}_2 \mid c_1 c_2 = 0\}$ is connected but not irreducible. Since every finite monoid M is an algebraic monoid, we assume throughout that M is irreducible. Every algebraic monoid M has a *dimension*, dim M, which we will not attempt to define [15, 3.1]. On the intuitive level, dim $\mathbf{M}_n = n^2$ and dim $\mathbf{D}_n = n$.

A *linear algebraic group*, for short an *algebraic group*, is a subgroup of $\mathbf{GL}_n$ which is the intersection of $\mathbf{GL}_n$ with a Zariski closed subset of $\mathbf{M}_n$. [3] For example $G = \mathbf{GL}_n$ is an algebraic group, as is the group $G = \mathbf{T}_n$ of invertible diagonal matrices. The *coordinate ring* or *affine ring* of G is the K-algebra of functions on G generated by the restrictions to G of the coordinate functions X_{ij} as well as the reciprocal of the determinant function det : $\mathbf{GL}_n \to K^*$, where $K^* \simeq \mathbf{GL}_1$ denotes the multiplicative group of K. Thus, for example, if $G = \mathbf{T}_n$ and X_i denotes the restriction of X_{ii} to G then $K[G] = K[X_1, \ldots, X_n, (X_1 \cdots X_n)^{-1}] = K[X_1, \ldots, X_n, X_1^{-1}, \ldots, X_n^{-1}]$ is a ring of Laurent polynomials in n indeterminates with coefficients in K. A subgroup of G is, by definition, closed if it is the intersection of G with a Zariski closed subset of $\mathbf{M}_n$; thus a closed subgroup of an algebraic group G is itself an algebraic group. A direct product of algebraic groups is an algebraic group. An algebraic group G is connected if and only if it is irreducible as algebraic set [15, 7.3]. [4] If $G = \langle Y_1, \ldots, Y_r \rangle$ where the Y_i are closed connected subgroups which generate G as an abstract group, then G is connected [15, Corollary 7.5]. This is a useful fact which we will apply several times.

It is very easy to construct algebraic monoids. If X is any subset of $\mathbf{M}_n$ let $\overline{X}$ denote the Zariski closure of X, the intersection of all Zariski closed sets which include X. The Zariski closure of any submonoid of $\mathbf{M}_n$ is a monoid [5] and hence an algebraic monoid. In particular, the Zariski closure $\overline{G}$ of any subgroup $G \subseteq \mathbf{GL}_n$ is

[2] The classical groups over finite fields were introduced by Jordan and were placed in a Lie theoretic context by Dickson and Chevalley. Steinberg [51, §10 – §15] studied the Chevalley groups, and certain variations on them which include the finite unitary groups, as fixed point sets $G_\sigma = \{g \in G \mid \sigma g = g\}$ where G is a semisimple algebraic group defined over $\mathbf{F}_q$. Renner introduced the analogous construction $M \rightsquigarrow M_\sigma$ for algebraic monoids [44],[46] and studied the M_σ when M is reductive; see Section 5 of this paper.

[3] We use Borel [3] and Humphreys [15] as references on algebraic groups; with occasional reference to Springer [50] and Séminaire Chevalley [6]. The boldface notation used here differs from the boldface notation in [3] and [15]. It is "dictated" if we let $\mathbf{D}_n$ denote the monoid of diagonal matrices in $\mathbf{M}_n$. Thus, here, $\mathbf{T}_n$ is the standard torus of invertible diagonal matrices and $\mathbf{B}_n$ is the Borel subgroup of invertible upper triangular matrices in $\mathbf{GL}_n$.

[4] In the theory of algebraic groups the term "irreducible" is usually suppressed in this context and is reserved for representation theory. By convention, this is also done for monoids in [31, Definition 1], so that "connected" in [31] means "irreducible" here.

[5] To prove this, argue as in Lemma 2.1 of [25].

an algebraic monoid. If $G \subseteq \mathbf{GL}_n(K)$ is an algebraic group then $\overline{G}$ is an algebraic monoid which has G as its group of units. In the opposite direction, if $M \subseteq \mathbf{M}_n$ is an algebraic monoid then $M \cap \mathbf{GL}_n$ is an algebraic group which is in fact the group of units of M, the set of elements of M with an inverse in M [10, Corollary II, §2, 3.5], [39, Corollary 2.2.2]. [6] Thus if $G \subseteq \mathbf{GL}_n(K)$ is an algebraic group, then $\overline{G}$ is an algebraic monoid which has G as its group of units. For example, if $G = \mathbf{GL}_n$ then $\overline{G} = \mathbf{M}_n$ and $\mathbf{GL}_n$ is the group of units of $\mathbf{M}_n$. If $G = \mathbf{T}_n$ then $\overline{G} = \mathbf{D}_n$ and $\mathbf{T}_n$ is the group of units of $\mathbf{D}_n$. Thus we have a method for constructing algebraic monoids with a prescribed algebraic group G of units. If G is connected then $\overline{G}$ is irreducible [31, p.48].

To construct interesting algebraic monoids we choose an algebraic group $G_0 \subseteq \mathbf{GL}_m$ and let $G = \rho(G_0)$ where where ρ is a *rational representation*. Recall that a representation of a group G_0 is a homomorphism $\rho : G_0 \to \mathbf{GL}_n$ for some n. A representation ρ of an algebraic group G_0 is a rational representation if, for each coordinate function X_{ij} on $\mathbf{M}_n$, the function $g \mapsto X_{ij}(\rho(g))$, $g \in G_0$, lies in the affine ring $K[G_0]$. For example, if $A = [a_{ij}]$ is any $m \times n$ matrix of integers then we may define a rational representation $\rho : \mathbf{T}_m \to \mathbf{GL}_n$ by

$$\rho(g) = \operatorname{diag}(t_1^{a_{11}} \cdots t_m^{a_{m1}}, \ldots, t_1^{a_{1n}} \cdots t_m^{a_{mn}}) \in \mathbf{T}_n \subseteq \mathbf{GL}_n \tag{1}$$

for $g = \operatorname{diag}(t_1, \ldots, t_m) \in \mathbf{T}_m$. The representation is rational because $X_{ij}(\rho(g)) = 0$ for $i \neq j$ and $X_{jj}(\rho(g)) = t_1^{a_{1j}} \cdots t_m^{a_{mj}}$ is a Laurent polynomial in the matrix entries of g. In fact every rational representation of $\mathbf{T}_m$ is equivalent to a representation of the form (1). [7] If $G_0 = \mathbf{GL}_m$ and $^\top$ means transpose then $\rho : G_0 \to \mathbf{GL}_m$ defined by $g \to (g^{-1})^\top$ is a rational representation because each matrix entry of g^{-1} is the product of $(\det g)^{-1}$ and a cofactor of g; the cofactor is a polynomial in the matrix entries of g. If G_0 is any algebraic group and ρ is a rational representation then $\rho(G_0)$ is an algebraic group [3, 1.4],[15, Proposition 7.4.B(b)]. Unfortunately, the proof of this theorem does not tell you how to construct polynomials in n^2 indeterminates X_{ij} which generate the ideal of polynomials which vanish on $\rho(G_0)$. It is hard to give an explicit description of M in terms of concretely given G_0 and ρ.

Let's look at an example. Let $G_0 = \mathbf{SL}_m = \mathbf{SL}_m(K)$ be the *special linear group*, which consists of all elements of $\mathbf{GL}_m$ with determinant 1. Let $\rho : G_0 \to \mathbf{GL}_n$ be a rational representation. Since $G = \rho(G_0)$ is an algebraic group, there is a Zariski closed subset X of $\mathbf{M}_n$ such that $G = X \cap \mathbf{GL}_n$. Since every $g \in G_0$ is a product of commutators $xyx^{-1}y^{-1}$ it follows that $\rho(g)$ is a product of commutators $\rho(x)\rho(y)\rho(x)^{-1}\rho(y)^{-1}$. Thus $\det \rho(g) = 1$ so $G \subseteq \mathbf{SL}_n$. Thus $G = X \cap \mathbf{SL}_n$. But $\mathbf{SL}_n$ is Zariski closed because it is the zero set of the polynomial $\det [X_{ij}] - 1$. Thus G is Zariski closed in $\mathbf{M}_n$. So the monoid $\overline{G} = G$ is a group and we have accomplished nothing.

---

[6] It is not clear without argument that if $a \in M$ is an invertible matrix then $a^{-1} \in M$. For example, if $a \in K^*$ has infinite order and $M = \{1, a, a^2, \ldots\}$ then M is a monoid, not an algebraic monoid, in which every element is invertible in K, but M has no units except 1.

[7] The key point here is the case $m = n = 1$ where one must check that every rational homomorphism $K^* \to K^*$ has the form $t \mapsto t^a$ for some $a \in \mathbf{Z}$. Then use the fact [3, Proposition 8.4] that $\rho(\mathbf{T}_m)$ is conjugate to a subgroup of $\mathbf{T}_n$.

This is just setting you up. We have learned something. To construct monoids which are not groups, G must contain matrices with determinant different from 1. [8] This suggests that we adjoin the scalar matrices to $\rho(G_0)$ and choose

$$G = \{c\rho(g) \mid c \in K^* \text{ and } g \in G_0\} = K^*\rho(G_0) \supseteq \rho(G_0). \tag{2}$$

We have taken our first attempt $\rho(G_0)$ and modified it by adjoining the nonzero scalar matrices. The group G is an algebraic group but it is *not* a closed subset of $\mathbf{M}_n$. [9] Let M be its Zariski closure. Then, as we have remarked, M is an algebraic monoid with G as its group of units. Write $M = M(\rho)$ to denote the dependence on ρ. Since G_0 and K^* are connected, so is $G = K^*\rho(G_0)$. Thus $M(\rho)$ is an irreducible algebraic monoid. Since Ker ρ is a normal subgroup of $G_0 = \mathbf{SL}_m$, it is either equal to G_0, a case we agree to ignore, or it is a finite central subgroup of G_0 of order equal to a divisor of m. Thus for any nontrivial representation ρ, the group $G = K^*\rho(G_0)$ of units of $M(\rho)$ is a close relative of $\mathbf{GL}_m$. [10] Any significant structural difference between $M(\rho)$ and $\mathbf{M}_m$ must lie in its set of idempotents. [11]

If M is a monoid, let $E(M)$ denote its set of idempotents. We give $E(M)$ the partial order

$$f \leq e \Leftrightarrow ef = f = fe\,. \tag{3}$$

We agree that the partial order on any subset of $E(M)$ is inherited from this one.

We give three examples of the construction $(G_0, \rho) \rightsquigarrow M(\rho)$ with $G_0 = \mathbf{SL}_m$ and various ρ. They presuppose no special knowledge about the representation theory of $\mathbf{SL}_m$. Even so, the monoid in the third example is quite different from familiar objects. Let K^m be the space of column vectors and let $v_1, \ldots, v_m$ be the standard basis for K^m. Let $T_0 = G_0 \cap \mathbf{T}_m$ be the group of diagonal matrices in G_0 and let $T = K^*\rho(T_0) \subseteq K^*\rho(G_0) = G$. Thus $\overline{T} \subseteq \overline{G} = M(\rho)$.

Example 2.1 Let $V = K^m$. Thus $n = \dim V = m$. Define $\rho : G_0 \to \mathbf{GL}(V)$ by $\rho(g) = g$. Then $G = \{cg \mid c \in K^*, g \in \mathbf{SL}_m\} = \mathbf{GL}_n$, $T = \mathbf{T}_n$, and $M(\rho) = \mathbf{M}_n$. We are on very familiar territory. This example is the prototype. Let $E_{ij} \in \mathbf{M}_n$ be the matrix units. The idempotents in $E(\overline{T})$ are of the form $e_I = \sum_{i \in I} E_{ii}$ where I ranges over the subsets of $\{1, \ldots, n\}$. The poset $E(\overline{T})$ is isomorphic to the Boolean

[8] Renner [39, Theorem 3.3.6] and Waterhouse [54] gave precise conditions under which an algebraic group G may be imbedded as the group of units of an algebraic monoid M which is not a group.

[9] Imbed $K^* \times G_0 \to \mathbf{GL}_{n+1}$ in the natural way. Then G is the image of the algebraic group $K^* \times G_0$ under the representation $(c, g) \mapsto c\rho(g)$ and hence is an algebraic group. Since the closure of $K^*\rho(1)$ contains the zero matrix, so does $\overline{G}$.

[10] To be precise, note that the homomorphism $K^* \times G_0 \to G$ is surjective with finite kernel. So is the homomorphism $K^* \times G_0 \to \mathbf{GL}_m$ given by $(c, g) \mapsto cg$.

[11] One should not conclude from this statement that the distinction between $K^* \times \mathbf{SL}_m$ and $\mathbf{GL}_m$ may be ignored in the context of monoids. For example, if $m = 2$ it enters the proof of Renner's classification theorem for semisimple algebraic monoids [40], [41] and into Renner's construction of an analogue for M of the Bruhat decomposition for G [43]. However, in the present context, the difference between groups is subtle while the difference between idempotent sets can be spectacular. Thus it seems best, with first examples, to concentrate on the idempotents.

lattice of subsets of $\{1,\ldots,n\}$. In view of Example 5.4, it is equally true, and closer to the general Theorem 5.4 which governs $E(\overline{T})$, to say that $E(\overline{T})$ is isomorphic to the lattice of faces of a simplicial cone in $\mathbf{R}^n$ generated by n rays through the origin.

Example 2.2 Let $V = K^m \otimes K^m$. Then V has a basis $\{v_i \otimes v_j \mid 1 \leq i < j \leq m\}$ so $n = \dim V = m^2$. Define $\rho : G_0 \to \mathbf{GL}(V)$ by $\rho(g)(v \otimes v') = gv \otimes gv'$. Then $G = \{c\rho(g) \mid c \in K^*, g \in \mathbf{SL}_m\} = \{g \otimes g \mid g \in \mathbf{GL}_m\}$. Note that ρ is a rational representation: if $g = [c_{ij}]$ then the matrix entries of $\rho(g)$ have the form $c_{ij}c_{kl}$ and are thus polynomials in the matrix entries of g. We may extend ρ to a representation of $\mathbf{M}_m$ by defining $\rho(a)(v \otimes v') = av \otimes av'$ for $a \in \mathbf{M}_m$. Then $\rho(\mathbf{M}_m) = \{a \otimes a \mid a \in \mathbf{M}_m\}$ is Zariski closed[12] and has G as its group of units, so $\overline{G} = \rho(\mathbf{M}_m)$ and thus $M(\rho) = \rho(\mathbf{M}_m)$. In particular, $E(M(\rho)) \simeq E(\mathbf{M}_m)$, isomorphism of posets, via $e \otimes e \leftrightarrow e$. Thus we get nothing new.

Example 2.3 Let $V = K^m \otimes K^m$ be as in Example 2.2 so again $n = \dim V = m^2$. Define $\rho : G_0 \to \mathbf{GL}(V)$ by $\rho(g)(v \otimes v') = gv \otimes (g^{-1})^\top v'$. Note that ρ is a rational representation: if $g = (c_{ij})$ then the matrix entries of $\rho(g)$ have the form $c_{ij}c'_{kl}$ where the c'_{kl} are cofactors of g and hence polynomials in the matrix entries of g. We will construct some idempotents in $E(\overline{T})$ which show that $E(M(\rho))$ is radically different from $E(\mathbf{M}_m)$. Our calculation here contains the germ of a general argument whose ancestry may be traced to Hilbert; see Theorem 5.4 for a formulation in the context of reductive monoids. If $g \in T_0$ write $g = \operatorname{diag}(t_1,\ldots,t_m)$ where $t_i \in K^*$ and $t_1 \cdots t_m = \det g = 1$. Then $\rho(g)(v_i \otimes v_j) = t_i t_j^{-1}(v_i \otimes v_j)$. Thus, using the basis of elements $v_i \otimes v_j$, the matrix $\rho(g)$ is diagonal:

$$\rho(g) = \operatorname{diag}(\ \ldots\ , t_i t_j^{-1},\ \ldots\ )\ . \tag{4}$$

Let's try to produce idempotents in $\overline{T}$. Here is the idea. Suppose we can find non-negative integers $a_1,\ldots,a_n$ such that

$$\operatorname{diag}(t^{a_1},\ldots,t^{a_n}) \in T \text{ for all } t \in K^*\ . \tag{5}$$

Then

$$\operatorname{diag}(t^{a_1},\ldots,t^{a_n}) \in \overline{T} \text{ for all } t \in K\ . \tag{6}$$

We may set $t = 0$ in (6). Under this substitution, $t^{a_i} \mapsto 1$ if $a_i = 0$ and $t^{a_i} \mapsto 0$ if $a_i > 0$. Thus the element in (6) is a diagonal matrix with entries in $\{0,1\}$ and hence is an idempotent $e \in \overline{T}$. It is tempting to write $e = \lim_{t\to 0} \operatorname{diag}(t^{a_1},\ldots,t^{a_n})$. See (77) for an elaboration of this limit argument. Set $t_i = t^{b_i}$ and $t_j = t^{b_j}$ in

[12] This must be checked. If $M \subseteq \mathbf{M}_m$ is an algebraic monoid and $\rho : M \to \mathbf{M}_n$ is a homomorphism of monoids such that the matrix entries of $\rho(a)$ are polynomials in the matrix entries of a, it does not follow that $\rho(M)$ is a closed submonoid of $\mathbf{M}_n$. This is quite different from the situation with algebraic groups [15, Proposition 7.4.B]. For example take $M = \mathbf{M}_m$ and let $\rho(a) = \det(a)a$.

(4) where $b_1, \ldots, b_m \in \mathbf{Z}$ are to be found. The condition $t_1 \cdots t_m = 1$ demands $b_1 + \cdots + b_m = 0$. Thus

$$\text{diag}(\ldots, t^{b_i - b_j}, \ldots) \in \rho(T_0) \subset T \tag{7}$$

for all $t \in K^*$ and all integers $b_1, \ldots, b_m$ such that $b_1 + \cdots + b_m = 0$. There is still some freedom left since $T = K^* \rho(T_0)$ contains the scalar matrices $t^{b_0}\rho(1)$ for $t \in K^*$ and $b_0 \in \mathbf{Z}$. Thus

$$\text{diag}(\ldots, t^{b_0 + b_i - b_j}, \ldots) \in T\,. \tag{8}$$

We want to choose integers $b_0, b_1, \ldots, b_m$ so that $b_1 + \cdots + b_m = 0$, all $b_0 + b_i - b_j$ are non-negative, and some $b_0 + b_i - b_j$ is positive. This is a problem in linear programming over $\mathbf{Z}$. Let's exhibit some solutions. Fix a pair of indices i, j with $1 \leq i \neq j \leq m$. Choose $b_0 = 2, b_i = -1, b_j = 1$, and let $b_k = 0$ for $k \neq i, j$. Then $b_0 + b_i - b_j = 0$ and all other $b_0 + b_p - b_q$ are positive. This gives us an idempotent, say $e_{ij} \in \overline{T}$.

Let $M = M(\rho)$. Since $e_{ij} \in \mathbf{D}_m \otimes \mathbf{D}_m \simeq \mathbf{D}_n$ is a diagonal matrix with just one nonzero entry, it has rank 1 and hence is a minimal element of the poset $E(M)$. Thus $E(M)$ has a set of $m(m-1)$ minimal elements which mutually commute. On the other hand, in $E(\mathbf{M}_m)$ any set of commuting minimal idempotents may be simultaneously diagonalized and thus has cardinality at most m. Thus $E(M)$ and $E(\mathbf{M}_m)$ are not isomorphic. The monoids M and $\mathbf{M}_m$ have closely related groups of units but they are radically different from one another. The poset $E(\overline{T})$ is in fact isomorphic to the lattice of faces of the polytope $\mathcal{P}$ in $\mathbf{R}^m$ which has its vertices at the points $\mathbf{e}_{ij} = \mathbf{e}_i - \mathbf{e}_j$, where $\mathbf{e}_1, \ldots, \mathbf{e}_m$ is the standard basis for the space $\mathbf{R}^m$ of column vectors. This fact may be deduced from (103). By contrast, the corresponding poset $E(\mathbf{T}_m)$ is isomorphic to the lattice of faces of a simplex.

Let's describe the polytope $\mathcal{P}$ explicitly in case $m = 4$. The calculation begun here which leads to this description will be completed in Example 5.6. We can draw the polytope $\mathcal{P}$ in $\mathbf{R}^3$ because its vertices $\mathbf{e}_{ij}$ lie in the hyperplane $H = \{\sum_{i=1}^4 x_i \mathbf{e}_i \mid \sum_{i=1}^4 x_i = 0\}$ of $\mathbf{R}^4$. Define an $\mathbf{R}$-linear map $\phi : H \to \mathbf{R}^3 \subset \mathbf{R}^4$ by $\phi(\mathbf{e}_1 - \mathbf{e}_2) = \mathbf{e}_2 - \mathbf{e}_3$, $\phi(\mathbf{e}_2 - \mathbf{e}_3) = \mathbf{e}_1 - \mathbf{e}_2$, $\phi(\mathbf{e}_3 - \mathbf{e}_4) = \mathbf{e}_2 + \mathbf{e}_3$. The $\phi(\mathbf{e}_{ij})$ are the 12 points $\sum_{i=1}^3 x_i \mathbf{e}_i$ such that $\{x_1, x_2, x_3\} \subseteq \{0, \pm 1\}$ and precisely one of x_1, x_2, x_3 is 0. These points are the midpoints of the edges of a cube with vertices at $\pm\mathbf{e}_1 \pm \mathbf{e}_2 \pm \mathbf{e}_3$. Thus the polytope $\mathcal{P}$ is the convex hull of the midpoints of a cube. This is a cuboctahedron. In Figure 2.1, $\phi(\mathbf{e}_{ij})$ is labeled ij and the "invisible" vertices 41, 42, 43 are omitted.

We may perform the construction $(G_0, \rho) \rightsquigarrow M(\rho)$ with any algebraic group $G_0 \subseteq \mathbf{GL}_m$ and rational representation ρ. If we want interesting monoids we start with interesting groups. By common agreement (?/!) these are the reductive groups. We postpone the definition of "reductive" to Section 3 and simply remark here that the family of reductive groups includes the special linear, general linear, symplectic and orthogonal groups, and that reductive groups have an extraordinary structure governed by finite group W called the Weyl group. One might hope that if we apply

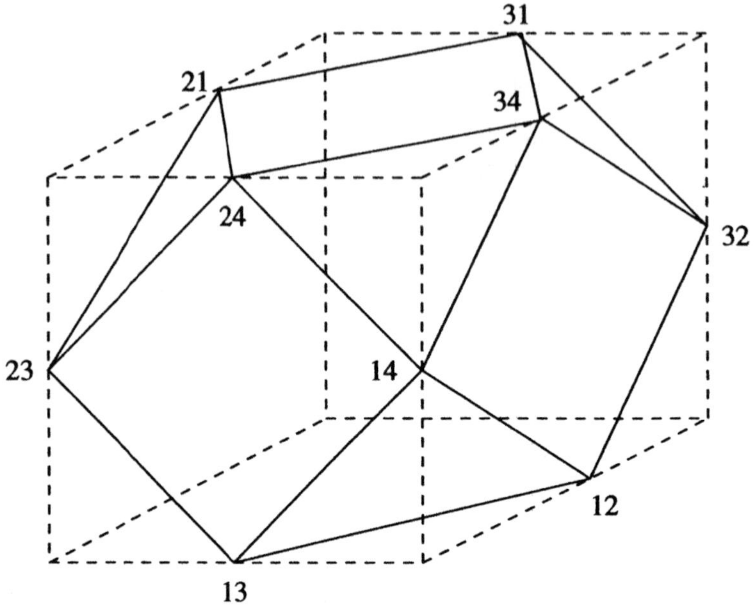

Figure 2.1. Cuboctahedron

the construction $(G_0, \rho) \rightsquigarrow M(\rho)$ to a reductive group, then the resulting monoid will also have extraordinary structure. This turns out to be the case; see Section 5 where we apply this construction in case G_0 is *semisimple* and ρ has finite kernel. Example 2.3 exhibits some of the appeal of the theory of reductive groups. At the core of an enormous edifice lie the regular polyhedra and their close relatives. In any event, since $G = \mathbf{GL}_n$ is reductive with Zariski closure $M = \mathbf{M}_n$, the hypothesis that a monoid is reductive can't lead us too far astray - any general theory must include the multiplicative aspects of matrix algebra over a field. [13]

An *algebraic torus*, or simply a *torus*, is an algebraic group T isomorphic to $\mathbf{T}_m$ for some integer m. The dimension of T is m. We often identify $\mathbf{T}_m$ with $(K^*)^m$. A closed connected subgroup of a torus T is itself a torus [3, Corollary 8.5]. For example $\mathbf{T}_m \cap \mathbf{SL}_m$ is a closed subgroup of $\mathbf{T}_m$ of dimension $m-1$, which is isomorphic to $(K^*)^{m-1}$ via $\mathrm{diag}(t_1, \ldots, t_m) \mapsto (t_1, \ldots, t_{m-1})$ because $t_1 \cdots t_m = 1$. A *rational character* of a torus T is a rational representation $\chi : T \to K^*$. For example if $g = \mathrm{diag}(t_1, \ldots, t_m) \in \mathbf{T}_m$ then

$$\varepsilon_i(g) = t_i \ , \ 1 \leq i \leq m \tag{9}$$

[13] There is no reason except for the infancy of the subject, to assume that K is a field; one might envision $K = \mathbf{Z}$ or a more general coefficient ring as in the theory of algebraic groups. In case K is a field Renner, [43] has given a monoid analogue of the row echelon form for square matrices.

defines a rational character ε_i of $\mathbf{T}_m$. The rational characters of a torus T form a group $X(T)$ under multiplication, which is free abelian of rank equal to dim T [3, 8.5]. For example $X(\mathbf{T}_m)$ has a basis $\{\varepsilon_1, \ldots, \varepsilon_m\}$; if $\chi \in X(\mathbf{T}_m)$ is a rational character of $\mathbf{T}_m$ then there exist unique $a_1, \ldots, a_m \in \mathbf{Z}$ such that $\chi = \varepsilon_1^{a_1} \cdots \varepsilon_m^{a_m}$. Thus, if $g = \mathrm{diag}(t_1, \ldots, t_m)$ then

$$\chi(g) = t_1^{a_1} \cdots t_m^{a_m} \ . \tag{10}$$

This is just (1) when $n = 1$.

Let $T_0 \subseteq \mathbf{T}_m$ be a torus and let $\rho : T_0 \to \mathbf{GL}(V) \simeq \mathbf{GL}_n$ be a rational representation. [14] A nonzero vector $v \in V$ is a ***weight vector*** for ρ if $\rho(T_0)v = Kv$. Thus v is an eigenvector for all $\rho(g)$ with $g \in T_0$. Define a function $\chi : T_0 \to K^*$ by

$$\rho(g)v = \chi(g)v \ g \in T_0 \ . \tag{11}$$

Since ρ is a representation, χ is a character of T_0. The characters χ which arise in this way are called ***weights*** of the representation ρ. Let $\Phi(\rho) \subset X(T_0)$ denote the set of weights of ρ. The set $\Phi(\rho)$ is finite, of cardinality at most n. If $\chi \in \Phi(\rho)$ then the space $V_\chi = \{v \in V \mid gv = \chi(g)v \text{ for all } g \in T_0\}$ is called the ***weight space*** corresponding to χ. The dimension dim V_χ is called the ***multiplicity*** of the weight χ. It is a fact [3, 8.17] that there exists a basis $\mathcal{B}$ for V which consists of weight vectors for ρ. We usually use $\mathcal{B}$ to identify $\mathbf{GL}(V)$ with $\mathbf{GL}_n$ so that $\rho(T_0) \subseteq \mathbf{T}_n$. If $\mathcal{B} = \{v_1, \ldots, v_n\}$ is the standard basis for K^n and $\rho(g)v_j = \chi_j(g)v_j$ for $g \in T_0$ then $\Phi(\rho) = \{\chi_1, \ldots, \chi_n\}$. It may happen that $\chi_j = \chi_k$ for $j \neq k$.

For example if $\rho : \mathbf{T}_m \to \mathbf{T}_n$ is the representation in (1), define $\chi_j \in X(\mathbf{T}_m)$ for $1 \leq j \leq n$ by $\chi_j(\mathrm{diag}(t_1, \ldots, t_m)) = t_1^{a_{1j}} \cdots t_m^{a_{mj}}$. Then $\Phi(\rho) = \{\chi_1, \ldots, \chi_n\}$. Let's describe $\Phi(\rho)$ for the representations ρ of $\mathbf{SL}_m$ in Examples 2.1-2.3. We view ρ as a representation of the torus $T_0 = \mathbf{T}_m \cap \mathbf{SL}_m$. Define $\varepsilon_i : T_0 \to K^*$ by

$$\varepsilon_i(g) = t_i \ , \ 1 \leq i \leq m \tag{12}$$

for $g = \mathrm{diag}(t_1, \ldots, t_m) \in T_0$. Although the formula in (12) is exactly as in (9), one must remember that in $X(T_0)$ we have $\varepsilon_1 \cdots \varepsilon_m = 1$ because $\det g = 1$ for $g \in T_0$. Thus the ε_i in (12) are the restrictions to T_0 of the ε_i in (9). The bases $\mathcal{B}$ of weight vectors and the corresponding sets $\Phi(\rho)$ are given in the following table. Here ε_i is defined as in (12) and $v_1, \ldots, v_m$ is the standard basis for K^m.

Example	$\mathcal{B}$	$\Phi(\rho)$
(2.1)	$\{v_i \mid 1 \leq i \leq m\}$	$\{\varepsilon_i \mid 1 \leq i \leq m\}$
(2.2)	$\{v_i \otimes v_j \mid 1 \leq i, j \leq m\}$	$\{\varepsilon_i\varepsilon_j \mid 1 \leq i, j \leq m\}$
(2.3)	$\{v_i \otimes v_j \mid 1 \leq i, j \leq m\}$	$\{\varepsilon_i\varepsilon_j^{-1} \mid 1 \leq i, j \leq m\}$.

In Example 2.3 the weights $\varepsilon_i\varepsilon_j^{-1}$ with $i \neq j$ have multiplicity 1 while the weight $1 = \varepsilon_1\varepsilon_1^{-1} = \cdots = \varepsilon_m\varepsilon_m^{-1}$ has multiplicity m.

[14] The change in notation from T to T_0 will be useful in Section 5 where $T_0 \subseteq \mathbf{T}_m$ and $\rho(T_0) \subseteq T \subseteq \mathbf{T}_n$. In Section 4 a torus will usually be written T.

3. The $\mathcal{J}$-class structure of an Algebraic Monoid

Let M be an algebraic monoid. How shall we begin a structure theory of M? Let G be the group of units of M. Since $GM \subseteq M$ and $MG \subseteq M$, the group $G \times G$ acts on M by $(g,h) \cdot a = gah^{-1}$ for $g,h \in G$ and $a \in M$. Let $G\backslash M/G$ denote the set of orbits $\mathcal{O} = GaG$ for this action. From the group theoretic point of view, the first main problem is to describe the set $G\backslash M/G$.

From the point of view of semigroup theory, the analogous problem is to describe the equivalence classes for Green's $\mathcal{J}$-relation, which is defined by $a\mathcal{J}b$ if and only if $MaM = MbM$ [31, 1.1]. Clearly $GaG = GbG \Rightarrow MaM = MbM$. Recall that we have agreed to assume throughout that M is irreducible. If M is irreducible then $MaM = MbM \Rightarrow GaG = GbG$ [31, 6.1], [24, Theorem 13]. Thus a, b lie in the same $\mathcal{J}$-class if and only if they lie in the same $G \times G$ orbit, so that the natural group theoretic and semigroup theoretic equivalence relations are the same. The $\mathcal{J}$-class of a is thus the orbit GaG. We will often write $G \times G$ orbit rather than $\mathcal{J}$-class because it suggests a connection with other parts of mathematics. In general, the set of $\mathcal{J}$-classes in a monoid M carries a natural partial order given by inclusion: $MaM \leq MbM \Leftrightarrow MaM \subseteq MbM$ [31, 1.1].

Example 3.1 Let $M = \mathbf{M}_n$. Then $G = \mathbf{GL}_n$. If $a, b \in M$ then $GaG = GbG$ if and only if rank a = rank b. Thus there is a bijection $G\backslash M/G \leftrightarrow \{0, 1, \ldots, n\}$ given by $GaG \to$ rank a. In particular, the number of $G \times G$ orbits is finite. The partial order is the natural linear order on $\{0, 1, \ldots, n\}$.

Example 3.2 ([24, Example 15]) Let $M \subset \mathbf{M}_{n+1}$ consist of all matrices

$$a = \begin{bmatrix} x & x_1 & x_2 & \ldots & x_n \\ 0 & x & 0 & \ldots & 0 \\ 0 & 0 & x & \ldots & 0 \\ \vdots & \vdots & \vdots & \ddots & \vdots \\ 0 & 0 & 0 & \ldots & x \end{bmatrix}$$

where $x, x_1, \ldots, x_n \in K$. For simplicity of notation write $a = (x, x_1, \ldots, x_n)$. The group G consists of those a with $x \neq 0$. The $G \times G$ orbits are G, $\{0 = (0, \ldots, 0)\}$ and orbits which contain matrices $(0, x_1, \ldots, x_n)$ with x_i not all 0. For the latter we have $(0, x_1, \ldots, x_n)$ and $(0, y_1, \ldots, y_n)$ in the same orbit if and only if there exists $c \in K^*$ with $y_i = cx_i$ for all i. Thus these orbits are in one to one correspondence with points in the projective space $\mathbf{P}^{n-1}(K)$. There is a bijection $G\backslash M/G \leftrightarrow G \cup \{0\} \cup \mathbf{P}^{n-1}(K)$. This example has some undesired features. The number of $G \times G$ orbits is infinite (unless $n = 1$) and M has no idempotents except $0, 1$. Thus we are led to the following definition [31, 1.5].

Definition 1 Let $\mathcal{U}(M) = \{\mathcal{O} \in G\backslash M/G \mid \mathcal{O} \cap E(M) \neq \emptyset\}$.

Thus $\mathcal{U}(M)$ is the set of $\mathcal{J}$-classes which contain an idempotent. It inherits a partial order from the set of all $\mathcal{J}$-classes. In Example 3.2 the poset $\mathcal{U}(M)$ is isomorphic to the poset $\{0, 1\}$. In general, if M is an irreducible algebraic monoid then $\mathcal{U}(M)$ is a finite lattice [31, Theorem 5.10], [21, Theorem 1.7], [22, Theorem 2.7] and any two maximal chains in $\mathcal{U}(M)$ have the same length [23, Theorem 1.9]. The following conjugacy theorem will be useful in Section 5.

Theorem 3.1 ([24, Theorem 9]) *Let M be an irreducible algebraic monoid. Suppose $e, f \in E(M)$. Then $GeG = GfG$ if and only if e and f are conjugate under G.*

Example 3.3 Suppose $M = \mathbf{M}_n$. Then $G = \mathbf{GL}_n$. If $e, f \in E(M)$ and $GeG = GfG$, then rank e = rank f so e, f are conjugate under G.

We will impose conditions on M which exclude the pathology in Example 3.2 and insure that $G\backslash M/G = \mathcal{U}(M)$. A monoid M is *unit regular* if $M = GE(M)$. A monoid M is *regular* if given $a \in M$ there exists $b \in M$ such that $a = aba$.

Proposition 3.2 *Let M be an irreducible algebraic monoid. Then the conditions (1) M is unit regular, (2) M is regular and (3) $G\backslash M/G = \mathcal{U}(M)$, are equivalent.*

For (1) $\Rightarrow$ (2) note that if $a = eg$ with $e \in E(M)$ and $g \in G$ then $ag^{-1}a = ea = a$. For (2) $\Rightarrow$ (3) note that, by our earlier remark, the $\mathcal{J}$-classes are the $G \times G$ orbits, and that if $aba = a$ then $ab = abab$ is an idempotent in the $\mathcal{J}$-class of a. To prove that (3) $\Rightarrow$ (1), given $a \in M$ choose $g, h \in G$ and $e \in E(M)$ with $a = geh$. Then $a = gh \cdot h^{-1}eh \in GE(M)$.

Thus, if M is regular then the set $G\backslash M/G$ is finite. If $a \in M$ then MaM is stable under the $G \times G$ action and is thus a finite union of orbits GbG. The closure $\overline{MaM}$ is thus a finite union of sets $\overline{GbG}$. On the other hand, if M regular then MaM is closed and irreducible as algebraic set. This is proved in [22, Corollary 2.5] under the hypothesis that a is idempotent, but this is no restriction since $G\backslash M/G = \mathcal{U}(M)$. Thus $MaM = \overline{GaG}$. It follows that if M is regular then for $a, b \in M$ we have $MaM \leq MbM \Leftrightarrow GaG \subseteq \overline{GbG}$. Thus the partial order on $\mathcal{U}(M)$ which is defined naturally in terms of the semigroup structure has an algebro-geometric interpretation in terms of the closures of group orbits. In Example 3.1, if rank $a = r$ then $\overline{GaG}$ consists of the matrices of rank at most r.

We may also exclude the undesired behavior in Example 3.2 by imposing the requirement that G be a *reductive group*. What is a reductive group? To define it we need some preliminaries. A matrix $u \in \mathbf{GL}_n$ is *unipotent* if all its eigenvalues are 1 [15, 15.1]. A subgroup of $\mathbf{GL}_n$ is unipotent if all its elements are unipotent [15, 17.5]. For example the group

$$\mathbf{U}_n = \{I + \sum_{1 \leq i < j \leq n} c_{ij}E_{ij} \mid c_{ij} \in K\} \tag{13}$$

of all upper unitriangular matrices is unipotent. In fact every unipotent subgroup of $\mathbf{GL}_n$ is conjugate to a subgroup of $\mathbf{U}_n$ [3, Theorem 4.8], [15, Corollary 17.5]. The set of all connected unipotent normal subgroups of a connected algebraic group G

has a unique maximal element $\mathcal{R}_u(G)$ called the *unipotent radical* [3, 11.21], [15, 19.5] of G. The group G is said to be *reductive* [3, 11.21], [15, 19.5] if $\mathcal{R}_u(G) = \{1\}$. Let $\mathbf{B}_n \subseteq \mathbf{GL}_n$ denote the group of all invertible upper triangular matrices.

Example 3.4 If $G = \mathbf{B}_n$ then $\mathcal{R}_u(G) = \mathbf{U}_n$. Thus $\mathbf{B}_n$ is not reductive for $n \geq 2$.

Example 3.5 Let $G = \mathbf{GL}_n$. Since any normal subgroup of G either contains $\mathbf{SL}_n$ or is included in the center $Z(G) = \{tI \mid t \in K^*\}$, G is reductive. Similarly $\mathbf{SL}_n$ is reductive.

If $e \in E(M)$ is idempotent, define the *right centralizer* $P(e)$ and *left centralizer* $P^-(e)$ of e in G to be the subgroups [31, p.47] [15]

$$(14) \qquad P(e) = \{g \in G \mid ge = ege\} \ , \ P^-(e) = \{g \in G \mid eg = ege\} \ .$$

Let $C_G(e) = \{g \in G \mid ge = eg\}$ be the centralizer of e in G. Then

$$(15) \qquad C_G(e) = P(e) \cap P^-(e) \ .$$

Sometimes we write $L(e) = C_G(e)$ since it turns out that $L(e)$ is a common Levi factor of $P(e)$ and $P^-(e)$; see (64) for the definition of Levi factor and see the case $\Gamma = \{e\}$ of Theorem 5.7.

Example 3.6 Let $G = \mathbf{GL}_n$. For $1 \leq r \leq n$ let $e = e_r = \text{diag}(1, \ldots, 1, 0, \ldots, 0) \in \mathbf{D}_n$ where there are r entries equal to 1. Then $P(e) = \left[\begin{smallmatrix} * & * \\ 0 & * \end{smallmatrix}\right]$ and $P^-(e) = \left[\begin{smallmatrix} * & 0 \\ * & * \end{smallmatrix}\right]$ where the $*$'s are arbitrary matrices of the appropriate size such that the diagonal $*$'s are invertible. These groups have unipotent radicals $U(e) = \left[\begin{smallmatrix} I & * \\ 0 & I \end{smallmatrix}\right]$ and $U^-(e) = \left[\begin{smallmatrix} I & 0 \\ * & I \end{smallmatrix}\right]$ respectively, where the $I's$ are identity matrices of the appropriate size. Note that the intersection $L(e) = \left[\begin{smallmatrix} * & 0 \\ 0 & * \end{smallmatrix}\right]$ is the *centralizer* of e in $\mathbf{GL}_n$. It is isomorphic to $\mathbf{GL}_r \times \mathbf{GL}_{n-r}$ and hence is a reductive group. Note also that $P(e)$ and $P^-(e)$ factor as semidirect products $P(e) = L(e)U(e)$ and $P^-(e) = L(e)U^-(e)$. Since $L(e)$ is reductive and $U(e)$ is a unipotent normal subgroup of $P(e)$ we have $U(e) = \mathcal{R}_u(P(e))$. Similarly $U^-(e) = \mathcal{R}_u(P^-(e))$. This is an important example in the general theory; see (64) and Example 4.30.

Definition 2 A *reductive monoid* is an algebraic monoid which is irreducible as algebraic set and has a connected reductive group of units. [16]

Thus, for example, $\mathbf{M}_n$ is reductive because $\mathbf{GL}_n$ is reductive. The monoid in Example 3.2 is not reductive. An element $g = (x, x_1, \ldots, x_n) \in G$ is unipotent if and only if $x = 1$. The set $U = \{(1, x_1, \ldots, x_n)\}$ of all unipotent elements in G is a normal subgroup of G which is Zariski closed because it is isomorphic to a product of n copies of K. Thus $U = \mathcal{R}_u(G)$ and G is not reductive.

It is truly remarkable, an unexpected confluence of group theory and semigroup theory, that M is reductive if and only if M is regular.

---

[15] These are sometimes written as $C_G^r(e)$ and $C_G^l(e)$. The notation $P(e), P^-(e)$ is more suggestive in the context of reductive monoids, since $P(e), P^-(e)$ are parabolic subgroups of G; see the remarks which precede Example 4.29 for the definition of parabolic subgroup.

[16] The connectivity of G is in fact a consequence of the irreducibility of M [25, p. 695].

Theorem 3.3 *Let M be an irreducible algebraic monoid with zero. Then M is reductive if and only if M is regular.* [17]

There is a clear signal here. Regular semigroups are a distinguished class in semigroup theory. Reductive groups form the centerpiece of the theory of algebraic groups. Two notions which, on the face of it, have nothing to do with one another are in fact closely related. From the point of view of semigroup theory, it might be a good idea to learn something about reductive groups. We will do this in the next section. From Theorem 3.3 and Proposition 3.2 we conclude:

Theorem 3.4 *Let M be a reductive monoid with zero. Let G be the group of units of M. Then $G\backslash M/G$ is finite, and every $G \times G$ orbit contains an idempotent.*

4. Reductive Groups

This is a big subject. Our aim is to maximize understanding in minimal time by study of some of the main examples. Thus the sequence of definitions and theorems given here is not the sequence one would ordinarily use to develop the theory. We have not in general used the formal notice **Theorem**. There are two exceptions. It would not be possible to prove the main theorems on reductive groups in the order given here. The Lie algebra has been suppressed, the roots are defined in terms of their corresponding root groups and introduction of the Borel subgroup is long postponed. We will make the necessary definitions, state some of the main theorems, and spend most of our time on examples. The examples will be $\mathbf{GL}_n$, $\mathbf{SL}_n$, and the symplectic group $\mathbf{Sp}_n$. To avoid trivial cases we assume throughout that $n \geq 2$. Note though that $\mathbf{GL}_1 \simeq K^*$.

This is a long section, which I have tried to make "user-friendly." A reader who cares about semigroups, but knows nothing about reductive groups, might turn to the last paragraph of this section. Here, in the correct general context, we find the right and left centralizers $P(e), P^-(e)$ of an idempotent $e \in \mathbf{M}_n$ which were defined in (14) and computed in Example 3.6.

Our first aim is to introduce the *roots* of a reductive group, relative to a chosen *maximal torus*. Let's begin in comfort with $\mathbf{GL}_n$. For $1 \leq i \neq j \leq n$ let

$$U_{ij} = \{I + cE_{ij} \mid c \in K\} \tag{16}$$

where E_{ij} is a matrix unit. Then U_{ij} is a unipotent subgroup of G and the map $x_{ij} : K^+ \to U_{ij}$ defined by

[17] The implication "regular $\Rightarrow$ reductive" was proved by Putcha in [26, Theorem 2.11]. The converse was proved by Putcha [28, Theorem 2.4] if K has characteristic zero, using complete reducibility of the rational representations of G. The converse was proved in arbitrary characteristic by Renner in his Thesis [39, Theorem 4.4.15], [42, Theorem 3.1]. Renner's proof uses a fair amount of algebraic geometry. Putcha [29, Theorem 1.1] gave a different proof, valid in arbitrary characteristic, which uses the structure of reductive groups and some geometry. This argument also appears in [31, Theorem 7.3]. The hypothesis concerning zero is "necessary". If M is a regular algebraic monoid and H is a connected algebraic group which is not reductive, then the monoid of all matrices diag(a, h) with $a \in M$ and $h \in H$ is regular but not reductive.

(17) $$x_{ij}(c) = I + cE_{ij}\ ,\ c \in K$$

is an isomorphism. Here K^+ denotes the additive group of K. It has been known since Jordan in the case of the finite field $\mathbf{F}_p$, and since Dickson in general, that the subgroups U_{ij} generate $\mathbf{SL}_n$ and hence, together with the invertible diagonal matrices, generate $\mathbf{GL}_n$. If $t = \mathrm{diag}(t_1, \ldots, t_n) \in \mathbf{T}_n$ then

(18) $$tx_{ij}(c)t^{-1} = x_{ij}(t_i t_j^{-1} c)\ ,\ t \in \mathbf{T}_n,\ c \in K\ .$$

For $i \neq j$ define $\alpha_{ij} \in X(\mathbf{T}_n)$ by $\alpha_{ij} = \varepsilon_i \varepsilon_j^{-1}$ where ε_i is as in (9). Then we may rewrite (18) as

(19) $$tx_{ij}(c)t^{-1} = x_{ij}(\alpha_{ij}(t)c)\ ,\ t \in \mathbf{T}_n,\ c \in K\ .$$

This formula suggests a definition, (20) below, in any connected reductive group G. First we define the analogue in G of the subgroup $\mathbf{T}_n$ of $\mathbf{GL}_n$. A *maximal torus* of G is a torus $T \subseteq G$ of maximal dimension. Since a closed connected subgroup of a torus is itself a torus [3, Corollary 8.5], a maximal torus is maximal in the set theoretic sense. Choose, once and for all, a maximal torus T. Since T is uniquely determined up to conjugacy in G [3, Corollary 11.3], [15, Corollary 21.3.A], our later constructions do not depend on the choice of T. A character $\alpha \in X(T)$ is a *root* of G relative to T if there exists a monomorphism $\varphi : K^+ \to G$ such that [18]

(20) $$t\varphi(c)t^{-1} = \varphi(\alpha(t)c)\ ,\ t \in T,\ c \in K\ .$$

Define

(21) $$U_\alpha = \varphi(K)\ .$$

In view of the preceding footnote, the group $\varphi(K)$ is uniquely determined by α and our usage of U_α is consistent with that in [3, 15]. The subgroups U_α are called *root*

[18] This is not the usual definition of "root" as given for example in [3, 8.17] or [15, 16.4] but is equivalent to it. The point of the definition given here is that it allows us to enter the subject with minimal prerequisites. Since roots are the essence of reductive groups we show the reader where to find a proof that the two definitions are equivalent. The group G and hence the maximal torus T has a representation on the Lie algebra of G - not defined in this paper - called the ***adjoint representation*** [15, 9.1], [3, 3.13] and written Ad. The set of roots of G relative to T is, according to the usual definition, the set $\Phi(Ad)$ of weights of the adjoint representation. If $\alpha \in \Phi(Ad)$ then [15, Theorem 26.3(c)] proves the existence of a monomorphism φ which satisfies (20). Conversely suppose $\alpha \in X(T)$ satisfies (20). Then T normalizes $\varphi(K)$. It follows from [3, Proposition 13.20] with $H = \varphi(K)$ that there exists $\beta \in \Phi(Ad)$ such that $\varphi(K) = U_\beta$ where the group U_β is defined in [3, Theorem 13.18(4)(d)] or [15, Theorem 26.3(a)]. It follows from [15, Theorem 26.3(c)] that there exists a monomorphism $\psi : K^+ \to G$ such that $t\psi(c)t^{-1} = \psi(\beta(t)c)$ for $t \in T$ and $c \in K$. Since $\psi^{-1}\varphi$ is an automorphism of K^+ it follows from [3, 10.10] that there exists $k \in K^*$ with $\psi(c) = \varphi(kc)$. Then $\varphi(k\alpha(t)c) = \varphi(k\beta(t)c)$ so $\alpha = \beta$. Thus $\alpha \in \Phi(Ad)$. This argument also shows that the group $\varphi(K)$ in (19) is uniquely determined by the root α as $\varphi(K) = U_\alpha$.

groups. The map φ is not uniquely determined by α. However, as remarked in the preceding footnote, the automorphisms of K^+ have the form $c \mapsto kc$, so the map φ is determined up to replacement of φ by $c \mapsto \varphi(kc)$ for some $k \in K^*$. For given α we make a fixed choice of φ.

The term "root" may puzzle a beginner. The Lie algebra of $\mathbf{GL}_n$ is $\mathbf{M}_n$ with the bracket operation $[a,b] = ab - ba$ for $a,b \in \mathbf{M}_n$. If $a = \mathrm{diag}(a_1, \cdots, a_n) \in \mathbf{D}_n \subseteq \mathbf{M}_n$ is a diagonal matrix, then the formula $[a, E_{ij}] = (a_i - a_j)E_{ij}$ shows that E_{ij} is an eigenvector for the K-linear map denoted ad $a : \mathbf{M}_n \to \mathbf{M}_n$ which is defined by $b \mapsto [a,b]$, for $b \in \mathbf{M}_n$ and that the corresponding eigenvalue, alias characteristic *root*, is $a_i - a_j$. This is the Lie algebraic version of (18) and is the source of the terminology.

The set $\Phi(G,T)$ of roots is finite and [19]

$$|\Phi(G,T)| = \dim G - \dim T \; . \tag{22}$$

Example 4.1 Let $G = \mathbf{GL}_n$. Then G is a connected reductive group and $T = \mathbf{T}_n$ is a maximal torus. [20] Let ε_i be as in (9). Then $\Phi(G,T) = \{\varepsilon_i\varepsilon_j^{-1} \mid 1 \le i \ne j \le n\}$ because (18) shows that $\varepsilon_i\varepsilon_j^{-1} \in \Phi(G,T)$ and $|\Phi(G,T)| = \dim G - \dim T = n^2 - n$ is the number of $\varepsilon_i\varepsilon_j^{-1}$ with $i \ne j$. If $\alpha = \varepsilon_i\varepsilon_j^{-1}$ then $U_\alpha = U_{ij}$ is as in (16).

Example 4.2 Let $G = \mathbf{SL}_n$. Then G is a connected reductive group and $T = G \cap \mathbf{T}_n$ is a maximal torus. Let ε_i be as in (12). Then $\Phi(G,T) = \{\varepsilon_i\varepsilon_j^{-1} \mid 1 \le i \ne j \le n\}$. The homomorphisms x_{ij} are exactly as in $\mathbf{GL}_n$ and $|\Phi(G,T)| = \dim G - \dim T = (n^2-1) - (n-1) = n^2 - n$ as in $\mathbf{GL}_n$. Again, if $\alpha = \varepsilon_i\varepsilon_j^{-1}$ then U_α is as in (16).

Example 4.3 Let $n = 2l$ be even and let $\{v_1, \dots, v_n\}$ be the standard basis for K^n. Let $J = \left[\begin{smallmatrix} 0 & I \\ -I & 0 \end{smallmatrix}\right] \in \mathbf{M}_n$ where I is the identity matrix of size l. Thus J is nonsingular and skew symmetric. The *symplectic group* is by definition

$$G = \mathbf{Sp}_n = \{g \in \mathbf{GL}_n \mid g^\top J g = J\} \; . \tag{23}$$

It is connected and reductive. [21] Let $T = G \cap \mathbf{T}_n$. Elements of T have the shape

$$t = \mathrm{diag}(t_1, \dots, t_l, t_1^{-1}, \dots, t_l^{-1}) \tag{24}$$

[19] This follows from [3, Proposition 13.20], [15, Corollary 26.2.B] and the fact that the dimension of G is equal to the dimension of its Lie algebra [3, 3.5], [15, 9.1].

[20] Since the torus T is its own centralizer in G it is a maximal torus. To prove that G is connected, note that $G = \langle T, U_{ij} \mid 1 \le i \ne j \le n \rangle$. Since $T \simeq (K^*)^n$ and the $U_{ij} \simeq K^+$ are connected we may apply [15, Corollary 7.5]. We have already remarked that G is reductive.

[21] The group G is generated by transvections [1, Theorem 3.21], each of which may be imbedded in a subgroup isomorphic to the connected group K^+ [50, 6.6]. Thus we may apply [15, Corollary 7.5] to prove that G is connected. To prove that G is reductive one may use the fact that the only proper normal subgroup of G, as abstract group, is its center of order at most 2 [1, Theorem 5.2]. Alternatively, one may exploit the fact that the natural representation of G on K^n is irreducible [6, Exposé 22, p.1].

where $t_1, \dots, t_l$ are arbitrary in K^*. Thus T is a torus of dimension l. As in the case of $\mathbf{GL}_n$, it is a maximal torus because it is its own centralizer in G. Define $\varepsilon_i : T \to K^*$ for $1 \le i \le l$ by

$$\varepsilon_i(\mathrm{diag}(t_1, \dots, t_l, t_1^{-1}, \dots, t_l^{-1})) = t_i. \tag{25}$$

Since dim $G = l(2l+1)$ [6, Exposé 22, p.3] we have $|\Phi(G,T)| = \dim G - \dim T = l(2l+1) - l = 2l^2$.[22] We will exhibit $2l^2$ homomorphisms $\varphi : K^+ \to G$ which satisfy (20) for suitable characters $\alpha \in X(T)$ [6, Exposé 22, p.2]. For $1 \le i \ne j \le l$ define $y_{ij} : K^+ \to \mathbf{SL}_l$ by $y_{ij}(c) = I + cE_{ij}$ for $c \in K$. These are precisely the maps used in (17) with a change of notation because the letter "x" is now reserved for $\mathbf{Sp}_n$. Define $x_{ij} : K^+ \to \mathbf{Sp}_n$ by

$$x_{ij}(c) = \begin{bmatrix} y_{ij}(c) & 0 \\ 0 & y_{ji}(-c) \end{bmatrix}. \tag{26}$$

To check that $x_{ij}(c) \in \mathbf{Sp}_n$ use $y_{ji}(c) = y_{ij}(c)^\top$. For $1 \le i \ne j \le l$ and $c \in K$, define x'_{ij}, x''_{ij} by

$$x'_{ij}(c) = \begin{bmatrix} I & c(E_{ij}+E_{ji}) \\ 0 & I \end{bmatrix}, \quad x''_{ij}(c) = \begin{bmatrix} I & 0 \\ c(E_{ij}+E_{ji}) & I \end{bmatrix}. \tag{27}$$

For $1 \le i \le l$ and $c \in K$, define x_i, x'_i by

$$x_i(c) = \begin{bmatrix} I & cE_{ii} \\ 0 & I \end{bmatrix}, \quad x'_i(c) = \begin{bmatrix} I & 0 \\ cE_{ii} & I \end{bmatrix}. \tag{28}$$

Here the $E_{ij} \in \mathbf{M}_l$ are matrix units and I is the identity matrix of size l. These maps $\varphi = x_{ij}, x'_{ij}, x''_{ij}, x_i, x'_i$ are all monomorphisms from $K^+ \to \mathbf{Sp}_n$.[23] We must compute $t\varphi(c)t^{-1}$ for these monomorphisms φ and check that there exist characters $\alpha \in X(T)$ which satisfy (20). This calculation may be done easily with block matrices of size l and we simply tabulate the results below. We have

(29)

φ	$t\varphi(c)t^{-1}$	α	
x_{ij}	$x_{ij}(t_i t_j^{-1} c)$	$\varepsilon_i \varepsilon_j^{-1}$	$1 \le i \ne j \le l$
x'_{ij}	$x'_{ij}(t_i t_j c)$	$\varepsilon_i \varepsilon_j$	$1 \le i < j \le l$
x''_{ij}	$x'_{ij}(t_i^{-1} t_j^{-1} c)$	$\varepsilon_i^{-1} \varepsilon_j^{-1}$	$1 \le i < j \le l$
x_i	$x_i(t_i^2 c)$	ε_i^2	$1 \le i \le l$
x'_i	$x'_i(t_i^{-2} c)$	ε_i^{-2}	$1 \le i \le l$

[22] The dimension of G is computed in [6, Exposé 22] by algebro-geometric argument; one may also compute it as the dimension of the Lie algebra Lie$(G) = \{a \in \mathbf{M}_n \mid a^\top J + Ja = 0\}$.

[23] They may be found, with slightly different notation, in [6, Exposé 22]

where the ε_i are as in (25) and t and the t_i are as in (24). Thus the set $\Phi(G,T)$ of roots of $G = \mathbf{Sp}_n$ relative to the chosen maximal torus T is the set of characters α in the right hand column of the table, namely

$$\Phi(G,T) = \{\varepsilon_i\varepsilon_j^{-1},\ \varepsilon_i\varepsilon_j,\ \varepsilon_i^{-1}\varepsilon_j^{-1},\ \varepsilon_i^2, \varepsilon_i^{-2}\} \tag{30}$$

with the restrictions on i,j given above. If $\alpha \in \Phi$ then U_α is the set of matrices which appear in the appropriate line (26)-(28).

Let G be a connected reductive group. Then

$$G = < T, U_\alpha \mid \alpha \in \Phi(G,T) > \tag{31}$$

is generated by the maximal torus T and the root groups U_α [3, Corollary 14.8.1], [15, Theorem 26.3(d)]. Thus the structure of G is determined in large part by the manner in which the various U_α are glued together. This is controlled by the *Weyl group*, a finite group with extraordinary properties. The setup is complex enough to be interesting but structured enough to be manageable. Our next objective is to study the Weyl group. The Weyl group of G relative to T is by definition

$$W(G,T) = N_G(T)/T \tag{32}$$

where $N_G(T)$ is the normalizer of T in G. [24] Let's simplify notation. Henceforth we fix a maximal torus T and write $\Phi = \Phi(G,T)$ for the set of roots, $N = N_G(T)$ for the normalizer and $W = W(G,T)$ for the Weyl group. The Weyl group is finite [3, 11.19], [15, Proposition 24.1.A].

Example 4.4 Let $G = \mathbf{GL}_n$. Let S_n be the symmetric group on $\{1, \dots, n\}$. Define a subgroup $\mathbf{N}_n$ of $\mathbf{GL}_n$ by

$$\mathbf{N}_n = \{\omega = \textstyle\sum_{j=1}^m t_j E_{\pi j,j} \mid t_j \in K^* \text{ and } \pi \in S_n\}\ . \tag{33}$$

The group $\mathbf{N}_n$ consists of all matrices which have just one nonzero entry in each row and column. These are sometimes called *monomial matrices*. The notation $\mathbf{N}_n$ is not standard but will be useful. Let's choose $T = \mathbf{T}_n$ as our maximal torus. Then $N = \mathbf{N}_n$. The map $\omega = \sum_{j=1}^n t_j E_{\pi j,j} \mapsto \pi$ is a homomorphism from N onto S_n with kernel T. Thus $W = N/T \simeq S_n$ so the Weyl group is isomorphic to the symmetric group on n letters. Let $P_n \subset G$ be the group of *permutation matrices*. By definition these are matrices of the form $p = \sum_{j=1}^n E_{\pi j,j}$ where $\pi \in S_n$. Then $N = TP_n$ splits as a semidirect product and we may realize W as a subgroup of G. This simplifies calculations in case $G = \mathbf{GL}_n$ but it is misleading. In general there is no semidirect product decomposition $N = TP$ with P isomorphic to W.

[24] If G is a connected algebraic group and $T \subseteq G$ is a torus, then the Weyl group is defined [3, 11.19], [15, 24.1] by $W(G,T) = N_G(T)/Z_G(T)$ where $Z_G(T)$ is the centralizer of T in G. If G is reductive and T is a maximal torus then $Z_G(T) = T$ [3, Corollary 13.17(2)], [15, Corollary 26.2.A] so $W(G,T)$ is as in (32).

Example 4.5 Let $G = \mathbf{SL}_n$ and $T = G \cap \mathbf{T}_n$. Then $N = G \cap \mathbf{N}_n$. The map $N \to S_n$ defined by $p = \sum_{j=1}^{n} t_j E_{\pi j,j} \mapsto \pi$ is surjective, as in the case of $\mathbf{GL}_n$ and has kernel T. Thus again $W = N/T \simeq S_n$. Note that permutation matrices may have determinant -1 and thus need not be in G. For example, if $n = 2$ and $\pi = (12)$, the corresponding coset in N may be represented by $\omega = \left[\begin{smallmatrix} 0 & 1 \\ -1 & 0 \end{smallmatrix}\right]$ but not by $p = \left[\begin{smallmatrix} 0 & 1 \\ 1 & 0 \end{smallmatrix}\right]$.

Example 4.6 Let $G = \mathbf{Sp}_n$ and let T be as in Example 4.3. Let's compute W. The description of N is complicated by the fact that $\mathbf{Sp}_n \subseteq \mathbf{SL}_n$ so that $\mathbf{Sp}_n$ does not contain the group P_n of permutation matrices; see the last sentence in Example 4.5. Let's ignore this difficulty for the moment. If $\pi \in S_n$ let $p_\pi = \sum_{i=1}^{n} E_{\pi i,i} \in P_n$ be the corresponding permutation matrix. Define an involutory permutation $i \mapsto \bar{i}$ of $\{1, \dots, n\}$ by $\bar{i} = i + l$ if $1 \leq i \leq l$ and $\bar{i} = i - l$ if $l + 1 \leq i \leq n$. Let C denote the centralizer of this involution in S_n. Then p_π normalizes T if and only if $\pi \in C$. The group C factors as a semidirect product $C = C_1 C_2$ where C_1 is a normal abelian subgroup of order 2^l generated by the transpositions $(1\bar{1}), \dots, (l\,\bar{l})$ and $C_2 \simeq S_l$ consists of all permutations $\pi \in S_n$ which stabilize $\{1, \dots, l\}$ and act on the complement $\{l+1, \dots, n\}$ in the unique manner consistent with the assertion that $\pi \in C$. For example, with $l = 5$, if the restriction of π to $\{1, \dots, 5\}$ is $(135)(24)$ in cycle notation, then $\pi = (135)(24)(\bar{1}\,\bar{3}\,\bar{5})(\bar{2}\,\bar{4})$. Note that $|C| = |C_1||C_2| = 2^l l!$. Now return to the difficulty mentioned at the beginning. It is possible to multiply p_π by a suitable diagonal matrix d_π with entries ± 1 such that $\omega_\pi = d_\pi p_\pi \in \mathbf{Sp}_n$. Since p_π normalizes T so does ω_π. Thus $\omega_\pi \in N$. This shows that $N \supseteq \sqcup_{\pi \in C} \omega_\pi T$, where $\sqcup$ means disjoint union. In fact $N = \sqcup_{\pi \in C} \omega_\pi T$. Thus $W = N/T \simeq C \subseteq S_n$. For $i = 1, 2$ let $N_i = \sqcup_{\pi \in C_i} \omega_\pi T$ and let $W_i = N_i/T$. Then $W = W_1 W_2 \simeq C_1 C_2 = C \subseteq S_n$ where the products are semidirect. There exists an isomorphism $\theta : W \to \mathbf{GL}_l(\mathbf{R})$ such that $\theta(W_1)$ is the group of diagonal matrices with entries in $\{\pm 1\}$ and $\theta(W_2) = P_l$ is the group of permutation matrices. The group $\theta(W)$ is the symmetry group of the ***hyperoctahedron*** or ***cross polytope*** in $\mathbf{R}^l$ with vertices at the $2l$ points $\pm \mathbf{e}_1, \dots, \pm \mathbf{e}_l$ where $\mathbf{e}_1, \dots, \mathbf{e}_l$ is the standard basis for $\mathbf{R}^l$. Thus W is sometimes called the ***hyperoctahedral group***.

Let G be a connected reductive group and let $T \subseteq G$ be a maximal torus. If $\chi \in X(T)$ and $\omega \in N$, define a function $\omega\chi : T \to K$ by $(\omega\chi)(t) = \chi(\omega^{-1} t \omega)$ for $t \in T$. Then $\omega\chi \in X(T)$. Since $\omega^{-1} t \omega$ depends only on $w = \omega T$ we write $w\chi = \omega\chi$. Thus W acts on $X(T)$. If $\chi, \chi' \in X(T)$ then $\omega(\chi\chi') = (\omega\chi)(\omega\chi')$, so $w(\chi\chi') = (w\chi)(w\chi')$. Thus W acts as a group of automorphisms of the abelian group $X(T)$. This action is faithful: if $w\chi = \chi$ for all $\chi \in X(T)$ then $w = 1$.[25]Let $\rho : G \to \mathbf{GL}(V)$ be a rational representation. View ρ as a representation of T and, as in Section 2, let $\Phi(\rho) \subset X(T)$ be the set of weights of ρ. Suppose $\chi \in \Phi(\rho)$ and $v \in V$ is a vector of weight χ. Thus $\rho(t)v = \chi(t)v$ for $t \in T$. If $\omega \in N$ then $\rho(t)\rho(\omega)v = \rho(\omega)\rho(\omega^{-1} t \omega)v = \rho(\omega)\chi(\omega^{-1} t \omega)v = (\omega\chi)(t)\rho(\omega)v$. Thus $\omega\chi \in \Phi(\rho)$. Thus

$$W\Phi(\rho) = \Phi(\rho)\ . \tag{34}$$

[25] Suppose $\omega \in N$ and $\omega\chi = \chi$ for all $\chi \in X(T)$. If $t \in T$ then $\chi(\omega^{-1} t \omega t^{-1}) = \chi(\omega^{-1} t \omega)\chi(t)^{-1} = 1$. Thus $\omega t \omega^{-1} t^{-1} = 1$ so ω centralizes T and thus $\omega \in T$; see the footnote which follows (32).

If we had defined roots using the adjoint representation - see the footnote to (20) - then it would follow in particular from (34) that $W\Phi = \Phi$. Let's check this directly using our definition (20). Note that N acts on $\mathrm{Hom}(K^+, G)$ by $(\omega\varphi)(c) = \omega\varphi(c)\omega^{-1}$ for $\omega \in N$, $\varphi \in \mathrm{Hom}(K^+, G)$ and $c \in K$. If $\alpha \in \Phi$ and $\varphi \in \mathrm{Hom}(K^+, G)$ satisfy (20) then $\beta = \omega\alpha$ and $\psi = \omega\varphi$ satisfy $t\psi(c)t^{-1} = \psi(\beta(t)c)$ for $t \in T$ and $c \in K$ so $\beta = \omega\alpha \in \Phi$. Thus indeed

$$W\Phi = \Phi\,. \tag{35}$$

If $\alpha \in \Phi$ then, since T normalizes U_α, the group $\omega U_\alpha \omega^{-1}$ depends only on $w = \omega T \in W$. We abuse notation and write $wU_\alpha w^{-1} = \omega U_\alpha \omega^{-1}$. Thus [15, Theorem 26.3(b)]

$$wU_\alpha w^{-1} = U_{w\alpha} \text{ for } w \in W\,. \tag{36}$$

Example 4.7 Let $G = \mathbf{GL}_n$ and let T, N be as in Example 4.4. If $\pi \in S_n$ let $p = \sum_{j=1}^n E_{\pi j,j} \in P_n \subseteq N$ be the corresponding permutation matrix. Let $\varepsilon_i \in X(T)$ be as in (9). Then $\{\varepsilon_1, \ldots, \varepsilon_n\}$ is a basis for the abelian group $X(T)$. Since W acts as a group of automorphisms of $X(T)$, the action of w on $X(T)$ is determined by the $w\varepsilon_i$. Let $w = pT$. Then $w\varepsilon_i = \varepsilon_{\pi i}$. Thus, as one might expect, $W \simeq S_n$ permutes $\varepsilon_1, \ldots, \varepsilon_n$ in the same way that S_n permutes $\{1, \ldots, n\}$. From Example 4.1 we have $\Phi = \{\varepsilon_i\varepsilon_j^{-1} \mid 1 \le i \ne j \le n\}$. Then $w(\varepsilon_i\varepsilon_j^{-1}) = (w\varepsilon_i)(w\varepsilon_j)^{-1} = \varepsilon_{\pi i}\varepsilon_{\pi j}^{-1} \in \Phi$. If U_{ij} is the root group corresponding to $\varepsilon_i\varepsilon_j^{-1}$ and $\pi \in S_n$ is the permutation corresponding to $w \in W$ then $wU_{ij}w^{-1} = U_{\pi i,\pi j}$.

Example 4.8 Let $G = \mathbf{SL}_n$. Let T, N be as in Example 4.5 and let $\varepsilon_1, \ldots, \varepsilon_n$ be as in (12). Remember in this case that $\varepsilon_1 \cdots \varepsilon_n = 1$. Here we may choose $\varepsilon_1, \ldots, \varepsilon_{n-1}$ as a basis for $X(T)$. In spite of this difference, the bottom line for $\mathbf{SL}_n$ is the same as for $\mathbf{GL}_n$. By Example 4.2 we have $\Phi = \{\varepsilon_i\varepsilon_j^{-1} \mid 1 \le i \ne j \le n\}$. If $w \in W$ corresponds to $\pi \in S_n$ then $w\varepsilon_i = \varepsilon_{\pi i}$ and $w(\varepsilon_i\varepsilon_j^{-1}) = \varepsilon_{\pi i}\varepsilon_{\pi j}^{-1} \in \Phi$ and $wU_{ij}w^{-1} = U_{\pi i,\pi j}$.

Example 4.9 Let $G = \mathbf{Sp}_n$ where $n = 2l$. Let T, N be as in Example 4.6 and let $\varepsilon_1, \ldots, \varepsilon_l$ be as in (25). Recall from Example 4.6 that $W = W_1W_2 \simeq C_1C_2 = C \subseteq S_n$. To describe the action of $w \in W$ on $X(T)$ it suffices to know the action of $w \in W_1$ and $w \in W_2$. If $w \in W_1$ and $\pi \in S_n$ is the corresponding permutation then there exists a subset I of $\{1, \ldots, l\}$ such that $\pi = \prod_{i \in I}(i\bar{i})$. Then for $1 \le i \le l$ we have $w\varepsilon_i = \varepsilon_i^{-1}$ if $i \in I$ and $w\varepsilon_i = \varepsilon_i$ if $i \notin I$. If $w \in W_2$ and $\pi \in S_n$ is the corresponding permutation then $\pi\{1, \ldots, l\} \subseteq \{1, \ldots, l\}$ and the action of w on $X(T)$ is given by $w\varepsilon_i = \varepsilon_{\pi i}$ for $1 \le i \le l$. It follows that the set Φ listed in (30) is stable under both W_1 and W_2 and is thus stable under W. To check that $wU_\alpha w^{-1} = U_{w\alpha}$ in $\mathbf{Sp}_n$ one must separate the cases (26), (27), (28) and one may choose $w \in W_1$ or $w \in W_2$.

Let G be a connected reductive group and let $T \subseteq G$ be a maximal torus. Since $X(T)$ is a free abelian group, we may use additive notation and write $\chi + \chi'$ rather than $\chi\chi'$. If $w \in W$ and $\chi, \chi' \in X(T)$ then $w(\chi + \chi') = w\chi + w\chi'$. Thus W acts faithfully as a group of $\mathbf{Z}$-linear mappings of $X(T)$. Identify $X(T)$ with a $\mathbf{Z}$-submodule of the real vector space $X = X(T) \otimes \mathbf{R}$ via $\chi = \chi \otimes 1$ [3, 14.8],[15,

26.1][26].The dimension $\dim_{\mathbf{R}} X$ of X as vector space over $\mathbf{R}$ is equal to the rank of $X(T)$ as free abelian group which is dim T. If $\{\chi_1, \dots, \chi_r\}$ is a basis for $X(T)$ as free abelian group, then $\{\chi_1, \dots, \chi_r\}$ is also a basis for X as vector space over $\mathbf{R}$. The Weyl group W may now be viewed as a group of $\mathbf{R}$-linear transformations of X, as well as a group of $\mathbf{Z}$-linear transformations of $X(T)$.

The roots Φ need not span the real vector space X. To describe the difference between $\dim_{\mathbf{R}} X$ and $\dim_{\mathbf{R}} \sum_{\alpha\in\Phi} \mathbf{R}\alpha$ we introduce the notions of rank and semisimple rank. The integer $r = \dim T = \dim_{\mathbf{R}} X$ is called the ***rank*** [15, 21.3] of G. [27] To define the semisimple rank of G we introduce the notions of radical and semisimple group. Any connected algebraic group G has a unique maximal connected solvable normal subgroup $\mathcal{R}(G)$ called the ***radical*** [3, 11.21], [15, 19.5] of G. The radical is automatically a closed subgroup of G. The group G is by definition ***semisimple*** [3, 11.21], [15, 19.5] if $G \neq \{1\}$ and $\mathcal{R}(G) = \{1\}$. If G is semisimple then the connected component $Z(G)^\circ$ of the center is $\{1\}$ so $Z(G)$ is finite. For example $G = \mathbf{SL}_n$ and $G = \mathbf{Sp}_n$ are semisimple; in either case a normal subgroup different from G is a subgroup of $Z(G)$. If G is semisimple then G is reductive. In the opposite direction, if G is reductive then $\mathcal{R}(G) = Z(G)^\circ$ is a torus [3, Proposition 11.21], [15, Lemma 19.5]. If G is reductive then the commutator subgroup (G,G) is semisimple [3, 14.2]. Furthermore $G = (G,G)\,Z(G)^\circ$ and $(G,G) \cap Z(G)$ is finite [3, Proposition 14.2], [15, 27.5]. For example, if $G = \mathbf{GL}_n$ then $\mathcal{R}(G) = Z(G) = \{tI \mid t \in K^*\}$ and $(G,G) = \mathbf{SL}_n$ so that $(G,G) \cap Z(G) = \{tI \mid t^n = 1\}$. If G is a connected reductive group and T is a maximal torus of G, the ***semisimple rank*** [3, 13.13], [15, 25.3] of G is by definition $l = \dim T - \dim\,(T \cap \mathcal{R}(G))$. Then $l = \dim T - \dim\, Z(G) = \dim_{\mathbf{R}} \sum_{\alpha\in\Phi} \mathbf{R}\alpha$ by [3, Theorem 14.8]. Thus the semisimple rank of G is equal to the dimension of the real vector space spanned by the roots. Let's translate the statements in Examples 4.7-4.9 into the additive language.

Example 4.10 Let $G = \mathbf{GL}_n$. Here $X(T) = \mathbf{Z}\varepsilon_1 \oplus \cdots \oplus \mathbf{Z}\varepsilon_n$ and $X = \mathbf{R}\varepsilon_1 \oplus \cdots \cdots \oplus \mathbf{R}\varepsilon_n$. Then $\Phi = \{\varepsilon_i - \varepsilon_j \mid 1 \le i \neq j \le n\}$. If $w \in W$ corresponds to the permutation $\pi \in S_n$ then $w\varepsilon_i = \varepsilon_{\pi i}$ and thus $w(\varepsilon_i - \varepsilon_j) = \varepsilon_{\pi i} - \varepsilon_{\pi j}$. The rank of G is $r = n$; the semisimple rank is $l = n-1$.

Example 4.11 Let $G = \mathbf{SL}_n$. Here $X(T) = \mathbf{Z}\varepsilon_1 + \cdots + \mathbf{Z}\varepsilon_n$ and $X = \mathbf{R}\varepsilon_1 + \cdots + \mathbf{R}\varepsilon_n$. The sums are not direct since $\varepsilon_1 + \cdots + \varepsilon_n = 0$. We have $X(T) = \mathbf{Z}\varepsilon_1 \oplus \cdots \oplus \mathbf{Z}\varepsilon_{n-1}$ and $X = \mathbf{R}\varepsilon_1 \oplus \cdots \oplus \mathbf{R}\varepsilon_{n-1}$. The bottom line for $\mathbf{SL}_n$ is the same as for $\mathbf{GL}_n$. We have $\Phi = \{\varepsilon_i - \varepsilon_j \mid 1 \le i \neq j \le n\}$ and $w\varepsilon_i = \varepsilon_{\pi i}$ so $w(\varepsilon_i - \varepsilon_j) = \varepsilon_{\pi i} - \varepsilon_{\pi j} \in \Phi$. The rank of G and the semisimple rank of G are given by $r = l = n - 1$.

Example 4.12 Let $G = \mathbf{Sp}_n$. Here $X(T) = \mathbf{Z}\varepsilon_1 \oplus \cdots \oplus \mathbf{Z}\varepsilon_l$ and $X = \mathbf{R}\varepsilon_1 \oplus \cdots \oplus \mathbf{R}\varepsilon_l$ where $n = 2l$. From (30) we get

$$\Phi = \{\varepsilon_i - \varepsilon_j \mid 1 \le i \neq j \le l\} \cup \{\pm(\varepsilon_i + \varepsilon_j) \mid 1 \le i < j \le l\} \cup \{\pm 2\varepsilon_i \mid 1 \le i \le l\}\,.$$

---

[26] In the rest of this section it would be possible, in most statements, to replace the real field $\mathbf{R}$ by the rational field $\mathbf{Q}$ and work in $X(T) \otimes \mathbf{Q}$ as in [3, 15].

[27] This differs from the definition of rank given in [3, 12.2] but the two definitions are equivalent when G is reductive, by [3, Corollary 13.17.2(c)].

In the notation of Example 4.9 the action of W is given as follows. If $w \in W_1$ corresponds to $\pi = \prod_{i \in I}(i\bar{i}) \in S_n$ then $w\varepsilon_i = -\varepsilon_i$ when $i \in I$ and $w\varepsilon_i = \varepsilon_i$ when $i \notin I$. If $w \in W_2$ corresponds to $\pi \in S_n$ then $w\varepsilon_i = \varepsilon_{\pi i}$ for $1 \leq i \leq l$. Thus W, in its action on X, permutes the set $\{\pm\varepsilon_1, \dots, \pm\varepsilon_l\}$. If we use the basis $\varepsilon_1, \dots, \varepsilon_l$ to identify X with Euclidean space $\mathbf{R}^l$ then the points $\pm\varepsilon_i$ are the vertices of the cross polytope, and this representation of W by linear transformations is the one mentioned in Example 4.6. The rank of G and the semisimple rank of G are given by $r = l = \frac{1}{2}n$.

Let X be a real vector space of finite dimension. An element $s \in \mathbf{GL}(X)$ is a *reflection* [3, 14.7], [15, p.229] if $s \neq 1$ and s fixes a subspace of codimension 1. Then s has just one eigenvalue different from 1 which, since the characteristic polynomial of s has real coefficients, must be -1. Thus $s^2 = 1$ and there is a nonzero vector α, determined uniquely up to a scalar multiple, such that $s\alpha = -\alpha$.

Theorem 4.1 *Let G be a connected reductive group and let T be a maximal torus. Let $\Phi = \Phi(G,T)$ be the set of roots and let $W = W(G,T) \hookrightarrow \mathbf{GL}(X)$ be the Weyl group. If $\alpha \in \Phi$ then $-\alpha \in \Phi$ and there exists a reflection $s_\alpha \in W$ such that $s_\alpha\alpha = -\alpha$. The group*

(37) $$W = < s_\alpha \mid \alpha \in \Delta >$$

is generated by the reflections s_α corresponding to the simple roots $\alpha \in \Delta$. Furthermore, if $\alpha, \beta \in \Phi$ then there exists $n_{\beta,\alpha} \in \mathbf{Z}$ such that

(38) $$s_\alpha\beta = \beta - n_{\beta,\alpha}\alpha \ .$$

This theorem is the most amazing and important fact in the structure theory of reductive groups. It makes the subject. The proof is long and difficult. [28] Note that $n_{\alpha,\alpha} = 2$ since $s_\alpha\alpha = -\alpha$.

For the moment, let X be any real vector space of finite dimension and let $W \subset \mathbf{GL}(X)$ be a finite group of linear transformations. There exists a positive definite inner product (symmetric bilinear form) on X which is W-invariant: [29] $(wx, wx') = (x, x')$ for $w \in W$ and $x, x' \in X$. Let $\mathbf{O}(X) \subset \mathbf{GL}(X)$ be the orthogonal group, the group of all linear transformations $g \in \mathbf{GL}(X)$ such that $(gx, gx') = (x, x')$ for all $x, x' \in X$. Then $W \subset \mathbf{O}(X)$. If $s \in \mathbf{O}(X)$ is a reflection and $\alpha \in V$ is a nonzero vector with $s\alpha = -\alpha$ then

(39) $$sx = x - \frac{2(x,\alpha)}{(\alpha,\alpha)}\,\alpha \ .$$

[28] The main effort is to show that Φ is a *root system* in the real vector space $\sum_{\alpha\in\Phi} \mathbf{R}\alpha$. See [3, 14.7], [9, 64.4], [15, 27.1], for the definition of root system, and [3, Theorem 14.8] for the last lines in the proof. There are important preliminary steps in the proof, which concern reductive groups of semisimple rank 1; these are used to prove the existence of the groups U_α. See [3, Theorem 13.18(4)], [15, 27.1] for the existence of a reflection s_α with $s_\alpha\alpha = -\alpha$. See [3, Corollary 14.6], [15, 27.2] for the proof of (38). Once one knows that Φ is indeed a root system, the assertion (37) follows from properties of "abstract root systems" in a real vector space [3, 14.7], [15, p.229].

[29] If $\langle x, x'\rangle$ is any positive definite inner product on X then $(x, x') = \sum_{w\in W}\langle wx, wx'\rangle$ is a W-invariant positive definite inner product.

Return to the case of a reductive group where W is the Weyl group. It is convenient in the examples, but not necesary, to choose a W-invariant inner product on $X = X(T) \otimes \mathbf{R}$. This inner product is not in general unique. It is unique up to scalar multiple in case W is the Weyl group of $\mathbf{SL}_n$, or $\mathbf{Sp}_n$ because W acts irreducibly on X in these cases.

Example 4.13 Suppose that $G = \mathbf{GL}_n$, $\mathbf{SL}_n$, or $\mathbf{Sp}_n$ and that the ε_i are as in Examples 4.10-4.12. If $G = \mathbf{GL}_n$ or $\mathbf{Sp}_n$ then the ε_i are an $\mathbf{R}$-basis for X and we may define an inner product on X by $(\varepsilon_i, \varepsilon_j) = \delta_{ij}$, Kronecker delta. If $G = \mathbf{SL}_n$ we must be careful because $\varepsilon_1 + \cdots + \varepsilon_n = 0$. In this case there is a well defined inner product on X such that $(\varepsilon_i, \varepsilon_j) = \delta_{ij} - \frac{1}{n}$. In all three cases, it is clear from the formulas for $w \in W$ in Examples 4.4-4.6 that the inner product is W-invariant.

If we compare (38) with (39) we see that

$$\frac{2(\beta, \alpha)}{(\alpha, \alpha)} = n_{\beta,\alpha} \in \mathbf{Z}\ . \tag{40}$$

Example 4.14 Let $G = \mathbf{GL}_n$. Use the notation of Example 4.7 and the inner product in Example 4.13. If $\alpha = \varepsilon_i - \varepsilon_j$, it follows from (39) that $s_\alpha \varepsilon_i = \varepsilon_j$, $s_\alpha \varepsilon_j = \varepsilon_i$, and $s_\alpha \varepsilon_k = \varepsilon_k$ for $k \neq i, j$. Thus the matrix for $s_\alpha \in \mathbf{O}(X)$ with respect to the basis $\{\varepsilon_1, \dots, \varepsilon_n\}$ is the permutation matrix corresponding to the transposition (ij). The assertion (37) amounts in this case to the fact that the transpositions (ij) generate the symmetric group. If $\alpha = \varepsilon_i - \varepsilon_j$ and $\beta = \varepsilon_p - \varepsilon_q$ then $(\alpha, \alpha) = 2$ and the $n_{\beta,\alpha}$ in (40) are given by

$$n_{\beta,\alpha} = \frac{2(\beta, \alpha)}{(\alpha, \alpha)} = \delta_{ip} - \delta_{iq} - \delta_{jp} + \delta_{jq} \in \mathbf{Z}\ . \tag{41}$$

Example 4.15 Let $G = \mathbf{SL}_n$. Use the notation of Example 4.8 and the inner product in Example 4.13. Although the inner product here is not the same as in $\mathbf{GL}_n$ the bottom line is the same. If $\alpha = \varepsilon_i - \varepsilon_j$ then again $s_\alpha \varepsilon_i = \varepsilon_j$, $s_\alpha \varepsilon_j = \varepsilon_i$, and $s_\alpha \varepsilon_k = \varepsilon_k$ for $k \neq i, j$. Since $\dim_{\mathbf{R}} X = n - 1$, the matrix $s_\alpha \in \mathbf{O}(X)$ with respect to the basis $\{\varepsilon_1, \dots, \varepsilon_{n-1}\}$ is not a permutation matrix. Nevertheless, the permutation in S_n which corresponds to s_α is again (ij). The integers $n_{\beta,\alpha}$ are exactly as in the case $\mathbf{GL}_n$.

Example 4.16 Let $G = \mathbf{Sp}_n$ where $n = 2l$. Use the notation of Example 4.9 and the inner product in Example 4.13. We compute the action of the $s_\alpha \in \mathbf{O}(X)$ on the basis $\{\varepsilon_1, \dots, \varepsilon_l\}$ as well as the corresponding permutations $\pi \in C \subseteq S_n$. If $\alpha = \varepsilon_i - \varepsilon_j$ then (39) gives $s_\alpha \varepsilon_i = \varepsilon_j$, $s_\alpha \varepsilon_j = \varepsilon_i$, and $s_\alpha \varepsilon_k = \varepsilon_k$ for $k \neq i, j$. The matrix for s_α is the permutation matrix corresponding to the transposition (ij). Remember here that $1 \leq i \neq j \leq l$ so that $s_\alpha \in P_l$. The corresponding permutation in C is $\pi = (ij)(\bar{i}\bar{j})$. If $\alpha = \varepsilon_i + \varepsilon_j$ then (39) gives $s_\alpha \varepsilon_i = -\varepsilon_j$, $s_\alpha \varepsilon_j = -\varepsilon_i$, and $s_\alpha \varepsilon_k = \varepsilon_k$ for $k \neq i, j$. The corresponding permutation in C is $\pi = (i\bar{j})(\bar{i}j)$. If $\alpha = 2\varepsilon_i$ then $s_\alpha \varepsilon_i = -\varepsilon_i$ and $s_\alpha \varepsilon_k = \varepsilon_k$ for $k \neq i$. The corresponding permutation in

C is $\pi = (i\bar{i})$. The permutation matrices of size l together with the diagonal matrices $\text{diag}(1, \dots, -1, \dots, 1)$ generate W when viewed as subgroup of $\mathbf{GL}(X)$ relative to the basis $\varepsilon_1, \dots, \varepsilon_l$. This is the assertion (37) in our present example. We omit the list of integers $2(\beta, \alpha)/(\alpha, \alpha)$, but some of them are given later in (4.24).

Let X be a real vector space of finite dimension, with positive definite inner product. Let $W \subset \mathbf{O}(X)$ be a finite group generated by reflections. Say that W is *crystallographic* if it satisfies the condition (40), called the *crystallographic restriction* [16, 2.8-2.9]. There are finite groups generated by reflections which are not crystallographic. For example if $\mathcal{P}$ is a regular n-gon in the Euclidean plane $\mathbf{R}^2$ then the symmetry group $W(\mathcal{P})$ is dihedral and hence generated by reflections, but $W(\mathcal{P})$ is crystallographic only for $n \in \{3, 4, 6\}$. If $\mathcal{P}$ is one of the regular polytopes in $\mathbf{R}^3$ then $W(\mathcal{P})$ is generated by reflections, but $W(\mathcal{P})$ is not crystallographic if $\mathcal{P}$ is the icosahedron or dodecahedron. In fact, a finite group generated by reflections is crystallographic if and only if it is the Weyl group of some reductive group. [30]

Now return to the context of roots in a reductive group. Since W permutes Φ and $s_\alpha \alpha = -\alpha$, it follows that if $\alpha \in \Phi$ then $-\alpha \in \Phi$. We will decompose the set of roots as a disjoint union $\Phi = \Phi^+ \cup \Phi^-$ in such a way that (i) $\alpha \in \Phi^+ \Leftrightarrow -\alpha \in \Phi^-$ and (ii) if $\alpha, \beta \in \Phi^+$ and $\alpha + \beta \in \Phi$ then $\alpha + \beta \in \Phi^+$. This may be done with an elementary geometric argument in the context of a real vector space X and finite subgroup of $\mathbf{GL}(X)$ generated by reflections [14, 4.1], [16, 1.3]. However, in the context of reductive groups, we must relate the decomposition $\Phi = \Phi^+ \cup \Phi^-$ to the structure of the group G. This depends on the notion of a Borel subgroup, which has been suppressed so far in this exposition, but is used in the proof of Theorem 4.1.

Let G be a connected reductive group. A *Borel subgroup* of G is a maximal element B in the set of closed connected solvable subgroups of G. Then B is maximal in the set of solvable subgroups of G [15, Corollary 23.1.A]. Borel subgroups exist and are uniquely determined up to conjugacy in G [3, Theorem 11.1], [15, Theorem 21.3]. [31] Let T be a maximal torus of G and let $\mathcal{B}^T$ denote the set of Borel subgroups of G which include T. Then $\mathcal{B}^T$ is finite, nonempty, and in fact [3, Proposition 11.19], [15, Proposition 24.1.A]

$$|\mathcal{B}^T| = |W(G,T)| \,. \tag{42}$$

If $B, B' \in \mathcal{B}^T$ then $B' = \omega B \omega^{-1}$ for some $\omega \in N$ and, for given B, the coset $w = \omega T$ is uniquely determined by B' [3, Proposition 11.19], [15, Proposition 24.1.A]. Since ωB and $B\omega$ depend only on w we may write $\omega B = wB$ and $B\omega = Bw$. Thus the Borel subgroups which include T have the form wBw^{-1} for some unique $w \in W$:

---

[30] It follows from (41) that the Weyl group of a reductive group is a crystallographic reflection group. For the converse, one must prove the existence of reductive groups. This can be done "case by case" using the classification of finite reflection groups. Such argument is of necessity long, since one must construct groups corresponding to the exceptional root systems E_6, E_7, E_8, F_4, G_2 [15, 33.6]. See [50, Chapter 12] for a case free argument.

[31] Borel subgroups may be defined in any connected algebraic group and are uniquely determined up to conjugacy [3, Theorem 11.1], [15, Theorem 21.3].

(43) $$\mathcal{B}^T = \{wBw^{-1} \mid w \in W\} \ .$$

Let $U = \mathcal{R}_u(B)$ be the unipotent radical of B. Since B is solvable, U is the set of unipotent elements of B, and B may be decomposed as a semidirect product [3, Theorem 10.6(4)], [15, Theorems 19.3(b) and 19.5]

(44) $$B = TU = T\mathcal{R}_u(B) \ .$$

Example 4.17 Let $G = \mathbf{GL}_n$ and $T = \mathbf{T}_n$ be as in Example 4.1. Then $B = \mathbf{B}_n$ is a Borel subgroup which includes T. We have remarked in Example 3.4 that $\mathcal{R}_u(\mathbf{B}_n) = \mathbf{U}_n$ where $\mathbf{U}_n$ is as in (13). Clearly $\mathbf{B}_n = \mathbf{T}_n\mathbf{U}_n$.

Example 4.18 Let $G = \mathbf{SL}_n$. Choose $T = G \cap \mathbf{T}_n$ as in Example 4.2. Then $B = G \cap \mathbf{B}_n$ is a Borel subgroup of G which includes T, and $U = \mathcal{R}_u(B) = G \cap \mathbf{U}_n$.

Example 4.19 Let $G = \mathbf{Sp}_n$ where $n = 2l$. Choose $T = G \cap \mathbf{T}_n$ as in Example 4.3. Define subgroups B_1, B_2 of G by

$$B_1 = \{ \begin{bmatrix} I & a \\ 0 & I \end{bmatrix} \mid a \in \mathbf{M}_l \text{ is symmetric } \}, \qquad B_2 = \{ \begin{bmatrix} b & 0 \\ 0 & (b^{-1})^{\mathsf{T}} \end{bmatrix} \mid b \in \mathbf{B}_l \ \} \ .$$

Then $B = B_1B_2$ is a Borel subgroup of $\mathbf{Sp}_n$ which includes T, and $U = \mathcal{R}_u(B) = B \cap \mathbf{U}_n$.

Let G be a connected reductive group, let T be a maximal torus, and let $B \supset T$ be a Borel subgroup. There exists a unique Borel subgroup $B^- \supset T$ such that $B \cap B^- = T$ [3, Theorem 14.1], [15, Theorem 26.2.B]. The Borel subgroups B, B^- are said to be *opposite*. For example if $G = \mathbf{GL}_n$ and $B = \mathbf{B}_n$ is the group of upper triangular matrices then B^- is the group $\mathbf{B}_n^-$ of lower triangular matrices. Define

(45) $$\Phi^+(B) = \{\alpha \in \Phi \mid U_\alpha \subset B\}, \qquad \Phi^-(B) = \{\alpha \in \Phi \mid U_\alpha \subset B^-\} \ .$$

The elements of $\Phi^+(B)$ are called *positive roots relative to B*; the elements of $\Phi^-(B)$ are called *negative roots relative to B*. In the rest of this paper we fix a Borel subgroup $B \supset T$ and write $\Phi^+ = \Phi^+(B)$ and $\Phi^- = \Phi^-(B)$. Elements of Φ^+ are called *positive roots*; elements of Φ^- are called *negative roots*. Then [3, Theorems 13.18(5)(a) and 14.1.III]

(46) $$\Phi = \Phi^+ \cup \Phi^- \text{ and } \Phi^- = -\Phi^+$$

where the union is disjoint. The group $X(T)$ may be given the structure of a totally ordered abelian group such that Φ^+ is the set of positive elements in Φ [3, Theorem 13.18(5)(d)], [15, 28.1]. Thus

(47) $$\alpha, \beta \in \Phi^+ \text{ and } \alpha + \beta \in \Phi \Rightarrow \alpha + \beta \in \Phi^+ \ .$$

Furthermore

$$U = \mathcal{R}_u(B) = \prod_{\alpha \in \Phi^+} U_\alpha \, . \tag{48}$$

The groups U_α in the product may be taken in any order [3, Proposition 14.4], [15, Proposition 28.1].

Example 4.20 Let $G = \mathbf{GL}_n$. Let B be as in Example 4.17. If $\alpha = \varepsilon_i - \varepsilon_j$ then $U_\alpha \subset B$ precisely when $i < j$. Thus $\Phi^+ = \{\varepsilon_i - \varepsilon_j \mid 1 \leq i < j \leq n\}$ and $\Phi^- = \{\varepsilon_j - \varepsilon_i \mid 1 \leq i < j \leq n\}$. Assertion (46) is clear. As for (47), if $\alpha = \varepsilon_i - \varepsilon_j \in \Phi^+$ and $\beta = \varepsilon_k - \varepsilon_l \in \Phi^+$ then $i < j$ and $k < l$ so $\alpha + \beta = \varepsilon_i - \varepsilon_j + \varepsilon_k - \varepsilon_l \in \Phi$ precisely when $j = k$, in which case $\alpha + \beta = \varepsilon_i - \varepsilon_l \in \Phi^+$ because $i < l$. Formula (48) asserts in this case that every upper unitriangular matrix may be written in the form $\prod_{1 \leq i < j \leq n}(I + c_{ij}E_{ij})$ for suitable $c_{ij} \in K$. The product may be taken in any order. Once the order is fixed, the coefficients c_{ij} are in fact uniquely determined. If $G = \mathbf{SL}_n$ and B is as in Example 4.18, the conclusions are exactly as in $\mathbf{GL}_n$.

Example 4.21 Let $G = \mathbf{Sp}_n$. Let B be as in Example 4.19. It follows from the shape of the matrices in (26)-(28) and the list Φ in Example 4.12 that

$$\Phi^+ = \{\varepsilon_i - \varepsilon_j \mid 1 \leq i < j \leq l\} \cup \{\varepsilon_i + \varepsilon_j \mid 1 \leq i < j \leq l\} \cup \{2\varepsilon_i \mid 1 \leq i \leq l\} \, .$$

Assertion (46) is clear. To verify (47), use $(\varepsilon_i - \varepsilon_j) + (\varepsilon_j - \varepsilon_k) = (\varepsilon_i - \varepsilon_k)$, $(\varepsilon_i - \varepsilon_j) + (\varepsilon_i + \varepsilon_j) = 2\varepsilon_i$ and $(\varepsilon_i - \varepsilon_j) + 2\varepsilon_j = \varepsilon_i + \varepsilon_j$. To verify (48) in this case, note that U has a semidirect product decomposition $U = U_1U_2$ analogous to the decomposition $B = B_1B_2$ in Example 4.19.

We say that a positive root is a *simple root* if it cannot be written as a sum of two positive roots [3, 14.7], [15, 27.3]. Let Δ denote the set of simple roots. Then Δ is a linearly independent subset of the real vector space X. Furthermore if $\alpha \in \Phi^+$ then α may be written uniquely as a $\mathbf{Z}$-linear combination of elements in Δ with non-negative coefficients [3, Corollary 14.8.1],

Example 4.22 Let $G = \mathbf{GL}_n$ with Φ^+ as in Example 4.20. Then

$$\Delta = \{\varepsilon_1 - \varepsilon_2, \, \varepsilon_2 - \varepsilon_3, \, \ldots, \, \varepsilon_{n-1} - \varepsilon_n\} \, .$$

This set is clearly linearly independent. If $\alpha = \varepsilon_i - \varepsilon_j \in \Phi^+$, so that $i < j$, then $\alpha = \sum_{k=i}^{j-1}(\varepsilon_k - \varepsilon_{k+1})$ is the desired linear combination. If $G = \mathbf{SL}_n$ the formulas are exactly as in $\mathbf{GL}_n$.

Example 4.23 Let $G = \mathbf{Sp}_n$ and let Φ^+ be as in Example 4.21. Then

$$\Delta = \{\varepsilon_1 - \varepsilon_2, \, \varepsilon_2 - \varepsilon_3, \, \ldots, \, \varepsilon_{l-1} - \varepsilon_l, \, 2\varepsilon_l\} \, .$$

The desired linear combinations may be found using the formulas (i) $\varepsilon_i - \varepsilon_j = \sum_{k=1}^{j-1}(\varepsilon_k - \varepsilon_{k+1})$ for $1 \leq i < j < l-1$, (ii) $\varepsilon_j + \varepsilon_l = (\varepsilon_j - \varepsilon_l) + 2\varepsilon_l$ for $1 \leq j < l$, (iii) $2\varepsilon_j = (\varepsilon_j - \varepsilon_l) + (\varepsilon_j + \varepsilon_l)$ for $1 \leq j < l$, (iv) $\varepsilon_i + \varepsilon_j = (\varepsilon_i - \varepsilon_j) + 2\varepsilon_j$ for $1 \leq i < j < l$.

It is understood in what follows that $\Phi = \Phi(G,T)$ is the set of roots of a reductive group, that Δ is the set of simple roots with respect to a chosen Borel subgroup $B \supset T$ and that $W = N/T$ is the Weyl group. We introduce the Cartan matrix, Coxeter diagram and Dynkin diagram. Let $\Delta = \{\alpha_1, \ldots, \alpha_l\}$ denote the set of simple roots. Thus $l = \dim \sum_{\alpha\in\Phi} \mathbf{R}\alpha$ is the semisimple rank of G. The *Cartan matrix* of Φ [5, 3.6], [15, p.230] is the $l \times l$ matrix with (i,j) entry

$$n_{ij} = \frac{2(\alpha_i,\alpha_j)}{(\alpha_j,\alpha_j)} = n_{\alpha_i,\alpha_j} \in \mathbf{Z}\ . \tag{49}$$

The n_{ij} are integers by (40). Clearly $n_{ii} = 2$. Clearly $n_{ij} = 0$ if and only if $n_{ji} = 0$. It happens that $n_{ij} \in \{0,-1,-2,-3\}$ and that $n_{ij}n_{ji} = \max\{|n_{ij}|,|n_{ji}|\}$. Thus if $n_{ij} \neq 0$ then at least one of n_{ij}, n_{ji} is equal to -1. In view of (38), the Cartan matrix determines the action of W on T and hence on $X(T)$.

Example 4.24 If $G = \mathbf{SL}_n$ with $l = n-1$ and Δ is as in (4.22), or $G = \mathbf{Sp}_n$ with $n = 2l$ and Δ is as in (4.23), then the Cartan matrices are given by

Cartan matrix

$\mathbf{SL}_n$

$$\begin{bmatrix} 2 & -1 & 0 & \cdots & 0 & 0 \\ -1 & 2 & -1 & \cdots & 0 & 0 \\ 0 & -1 & 2 & \cdots & 0 & 0 \\ \vdots & \vdots & \vdots & & \vdots & \vdots \\ 0 & 0 & 0 & \cdots & 2 & -1 \\ 0 & 0 & 0 & \cdots & -1 & 2 \end{bmatrix}$$

$\mathbf{Sp}_n$

$$\begin{bmatrix} 2 & -1 & 0 & \cdots & 0 & 0 \\ -1 & 2 & -1 & \cdots & 0 & 0 \\ 0 & -1 & 2 & \cdots & 0 & 0 \\ \vdots & \vdots & \vdots & & \vdots & \vdots \\ 0 & 0 & 0 & \cdots & 2 & -1 \\ 0 & 0 & 0 & \cdots & -2 & 2 \end{bmatrix}$$

The Dynkin diagram is a graph with vertex set $\Delta = \{\alpha_1, \ldots, \alpha_l\}$, no loops, and directed possibly multiple edges, defined as follows. Suppose $i \neq j$. If $n_{ij} = 0$ then $n_{ji} = 0$. The multiplicity of the edge connecting α_i to α_j is $n_{ij}n_{ji}$; if this number is 0 it is understood that there is no edge. If $(\alpha_i,\alpha_i) > (\alpha_j,\alpha_j)$ then the edge is directed from α_i to α_j. If $(\alpha_i,\alpha_i) = (\alpha_j,\alpha_j)$ then the edge is undirected. These diagrams, with undirected edges and slightly different notation for the multiplicities (see below) were first introduced by Coxeter [8] in his enumeration of finite subgroups of $\mathbf{GL}_l(\mathbf{R})$ generated by reflections; then Dynkin introduced them in the classification of simple Lie algebras over $\mathbf{C}$ [11, p.1]. Their ancestry may be traced to Schläfli who used a similar notation in his enumeration of the regular polytopes in $\mathbf{R}^l$. [32]

[32] The labeling of multiple edges in the Dynkin diagram exhibits the integers $n_{ij}n_{ji}$. The directed edges exhibit the n_{ij} as well. Tits [52, p.24, p.243] was the first to use the directed diagrams; " ... à un détail près, celui utilisé par Dynkin ... ". Tits called them "Figures de Schläfli" following a terminology introduced by Borel and de Siebenthal. Plus ça change, plus c'est la même chose.

Example 4.25 The diagrams for $\mathbf{SL}_n$ and $\mathbf{Sp}_n$ are

$\mathbf{SL}_n$ $\qquad\qquad$ $\mathbf{Sp}_n$

Now let's describe Coxeter's presentation of W as a *Coxeter group*, given in (53) below. Define integers m_{ij} for $1 \leq i, j \leq l$ as follows. Let $m_{ii} = 1$. For $i \neq j$ let m_{ij} be the order of $s_i s_j \in W$. Thus $m_{ij} = m_{ji}$. The $l \times l$ matrix with (i, j) entry equal to m_{ij} is called the *Coxeter matrix* [4, n° 1.9].

Example 4.26 The Coxeter matrices for the Weyl groups of $\mathbf{SL}_n$ and $\mathbf{Sp}_n$ are

Coxeter matrix

$\mathbf{SL}_n$ $\qquad\qquad$ $\mathbf{Sp}_n$

$$\begin{bmatrix} 1 & 3 & 2 \cdots & 2 & 2 \\ 3 & 1 & 3 \cdots & 2 & 2 \\ 2 & 3 & 1 \cdots & 2 & 2 \\ \vdots & \vdots & \vdots & \vdots & \vdots \\ 2 & 2 & 2 \cdots & 1 & 3 \\ 2 & 2 & 2 \cdots & 3 & 1 \end{bmatrix} \qquad \begin{bmatrix} 1 & 3 & 2 \cdots & 2 & 2 \\ 3 & 1 & 3 \cdots & 2 & 2 \\ 2 & 3 & 1 \cdots & 2 & 2 \\ \vdots & \vdots & \vdots & \vdots & \vdots \\ 2 & 2 & 2 \cdots & 1 & 4 \\ 2 & 2 & 2 \cdots & 4 & 1 \end{bmatrix}$$

The connection between the Coxeter matrix and the Cartan matrix is given by the formula

$$4\cos^2(\pi/m_{ij}) = n_{ij}n_{ji} \ . \tag{50}$$

The explicit values of the m_{ij} in terms of the $n_{ij}n_{ji}$ are given by the table

(51)

m_{ij}	2	3	4	6
$n_{ij}n_{ji}$	0	1	2	3

To get the Coxeter diagram from the Dynkin diagram proceed as follows: if $i \neq j$ and $n_{ij}n_{ji} > 1$, replace the multiple edge in the Dynkin diagram, and its arrow, by a single edge adorned with the integer m_{ij}. Thus the Coxeter diagram and Dynkin diagram for $\mathbf{SL}_n$ are identical and the Coxeter diagram for $\mathbf{Sp}_n$ is obtained from the Dynkin diagram by replacing its unique multiple edge by $\circ\overset{4}{\text{—}}\circ$.

Let $s_i = s_{\alpha_i} \in \mathbf{O}(X)$ be the reflection corresponding to $\alpha_i \in \Delta$. Let

$$S = \{s_1, \dots, s_l\} . \tag{52}$$

Then S generates W, as we have remarked in (37). The set S is called a set of *Coxeter generators* of W. It depends on Φ^+ and hence on choice of the Borel subgroup B. The s_i satisfy the relations

$$(s_i s_j)^{m_{ij}} = 1 \ , \ 1 \le i, j \le l \tag{53}$$

where $[m_{ij}]$ is the Coxeter matrix. It is a theorem of Coxeter that these relations (53) are a set of defining relations for W [8], [9, Theorem 64.26], [15, Theorem 29.4].

Example 4.27 Let W be the Weyl group of $\mathbf{SL}_n$. Then $\alpha_i = \varepsilon_i - \varepsilon_{i+1}$ for $1 \le i \le n-1$. It follows from the formula for s_α given in Example 4.15 that the permutation corresponding to the reflection s_i under the isomorphism $W \simeq S_n$ is the transposition $\tau_i = (i, i+1)$. These transpositions generate S_n. It was shown by E.H. Moore in 1897 that S_n has a presentation in terms of these transpositions with defining relations

$$\begin{array}{lll} \tau_i^2 & = 1 & 1 \le i \le n-1 \\ (\tau_i\tau_{i+1})^3 & = 1 & 1 \le i \le n-2 \\ (\tau_i\tau_j)^2 & = 1 & 1 \le i < j-1 \le n-1 \ . \end{array}$$

This is the presentation in (53) with $S = \{s_1, \dots, s_n\}$ and s_i in place of τ_i.

Let $w \in W$. Since S generates W we may write $w = s_{i_1} \cdots s_{i_p}$ where $s_{i_1}, \dots, s_{i_p} \in S$. Define the *length* of w to be the least such p:

$$l(w) = \min\{p \mid w = s_{i_1} \cdots s_{i_p}\}. \tag{54}$$

For example $l(1) = 0$ and $l(s_i) = 1$. This length function appears in various parts of the theory of reductive groups, from combinatorics to cohomology. In general, if a group is given by generators and relations, it is hard to determine the length of a group element as a word in the given set of generators. For the Weyl group we have the following combinatorial description of length due to Iwahori [17], [5, Theorem 2.2.2]. If $w \in W$ let

$$\Phi_w^- = \{\alpha \in \Phi^+ \mid w\alpha \in \Phi^-\} \tag{55}$$

be the set of positive roots mapped to negative roots by w. Then [17, Lemma 2.2], [9, Theorem 64.16(iii)]

$$l(w) = |\Phi_w^-| \ . \tag{56}$$

Example 4.28 Let $G = \mathbf{GL}_n$. If $\alpha \in \Phi^+$ then $\alpha = \varepsilon_i - \varepsilon_j$ where $i < j$. If $w \in W$ then $w\alpha = \varepsilon_{\pi i} - \varepsilon_{\pi j}$ where $\pi \in S_n$ is the corresponding permutation. If $w\alpha \in \Phi^-$ then $\pi i > \pi j$. Thus $n(w)$ is the number of pairs (i, j) with $1 \leq i < j \leq n$ and $\pi i > \pi j$. This is the number of ***inversions*** in the permutation π. Thus $l(w)$ is the number of inversions in the permutation $\pi \in S_n$ which corresponds to w. In particular, the permutation $\pi_0 : 123\cdots n \mapsto n \cdots 321$ has order 2 and is the unique permutation in S_n with the maximum number of inversions. The corresponding Weyl group element w_0 maps Φ^+ to Φ^- and is the unique element with this property. If $B = \mathbf{B}_n$ is our chosen Borel subgroup, then $w_0 B w_0$ is the group $\mathbf{B}_n^-$ of *lower* triangular matrices, the opposite Borel subgroup.

The remarks in Example 4.28 are a special case of a general theorem about finite reflection groups: If X is a real vector space and $W \subset \mathbf{GL}(X)$ is a finite group generated by reflections, then W contains a unique element w_0, called the ***opposition element***, with the property $w_0 \Phi^+ = \Phi^-$ [5, Proposition 2.2.6], [9, Theorem 64.16(vi)], [16, 1.8]. It follows from (56) that w_0 is the unique element in W of maximal length, and that if $w \in W$ maps no positive root into a negative root then $w = 1$. Since $\Phi^- = -\Phi^+$ we have $w_0^2 \Phi^+ = \Phi^+$, so w_0^2 maps no positive root into a negative root, and thus $w_0^2 = 1$. Since $w_0 \Phi^+ = \Phi^-$, it follows from (36), (45) and (48) that

$$w_0 B w_0 = B^- \tag{57}$$

is the Borel subgroup opposite to our chosen Borel subgroup B.

Let G be a connected reductive group. A subgroup of G is a ***parabolic subgroup*** [3, Corollary 11.2], [15, 29.3] if it includes some Borel subgroup. A parabolic subgroup is closed [3, 14.16], [15, 21.3] and connected [3, Theorem 11.16], [15, Corollary 23.1.B]. A parabolic subgroup is a ***standard parabolic subgroup*** [3, 14.17], [15, 30.1] if it includes our chosen Borel subgroup B.

Example 4.29 Let $G = \mathbf{GL}_n$. A ***composition*** or ***ordered partition*** γ of n, is a k-tuple $(\gamma_1, \ldots, \gamma_k)$ of positive integers with $\gamma_1 + \cdots + \gamma_k = n$ where $1 \leq k \leq n$. The γ_i are called the ***parts*** of γ. There are 2^{n-1} compositions of n. Let P_γ be the group of all block diagonal matrices of the shape

$$\begin{bmatrix} A_{11} & A_{12} & \ldots & A_{1k} \\ 0 & A_{22} & \ldots & A_{2k} \\ \vdots & \vdots & \ddots & \vdots \\ 0 & 0 & \ldots & A_{kk} \end{bmatrix} \tag{58}$$

where A_{ij} is a matrix of size $\gamma_i \times \gamma_j$ and A_{ii} is invertible. Then $P_\gamma \supseteq B$ so P_γ is a standard parabolic subgroup of G. It is a consequence of Theorem 4.2 that each standard parabolic subgroup has the form P_γ for some composition γ of n. Thus the mapping $\gamma \to P_\gamma$ is bijective from the set of compositions of n to the set of standard parabolic subgroups of G. In particular, G has 2^{n-1} standard parabolic subgroups.

Let's examine, in the example $G = \mathbf{GL}_n$, the connection between the standard parabolic subgroups of G and certain subgroups of the Weyl group W. To avoid notational clutter, we agree to identify W with the symmetric group and with the group of permutation matrices. Thus $W = S_n \subset \mathbf{GL}_n = G$. With this identification let $W_\gamma = W \cap P_\gamma$. The elements of W_γ have the form $\mathrm{diag}(\pi_1, \dots, \pi_k)$ where π_i is a permutation matrix of size γ_i so that $W_\gamma \simeq S_{\gamma_1} \times \cdots \times S_{\gamma_k}$. Thus the groups W_γ, when viewed as groups of permutations, are the *partition subgroups* or *Young subgroups* of S_n. It is an elementary but important fact that

$$P_\gamma = BW_\gamma B \tag{59}$$

where $B = \mathbf{B}_n$ is the group of upper triangular matrices. Let's see why this is so. Let $B^- = \mathbf{B}_n^-$ be the group of lower triangular matrices. Recall from linear algebra, that if $a \in \mathbf{M}_n$ and the process of Gaussian elimination has no zero pivots, then a is the product of a lower triangular matrix and an upper triangular matrix. Thus if $g \in G$ has no zero pivots then $g \in B^-B$. A generalization of this fact, which takes zero pivots into account, is the formula $G = B^-WB$. This is sometimes called the *Birkhoff decomposition* of G. As in Example 4.28 let $\pi_0 \in W$ be the permutation with the maximum number of inversions. Since $\pi_0 G = G$, $\pi_0 W = W$ and $\pi_0 B \pi_0 = B^-$, it follows from the Birkhoff decomposition that $G = BWB$. This is called the *Bruhat decomposition* of G, and is the desired formula (59) when $\gamma = (n)$ has a single part. The general formula may be deduced from this special case by a matrix calculation. For example suppose $k = 2$ and $\gamma = (\gamma_1, \gamma_2)$. Given A_{ij} of size $\gamma_i \times \gamma_j$ with A_{ii} invertible, and $1 \le i \le j \le 2$ we may use the Bruhat decomposition for matrices of size γ_1, γ_2 to write $A_{ii} = B_{ii}\pi_i B_{ii}'$ where B_{ii}, B_{ii}' are invertible upper triangular matrices of size γ_i and π_i is a permutation matrix of size γ_i. Then the equation

$$\begin{bmatrix} A_{11} & A_{12} \\ 0 & A_{22} \end{bmatrix} = \begin{bmatrix} B_{11} & B_{12} \\ 0 & B_{22} \end{bmatrix} \begin{bmatrix} \pi_1 & 0 \\ 0 & \pi_2 \end{bmatrix} \begin{bmatrix} B_{11}' & 0 \\ 0 & B_{22}' \end{bmatrix} \tag{60}$$

may be solved for a matrix B_{12} of size $\gamma_1 \times \gamma_2$.

To get the correct general formulation of these facts about $G = \mathbf{GL}_n$, let S be the set of Coxeter generators for W as in Example 4.27. If $I \subseteq S$ let W_I be the subgroup of W generated by I. If γ is a composition of n there exists a unique subset $I \subseteq S$ such that $W_I = W_\gamma$. Thus $I \mapsto BW_IB$ is a bijection between subsets of S and standard parabolic subgroups of $G = \mathbf{GL}_n$.

In 1962, J.Tits [53] introduced the notion of a group G with BN-pair. His intent, inspired by arguments of Chevalley [7, Théorème 2], was to study certain properties of the parabolic subgroups of a reductive group from an axiomatic point of view. The axioms are few and the consequences are many. The axioms concern a group G, two subgroups B, N and a set S of generators for $N/B \cap N$. The given data G, B, N, S together with the prescribed axioms are now called a *Tits system* [3, 14.15], [9, 65.1], [15, 29.1]. The axioms are satisfied in case G is a reductive group, B is a Borel subgroup, N is the normalizer of a maximal torus $T \subset B$, and S is the set of Coxeter generators for the Weyl group $W = N/T$ [3, 14.15], [15, 29.1]. They have had profound impact on the theory of groups, in particular on the classification of finite simple groups, and on geometry. They have also influenced the theory of reductive monoids see for example [32], [43], [48], and Section 5 of this paper.

Here are the axioms for a Tits system $\mathcal{T} = (G, B, N, S)$. Let G be a group and let B, N be two subgroups which generate G. Let $T = B \cap N$. Assume that $T \lhd N$ and that $W := N/T$ is generated by a set $S = \{s_1, \dots, s_l\}$ of involutory elements, assumed distinct. If $\omega \in N$ represents $w \in W$ we write $\omega B = wB$ and $B\omega = Bw$ as in (57). The group W is called the *Weyl group* of $\mathcal{T}$. Assume that

$$\text{(61)} \qquad \begin{aligned} &\text{Axiom 1: If } s \in S \text{ and } w \in W \text{ then } sBw \subseteq BwB \cup BswB\ . \\ &\text{Axiom 2: If } s \in S \text{ then } sBs \neq B\ . \end{aligned}$$

Axiom 2 excludes, for example, the possibility that $G = S_n$, $B = \{1\}$, and $N = S_n$. It takes some work to check Axiom 1 even in the prototype $G = \mathbf{GL}_n$, $B = \mathbf{B}_n$, $N = \mathbf{N}_n$ [9, Theorem 65.10]. It is remarkable how much follows from these Axioms. For example:

Theorem 4.2 ([5, 8.3], [9, §65], [15, Theorem 29.3]) *Let $\mathcal{T} = (G, B, N, S)$ be a Tits system. If $I \subseteq S$ let W_I be the subgroup of W generated by I. Then*
(1) $P_I = BW_IB$ is a group.
(2) P_I is equal to its normalizer in G.
(3) If $I \neq J$ then P_I and P_J are not conjugate in G; in particular $P_I \neq P_J$.
(4) $P_I \cap P_J = P_{I \cap J}$.
(5) The map $I \to P_I$ is bijective from the set of subsets of S to the set of subgroups of G which include B.

Since the subgroups P_I are the standard parabolic subgroups of G, the subgroups W_I are often called ***standard parabolic subgroups*** of W. Thus we have a one-to-one correspondence between standard parabolic subgroups P_I of G and standard parabolic subgroups W_I of W. The index set I is called the *type* of the parabolic subgroup P_I. It follows from Theorem 4.2(1) with $I = S$ that

$$\text{(62)} \qquad G = BWB = \bigsqcup_{w \in W} BwB \quad \text{and} \quad BwB = Bw'B \Rightarrow w = w'.$$

This double coset decomposition is called the *Bruhat decomposition* of G. As in (54) we define the length $l(w)$ for $w \in W$ to be the least integer p such that w may be written as a word of length p in the generating set S. It follows from the axioms that if $s \in S$ and $w \in W$ then $l(sw) = l(w) \pm 1$ and [15, Lemma 29.3.A]

$$\text{(63)} \qquad BsB \cdot BwB = \begin{cases} BswB & \text{if } l(sw) = l(w) + 1 \\ BswB \sqcup BwB & \text{if } l(sw) = l(w) - 1\ . \end{cases}$$

We emphasize that Theorem 4.2 and formulas (62) and (63) apply to any reductive group G where G, B, N, S have their usual meaning. In this case Theorem 4.2 shows that every standard parabolic subgroup of G, as defined prior to Example 4.29 has the form P_I for some subset I of the set S of Coxeter generators. In the context of reductive groups we often use the same index I for the corresponding subset $\{\alpha \in \Delta \mid s_\alpha \in I\}$ of the set of simple roots. If $G = \mathbf{GL}_n$ the correspondence $I \mapsto P_I$ is given explicitly in Example 4.29. See (122) for a monoid analogue of (63).

We have already remarked that a parabolic subgroup P of G is closed and connected. It is not in general reductive. It splits as a semidirect product

$$P = L\mathcal{R}_u(P) \tag{64}$$

where L is a closed connected reductive subgroup called a *Levi subgroup* or *Levi factor* of P [3, 11.22, Corollary 14.19], [15, Theorem 30.2]. The Levi factor is a connected reductive group [3, 11.22]. It is not in general unique, but any two Levi factors are conjugate by an element of $\mathcal{R}_u(P)$ [3, Proposition 14.21], [15, Theorem 30.2].

Example 4.30 Let $G = \mathbf{GL}_n$ and let $P = P_\gamma$ be as in Example 4.29. Then

$$\mathcal{R}_u(P) = \{ \begin{bmatrix} I & A_{12} & \dots & A_{1k} \\ 0 & I & \dots & A_{2k} \\ \vdots & \vdots & \ddots & \vdots \\ 0 & 0 & \dots & I \end{bmatrix} \}, \qquad L = \{ \begin{bmatrix} A_{11} & 0 & \dots & 0 \\ 0 & A_{22} & \dots & 0 \\ \vdots & \vdots & \ddots & \vdots \\ 0 & 0 & \dots & A_{kk} \end{bmatrix} \}$$

where A_{ij} has size $\gamma_i \times \gamma_j$. Although L is not uniquely determined, the L pictured here is the most natural choice. Note that $L \simeq \mathbf{GL}_{\gamma_1} \times \cdots \times \mathbf{GL}_{\gamma_k}$.

Two parabolic subgroups of a connected reductive group G are said to be *opposite parabolic subgroups* if their intersection is a Levi subgroup of both [3, 14.20]. This generalizes the notion of opposite Borel subgroups B, B^- in which case $B = T\mathcal{R}_u(B)$ as in (44), $B^- = T\mathcal{R}_u(B^-)$ and $B \cap B^- = T$ is the common Levi subgroup. If P is a parabolic subgroup with Levi factor L there is a unique opposite parabolic subgroup P^- with Levi factor L [3, Proposition 14.21]. If $P = P_\gamma$ is as in Example 4.29 then the group P_γ^- consisting of transposes of elements in P_γ is an opposite parabolic subgroup with common Levi subgroup pictured in (4.30).

Now return to Example 3.6 and re-examine it in current terms. We have a group $G = \mathbf{GL}_n$ and an idempotent $e = e_r$ in the Zariski closure $M = \overline{G} = \mathbf{M}_n$. Let γ be the composition $\gamma = (r, n-r)$ of n. Then the right centralizer $P(e) = P_\gamma$ of e in $\mathbf{GL}_n$ is a maximal parabolic subgroup of $\mathbf{GL}_n$, and the left centralizer $P^-(e) = P_\gamma^-$ of e in $\mathbf{GL}_n$ is also a maximal parabolic subgroup of $\mathbf{GL}_n$. These two parabolic subgroups are opposite. The common Levi subgroup is $L(e) = P(e) \cap P^-(e)$. It is this elementary observation which connects the idempotents in $\mathbf{M}_n$ to the group structure of $\mathbf{GL}_n$ and allows one to build a theory of reductive monoids on the theory of reductive groups.

5. Reductive Monoids

Let M be a reductive monoid with unit group G. Thus G is by definition a connected reductive group. In the theory of connected reductive groups, a main guiding principle is to reduce problems, as far as possible, to a study of the Weyl group W. In the theory of reductive monoids, the analogous guiding principle is to reduce to a study of W and its action on $E(\overline{T})$, where T is a maximal torus. Since T is conjugate to a subgroup of $\mathbf{T}_n$ [3, Theorem 8.5] we may assume that $T \subseteq \mathbf{T}_n$. Thus $\overline{T} \subseteq \mathbf{D}_n$, the monoid of diagonal matrices.

The group G acts on M by conjugation: $a \mapsto gag^{-1}$ for $a \in M$ and $g \in G$. The set $E(M)$ is stable under conjugation. The first main problem is to determine the G-orbits on $E(M)$. This is simplified by Theorem 5.2 which shows that every G-orbit on $E(M)$ meets $E(\overline{T})$. Theorems 5.2 and 5.3 allow us to reduce the study of $E(M)$ to that of W on $E(\overline{T})$. Theorem 5.1 shows that it suffices to prove Theorem 5.2 in case G is solvable. Note that Theorem 5.1 and Theorem 5.2 concern an irreducible algebraic monoid M which is not assumed to be reductive.

Theorem 5.1 ([23, Theorem 1.4], [31, Proposition 6.3]) *Let M be an irreducible algebraic monoid with unit group G. Let $B \subseteq G$ be a Borel subgroup of G. Then*

$$M = \bigcup_{g \in G} g\overline{B}g^{-1}.$$

Theorem 5.2 ([23, Corollary 1.6], [31, Corollary 6.10]) *Let M be an irreducible algebraic monoid with unit group G. Let $T \subseteq G$ be a maximal torus. Then*

$$E(M) = \bigcup_{g \in G} gE(\overline{T})g^{-1} .$$

Example 5.1 Suppose $M = \mathbf{M}_n$ so that $G = \mathbf{GL}_n$. Let $B = \mathbf{B}_n$ be the group of invertible upper triangular matrices, let $T = \mathbf{T}_n$ be the group of invertible diagonal matrices, and let $\mathbf{D}_n$ be the monoid of diagonal matrices. Theorem 5.1 asserts that every matrix is conjugate under $\mathbf{GL}_n$ to an element of $\mathbf{B}_n$. Theorem 5.2 asserts that every idempotent matrix is conjugate under $\mathbf{GL}_n$ to an element of $\mathbf{D}_n$.

Assume now that M is a reductive monoid with group of units G. As in Section 4, choose a maximal torus T and Borel subgroup B of G with $T \subset B$. Let $N = N_G(T)$ be the normalizer of T and let W be the Weyl group. If $\omega \in N$ and $e \in E(\overline{T})$, then $\omega e \omega^{-1} \in \overline{T}$. If $w = \omega T \in W$ then $\omega e \omega^{-1}$ depends only on w so we may define $wew^{-1} = \omega e \omega^{-1}$. Thus W acts, by conjugation, on $E(\overline{T})$.

Theorem 5.3 ([31, Theorem 6.25]) *Let M be a reductive monoid. If $e, f \in E(\overline{T})$ are conjugate under G they are conjugate under W.* [33]

Example 5.2 Suppose $M = \mathbf{M}_n$, $G = \mathbf{GL}_n$ and $T = \mathbf{T}_n$. Suppose $e, f \in E(\overline{T})$. Write $e = e_I$ and $f = e_J$ as in Example 2.1 where I, J are subsets of $\{1, \ldots, n\}$. If e, f are conjugate under G then $|I| = |J|$. Choose a permutation $\pi \in S_n$ such

[33] This theorem is true for any irreducible algebraic monoid. The Weyl group $W(G,T)$ is by definition $N_G(T)/Z_G(T)$; see the footnote which follows (32).

that $\pi I = J$. If $w \in W$ corresponds to π under the isomorphism $W \simeq S_n$ then $f = wew^{-1}$.

In view of Theorems 5.2, 5.3 and 3.1 we have bijections

$$G\backslash M/G = \mathcal{U}(M) \leftrightarrow E(M)/G \leftrightarrow E(\overline{T})/W \tag{65}$$

given by

$$GeG \leftrightarrow \{geg^{-1} \mid g \in G\} \leftrightarrow \{wew^{-1} \mid w \in W\} \tag{66}$$

for $e \in E(\overline{T})$ where $E(M)/G$ denotes the set of G-conjugacy classes in $E(M)$ and $E(\overline{T})/W$ denotes the set of W-conjugacy classes in $E(\overline{T})$. This gives us an explicit description of the $\mathcal{J}$-classes, alias $G \times G$ orbits, provided we can compute $E(\overline{T})$ and the Weyl group action.

In Example 2.3 we computed some idempotents in $E(\overline{T})$ for a monoid $M = M(\rho)$ defined by a representation ρ of $\mathbf{SL}_m$. At that time we did not have the terms "reductive group" and "maximal torus" at our disposal. Now we can put that computation in its proper context. It is a remarkable fact that if M is any reductive monoid with unit group G and T is a maximal torus of G then $E(\overline{T})$ is anti-isomorphic, by a W-equivariant map, to the lattice of faces of a rational convex cone. If $G = K^*\rho(G_0)$ where G_0 is a semisimple group and $\rho : G_0 \to \mathbf{GL}(V) = \mathbf{GL}_n$ is a rational representation then the cone may be replaced by a rational convex polytope.

To see how this comes about we need some general remarks about one-parameter subgroups of tori. Let G be a connected reductive group of rank r and let T be a maximal torus of G. A *one-parameter subgroup* of T is a homomorphism $\lambda : K^* \to T$ of algebraic groups [15, 16.1], [3, 8.6]. For example if $a_1, \dots, a_n \in \mathbf{Z}$ then the map $K^* \to \mathbf{T}_n$ defined in (5) by $t \mapsto \mathrm{diag}(t^{a_1}, \dots, t^{a_n})$ is a one-parameter subgroup of $\mathbf{T}_n$. Let $X_*(T)$ denote the set of one-parameter subgroups of T. Since T is abelian, $X_*(T)$ has the structure of abelian group with product defined by $(\lambda\lambda')(t) = \lambda(t)\lambda'(t)$ for $t \in K^*$. The group $X_*(T)$ is free abelian of rank $r = \dim T$. If T' is a subtorus of T there is a natural injection $X_*(T') \hookrightarrow X_*(T)$ which we view as inclusion. As with $X(T)$, we often use additive notation and write $\lambda + \lambda'$ rather than $\lambda\lambda'$. Let $X_* = X_*(T) \otimes \mathbf{R}$. We identify $X_*(T)$ with a $\mathbf{Z}$-submodule of the real vector space X_* via $\lambda = \lambda \otimes 1$. Then $\dim_{\mathbf{R}} X_* = \dim T$. If $\chi \in X(T)$ and $\lambda \in X_*(T)$ then $\chi \circ \lambda : K^* \to K^*$ is a rational homomorphism and thus has the form $t \mapsto t^a$ for some $a \in \mathbf{Z}$ and all $t \in K^*$. Write $\langle \chi, \lambda \rangle = a$. Thus we have a $\mathbf{Z}$-bilinear dual pairing [3, Proposition 8.6]

$$\langle\ ,\ \rangle : X(T) \times X_*(T) \to \mathbf{Z} \tag{67}$$

defined by

$$(\chi \circ \lambda)(t) = t^{\langle \chi, \lambda \rangle} \tag{68}$$

for $\chi \in X(T)$, $\lambda \in X_*(T)$ and $t \in K^*$ This pairing extends to a nondegenerate $\mathbf{R}$-bilinear pairing $X \times X_* \to \mathbf{R}$ of vector spaces where $X = X(T) \otimes \mathbf{R}$. Define a $\mathbf{Z}$-linear action of W on $X_*(T)$ as follows. If $w = \omega T$ with $\omega \in N(T)$ and $\lambda \in X_*(T)$ set

$$(w\lambda)(t) = \omega\lambda(t)\omega^{-1} \tag{69}$$

for $t \in K^*$. Then $w\lambda \in X_*(T)$ and

$$\langle w\chi, w\lambda\rangle = \langle \chi, \lambda\rangle \tag{70}$$

for $w \in W$, $\chi \in X(T)$ and $\lambda \in X_*(T)$. The action of W on $X_*(T)$ extends to an $\mathbf{R}$-linear action on X_* such that (70) holds for $\chi \in X$ and $\lambda \in X_*$. Let $\mathbf{Z}^n_*$ denote the set of *row* vectors over $\mathbf{Z}$.

Example 5.3 Suppose $G = \mathbf{GL}_n$ and $T = \mathbf{T}_n$. If $\mathbf{a} = (a_1, \dots, a_n) \in \mathbf{Z}^n_*$ define $\lambda^{\mathbf{a}} \in X_*(\mathbf{T}_n)$ by

$$\lambda^{\mathbf{a}}(t) = \operatorname{diag}(t^{a_1}, \dots, t^{a_n}) \tag{71}$$

for $t \in K^*$. The map $\mathbf{a} \mapsto \lambda^{\mathbf{a}}$ is an isomorphism $\mathbf{Z}^n_* \to X_*(\mathbf{T}_n)$ of abelian groups. For $1 \le i \le n$ define $\eta_i \in X_*(\mathbf{T}_n)$ by $\eta_i(t) = \operatorname{diag}(1, \dots, t, \dots, 1)$ with t in the i-th position. Then $\lambda^{\mathbf{a}} = \eta_1^{a_1} \cdots \eta_n^{a_n}$ and $\{\eta_1, \dots, \eta_n\}$ is a basis for $X_* = X_*(\mathbf{T}_n) \otimes \mathbf{R}$. Let $\varepsilon_j \in X(\mathbf{T}_n)$ be as in (9). Since $\varepsilon_j(\eta_i(t)) = t^{\delta_{ij}}$ we have $\langle \varepsilon_j, \eta_i\rangle = \delta_{ij}$. Thus $\{\varepsilon_1, \dots, \varepsilon_n\}$ and $\{\eta_1, \dots, \eta_n\}$ are dual bases. The Weyl group actions on $X(T)$ and $X_*(T)$ are given by $w\varepsilon_j = \varepsilon_{\pi j}$ and $w\eta_i = \eta_{\pi i}$ where $\pi \in S_n$ corresponds to $w \in W$.

In general we have a commutative diagram

$$\begin{array}{ccc} T & \to & \mathbf{T}_n \\ \downarrow & & \downarrow \\ \overline{T} & \to & \mathbf{D}_n \end{array} \tag{72}$$

where all the maps are inclusions. A *rational character* of $\overline{T}$ is a morphism $\overline{T} \to K^\times$ of algebraic monoids where $K^\times$ denotes the field K viewed as monoid under multiplication. The set $X(\overline{T})$ of rational characters of $\overline{T}$ is a commutative monoid under multiplication. Restriction defines a monoid homomorphism $X(\overline{T}) \to X(T)$ which is injective [31, p.80]. If $\chi \in X(T)$ extends to a character of $\overline{T}$, let $\overline{\chi}$ denote the unique extension. We have an injection

$$X(\mathbf{D}_n) = \mathbf{N}\overline{\varepsilon}_1 \oplus \cdots \oplus \mathbf{N}\overline{\varepsilon}_n \to \mathbf{Z}\varepsilon_1 \oplus \cdots \oplus \mathbf{Z}\varepsilon_n = X(\mathbf{T}_n) \tag{73}$$

where $\mathbf{N}$ is the set of non-negative integers and ε_j is the j-th coordinate function on $\mathbf{T}_n$. It is tempting to write ε_j rather than $\overline{\varepsilon}_j$ on the left side of (73) and view the map (73) as inclusion, but it is in fact restriction. Let χ_j denote the restriction of ε_j to T and let $\overline{\chi}_j$ denote the restriction of $\overline{\varepsilon}_j$ to $\overline{T}$. We have a commutative diagram

$$\begin{array}{ccc} X(\mathbf{D}_n) & \to & X(\overline{T}) \\ \downarrow & & \downarrow \\ X(\mathbf{T}_n) & \to & X(T) \end{array} \qquad \begin{array}{ccc} \overline{\varepsilon}_j & \mapsto & \overline{\chi}_j \\ \downarrow & & \downarrow \\ \varepsilon_j & \mapsto & \chi_j \end{array} \tag{74}$$

where all the maps are restrictions. We have already remarked that the vertical maps are injective. The horizontal map in the bottom row is surjective by [3, Proposition 8.2(c)]. The horizontal map in the top row is also surjective; this is shown in the proof of Lemma 2.2 in [23]. The inclusion $T \subseteq \mathbf{T}_n$ thus provides us with a set $\{\overline{\chi}_1, \dots, \overline{\chi}_n\}$ of generators for the monoid $X(\overline{T})$ of characters of $\overline{T}$ and we have an injection

$$X(\overline{T}) = \mathbf{N}\overline{\chi}_1 + \cdots + \mathbf{N}\overline{\chi}_n \to \mathbf{Z}\chi_1 + \cdots + \mathbf{Z}\chi_n = X(T) \ . \tag{75}$$

Since the coordinate ring of $\mathbf{D}_n$ is $K[\mathbf{D}_n] = K[\overline{\varepsilon}_1, \dots, \overline{\varepsilon}_n]$ we see that the coordinate ring of $\overline{T}$ is $K[\overline{T}] = K[\overline{\chi}_1, \dots, \overline{\chi}_n]$. Note that $\mathbf{N}\chi_1 + \cdots + \mathbf{N}\chi_n \subseteq X(T)$ is the set of characters of T which may be extended to characters of $\overline{T}$. Define a cone σ in X_* by

$$\sigma = \{\lambda \in X_* \mid \langle \chi_j, \lambda \rangle \geq 0 \text{ for } 1 \leq j \leq n\} \ . \tag{76}$$

The cone σ is the key to the structure of $E(\overline{T})$. To see why this might be so, suppose that $\mathbf{a} = (a_1, \dots, a_n) \in \mathbf{Z}^n_*$ and let $\lambda^{\mathbf{a}}$ be as in (71). If $\lambda^{\mathbf{a}} \in \sigma$ then $\langle \chi_j, \lambda^{\mathbf{a}} \rangle \geq 0$ for all $j = 1, \dots, n$ so $\mathbf{a} \in \mathbf{N}^n_*$. Thus we may set $t = 0$ in $\operatorname{diag}(t^{a_1}, \dots, t^{a_n})$. This gives an element of $\overline{T} \subseteq \mathbf{D}_n$, which is suggestively written as $\lambda(0)$ or $\lim_{t \to 0} \lambda(t)$. Since the diagonal entries of $\lambda(0)$ are 0 or 1 we have

$$\lambda(0) = \lim_{t \to 0} \lambda(t) \in E(\overline{T}) \ . \tag{77}$$

This is the construction we used in Example 2.3.

To proceed further, we need some elementary facts about convex polyhedral cones. Fulton [12, Section 1.2] gives a summary of these facts, with proofs, which is made to order for the present setup. Let M, N be free (additively written) abelian groups of rank r. Assume there is given a $\mathbf{Z}$-bilinear dual pairing $M \times N \to \mathbf{Z}$. Imbed $M \subseteq M_{\mathbf{R}} = M \otimes \mathbf{R}$, imbed $N \subseteq N_{\mathbf{R}} = N \otimes \mathbf{R}$, and extend the pairing to a nondegenerate $\mathbf{R}$-bilinear pairing $M_{\mathbf{R}} \times N_{\mathbf{R}} \to \mathbf{R}$. Let $\mathbf{R}^+$ be the set of non-negative real numbers. A *convex polyhedral cone* in $N_{\mathbf{R}}$ is a set of the form

$$\sigma = \mathbf{R}^+\nu_1 + \cdots + \mathbf{R}^+\nu_s \tag{78}$$

where $\nu_1, \dots, \nu_s \in N_{\mathbf{R}}$ are called *generators* of σ. We usually omit "convex polyhedral" and simply say that σ is a *cone*. Then

(79) $$\sigma^\vee = \{\chi \in M_{\mathbf{R}} \mid \langle \chi, \lambda \rangle \geq 0 \text{ for } \lambda \in \sigma\}$$

is a cone in $M_{\mathbf{R}}$ called the *dual* of σ [12, 1.2.9]. We have $(\sigma^\vee)^\vee = \sigma$ [12, 1.2.1]. Since we may interchange $N_{\mathbf{R}}$ and $M_{\mathbf{R}}$ in (78) and (79), it follows that if σ is a cone in $N_{\mathbf{R}}$ then there exist $\chi_1, \ldots, \chi_n \in M_{\mathbf{R}}$ such that

(80) $$\sigma = \{\lambda \in N_{\mathbf{R}} \mid \langle \chi_j, \lambda \rangle \geq 0 \text{ for } 1 \leq j \leq n\} .$$

Let $\mathcal{F}(\sigma)$ denote the set of faces of σ partially ordered by inclusion. The poset $\mathcal{F}(\sigma)$ is finite [12, 1.2.2]. If $\tau \in \mathcal{F}(\sigma)$ then τ is itself a cone [12, 1.2.2]. If $\tau, \tau' \in \mathcal{F}(\sigma)$ then $\tau \cap \tau' \in \mathcal{F}(\sigma)$ [12, 1.2.3]. Thus $\mathcal{F}(\sigma)$ is a finite lattice. The minimal element of $\mathcal{F}(\sigma)$ is $\sigma \cap (-\sigma)$. A cone σ is *strongly convex* if $\sigma \cap (-\sigma) = \{0\}$ or, equivalently, if $\sigma^\vee$ spans $M_{\mathbf{R}}$ [12, 1.2.13]. If σ is strongly convex and different from $\{0\}$ it has a minimal set $\{\nu_1, \ldots, \nu_s\}$ of generators which are unique up to multiplication by positive real numbers; the rays $\mathbf{R}^+\nu_i$ are the one-dimensional faces of σ [12, 1.2.13].

A convex polyhedral cone σ in $N_{\mathbf{R}}$ is by definition *rational* if it has the form (80) where $\chi_1, \ldots, \chi_n \in M$ [12, p.12]. If σ is rational then $\sigma^\vee$ is rational [12, p.12]. If σ is rational then it may be written in the form (78) where $\nu_i \in N$ for $i = 1, \ldots, s$. Thus if σ is rational then $\sigma = \mathbf{R}^+(\sigma \cap N)$. A face of a rational cone is rational [12, p.13]. Let relint(σ) denote the relative interior of σ. If σ is rational then relint$(\sigma) \cap N$ is non-empty. Thus, since a face of a rational cone is rational, relint$(\tau) \cap N$ is non-empty for every $\tau \in \mathcal{F}(\sigma)$.

Henceforth we assume that $M = X(T)$ and $N = X_*(T)$ where T is a maximal torus in a connected reductive group G. Thus $M_{\mathbf{R}} = X$ and $N_{\mathbf{R}} = X_*$ in our earlier notation. Let σ be as in (76). Since $\sigma = (\sigma^\vee)^\vee$ it follows from (76) that

(81) $$\sigma^\vee = \mathbf{R}^+\chi_1 + \cdots + \mathbf{R}^+\chi_n .$$

In particular, we see that $\sigma^\vee$ spans X so σ is strongly convex and thus the minimal element of $\mathcal{F}(\sigma)$ is $\{0\}$. Suppose $\tau \in \mathcal{F}(\sigma)$. Since τ is a rational cone the set relint$(\tau) \cap X_*(T)$ is non-empty. Suppose that $\lambda, \lambda' \in$ relint$(\tau) \cap X_*(T)$. If $1 \leq j \leq n$ then $\langle \chi_j, \lambda \rangle > 0$ if and only if $\langle \chi_j, \lambda' \rangle > 0$. Thus $\lambda(0) = \lambda'(0)$ depends only on τ. Let $e_\tau \in E(\overline{T})$ be the common value of the $\lambda(0)$ as λ ranges over relint$(\tau) \cap X_*(T)$. Note that $T\overline{T} \subseteq \overline{T}$ so that T acts by multiplication on $\overline{T}$. The following theorem describes the orbits for this action, and the orbit closures, in terms of $\mathcal{F}(\sigma)$. [34]

[34] Theorem 5.4 is part of the theory of toric varieties [18, Theorem 2] and [12, Section 3.1]. If the toric variety is $\overline{T}$, the word "idempotent" is implicit, but not explicit in the general theory. In the early stages of his work on algebraic monoids Putcha [23], [31, Chapter 8] observed the connection between idempotents in $\overline{T}$ and rational convex polytopes; he proved what he needed from first principles. Renner [39, Chapter 4], [41, Section 3] remarked that the structure of $E(\overline{T})$ follows from known facts about toric varieties. A reader who checks these references should note that our $\sigma \subseteq X_*(T) \otimes \mathbf{R}$ and $\sigma^\vee \subseteq X(T) \otimes \mathbf{R}$ agree with [18] and [12] but do not agree with [41, p.201] where $\sigma \subseteq X(T) \otimes \mathbf{R}$ means the dual cone.

Theorem 5.4 **([18, Theorem 2(c,d)])** *Let $T \subseteq \mathbf{T}_n$ be a torus and let $\chi_1, \cdots, \chi_n \in X(T)$ be the restrictions of the coordinate functions on $\mathbf{T}_n$ to T. Let*

$$\sigma = \{\lambda \in X_* \mid \langle \chi_j, \lambda \rangle \geq 0 \text{ for } 1 \leq j \leq n\} .$$

If $\tau \in \mathcal{F}(\sigma)$ let $e_\tau \in E(\overline{T})$ be the corresponding idempotent. Then (1) the map $\tau \mapsto Te_\tau$ is bijective from $\mathcal{F}(\sigma)$ to the set of T-orbits on $\overline{T}$, and (2) if $\tau, \tau' \in \mathcal{F}(\sigma)$ then $\tau \subseteq \tau'$ if and only if $\overline{Te_\tau} \supseteq Te_{\tau'}$.

It follows from (1) that each T-orbit on $\overline{T}$ contains an idempotent e. This idempotent is unique. For suppose $te \in E(\overline{T})$ for some $t \in T$. Since e is idempotent we have $\varepsilon_j(e) \in \{0, 1\}$ for $1 \leq j \leq n$. In either case $\varepsilon_j(te) = \varepsilon_j(e)$ so $te = e$. Thus e is unique. Suppose that $\tau, \tau' \in \mathcal{F}(\sigma)$ and $\tau \subseteq \tau'$. The set $\{a \in \overline{T} \mid ae_\tau = a\}$ is closed and includes Te_τ because e_τ is idempotent, so it includes $\overline{Te_\tau}$ and hence contains $e_{\tau'}$ by (2). Thus $e_{\tau'} e_\tau = e_{\tau'}$ so that $e_\tau \geq e_{\tau'}$ in the partial order (3) on $E(\overline{T})$. Thus

Corollary 5.5 *The map $\tau \to e_\tau$ is a lattice anti-isomorphism from $\mathcal{F}(\sigma)$ to $E(\overline{T})$.*

We can give a more explicit formula for e_τ. Choose $\lambda_\tau \in X_*(T) \cap \operatorname{relint}(\tau)$. Since $\lambda_\tau(t) = \operatorname{diag}(t^{\langle \chi_1, \lambda_\tau \rangle}, \ldots, t^{\langle \chi_n, \lambda_\tau \rangle})$,

$$e_\tau = \sum_{\langle \chi_j, \lambda_\tau \rangle = 0} E_{jj} . \tag{82}$$

The vertex $\{0\}$ of σ corresponds to the idempotent $1 \in E(\overline{T})$. The one-dimensional faces of σ correspond to the maximal idempotents of $E(\overline{T})$. Since every idempotent is a product of maximal idempotents, it suffices, in computing $E(\overline{T})$, to find the idempotents corresponding to the one-dimensional faces of σ.

The group W permutes $\mathcal{F}(\sigma)$ and $\mathcal{F}(\sigma^\vee)$. To see this, note that the inclusion map $T \to \mathbf{T}_n$ is a representation of T with weights $\chi_1, \cdots, \chi_n$ where χ_j is the restriction to T of the j-th coordinate function ε_j on $\mathbf{T}_n$. It follows from (34) that W permutes $\{\chi_1, \cdots, \chi_n\}$. Thus, from (81) we have $W\sigma^\vee = \sigma^\vee$. If $\lambda \in \sigma$ and $w \in W$ then, by (76) and (70) we have $\langle \chi_j, w\lambda \rangle = \langle w^{-1}\chi_j, \lambda \rangle \geq 0$ so $w\lambda \in \sigma$. Thus $W\sigma = \sigma$. It follows that

$$W\mathcal{F}(\sigma) = \mathcal{F}(\sigma) \text{ and } W\mathcal{F}(\sigma^\vee) = \mathcal{F}(\sigma^\vee) . \tag{83}$$

Suppose $\tau \in \mathcal{F}(\sigma)$ and $w \in W$. Write $w = \omega T$ for some $\omega \in N$ and write $e_\tau = \lambda(0)$ for some $\lambda \in \operatorname{relint}(\tau) \cap X_*(T)$. Then $\lambda' = \omega\lambda\omega^{-1} \in \operatorname{relint}(w\tau) \cap X_*(T)$. Since $\lambda'(0) = e_{w\tau}$

$$we_\tau w^{-1} = \omega e_\tau \omega^{-1} = e_{w\tau} . \tag{84}$$

Example 5.4 Suppose $G = \mathbf{GL}_n$ and $T = \mathbf{T}_n$ as in Example (5.3). Then $\chi_j = \varepsilon_j$. The faces of $\sigma = \mathbf{R}^+\eta_1 + \cdots + \mathbf{R}^+\eta_n$ have the form $\tau_I = \sum_{i \in I} \mathbf{N}\eta_i$ where I runs over the subsets of $\{1, \ldots, n\}$. Thus $\mathcal{F}(\sigma)$ is isomorphic to the lattice of subsets of

$\{1, \ldots, n\}$. Since elements of $X_*(\mathbf{T}_n)$ have the form $\eta_1^{a_1} \cdots \eta_n^{a_n}$ with $a_i \in \mathbf{N}$ we have $\sigma \cap X_*(T) = \mathbf{N}\eta_1 + \cdots + \mathbf{N}\eta_n$ in additive notation. If $\lambda \in \text{relint}(\tau_I)$ then, since $\text{relint}(\tau_I) \cap X_*(T) = \{\prod_{i \in I} \eta_i^{a_i} \mid a_i \in \mathbf{N} \text{ and } a_i > 0\}$, we have $\lambda(0) = \sum_{k \notin I} E_{kk}$. Maximal idempotents in $\overline{T}$ correspond to one-dimensional faces $\mathbf{R}^+\eta_1, \ldots, \mathbf{R}^+\eta_n$ of σ and are thus given by $\eta_i(0) = \sum_{k \neq i} E_{kk}$. The $\mathbf{T}_n$-orbit on $\mathbf{D}_n$ corresponding to the face τ_I is $\{\sum_{k \notin I} t_k E_{kk} \mid t_k \in K^*\}$. This orbit has closure $\{\sum_{k \notin I} t_k E_{kk} \mid t_k \in K\}$. The Weyl group action is given by permutation of coordinates.

Assume until further notice that $\dim Z(G) = 1$ and that $0 \in M$. These monoids are said to be ***semisimple*** [38, p.313]. Renner [41] has classified the semisimple monoids in the spirit of the classification of algebraic groups by root systems Φ [50, Chapters 11,12], under the additional hypothesis that M is a normal algebraic variety. The replacement for Φ is a combinatorial datum called a ***polyhedral root system*** [41, Definition 3.6] which is an abstraction for the triple $(X(T), \Phi, X(\overline{T}))$. Our assumptions insure that $0 \in E(\overline{T})$ and that the semisimple rank l of G is given by $l = r - 1$. [35] Let $Z = Z(G)^\circ$. Write $G' = (G, G)$. Since G is reductive, G' is semisimple and $G = G'Z$ where $G' \cap Z$ is finite. Let $T' = G' \cap T$. Since $Z \subseteq T$ we have $T = T'Z$ where $T' \cap Z$ is finite. Thus $\dim T' = \dim T - \dim Z = r - 1 = l$. Thus T' is a maximal torus of G' and, since $T' \cap Z$ is finite, $X_*(Z) \cap X_*(T') = \{0\}$, in additive notation. Let $N' = N_{G'}(T')$. The inclusion $N' \subset N$ induces an isomorphism $W(G', T') \simeq W(G, T)$.

Let $X'_* = X_*(T') \otimes \mathbf{R}$. Then $\dim_{\mathbf{R}} X_* = 1 + \dim_{\mathbf{R}} X'_*$. Since $X_*(T') \subset X_*(T)$ we have $X'_* \subseteq X_*$. Since $\dim Z(G) = 1$ we may choose a non-trivial element $\lambda_0 \in X_*(Z) \subseteq X_*(T)$. Then $\lambda_0 \notin X'_*$ so

$$X_* = \mathbf{R}\lambda_0 \oplus X'_* \ . \tag{85}$$

Since G' is semisimple we have $G' = (G', G')$ so that $G' \subseteq \mathbf{SL}_n$. Thus, if $\lambda' \in X_*(T')$ and $t \in K^*$ then $\prod_{j=1}^n \chi_j(\lambda'(t)) = \det \lambda'(t) = 1$. Thus $\sum_{j=1}^n \langle \chi_j, \lambda' \rangle = 0$ for $\lambda' \in X_*(T')$. By $\mathbf{R}$-linearity the same holds for $\lambda' \in X'_*$. Suppose $\lambda \in X_*$. Write $\lambda = r\lambda_0 + \lambda'$ with $r \in \mathbf{R}$ and $\lambda' \in X'_*$. The map $f : \lambda \mapsto r$ is an $\mathbf{R}$-linear form on X_* and we have $\sum_{j=1}^n \langle \chi_j, \lambda \rangle = f(\lambda) \sum_{j=1}^n \langle \chi_j, \lambda_0 \rangle$. Since $0 \in E(\overline{T})$ it follows from Theorem 5.4 that we may choose our $\lambda_0 \in X_*(Z)$ so that $\langle \chi_j, \lambda_0 \rangle > 0$ for $1 \leq j \leq n$ and hence $\sum_{j=1}^n \langle \chi_j, \lambda_0 \rangle > 0$. It follows from the definition of σ that if $\lambda \in \sigma$ then $f(\lambda) \geq 0$ with equality if and only if $\lambda = 0$. Thus, if we set $\sigma(1) = \{\lambda \in \sigma \mid f(\lambda) = 1\}$ then $\sigma = \mathbf{R}^+\sigma(1)$ is a cone over $\sigma(1)$. In Example 5.4 $\sigma(1)$ is a simplex with vertices at $\eta_1, \ldots, \eta_n$. In general, since $\sigma(1)$ is the convex hull of its intersections with the one-dimensional faces of σ it is a ***rational polytope***: its vertices are in $X_*(T) \otimes \mathbf{Q}$. [36] If $\lambda' \in X'_*$ and $|\langle \chi_j, \lambda' \rangle| < 1$ for all j then $\langle \chi_j, \lambda_0 + \lambda' \rangle = 1 + \langle \chi_j, \lambda' \rangle > 0$ so that $\lambda_0 + \lambda' \in \sigma(1)$. Thus $\dim_{\mathbf{R}} \sigma(1) = l$ so that $\dim_{\mathbf{R}} \sigma = l + 1 = r$. Let $\mathcal{F}(\sigma(1))$ be the lattice of faces of $\sigma(1)$ including the empty face. It follows from Corollary 5.5 that there is a lattice anti-isomorphism $\mathcal{F}(\sigma(1)) \simeq E(\overline{T})$.

Now consider the action of the Weyl group. Suppose $w = \omega T \in W$ where $\omega \in N = N_G(T)$. Then ω normalizes T' because $G' \lhd G$. Suppose $\lambda \in X^*(T')$.

[35] In what follows we use the remarks about semisimple groups which precede Example 4.10.

[36] See [12, Section 1.5] for a discussion of rational polytopes in the present context.

Then $(w\lambda)(t) = \omega\lambda(t)\omega^{-1} \in T'$ for $t \in K^*$ so that $w\lambda \in X_*(T')$. Thus $WX_*(T') = X_*(T')$, so $WX'_* = X'_*$ and the second summand in (85) is stable under W. Since $\lambda_0(t) \in Z(G)$ we have $\omega\lambda_0(t)\omega^{-1} = \lambda_0(t)$ so that $w\lambda_0 = \lambda_0$ and the first summand in (85) is fixed by W. Thus $W\sigma(1) = \sigma(1)$. The lattice anti-isomorphism between $\mathcal{F}(\sigma(1))$ and $E(\overline{T})$ is thus W-equivariant.

The hypothesis that $0 \in M$ and dim $Z(G) = 1$ is satisfied in the following important special case. Let G_0 be a connected semisimple group of rank l and let $\rho : G_0 \to \mathbf{GL}_n$ be a rational representation with *finite kernel*. Let $G = K^*\rho(G_0)$ and let $M = M(\rho) = \overline{G}$. Since G_0 and K^* are connected, so is G. The group $\rho(G_0)$ is semisimple by [3, Proposition 14.10.1(c)] so G is reductive. The rank of $\rho(G_0)$ is l because ρ has finite kernel. Since $\rho(G_0)$ is semisimple its center $Z(\rho(G_0))$ is finite. Since $Z(G) = K^*Z(\rho(G_0))$ it follows that dim $Z(G) = 1$. In fact $Z = \{tI \mid t \in K^*\}$ so that we may choose $\lambda_0(t) = \mathrm{diag}(t, \dots, t)$ for $t \in K^*$. Let T_0 be a maximal torus of G_0. If T', G' are as in the preceding paragraphs then $T' = \rho(T_0)$ and $G' = \rho(G_0)$. Let $N_0 = N_{G_0}(T_0)$. Then $\rho(N_0) \subset N$. The homomorphism $W(G_0, T_0) \to W(G, T)$ defined by $\omega_0 T_0 \mapsto \rho(\omega_0)T$ for $\omega_0 \in N_0$ is an isomorphism since ρ has finite kernel. In what follows we identify $W(G_0, T_0) = W(G, T)$ and let W denote either of these groups. With this identification we have $w(\rho \circ \theta) = \rho \circ (w\theta)$ for $w \in W$ and $\theta \in X_*(T_0)$. This is the general construction $(G_0, \rho) \rightsquigarrow M(\rho)$ promised in Section 2, where we chose $G_0 = \mathbf{SL}_n$ and considered $M(\rho)$ for various representations ρ.

To compute σ and $\sigma(1)$ in concrete cases choose $\lambda_0, \lambda_1, \dots, \lambda_l \in X_*(T)$ such that

$$T = \{\lambda_0(t_0)\lambda_1(t_1)\cdots\lambda_l(t_l) \mid t_0, t_1, \dots, t_l \in K^*\} \subseteq \mathbf{T}_n \tag{86}$$

where, as before, $\lambda_0(t) = \mathrm{diag}(t, \dots, t)$. Such λ_i exist. Since dim $T = r = l + 1$ it follows that $\lambda_0, \lambda_1, \cdots, \lambda_l$ are linearly independent over $\mathbf{Z}$ and hence over $\mathbf{R}$ when viewed in X_*. If $G = K^*\rho(G_0)$ we choose $\lambda_1, \dots, \lambda_l$ as follows. Let $\theta_1, \dots, \theta_l$ be a $\mathbf{Z}$-basis for $X_*(T_0)$. [37] Then $T_0 = \{\theta_1(t_1)\cdots\theta_l(t_l) \mid t_1, \dots, t_l \in K^*\}$. Define $\lambda_i = \rho \circ \theta_i$. Then $\rho(T_0) = \{\lambda_1(t_1)\cdots\lambda_l(t_l) \mid t_1, \dots, t_l \in K^*\}$ so the conditions in (86) are satisfied. We do not assume uniqueness of expression in (86) since this may not hold if $G = K^*\rho(G_0)$. [38] In the rest of this section we distinguish row and column vectors over $\mathbf{R}$. Let $\mathbf{R}^r = \mathbf{R}^{1+l}$ denote the space of column vectors with standard basis written $\mathbf{e}_0, \mathbf{e}_1, \dots, \mathbf{e}_l$. Let $\mathbf{R}^r_* = \mathbf{R}^{1+l}_*$ denote the space of row vectors with standard basis written $\mathbf{e}_{0*}, \mathbf{e}_{1*}, \dots, \mathbf{e}_{l*}$ where $\mathbf{e}_{i*} = \mathbf{e}_i^\top$. We will see that choice of λ_i as in (86) leads to vector space isomorphisms $\chi \mapsto \chi^*$ from $X \to \mathbf{R}^r$ and $\beta \mapsto \beta_*$ from $\mathbf{R}^r_* \to X_*$. The notation is chosen so that $\langle\chi, \beta_*\rangle = \langle\chi^*, \beta\rangle$ where the pairing on the left is given by (67) and the pairing on the right is the natural pairing of column and row vectors $\mathbf{R}^r \times \mathbf{R}^r_* \to \mathbf{R}$; see (100).

Since $T \subseteq \mathbf{T}_n$ there exist $\mathbf{a}_i = (a_{i1}, \dots, a_{in}) \in \mathbf{Z}^n_*$ for $0 \le i \le l$ such that

$$\lambda_i(t) = \lambda^{\mathbf{a}_i}(t) = \mathrm{diag}(t^{a_{i1}}, \dots, t^{a_{in}}) \; . \tag{87}$$

[37] There is a natural choice for $\theta_1, \dots, \theta_l$ namely the *co-roots* $\alpha_1^\vee, \dots, \alpha_l^\vee$ as defined, for example, in [50, Lemma 9.1.5]. If $G_0 = \mathbf{SL}_m$ then $l = m - 1$ and $\alpha_i^\vee(t) = \mathrm{diag}(1, \dots, t, t^{-1}, \dots, 1)$ where t, t^{-1} are in positions $i, i+1$. We do not define the co-roots here.

[38] This happens in the simplest example $G_0 = \mathbf{SL}_2$ with $\rho(g) = g$.

Note that $\mathbf{a}_0 = (1, \dots, 1)$. Let

$$A = \begin{bmatrix} a_{11} & \dots & a_{1n} \\ \vdots & & \vdots \\ a_{l1} & \dots & a_{ln} \end{bmatrix} \in \mathbf{M}_{l,n}(\mathbf{Z}) \tag{88}$$

be the $l \times n$ integer matrix with rows $\mathbf{a}_1, \dots, \mathbf{a}_l$. Since $\lambda_0, \lambda_1, \dots, \lambda_l$ are linearly independent, so are $\mathbf{a}_0, \mathbf{a}_1, \dots, \mathbf{a}_l$. Thus the $r \times n$ matrix A_0 with rows $\mathbf{a}_0, \mathbf{a}_1, \dots, \mathbf{a}_l$ has rank r. In particular, rank $A = l$. The general element of T has the form

$$\lambda_0(t_0)\lambda_1(t_1)\cdots\lambda_l(t_l) = \operatorname{diag}(\ \dots,\ \textstyle\prod_{i=0}^{l} t_i^{a_{ij}}\ ,\ \dots) \,. \tag{89}$$

Thus $\chi_j(\lambda_i(t)) = t^{a_{ij}}$ for $t \in K^*$, so that $\langle \chi_j, \lambda_i \rangle = a_{ij}$. If $\beta = (b_0, b_1, \dots, b_l) \in \mathbf{Z}_*^r = \mathbf{Z}_*^{1+l}$ define $\beta_* \in X_*(T)$ by $\beta_* = \prod_{i=0}^{l} \lambda_i^{b_i}$. The map $\beta \mapsto \beta_*$ is a homomorphism $\mathbf{Z}_*^r \to X_*(T)$ of abelian groups. Extend this map to an $\mathbf{R}$-linear map $\beta \mapsto \beta_*$ from $\mathbf{R}_*^r \to X_*$. In additive notation

$$\beta_* = \sum_{i=0}^{l} b_i \lambda_i \tag{90}$$

for $\beta = (b_0, b_1, \dots, b_l) \in \mathbf{R}_*^r$. Define $\mathbf{R}$-linear forms $\xi_1, \dots, \xi_n$ on $\mathbf{R}_*^r$ by

$$\xi_j(\beta) = \langle \chi_j, \beta_* \rangle = \sum_{i=0}^{l} a_{ij} b_i \,. \tag{91}$$

Define a rational convex polyhedral cone $\gamma \subset \mathbf{R}_*^r$ by

$$\gamma = \{\beta \in \mathbf{R}_*^r \mid \xi_j(\beta) \geq 0 \text{ for } 1 \leq j \leq n\} \,. \tag{92}$$

Since $\beta \mapsto \beta_*$ is an isomorphism of vector spaces, it follows from (76) that it defines an isomorphism $\gamma \simeq \sigma$ of rational convex polyhedral cones. Let $\gamma(1) = \{(b_0, b_1, \dots, b_l) \in \gamma \mid b_0 = 1\}$ be the preimage of $\sigma(1)$ under the isomorphism. Then $\gamma(1) \subset \mathbf{R}_*^r$ is a rational convex polytope isomorphic to $\sigma(1)$.

Example 5.5 Let $G_0 = \mathbf{Sp}_n$. Then $T_0 = G_0 \cap \mathbf{T}_n$ is a maximal torus of G_0. [39] Let $G = K^* G_0 \subset \mathbf{GL}_n$. Then G is a connected reductive group and $T = K^* T_0$ is a maximal torus of G. The group G bears the same relation to $\mathbf{Sp}_n$ that $\mathbf{GL}_n$ does to $\mathbf{SL}_n$. Recall from (24) that elements of T_0 have the shape $\operatorname{diag}(t_1, \dots, t_l, t_1^{-1}, \dots, t_l^{-1})$ where $n = 2l$ and $t_1, \dots, t_l$ are arbitrary in K^*. The group G has rank $r = l + 1$ and

[39] The change in notation from G, T in Example 4.3 to G_0, T_0 here is forced by the following definitions of G, T.

semisimple rank l. For $1 \le i \le l$ define $\lambda_i \in X_*(T_0) \subset X_*(T)$ by $\lambda_i = \eta_i \eta_{l+i}^{-1}$ where η_i is as in Example 5.3. Then T is given by (86). In the notation of (71) and (87) we have $\mathbf{a}_i = \mathbf{e}_{i*} - \mathbf{e}_{l+i,*}$ for $1 \le i \le l$. The $l \times n$ matrix A is given by

$$A = \begin{bmatrix} 1 & 0 & \cdots & 0 & -1 & 0 & \cdots & 0 \\ 0 & 1 & \cdots & 0 & 0 & -1 & \cdots & 0 \\ \vdots & \vdots & & \vdots & \vdots & \vdots & & \vdots \\ 0 & 0 & \cdots & 1 & 0 & 0 & \cdots & -1 \end{bmatrix}. \tag{93}$$

The linear forms $\xi_j : \mathbf{Z}_*^r \to \mathbf{Z}$ are given for $1 \le j \le n$ by

$$\begin{aligned} \xi_i(\beta) &= b_0 + b_i \quad \text{for } 1 \le i \le l \\ \xi_{l+i}(\beta) &= b_0 - b_i \quad \text{for } 1 \le i \le l\,. \end{aligned} \tag{94}$$

To describe the cone γ, note that $\gamma(1)$ is defined by the inequalities $|b_i| \le 1$ for $1 \le i \le l$ and is thus a cube with 2^l vertices at the points $\beta = (1, \pm 1, \dots, \pm 1) \in \mathbf{Z}_*^r$. Since the maximal idempotents in $\overline{T}$ are in one-to-one correspondence with the 1-dimensional faces of σ it follows that $E(\overline{T})$ has 2^l maximal idempotents. If $\beta = (1, b_1, \dots, b_l)$ is a vertex of $\gamma(1)$ then $\beta_*(t) = \mathrm{diag}(t^{1+b_1}, \cdots, t^{1+b_l}, t^{1-b_1}, \dots, t^{1-b_l})$. The corresponding maximal idempotent is

$$\beta_*(0) = \sum_{j \in I(\beta)} E_{jj} \tag{95}$$

where $I(\beta) = \{1 \le j \le l \mid b_j = -1\} \cup \{l+1 \le j \le n \mid b_{j-l} = +1\}$. For example if $l = 3$ and $(b_1, b_2, b_3) = (-1, +1, +1)$ then $\beta_*(0) = \mathrm{diag}(1, t^2, t^2, t^2, 1, 1)$ so $I(\beta) = \{1, 5, 6\}$ and $\beta_*(0) = E_{11} + E_{55} + E_{66}$. The set $E(\overline{T})$ may be described as follows. Say that a subset $I \subseteq \{1, \dots, n\}$ is *admissible* if $j \in I$ implies $\bar{j} \notin I$ where $j \mapsto \bar{j}$ is the involutory permutation of $\{1, \dots, n\}$ introduced in Example 4.6. The sets $I(\beta)$ for β a vertex of $\gamma(1)$, are the maximal admissible subsets of $\{1, \dots, n\}$. The map

$$I \mapsto e_I := \sum_{j \in I} E_{jj} \tag{96}$$

is bijective from admissible subsets of $\{1, \dots, n\}$ to $E(\overline{T}) - \{1\}$. It remains to describe the Weyl group action on $E(\overline{T})$. Use the notation in Examples 4.6 and 4.9. If $w \in W_1$ and $\pi \in S_n$ corresponds to w, there exists a subset I of $\{1, \dots, l\}$ such that $\pi = \prod_{i \in I}(i\bar{i})$. Then for $1 \le i \le l$ we have $w\lambda_i = \lambda_i^{-1}$ if $i \in I$ and $w\lambda_i = \lambda_i$ if $i \notin I$. If $w \in W_2$ and $\pi \in S_n$ corresponds to w then $\pi\{1, \dots, l\} \subseteq \{1, \dots, l\}$ and the action of w on $X_*(T)$ is given by $w\lambda_i = \lambda_{\pi i}$ for $1 \le i \le l$. Thus the action of W on $X_*(T)$ is like that on $X(T)$. Since W centralizes the involution $i \mapsto \bar{i}$ it maps admissible sets to admissible sets. The action of W on $E(\overline{T})$ is given by $we_I w^{-1} = e_{wI}$. The group W acts transitively on faces of σ of any given dimension. Thus the W-orbits on $E(\overline{T})$, excluding the orbits of 0 and 1, are in one-to-one correspondence with the set $\{1,\dots,l\}$. The number of $\mathcal{J}$-classes in the reductive monoid $M(\rho)$, excluding $\{0\}$ and $\{1\}$, is thus l. The poset $\mathcal{J}$ is linearly ordered.

It is sometimes convenient to work with the dual cone $\sigma^\vee \subset X$. If $\tau \in \mathcal{F}(\sigma)$ define a subset τ^* of X by

$$(97) \qquad \tau^* = \{\chi \in X \mid \langle \chi, \lambda \rangle \geq 0 \text{ for } \lambda \in \sigma \text{ and } \langle \chi, \lambda \rangle = 0 \text{ for } \lambda \in \tau\} .$$

The map $\tau \mapsto \tau^*$ is a lattice anti-isomorphism $\mathcal{F}(\sigma) \to \mathcal{F}(\sigma^\vee)$ [12, 1.2.10] which is W-equivariant by (70). It follows from Corollary 5.5 that the map $\tau^* \mapsto e_\tau$ is a lattice isomorphism $\mathcal{F}(\sigma^\vee) \to E(\overline{T})$. Define $\chi_0 \in X$ by $\chi_0 = \frac{1}{n}(\chi_1 + \cdots + \chi_n)$ where χ_j is the restriction of ε_j to T. Thus, in multiplicative notation, $\chi_0^n \in X(T)$ is the determinant. Define a subspace X' of X by $X' = \{\chi \in X \mid \langle \chi, \lambda_0 \rangle = 0\}$. Since $\dim_{\mathbf{R}} X' = l$ and $\langle \chi_0, \lambda_0 \rangle = 1$ we have

$$(98) \qquad X = \mathbf{R}\chi_0 \oplus X'.$$

As in (85) the Weyl group fixes the first summand and stabilizes the second. If $1 \leq j \leq n$ then $\langle \chi_j - \chi_0, \lambda_0 \rangle = 1 - 1 = 0$ so the decomposition of χ_j according to (98) is $\chi_j = \chi_0 + (\chi_j - \chi_0)$. Since $G' \subseteq \mathbf{SL}_n$ we have $\langle \chi_0, \lambda \rangle = 0$ for $\lambda \in X'_*$. Thus $X'_* = \{\lambda \in X_* \mid \langle \chi_0, \lambda \rangle = 0\}$. The restriction of the pairing $\langle \ , \ \rangle$ to $X' \times X'_*$ is nondegenerate. If $\chi \in X$ write $\chi = r\chi_0 + \chi'$ with $r \in \mathbf{R}$ and $\chi' \in X'$. The map $f : \chi \mapsto r$ is an $\mathbf{R}$-linear form on X. The argument which follows (85) shows that if we set $\sigma^\vee(1) = \{\chi \in \sigma^\vee \mid f(\chi) = 1\}$, then $\sigma^\vee(1)$ is stable under W and $\sigma^\vee = \mathbf{R}^+\sigma^\vee(1)$ is a cone over $\sigma^\vee(1)$. Let $\chi \in X$. By (81) we may write $\chi = \sum_{j=1}^n r_j\chi_j$ where $r_j \in \mathbf{R}^+$. Then $\chi = (\sum_{j=1}^n r_j)\chi_0 + \sum_{j=1}^n r_j(\chi_j - \chi_0)$ so $f(\chi) = \sum_{j=1}^n r_j$. Thus $\sigma^\vee(1) = \{\sum_{j=1}^n r_j\chi_j \mid r_j \in \mathbf{R}^+ \text{ and } \sum_{j=1}^n r_j = 1\}$ is the convex hull of the χ_j. [40]

Define an $\mathbf{R}$-linear map $\chi \mapsto \chi^*$ from $X \to \mathbf{R}^r$ by

$$(99) \qquad \chi^* = (\langle \chi, \lambda_0 \rangle, \langle \chi, \lambda_1 \rangle, \dots, \langle \chi, \lambda_l \rangle)^\top .$$

Then $\chi \mapsto \chi^*$ is an isomorphism of vector spaces. It follows from (90) and (99) that

$$(100) \qquad \langle \chi, \beta_* \rangle = \sum_{i=0}^{l} b_i \langle \chi, \lambda_i \rangle = \langle \chi^*, \beta \rangle$$

for $\chi \in X$ and $\beta \in \mathbf{R}_*$, where the pairing on the right is the natural pairing of column and row vectors $\mathbf{R}^r \times \mathbf{R}^r_* \to \mathbf{R}$. We may use the isomorphisms (90) and (99) to transport the action of W on X, X_* to $\mathbf{R}^r, \mathbf{R}^r_*$. Thus we have

[40] The χ_j are the weights of the representation $T \hookrightarrow \mathbf{T}_n$. Since $T_0 \to \rho(T_0) \to T$ where the first map is ρ and the second map is inclusion, there is a homomorphism $\pi : X(T) \to X(T_0)$ of abelian groups. The functions $\pi \circ \chi_j$ are the weights of the representation ρ. Since the restriction of π to $X(T) \cap X'$ is injective we may say, with slight abuse of terminology that $\sigma^\vee(1)$ is the convex hull of the set $\Phi(\rho)$ of weights of ρ; compare [41, Section 3].

$$w(\chi^*) = (w\chi)^*, \qquad (w\beta)_* = w(\beta_*) \tag{101}$$

for $\chi \in X$ and $\beta \in \mathbf{R}^r_*$. It follows from these formulas and (70) that

$$\langle w(\chi^*), w\beta \rangle = \langle \chi^*, \beta \rangle \tag{102}$$

so that W respects the pairing on $\mathbf{R}^r \times \mathbf{R}^r_*$. To compute matrix representations using these actions note that if $\beta = \mathbf{e}_{i*}$ then $\beta_* = \lambda_i$. Thus the matrix, say $\theta(w)$, for the action of w on $\mathbf{R}^r_*$ is precisely the matrix for its action on X_* with respect to the basis $\lambda_0, \dots, \lambda_l$. It follows from (102) that the matrix for the action of w on $\mathbf{R}^r$ is the contragredient $\theta(w^{-1})^\top$. The columns of A_0 are $\chi_1^*, \dots, \chi_n^*$. Thus left multiplication of A_0 by $\theta(w^{-1})^\top$ permutes the columns of A_0 as w permutes the set $\Phi(\rho)$ of weights.

We have already remarked that the polytope $\sigma^\vee(1)$ is the convex hull of the set $\{\chi_1, \dots, \chi_n\}$. Thus its image, say $\mathcal{P}_0 \subset \mathbf{R}^r$, under the map $\chi \mapsto \chi^*$ is the convex hull of the columns of A_0 and is thus isomorphic, by a translation, to the convex hull $\mathcal{P}$ in $\mathbf{R}^l$ of the columns of A. The Weyl group action on $\mathcal{P}_0$ and on $\mathcal{P}$ is given by $w \mapsto \theta(w^{-1})^\top$. The lattice anti-isomorphism $\mathcal{F}(\sigma) \to \mathcal{F}(\sigma^\vee)$ induces a lattice anti-isomorphism $\mathcal{F}(\sigma(1)) \to \mathcal{F}(\sigma^\vee(1))$ [12, p.24] and hence a lattice anti-isomorphism $\mathcal{F}(\sigma) \to \mathcal{F}(\mathcal{P})$. We conclude from Corollary 5.5 and formula (82) that the map

$$F \mapsto e_F := \sum_{\chi_j^* \in F} E_{jj}, \tag{103}$$

is a lattice isomorphism $\mathcal{F}(\mathcal{P}_0) \to E(\overline{T})$. Note that $\chi_j^* = (a_{1j}, \dots, a_{lj})^\top$ is the j-th column of the matrix A. This is the form in which Putcha [23], [31, Chapter 8] proved Corollary 5.5.

The polytopes $\sigma(1)$ and $\sigma^\vee(1)$ are *dual polytopes* [12, p.24]. For example, if $G = K^*\mathbf{Sp}_n$ where $n = 2l$ and $\mathbf{e}_1, \dots, \mathbf{e}_l$ is the standard basis for $\mathbf{R}^l$ then (93) shows that the set of columns of A is $\{\mathbf{e}_1, \dots, \mathbf{e}_l, -\mathbf{e}_1, \dots, -\mathbf{e}_l\}$. The polytope $\mathcal{P} \simeq \sigma^\vee(1)$ is thus the cross polytope, which is dual to the cube we found in Example 5.5.

Example 5.6 Let's reconsider Example 2.3 in case $m = 4$ and complete the calculation begun there. We will compute in detail, so that the reader may enjoy reproducing these calculations in other examples. Let $G_0 = \mathbf{SL}_4$ and let $T_0 = G_0 \cap \mathbf{T}_4$. Let $V = K^4 \otimes K^4$. Define $\rho : G_0 \to \mathbf{GL}(V)$ by $\rho(g)(v \otimes v') = gv \otimes (g^\top)^{-1}v'$. Define $\theta_1, \theta_2, \theta_3 \in X_*(T_0)$ by $\theta_1(t) = \mathrm{diag}(t, t^{-1}, 1, 1)$, $\theta_2(t) = \mathrm{diag}(1, t, t^{-1}, 1)$, $\theta_3(t) = \mathrm{diag}(1, 1, t, t^{-1})$. The general element of T_0 has the shape $t_0 = \theta_1(t_1)\theta_2(t_2)\theta_3(t_3)$. Let $\lambda_i = \rho \circ \theta_i \in X_*(\rho(T_0)) \subseteq X_*(\rho(T))$. The vectors $v_j \otimes v_k$ where $1 \le j, k \le 4$, are a basis of weight vectors. We view $\rho(T_0) \subseteq \mathbf{T}_4 \times \mathbf{T}_4$. Let $\varepsilon_{jk} \in X(\mathbf{T}_4 \times \mathbf{T}_4)$ denote the coordinate functions on $\mathbf{T}_4 \times \mathbf{T}_4$ and let χ_{jk} denote the restriction of ε_{jk} to $T = K^*\rho(T_0)$ or to $\rho(T_0)$. Since $t_0(v_j \otimes v_k) = (\chi_{jk} \circ \rho)(t_0)(v_j \otimes v_k)$ the functions $\chi_{jk} \circ \rho \in X(T_0)$ are the weights of ρ. For example since $\rho(t_0)(v_1 \otimes v_2) =$

$t_0v_1 \otimes t_0^{-1}v_2 = t_0v_1 \otimes (t_1t_2^{-1})^{-1}v_2 = t_1^2t_2^{-1}(v_1 \otimes v_2)$ we have $(\chi_{12} \circ \rho)(t_0) = t_1^2t_2^{-1}$. Write $(\chi_{jk} \circ \rho)(t_0) = t_1^{a_1}t_2^{a_2}t_3^{a_3}$ for suitable integers $a_i = a_{ijk}$. The list of weights is:

j k	a_1	a_2	a_3	j k	a_1	a_2	a_3
1 1	0	0	0	3 1	−1	−1	1
1 2	2	−1	0	3 2	1	−2	1
1 3	1	1	−1	3 3	0	0	0
1 4	1	0	1	3 4	0	−1	2
2 1	−2	1	0	4 1	−1	0	−1
2 2	0	0	0	4 2	1	−1	−1
2 3	−1	2	−1	4 3	0	1	−2
2 4	−1	1	1	4 4	0	0	0

The matrix A of size 3×16 has the vectors $(a_1, a_2, a_3)^\top$ as its columns. The polytope $\mathcal{P}$ is the convex hull of these columns. The 12 nonzero columns are not the midpoints of the edges of a cube with vertices $\pm\mathbf{e}_1 \pm \mathbf{e}_2 \pm \mathbf{e}_3$ that we found in Example 2.3. We haven't lost it though. For $1 \leq j \neq k \leq 4$ let $\mathbf{e}_{jk} = \mathbf{e}_j - \mathbf{e}_k \in \mathbf{R}^4$. The $\mathbf{e}_{jk}$ lie in the hyperplane $H = \{\sum_{i=1}^4 x_i\mathbf{e}_i \mid \sum_{i=1}^4 x_i = 0\}$. Map $H \to \mathbf{R}^3$ by $(x_1, x_2, x_3, x_4)^\top \mapsto (x_1 - x_2, x_2 - x_3, x_3 - x_4)^\top$; this is not the map ϕ we used in Example 2.3. The $\mathbf{e}_{jk}$ map to the vectors $(a_1, a_2, a_3)^\top$ so $\mathcal{P}$ is the cuboctahedron after all.

We compute the W-conjugacy classes in $E(\overline{T})$. Since the correspondence $\mathcal{F}(\mathcal{P}) \leftrightarrow E(\overline{T})$ respects the W-action, we may compute the orbits for the W-action on vertices, edges, and faces of $\mathcal{P}$. The reader should look at the picture of the cuboctahedron given in Example 2.3. The symmetry group of the cube, and of the inscribed cuboctahedron, has order 48. The Weyl group W of $\mathbf{SL}_4$ has order 24, and is isomorphic as abstract group to the symmetric group on four letters. The group W has two inequivalent irreducible representations in $\mathbf{R}^3$, one as the group of rotations of the cube, the other as the group generated by reflections in the six symmetry planes of the cube which are different from the coordinate planes. Our representation of W in Example 2.3 is the one generated by reflections. Thus W is transitive on vertices of the cuboctahedron, has 2 orbits on edges, and has 3 orbits on faces of dimension 2, one orbit consisting of squares and two orbits consisting of triangles. We conclude that the number of W-congugacy classes of idempotents in $E(\overline{T})$ is, excluding the classes $\{0\}, \{1\}$, equal to $1+2+3$. In view of (65) this is the number of $G \times G$ orbits on M or the number of G orbits on $E(M)$ excluding $\{0\}, \{1\}$.

Example 5.7 Let $G_0 = \mathbf{SL}_4$ and let $T_0 = G_0 \cap \mathbf{T}_4$. Let $V_p = \bigwedge^p K^4$ be the p-th exterior power of K^4. Thus V_1 has basis $\{v_i \mid 1 \leq i \leq 4\}$, V_2 has basis $\{v_j \wedge v_k \mid 1 \leq j < k \leq 4\}$, and V_3 has basis $\{v_p \wedge v_q \wedge v_r \mid 1 \leq p < q < r \leq 4\}$. Let $V = V_1 \otimes V_2 \otimes V_3$. Then dim $V = 96$. Define $\theta_1, \theta_2, \theta_3$ as in Example 5.6 and let $\lambda_i = \rho \circ \theta_i$. The matrix A has size 3×96. It is hard to get a picture of $\mathcal{P}$ from this matrix. However, there is a trick we can use which is implicit in the computation of Example 5.6. Write the elements of $\mathbf{SL}_4$ as diag(t_1, t_2, t_3, t_4) where $t_1t_2t_3t_4 = 1$. Ignore, for a moment, the restriction $t_1t_2t_3t_4 = 1$, compute a congruent polytope in $\mathbf{R}^4$ and return to the desired polytope in $\mathbf{R}^3$ via the mapping $(x_1, x_2, x_3, x_4) \mapsto (x_1 - x_2, x_2 - x_3, x_3 - x_4)$. The vector v_i has weight t_i, the vector $v_j \wedge v_k$ has weight t_jt_k and the vector $v_p \wedge v_q \wedge v_r$ has weight $t_pt_qt_r$. Thus $v_i \otimes (v_j \wedge v_k) \otimes (v_p \wedge v_q \wedge v_r)$ has weight $t_it_jt_kt_pt_qt_r$. This gives us a matrix of size 4×96. If we write $t_it_jt_kt_pt_qt_r = t_1^a t_2^b t_3^c t_4^d$ we find that the only column vectors $(a, b, c, d)^\top$ which occur in this matrix are gotten by permuting the components of a vector in the list: $(3,2,1,0)^\top$, $(3,1,1,1)^\top$, $(2,2,2,0)^\top$, $(2,2,1,1)^\top$. These all live in the hyperplane $x_1 + x_2 + x_3 + x_4 = 6$. When we pass to the convex hull, only the vector $(3,2,1,0)^\top$ and its permutations survive. The polytope $\mathcal{P}$ is the convex hull of the 24 points which are permutations of $(3,2,1,0)^\top$. This polytope is often called the *permutohedron* because its vertices correspond to permutations on 4 letters. Here is a picture (Figure 5.1), adapted from [2, p. 136].

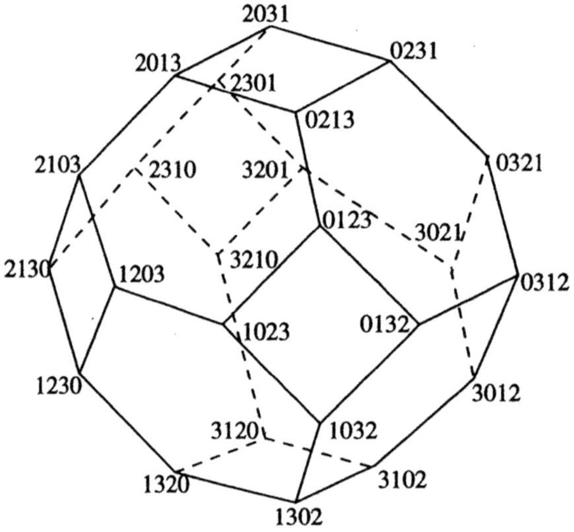

Figure 5.1. Permutohedron

In this example W is transitive on vertices, has 3 orbits on edges, and 3 orbits on faces of dimension 2. Thus the number of $\mathcal{J}$-classes in the corresponding monoid $M(\rho)$, excluding the classes $\{0\}$, $\{1\}$, is $1+3+3 = 2^3 - 1$. If ρ is *any* representation of $\mathbf{SL}_4$ such that W is transitive on the vertices of the corresponding polytope $\mathcal{P}$ then the number of $\mathcal{J}$-classes, excluding the classes $\{0\}$, $\{1\}$, is at most $2^3 - 1$. The proper context for this fact lies in the notion of a *canonical monoid* for $\mathbf{SL}_4$; see [38], [35] and the remarks which follow Theorem 5.9.

Now return to the general theory of a reductive monoid M with unit group G. We drop the assumption that dim $Z(G) = 1$ but, since some of the theorems we state require that $0 \in M$, we *assume henceforth* that $0 \in M$. Note that $0 \in M(\rho)$ for any representation ρ. Theorems 5.6 and 5.7 show how to recapture some of the standard constructs in the theory of a reductive group, for example Borel subgroup, maximal torus, parabolic subgroup, Levi factor, in terms of the idempotent set $E(M)$. A *chain of idempotents* is a linearly ordered subset $\Gamma = \{e_1 < \cdots < e_k\}$ of the poset $E(M)$. If Γ is a chain of idempotents then there exists a maximal torus T of G such that $\Gamma \subset \overline{T}$ [31, Corollary 6.10]. [41] The length of a maximal chain in $E(M)$ or $E(\overline{T})$ is dim T [31, Theorem 6.20] which is also the length of a maximal chain in $G\backslash M/G$ [31, Theorem 6.20]. If $\Gamma \subseteq E(M)$, define the right and left centralizers $P(\Gamma)$, $P^-(\Gamma)$ by

$$P(\Gamma) = \bigcap\nolimits_{e\in\Gamma} P(e), \qquad P^-(\Gamma) = \bigcap\nolimits_{e\in\Gamma} P^-(e)\ , \tag{104}$$

where $P(e), P^-(e)$ are as in (14). Define the centralizer $C_G(\Gamma)$ by

$$C_G(\Gamma) = P(\Gamma) \cap P^-(\Gamma) = \bigcap\nolimits_{e\in\Gamma} C_G(e)\ . \tag{105}$$

Theorem 5.6 ([31, Theorem 7.1]) *Let M be a reductive monoid. Let Γ be a maximal chain in $E(M)$. Then $C_G(\Gamma)$ is a maximal torus of G and $P(\Gamma), P^-(\Gamma)$ are a pair of opposite Borel subgroups relative to $C_G(\Gamma)$.*

Every maximal torus and Borel subgroup of G can be obtained in this way [26, Theorem 4.5]. Thus the Borel subgroups and maximal tori in G may be recovered from $E(M)$. How about the parabolic subgroups?

Theorem 5.7 ([26, Theorem 4.6], [27, Theorem 4]) *Let M be a reductive monoid. Let Γ be a chain in $E(M)$. Then $P(\Gamma)$ and $P^-(\Gamma)$ are a pair of opposite parabolic subgroups with common Levi factor $C_G(\Gamma)$.*

Every parabolic subgroup P of G has the form $P = P(\Gamma)$ for some chain $\Gamma \subset E(M)$ [30, Theorem 7.2]. [42] The simple statement of Theorem 5.7 contains implicit information. For example it says, by the definition (64) of Levi subgroup, that $C_G(\Gamma)$ is a connected reductive group. In particular, if $\Gamma = \{e\}$ then $C_G(e)$ is a connected reductive group and $P(e), P^-(e)$ are a pair of opposite parabolic subgroups. If Γ is a maximal chain then Theorem 5.7 reduces to Theorem 5.6.

Example 5.8 Let $M = \mathbf{M}_n$ so that $G = \mathbf{GL}_n$. Let $e_r \in \mathbf{D}_n$ be as in Example 3.6. Let $\gamma = (\gamma_1, \dots, \gamma_k)$ be a composition of n as in (4.29). Then $\Gamma = \{e_{\gamma_1}, e_{\gamma_1+\gamma_2}, \dots, e_{\gamma_1+\gamma_2+\cdots+\gamma_k}\}$ is a chain of idempotents and $P(\Gamma)$ is the parabolic subgroup pictured in (58). The opposite parabolic subgroup $P^-(\Gamma)$ is the transpose of $P(\Gamma)$. The intersection $C_G(\Gamma) = P(\Gamma) \cap P^-(\Gamma)$ is the Levi subgroup pictured in Example 4.30. If $\Gamma = \{e_r\}$ then we are back in Example 3.6.

---

[41] If $\Gamma = \{e\}$ this is Theorem 5.2, which serves to start an induction.

[42] The proof uses the notion of cross section lattice, introduced below, as well as an algebro-geometric result of Renner [39].

Let M be a reductive monoid with unit group G and maximal torus T. It follows from (66) that if $\Lambda \subseteq E(\overline{T})$ represents the set of W-orbits on T then the map $\Lambda \to G\backslash M/G$ defined by $e \mapsto GeG$ is bijective. Since both $E(\overline{T})$ and $G\backslash M/G$ are partially ordered sets it is natural to ask if one can choose Λ so that the bijection respects the partial orders. This question leads to the following important definition [26, §6], [31, Definition 9.1]:

Definition 3 A set $\Lambda \subseteq E(\overline{T})$ is a *cross section lattice* if Λ is a set of representatives for the W-orbits on $E(\overline{T})$ *and* the bijection $\Lambda \to G\backslash M/G$ is order preserving.

Since $G\backslash M/G$ is a lattice so is Λ. It is not clear that cross section lattices exist. In fact they do exist by a remarkable theorem of Putcha, which shows in addition that there is a bijection between the set of cross section lattices and the set $\mathcal{B}^T$ of Borel subgroups of G which include T. Recall that if $e \in E(\overline{T})$ then the right centralizer $P(e)$ is a parabolic subgroup of G. Since $\overline{T}$ is commutative we have $P(e) \supset T$.

Theorem 5.8 ([30, Theorem 1.1], [31, Theorem 9.10]) *If $B \in \mathcal{B}^T$, define a subset $\Lambda(B)$ of $E(\overline{T})$ by $\Lambda(B) = \{e \in E(\overline{T}) \mid P(e) \supseteq B\}$. The map $B \mapsto \Lambda(B)$ is bijective from $\mathcal{B}^T$ to the set of cross section lattices in $E(\overline{T})$.*

We may reconstruct the corresponding Borel subgroup from a given cross section lattice Λ according to the recipe $B = \bigcap_{e\in\Lambda} P(e)$. Since $P(wew^{-1}) = wP(e)w^{-1}$ we have $\Lambda(wBw^{-1}) = w\Lambda(B)w^{-1}$. It follows, in view of (43), that if $B \in \mathcal{B}^T$ is a fixed Borel subgroup then the map $w \to w\Lambda(B)w^{-1}$ is bijective from W to the set of all cross section lattices in $E(\overline{T})$.

Example 5.9 Suppose $M = \mathbf{M}_n$, $G = \mathbf{GL}_n$ and $T = \mathbf{T}_n$. Let $e \in E(\overline{T})$. Then $P(e) \supseteq \mathbf{B}_n$ if and only if $e \in \{e_0, e_1, \ldots, e_n\}$ with e_r as in Example 3.6. Thus the cross section lattice corresponding to the standard Borel subgroup $\mathbf{B}_n$ is $\Lambda(\mathbf{B}_n) = \{e_0, e_1, \ldots, e_n\}$ where $e_0 = 0$ and $e_n = 1$. If $I \subseteq \{1, \ldots, n\}$ write $e_I = \sum_{i\in I} E_{ii}$ as in Example 2.1. Thus $e_r = e_{\{1,\ldots,r\}}$. Since $we_Iw^{-1} = e_{\pi I}$ where $\pi \in S_n$ is the permutation corresponding to w, we see that the cross section lattices in $E(\overline{T})$ have the shape $\{e_{I_0}, e_{I_1}, e_{I_2}, \ldots, e_{I_n}\}$ where $\emptyset = I_0 \subset I_1 \subset I_2 \subset \cdots \subset I_n = \{1, \ldots, n\}$ and $|I_k| = k$.

Theorem 5.8 allows us to describe the partial order in $G\backslash M/G$ in terms of idempotents which may be computed using the geometric method in (103). The partial order structure in $G\backslash M/G$ in Examples 5.5, 5.6, 5.7 is, omitting the orbits $\{0\}, \{1\}$, given by:

cuboctahedron

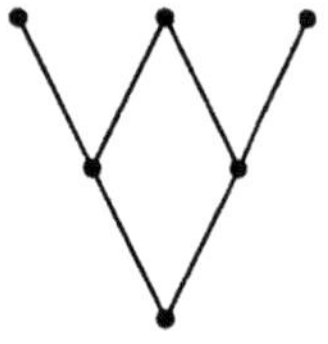

cross polytope

permutohedron

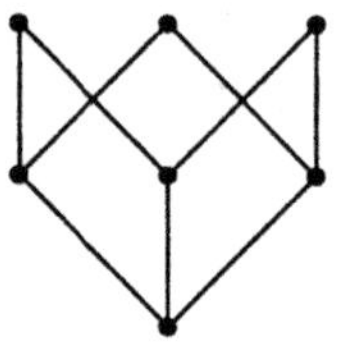

Now fix a Borel subgroup $B \supset T$ and let $\Lambda = \Lambda(B)$. If $e \in \Lambda$ then $P(e) \supseteq B$ is a standard parabolic subgroup and hence, by Theorem 4.2, has the form P_I for some $I \subseteq S$, called the type of $P(e)$. This gives a map

(106) $$\lambda : \Lambda \to 2^S$$

called the *type map*, defined by $\lambda(e) =$ type $P(e)$. In fact [30, Lemma 2.4], [35, p.645]

(107) $$\text{type } P(e) = \{s \in S \mid ses^{-1} = e\} ,$$

is the set of Coxeter generators which fix e. See [36, Section 2] and the references given there for more about the type map. In Example 5.9 the image of the type map consists, aside from S and $\emptyset$, of the sets $\{s_1, \dots, s_{r-1}, s_{r+1}, \dots, s_{n-1}\}$ for $1 \leq r \leq n-1$. The corresponding parabolic subgroups are, in view of Theorem 4.2(b), the maximal parabolic subgroups given in Example 3.6. On the other hand, in Example 5.7 the type map $\lambda : \Lambda \to 2^S$ is surjective, and is in fact bijective when restricted to $\Lambda - \{0\}$. Note in both examples that the type map is injective, and that W acts transitively on the set of vertices of $\mathcal{P}$ and hence acts transitively on the set of minimal nonzero idempotents of $E(\overline{T})$.

Theorem 5.9 ([41, Corollary 8.3.3], [31, Corollary 15.3]) *Let M be a reductive monoid. Then W acts transitively on the set of minimal nonzero idempotents of $E(\overline{T})$ if and only if M has an irreducible idempotent separating representation ρ.*[43]

Monoids with the property of Theorem 5.9 are said to be $\mathcal{J}$-*irreducible* because transitivity of W on the set of minimal nonzero idempotents of $E(\overline{T})$ is equivalent, in view of Theorem 65, to the assertion that M has a unique minimal non-zero $\mathcal{J}$-class. If M is $\mathcal{J}$-irreducible then dim $Z(G) = 1$ so that M is a semisimple monoid [31, Corollary 15.3], [41, Lemma 8.3.2]. Thus the remarks which follow 5.4 may be applied to M. If $\rho : G_0 \to \mathbf{GL}_n$ is an irreducible representation of a semisimple group G_0, then the inclusion map $M(\rho) \to \mathbf{M}_n$ is a faithful irreducible representation of $M(\rho)$. Thus, if ρ is an irreducible representation of G_0 then $M(\rho)$ is a $\mathcal{J}$-irreducible monoid. All the examples we have given to illustrate the geometry of the cone σ and the polytope $\mathcal{P}$ are $\mathcal{J}$-irreducible. If M is $\mathcal{J}$-irreducible and $e, f \in E(M)$ then $P(e) = P(f)$ if and only if $eG = fG$ [37, Proposition 4.3]. Thus if M is $\mathcal{J}$-irreducible then the type map is injective. If M is $\mathcal{J}$-irreducible, there is an effective method for computing the type map [37, Corollary 4.11,Theorem 4.16] which by-passes the geometric arguments given earlier in this Section. This argument has been used to compute the posets $\Lambda \simeq G\backslash M/G$ if G_0 is a simple algebraic group, in case ρ is the adjoint representation, in case the highest weight of ρ is a fundamental weight, and

[43] A representation is ***idempotent separating*** if it is faithful on $E(M)$. The condition that ρ be idempotent separating may be replaced by the condition that ρ is a finite morphism in the sense of algebraic geometry [41, Corollary 8.3.3]; the hypothesis that M is a normal variety is not used in the proof of [41, Lemma 8.3.2].

in some other cases [37, Section 6]. [44] If the highest weight is fixed by no nonidentity element of W, then the type map $\lambda : \Lambda - \{0\} \to 2^S$ is surjective and hence bijective as in Example 5.7. Monoids with this property play a special role in the representation theory of "finite monoids of Lie type"; see the remarks which follow Example 5.11

The Bruhat decomposition and Tits system in a reductive group G are central in the structure theory. They may be used to reduce many questions about G to questions about the Weyl group. Renner [43] found an analogous decomposition for reductive monoids with striking consequences. [45] Let M be a reductive monoid and let $G, B \supset T, N, W$ be as before. Let $\overline{N}$ be the Zariski closure of N in M. Then $\overline{N}$ is a monoid which normalizes T so $R = \overline{N}/T$ is a monoid. Thus

$$R = \overline{N}/T \supset N/T = W. \tag{108}$$

Example 5.10 Let $M = \mathbf{M}_n$. Then, as in Example 4.4, N consists of all monomial matrices, while $\overline{N}$ consists of matrices with *at most* one nonzero entry in each row and column. Thus R may be identified with the monoid of all zero-one matrices which have at most one entry equal to 1 in each row and column. This is a finite monoid isomorphic to the *symmetric inverse semigroup* I_n which consists, by definition, of all injective partial maps from $\{1, \ldots, n\} \to \{1, \ldots, n\}$. It is sometimes useful, from a combinatorial point of view, to think of these matrices as placements of non-attacking rooks on an $n \times n$ board. In [48] R is called the rook monoid. The order of R is $|R| = \sum_{r=0}^{n} \binom{n}{r}^2 r!$. The group of units of I_n is the group P_n of permutation matrices.

Theorem 5.10 *Let M be a reductive monoid. Let $\Lambda = \Lambda(B)$ be a cross section lattice for $E(\overline{T})$ and let $e \in \Lambda$. Then*
(1) R is a finite inverse monoid.
(2) The group of units of R is W, and $R = WE(R)$.
(3) $E(R) \simeq E(\overline{T})$.
(4) $M = \bigsqcup_{\rho \in R} B\rho B$, and $B\rho B = B\rho' B \Rightarrow \rho = \rho'$.
(5) If $s \in S$ is a Coxeter generator then $BsB \cdot B\rho B \subseteq Bs\rho B \cup B\rho B$.
(6) $GeG = \bigsqcup_{\rho \in WeW} B\rho B$.
(7) If $w_0 \in W$ is the opposition element then Bw_0eB is open and dense in GeG.

When Renner proved these facts, they were new even in case $M = \mathbf{M}_n(K)$; no-one had noticed the significance of the symmetric inverse semigroup for the structure theory of $\mathbf{M}_n(K)$, although it is easy to prove (4) using elementary row and column operations in this case: just note that for any field K a square matrix over K is equivalent to a rook matrix by elementary row and column operations in which *rows* are moved *up* and *columns* are moved to the *right*. As an analogue of the Weyl

[44] We have not defined "highest weight" or "fundamental weight" in this paper; see for example [15, Theorem 31.3]. If G_0 is a semsimple group and $\rho : G_0 \to \mathbf{GL}_n$ is an irreducible representation with highest weight μ then the polytope $\mathcal{P}$ is the convex hull of the set $W\mu$. This is shown, with slightly different formulation, in [41, Proposition 3.5].

[45] Monoids corresponding to classical groups in their natural representations were first studied by Grigor'ev [13], who looked for an analogue of the Bruhat decomposition.

group, the *Renner monoid* R plays a central role in the theory of reductive monoids. Note that (4) may be viewed as an analogue of the Bruhat decomposition of G, that (5) gives M a structure like a Tits system, that (6) in case $e = 1$ is the Bruhat decomposition of G and that (7) in case $e = 1$ asserts the existence of Chevalley's "big cell" Bw_0B where w_0 is the longest element of W. The "big cell" should be familiar in case $G = \mathbf{GL}_n$ under a different name, since $Bw_0B = w_0B^-B$ and the set B^-B consists of the matrices for which the process of Gaussian elimination involves no zero pivots; see the remarks which follow (62).

Let K be the algebraic closure of $\mathbf{F}_q$. The rest of this paper concerns those finite submonoids of $\mathbf{M}_n(K)$ which are fixed points under the Frobenius map $\sigma : \mathbf{M}_n(K) \to \mathbf{M}_n(K)$ defined by $\sigma : [c_{ij}] \mapsto [c_{ij}^q]$. These monoids and more general fixed point monoids were introduced by Renner in [44]; see also [38, Section 4], [46]. To agree with Renner's paper in these Proceedings we *change notation*; this looks unnecessarily complicated to start but is ultimately simpler. Henceforth we write $\mathbf{M}_n(K)$ rather than $\mathbf{M}_n$ and let $\underline{M} \subseteq \mathbf{M}_n(K)$ denote a reductive algebraic monoid with 0 such that $\sigma\underline{M} = \underline{M}$. We let $M = \{a \in \underline{M} \mid \sigma a = a\}$ denote the finite submonoid of fixed points. Thus, for example, if $\underline{M} = \mathbf{M}_n(K)$ then $M = \mathbf{M}_n(\mathbf{F}_q)$.[46] We call M a *finite reductive monoid*. The remarks in the following paragraph say, roughly, that many of the constructs in this paper descend from the reductive monoid $\underline{M}$ to the finite reductive monoid M.

Let $\underline{G}$ be the group of units of $\underline{M}$. According to Steinberg [51, Corollary 10.10] we may choose a Borel subgroup $\underline{B}$ and maximal torus $\underline{T} \subset \underline{B}$, both stable under σ. Let $\underline{N}$ denote the normalizer of $\underline{T}$ in $\underline{G}$. Let $\underline{W}$ denote the Weyl group, let $\underline{R}$ denote the Renner monoid, and let $\underline{\Lambda}$ denote the cross section lattice defined in Theorem 5.8. Let $G = \underline{G} \cap M$ denote the finite group of units of M, let $B = \underline{B} \cap M$, let $T = \underline{T} \cap M$, and let $N = \underline{N} \cap M$. The Frobenius automorphism acts on $\underline{W}, \underline{R}$ and $\underline{\Lambda}$. Let W, R, Λ denote the corresponding fixed point sets. Then $R = \underline{R}, \Lambda = \underline{\Lambda}$ and $W = \underline{W} \simeq N/T$ [44], [46]. [47] We identify $N/T = W$. Then (G, B, N, S) is a Tits system with Weyl group W. There is a decomposition

$$M = BRB = \bigsqcup_{\rho \in R} B\rho B \tag{109}$$

analogous to the Bruhat decomposition $G = \sqcup_{w \in W} BwB$ of the finite reductive group G. The union in (109) is disjoint and $B\rho B = B\rho' B$ implies $\rho = \rho'$.

Example 5.11 If $\underline{M} = \mathbf{M}_n(K)$ then $M = \mathbf{M}_n(\mathbf{F}_q)$, $G = \mathbf{GL}_n(\mathbf{F}_q)$, B is the group of invertible upper triangular matrices over $\mathbf{F}_q$, and T is the group of invertible diagonal matrices over $\mathbf{F}_q$. Here R is isomorphic to the symmetric inverse semigroup (rook monoid), W is isomorphic to the symmetric group S_n, and Λ is as in Example 5.9.

---

[46] Renner allows for a more general setup which includes, for example, monoid analogues of the finite unitary groups. His endomorphisms σ are analogous to Steinberg's endomorphisms [51] in the theory of reductive groups. This setup is more subtle than the one we consider here.

[47] The simplicity of this statement depends on our assumption that σ is the q-th power map; see [46, 3.3] for an example in which G is a finite unitary group.

There is a close connection between the modular representation theories of G and M. If G is any finite reductive group over $\mathbf{F}_q$, there exists [38, Section 4] a reductive monoid $\underline{M}$, over the algebraically closure K of $\mathbf{F}_q$ such that (1) $G(M)$ is a finite central extension of G and (2) the type map $\lambda : \Lambda - \{0\} \to 2^S$ is an order preserving bijection. The finite monoid M has a central extension $\mathcal{M}$ whose representation theory over K is intimately related to the modular representation theory of G. In particular, [38, Theorem 2.2], every irreducible modular representation of $\mathcal{M}$ restricts to an irreducible representation of G and the number of extensions of a modular representation of G to $\mathcal{M}$ is determined explicitly in terms of the Curtis-Richen representation theory of G. See Renner's paper [47] in these Proceedings for an explicit statement and further references.

Since the Bruhat decomposition is at the heart of the theory of finite reductive groups, one might expect the same of Renner's decomposition (109) for finite reductive monoids. With this in mind, we sketch the analogue for a finite reductive monoid M of Iwahori's double coset ring construction for G [9, §67], [17]. Let $\mathbf{C}M = \bigoplus_{a \in M} \mathbf{C}a$ be the monoid algebra, a finite dimensional associative algebra with identity which contains the group algebra $\mathbf{C}G$. If X is any subset of M write $[X] = \sum_{a \in X} a \in \mathbf{Z}M$. Thus $\epsilon = |B|^{-1}[B] \in \mathbf{C}G$ is the idempotent corresponding to the subgroup B. The Iwahori algébra [48] is by definition

$$\mathcal{H}_{\mathbf{C}}(G,B) = \epsilon \mathbf{C}G\epsilon = \bigoplus_{w \in W} \mathbf{C}[BwB] \,. \tag{110}$$

It is semisimple because $\mathbf{C}G$ is semisimple. If we define

$$T_w = \frac{1}{|B|}[BwB] \tag{111}$$

then, on general group theoretic grounds [9, 11.30],

$$\mathcal{H}(G,B) = \bigoplus_{w \in W} \mathbf{Z}T_w \tag{112}$$

is a ring. Iwahori studied this ring [17] because $\mathcal{H}_{\mathbf{C}}(G,B)$ controls the decomposition of the permutation representation of G on G/B. In particular, he found the multiplication table for the basis elements T_w [17, Corollary 4.2], [9, Theorem 67.2]:

Theorem 5.11 *Let G be a finite reductive group. The ring $\mathcal{H}(G,B)$ is generated by the T_s for $s \in S$. The multiplication of the T_w is determined by the formulas*

[48] This is usually called the Iwahori Hecke algebra or simply Hecke algebra; see [9, 11.22] and the Introduction to [48] for historical remarks on the terminology. See [9, §11] for the theory of Hecke algebras or double coset rings in the context of general representation theory of finite groups and [9, §§67,68] for these algebras in the context of finite groups with a Tits system. Iwahori [17] was the first to notice that the double coset ring of a finite reductive group G with respect to a Borel subgroup B has a remarkable structure. The examples he studied are the ones which have surfaced as "Hecke algebras" in other parts of mathematics.

(113) $$T_s T_w = \begin{cases} T_{sw} & \textit{if } l(sw) = l(w) + 1 \\ qT_{sw} + (q-1)T_w & \textit{if } l(sw) = l(w) - 1 \end{cases}$$

where $s \in S$ and $w \in W$.

Here $w \mapsto l(w)$ is the length function given by (54). If $w = s_{i_1} \cdots s_{i_p}$ where $s_{i_1}, \ldots, s_{i_p} \in S$ and $l(w) = p$, it follows from (113) that $T_w = T_{s_{i_1}} \cdots T_{s_{i_p}}$. Thus, in view of (113), if $w', w'' \in W$ we have $T_w T_{w'} = \sum_{w'' \in W} c_{w,w',w''} T_{w''}$ where the $c_{w,w',w''}$ are polynomials in q with integer coefficients.

Is there an analogous theorem for a reductive monoid M? If M is *any* finite monoid with group of units G, and B is a subgroup of G, then we may define a $\mathbf{C}$-algebra $\mathcal{H}_{\mathbf{C}}(M, B)$ by

(114) $$\mathcal{H}_{\mathbf{C}}(M, B) = \epsilon\mathbf{C}[M]\epsilon = \bigoplus_D \mathbf{C}[D] \supseteq \mathcal{H}_{\mathbf{C}}(G, B)$$

where D ranges over the $B \times B$ orbits on M. These orbits are the natural replacement for the double cosets. The algebra $\mathcal{H}_{\mathbf{C}}(M, B)$ won't, in general, be semisimple. So let's suppose that M is a finite reductive monoid, and let G, B be as before. Then $\mathbf{C}[M]$ is semisimple and hence so is $\mathcal{H}_{\mathbf{C}}(M, B)$. This is a theorem of Okniński and Putcha [20]. The proof is not easy; it uses the Harish-Chandra induction theory for characters of G. To gauge the depth of this result one should note that there are no general theorems about semisimplicity of monoid algebras which will prove the semisimplicity of $\mathbf{C}[M]$ in case $M = \mathbf{M}_n(\mathbf{F}_q)$; Kovaćs [19] has argued this special case from first principles. Since $\mathbf{C}[M]$ is semisimple, so is $\mathcal{H}_{\mathbf{C}}(M, B)$. What can we say about the structure of $\mathcal{H}_{\mathbf{C}}(M, B)$ in the direction of Iwahori's work for G? Renner's decomposition (109) gives

(115) $$\mathcal{H}_{\mathbf{C}}(M, B) = \bigoplus_{\rho \in R} \mathbf{C}[B\rho B] \, .$$

Putcha [33], [36, Section 4] studied $\mathcal{H}_{\mathbf{C}}(M, B)$ with remarkable results. Fix an orbit, alias $\mathcal{J}$-class, $\mathcal{O} \in G\backslash M/G$ and consider the *local monoid* $M(\mathcal{O}) = G \cup \mathcal{O} \cup \{0\}$ [35, p.638] where the product in $M(\mathcal{O})$ of two elements of $\mathcal{O}$ is defined to be 0 if their product in M is not in $\mathcal{O}$. Since $M(\mathcal{O}) \supset G$ we may consider the monoid algebra $\mathcal{H}_{\mathbf{C}}(M(\mathcal{O}), B) \supset \mathcal{H}_{\mathbf{C}}(G, B)$. Putcha determined the structure of these algebras $\mathcal{H}_{\mathbf{C}}(M(\mathcal{O}), B)$, and, by piecing them together over the various orbits $\mathcal{O}$, proved an isomorphism [33, Theorem 4.1]

(116) $$\mathcal{H}_{\mathbf{C}}(M, B) \simeq \mathbf{C}[R]$$

analogous to Tits' *deformation theorem* $\mathcal{H}_{\mathbf{C}}(G, B) \simeq \mathbf{C}[W]$ in the theory of finite groups with a Tits system [9, Theorem 68.21]. The proof involves an interesting application of the Kazhdan-Lusztig "R-polynomials."

There is no known general theorem for finite monoids, like [9, Theorem 11.30] for finite groups, which guarantees the existence of constants $a_D \in \mathbf{Q}$ corresponding to $B \times B$ orbits D on M, such that $\oplus_D \mathbf{Z}a_D[D]$ is a ring. If M is a finite reductive monoid one may hope for special dispensation. Thus the question is: are there elements $T_\rho = a_\rho[B\rho B]$ with $\rho \in R$ and $a_\rho \in \mathbf{Q}$, and formulas analogous to those in (113), which show that the structure constants for the multiplication table in $\mathcal{H}_{\mathbf{C}}(M, B)$ relative to the basis T_ρ are integers given by polynomials in q? The answer is "yes". To see how this comes about, we recall Iwahori's setup in $\mathcal{H}_{\mathbf{C}}(G, B)$. The trivial representation $g \mapsto 1$ for $g \in G$ of $\mathbf{C}G$ restricts to a representation ind : $\mathcal{H}_{\mathbf{C}}(G, B) \to \mathbf{C}$. The double coset BwB may be written in Chevalley normal form [9, 69.7]

(117) $$BwB = BwU_w^-$$

where

(118) $$U_w^- = \prod_{\alpha \in \Phi_w^-} U_\alpha$$

with U_α as in (21) and $\Phi_w^- = \{\alpha \in \Phi^+ \mid w\alpha \in \Phi^-\}$ as in (55). The groups U_w^- were defined by Chevalley in [7, p.42]. The expression of an element of BwB as bwu with $b \in B$, $w \in W$, and $u \in U_w^-$ is unique. Thus, in view of (56) we have

(119) $$|BwB| = |B|q^{l(w)}$$

where $l(w)$ is the length function on W. Thus

(120) $$\mathrm{ind}(T_w) = q^{l(w)} .$$

We would like monoid analogues of (117)-(120). If we find them, we have a good shot at the structure of $\mathcal{H}(M, B)$. To begin, we need a definition of length. In the case of reductive groups there are various ways to view the length function on W, all of them closely related:

(L1) Words: $l(w) = \min \{p \mid w = s_1 \dots s_p \text{ with } s_i \in S\}$ as in (54) .

(L2) Roots: $l(w) = |\Phi_w^-|$ as in (56).

(L3) Geometry of G/B: $l(w) = \dim BwB/B = \dim BwB - \dim B$ is the dimension of a Bruhat cell; in this formulation we are in the context of algebraic groups over an algebraically closed ground field K.

The monoid analogue of Iwahori's theorem (113) and of the deformation theorem (116) was proved in [48] in case $M = \mathbf{M}_n(\mathbf{F}_q)$. The argument there, defines a length function $\rho \mapsto l(\rho)$ in the spirit of (L1) and (L2). Each $W \times W$ orbit on R has its own length function. If the orbit is $W \cdot 1 \cdot W$ then the length function agrees with the length function $w \mapsto l(w)$ on W. The construction depends on the set

$\mathcal{N} = \{1, \nu, \nu^2, \ldots, \nu^{n-1}, \nu^n = 0\}$ where $\nu = E_{12} + E_{23} + \ldots + E_{n-1,n}$ is a Jordan block. The set $\mathcal{N}$ is a cross section for $G\backslash M/G$ and for $W\backslash R/W$; the elements of $G\nu^iG$ have rank $n-i$. If $\rho \in W\nu^iW$ write $\rho = w\nu^iw'$ with $w, w' \in W$ and define $l(\rho) = \min(l(w) + l(w'))$ over all such expressions. Thus the element ν^i is the unique element of length zero in its $W \times W$-orbit. The main combinatorial idea in [48] is to find a substitute for Chevalley's two part partitions of the set of positive roots: those positive roots which change sign under a given element $w \in W$ and those which don't. The corresponding partitions for $\rho \in R$ have five (some possibly empty) parts. In view of later developments [49] these partitions are a piece of the general theory. Renner [45] discovered the following marvelous fact, which proves the existence of a set analogous to $\mathcal{N} \subseteq \mathbf{M}_n(\mathbf{F}_q)$ in any reductive monoid.

Theorem 5.12 *Let M be a reductive monoid.* [49] *Let $\mathcal{N} = \dot{\mathcal{N}}(B) = \{\nu \in R \mid B\nu = \nu B\}$. The set $\mathcal{N}$ is a cross section for $G\backslash M/G$ as well as $W\backslash R/W$.*

If M is the monoid of all matrices over *any* field, and B is the group of upper triangular matrices, and $\nu = E_{12} + E_{23} + \ldots + E_{n-1,n}$ is a Jordan block then $\mathcal{N} = \{1, \nu, \ldots, \nu^{n-1}\}$.

Renner used his set $\mathcal{N}$ to define length in the spirit of (L3) as follows. Suppose $\rho \in R$. Choose $\nu \in \mathcal{N}$ with $\rho \in W\nu W$ and define

$$l(\rho) = \dim B\rho B - \dim B\nu B \ . \tag{121}$$

Thus the elements of $\mathcal{N}$ are, by definition, the elements of length zero. With this definition Renner [44] proved the monoid analogue of Tits' formulas (63) for the multiplication of double cosets, namely,

$$BsB \cdot B\rho B = \begin{cases} B\rho B & \text{if } l(s\rho) = l(\rho) \\ Bs\rho B & \text{if } l(s\rho) = l(\rho) + 1 \\ Bs\rho B \sqcup B\rho B & \text{if } l(s\rho) = l(\rho) - 1 \end{cases} \tag{122}$$

where $s \in S$ and $\rho \in R$. These formulas agree with those in [48, Proposition 3.14] if $M = \mathbf{M}_n(\mathbf{F}_q)$ which were proved using (L1) and (L2). The case $l(s\rho) = l(\rho)$ where $BsB \cdot B\rho B = B\rho B$ is impossible for groups. It turns out that one can also define length for the Renner monoid R of any reductive monoid M in the spirit of (L1) and (L2). One also has (L3) since there is a Chevalley normal form for the orbits $B\rho B$ with $\rho \in R$ which is analogous to (117) [49]. This leads, in particular, to the formula

$$|B\rho B| = |eBe| q^{l(\rho)} \tag{123}$$

[49] In this theorem, and in (122) one may choose M to be a reductive monoid over an algebraically closed field K *or* the finite monoid of fixed points under the Frobenius map in case K is the algebraic closure of $\mathbf{F}_q$. In (121) it is understood that the field is algebraically closed.

where $e \in \Lambda$ is the idempotent in the $\mathcal{J}$-class, alias $G \times G$ orbit, of ρ. As in the case of groups the restriction of the trivial representation $a \mapsto 1$ for $a \in M$ of $\mathbf{C}M$ restricts to a representation $\mathrm{ind} : \mathcal{H}_{\mathbf{C}}(M, B) \to \mathbf{C}$. Thus if we define

$$T_\rho = \frac{1}{|eBe|} [B\rho B] \tag{124}$$

we have a rational multiple of the double coset sum $[B\rho B]$ which satisfies the analogue of (120), namely

$$\mathrm{ind}(T_\rho) = q^{l(\rho)} . \tag{125}$$

This leads to a theorem [49] on the Iwahori ring of a finite reductive monoid:

Theorem 5.13 *Let M be a finite reductive monoid. Define a free $\mathbf{Z}$-module of rank $|R|$ by*

$$\mathcal{H}(M, B) = \bigoplus_{\rho \in R} \mathbf{Z} T_\rho .$$

Then $\mathcal{H}(M, B)$ is a ring generated by the T_s for $s \in S$, and T_ν for $\nu \in \mathcal{N}$. The multiplication table is determined by the formulas

$$T_s T_\rho = \begin{cases} qT_\rho & \text{if } l(s\rho) = l(\rho) \\ T_{s\rho} & \text{if } l(s\rho) = l(\rho) + 1 \\ qT_{s\rho} + (q-1)T_\rho & \text{if } l(s\rho) = l(\rho) - 1 \end{cases} \tag{126}$$

and

$$T_\nu T_\rho = q^{l(\rho) - l(\nu\rho)} T_{\nu\rho} \tag{127}$$

for $s \in S$, $\rho \in R$, and $\nu \in \mathcal{N}$. There are analogous formulas for right multiplication by T_s and the T_ν.

The structure constants in (127) are integers because $l(\rho) \geq l(\nu\rho)$ for all $\rho \in R$ and $\nu \in \mathcal{N}$. If $M = \mathbf{M}_n(\mathbf{F}_q)$ this inequality was proved in [48, Theorem 4.12]. The author first proved Theorem 5.13 with $\mathcal{H}(M, B)$ replaced by $\mathcal{H}_A(M, B) = \oplus_{\rho \in R} AT_\rho$ where A is any (commutative) coefficient ring in which q is invertible. The obstacle to replacement of A by $\mathbf{Z}$ lay in the conjectured inequality $l(\rho) \geq l(\nu\rho)$ for $\rho \in R$ and $\nu \in \mathcal{N}$, which one needs to control the exponent in (127). The inequality was proved by Putcha in a letter to the author. [50] Putcha [34] described the irreducible representations of $\mathcal{H}_{\mathbf{C}}(M, B)$, and found an explicit isomorphism $\mathcal{H}_{\mathbf{C}}(M, B) \simeq \mathbf{C}[R]$.

[50] The author would like to express his thanks.

The theory developed by Putcha and Renner seems to me to be the beginning of a long story. There is much to be done, on "Hecke rings", on finite reductive monoids, on reductive monoids over more general coefficient rings, and over special ones like $\mathbf{C}$. If you have seen this subject here for the first time, want to work on it, and come to it from outside the theory of reductive groups, my advice would be: compute one difficult example in great detail and bring to bear the part of mathematics (for example semigroup theory) that you know best. If you are already comfortable with reductive groups, keep in mind the diagram:

$$\begin{array}{ccc} M\ (=\mathbf{M}_n) & \rightsquigarrow & R\ (=I_n) \\ \cup & & \cup \\ G\ (=\mathbf{GL}_n) & \rightsquigarrow & W\ (=S_n) \end{array}$$

Many aspects of the bottom row are well understood. Use what you know and work on the top.

References

1. E. Artin, *Geometric algebra*, John Wiley, 1966.
2. C. Berge, *Principles of combinatorics*, Academic Press, 1971.
3. A. Borel, *Linear algebraic groups*, 2nd ed. Springer-Verlag, 1991.
4. N. Bourbaki, *Groupes et algèbres de Lie*, Chapitre 4, 5 et 6, Hermann, Paris, 1968.
5. R. Carter, *Simple groups of Lie type*, John Wiley, 1972.
6. C. Chevalley, *Séminaire: Classification des groupes de Lie algébriques*, Ecole Normale Supérieure, Paris, 1958.
7. C. Chevalley, *Sur certains groupes simples*, Tôhoku Math. J. **7** (1955), 14–66.
8. H. S. M. Coxeter, *Discrete groups generated by reflections*, Annals of Math. **35** (1934), 588–621.
9. C. W. Curtis and I. Reiner, *Methods of representation theory*, Volume I, John Wiley, 1981 and Volume II, John Wiley, 1987.
10. M. Demazure and P. Gabriel, *Groupes algébriques*, Tome I, North-Holland, 1970.
11. E. B. Dynkin, *The structure of semi-simple algebras*, American Math. Soc. Translation 17 (1950); Uspehi Matematicheskih Nauk (N.S.) **20** (1947) 59–127.
12. W. Fulton, *Introduction to toric varieties*, Annals of Math. Studies 131, Princeton Univ. Press, 1993.
13. D. Ju. Grigor'ev, *An analogue of the Bruhat decomposition for the closure of the cone of a Chevalley group over a finite field*, Soviet Math. Doklady **23** (1981), 393–397.
14. L. C. Grove and C. T. Benson, *Finite Reflection Groups*, Second Edition, Springer-Verlag, 1985.
15. J. E. Humphreys, *Linear Algebraic Groups*, Springer-Verlag, 1975.
16. J. E. Humphreys, *Reflection groups and Coxeter groups*, Cambridge Univ. Press, 1992.
17. N. Iwahori, *On the structure of a Hecke ring of a Chevalley group over a finite field*, J. Fac. Sci. Univ. Tokyo, Sec. I, **10** (1964), 215–236.
18. G. Kempf, F. Knudsen, D. Mumford and B. Saint-Donat, *Toroidal Embeddings*, Lecture Notes in Math. 339, Springer-Verlag, 1973.
19. L. G. Kovaćs, *Semigroup algebras of the full matrix semigroup over a finite field*, Proc. Amer. Math. Soc. **116** (1992), 911–919.
20. J. Okniński and M. S. Putcha, *Complex Representations of Matrix Semigroups*, Trans. Amer. Math. Soc. **323** (1991), 563–581.
21. M. S. Putcha, *On linear algebraic semigroups*, Trans. Amer. Math. Soc. **259** (1980), 457–469.
22. M. S. Putcha, *On linear algebraic semigroups II*, Trans. Amer. Math. Soc. **259** (1980), 471–491.
23. M. S. Putcha, *On linear algebraic semigroups III*, Internat. J. Math. & Math. Sci. **4** (1981) 667–690.

24. M. S. Putcha, *Green's relations on a connected algebraic monoid*, Linear and Multilinear Algebra **12** (1982), 205–214.
25. M. S. Putcha, *Connected algebraic monoids*, Trans. Amer. Math. Soc. **272** (1982), 693–709.
26. M. S. Putcha, *A semigroup approach to linear algebraic groups*, J. Algebra **80** (1983), 164–185.
27. M. S. Putcha, *Determinant functions on algebraic monoids*, Comm. in Algebra, **11** (1983), 695–710.
28. M. S. Putcha, *Reductive groups and regular semigroups*, Semigroup Forum **30** (1984), 253–261.
29. M. S. Putcha, *Regular linear algebraic monoids*, Trans. Amer. Math. Soc. **290** (1985), 615–626.
30. M. S. Putcha, *A semigroup approach to linear algebraic groups II. Roots*, J. Pure Appl. Algebra **39** (1986), 153–163.
31. M. S. Putcha, *Linear Algebraic Monoids*, London Math. Soc. Lecture Note Series 133, Cambridge Univ. Press 1988.
32. M. S. Putcha, *Monoids on groups with BN-pairs*, J. Algebra **120** (1989), 139–169.
33. M. S. Putcha, *Sandwich Matrices, Solomon algebras, and Kazhdan-Lusztig polynomials*, Trans. Amer. Math. Soc. **340** (1993), 415–428.
34. M. S. Putcha, *Monoid Hecke Algebras*, preprint, 1993
35. M. S. Putcha, *Classification of Monoids of Lie Type*, J. Algebra **163** (1994), 636–662.
36. M. S. Putcha, *Monoids of Lie type* (in these Proceedings).
37. M. S. Putcha and L. E. Renner, *The system of idempotents and the lattice of $\mathcal{J}$-classes of reductive algebraic monoids*, J. Algebra **116** (1988), 385–399.
38. M. S. Putcha and L. E. Renner, *The canonical compactification of a finite group of Lie type*, Trans. Amer. Math. Soc. 337 (1993), 305–319.
39. L. E. Renner, *Algebraic Monoids*, Thesis, The University of British Columbia, 1982.
40. L. E. Renner, *Classification of semisimple rank one monoids*, Trans. Amer. Math. Soc. **287** (1985), 457–473.
41. L. E. Renner, *Classification of semisimple algebraic monoids*, Trans. Amer. Math. Soc. **292** (1985), 193–223.
42. L. E. Renner, *Reductive monoids are von Neumann regular*, J. Algebra **93** (1985), 237–245.
43. L. E. Renner, *Analogue of the Bruhat decomposition for algebraic monoids*, J. Algebra **101** (1986), 303–338.
44. L. E. Renner, *Finite monoids of Lie type, Monoids and Semigroups with Applications*, J. Rhodes (ed.), World Scientific, 1991, 278–287.
45. L. E. Renner, *Analogue of the Bruhat decomposition for algebraic monoids, II. The length function and the trichotomy*, preprint, 1991.
46. L. E. Renner, *Finite reductive monoids*, these Proceedings.
47. L. E. Renner, *Modular representations of finite monoids of Lie type*, these Proceedings.
48. L. Solomon, *The Bruhat decomposition, Tits system and Iwahori ring for the monoid of matrices over a finite field*, Geom. Dedicata **36** (1990), 15–49.
49. L. Solomon, *The Iwahori ring of a finite reductive monoid*, (manuscript in preparation).
50. T. A. Springer, *Linear algebraic groups*, Birkhäuser, 1983.
51. R. Steinberg, *Endomorphisms of linear algebraic groups*, Memoirs Amer. Math. Soc. **80**, 1968.
52. J. Tits, *Sur certaines classes d'espaces homogènes de groupes de Lie*, Mémoires Académie royale de Belgique, Classe des Sciences, Tome XXIX, Fascicule 3, N° 3, 1955.
53. J. Tits, *Théorème de Bruhat et sous-groupes paraboliques*, C. R. Acad. Sci. Paris **254** (1962), 2910–2912.
54. W. C. Waterhouse, *The unit groups of affine algebraic monoids*, Proc. Amer. Math. Soc. **85** (1982), 506–508.

MONOIDS OF LIE TYPE

MOHAN S. PUTCHA*
Department of Mathematics
North Carolina State University
Raleigh, NC 27695-8205
U. S. A.

1. Introduction

The theory of semigroups has been intensely developed for over 50 years. However the subject has strayed away from the major areas of algebra. A clear evidence of this is the fact that linear representation theory plays no role in semigroup theory at a time when representation theory has become *the* central force in all other areas of algebra. The theory of reductive monoids and monoids of Lie type is a strong attempt to steer semigroup theory into the mainstream.

As with the original semigroup theorists (see for example [3]) we draw our inspiration from group theory, but in a radically different way. The class of monoids to be studied is decided, not by a random choice of made up equations but by geometric considerations.

Consider a reductive group $G \subseteq GL_n(F)$ where F is an algebraically closed field. Then the Zariski closure M of G in $M_n(F)$ is a regular monoid that is called a *reductive monoid.* Let B_0, B_0^- be opposite Borel subgroups of G, $T = B_0 \cap B_0^-$ and $W = N_G(T)/T$ the Weyl group. The set of diagonal idempotents $E(\overline{T})$ is isomorphic to the face lattice of a rational polytope. There is a cross-section lattice $\Lambda \subseteq E(\overline{T})$ such that

$$M = \bigsqcup_{e \in \Lambda} GeG.$$

The $G \times G$ orbits GeG are the $\mathcal{J}$-classes of M and hence we can form for $J = GeG$ a *local monoid* or *orbit monoid,*

$$M(J) = G \cup J^0 = G \cup J \cup \{0\}$$

where for $a, b \in J$

$$a \cdot b = \begin{cases} ab & \text{if } ab \in J, \\ 0 & \text{otherwise.} \end{cases}$$

* Partially supported by NSF Grant DMS9200077

J. Fountain (ed.), Semigroups, Formal Languages and Groups, 353–367.

The subgroups

$$P \;\; = P(e) = \{x \in G | xe = exe\},$$

$$P^- \; = P^-(e) = \{x \in G | ex = exe\}$$

are opposite parabolic subgroups of G by [12]. If $L = P \cap P^-$ then $eL = Le$ is the unit group of eMe and we have natural surjective homomorphisms

$$\delta_+ : P \to eL,\; \delta_+(a) = ae,$$

$$\delta_- : P^- \to eL,\; \delta_-(a) = ea.$$

These homomorphisms agree on L and factor through L. So $eL \cong L/K$ for some $K \triangleleft L$. Any two idempotents in J are conjugate and this implies by [19], [21] that the above data uniquely determines the monoid $M(J)$. Hence we write this monoid as $M(G, P, P^-, L/K)$ with the homomorphisms δ_+, δ_- being understood to be part of the data. The *Renner monoid*

$$R = R(M) = \overline{N_G(T)}/T = \langle \Lambda, W \rangle$$

is a finite inverse monoid and by [27]

$$M = \bigsqcup_{\sigma \in R} B\sigma B.$$

What miracles does taking the Zariski closure produce? Consider the multiplicative monoid $M = M_n(F)$, where F is an algebraically closed field. The unit group $G = GL_n(F)$ and the Weyl group W is the symmetric group with a generating set of simple reflections

$$S = \{(12), (23), \ldots, (n-1\; n)\}.$$

The cross-section lattice

$$\Lambda = \{0 < e_1 < e_2 < \ldots < e_{n-1} < 1\}$$

where $e_r = \left[\begin{smallmatrix} I_r & 0 \\ 0 & 0 \end{smallmatrix}\right]$. Then $P_r = P(e_r), P_r^- = P^-(e_r)$ are opposite maximal parabolic subgroups. If $J_r = Ge_rG$ then $M(J_r)$ is not a submonoid of M for $1 < r < n$. Now consider the representation θ of M given by

$$\theta(a) = \wedge^1(a) \otimes \wedge^2(a) \otimes \cdots \otimes \wedge^n(a),$$

the tensor product of all the exterior powers of a. This representation has degree $\prod_{r=1}^{n} \binom{n}{r}$. θ has the effect of killing the singular part of M since $\theta(M) = \theta(G) \cup \{0\}$. However something amazing happens when we take the Zariski closure $\mathcal{M}$ of $\theta(M)$. The cross-section lattice is now

$$\Lambda = \{0\} \cup \{e_I | I \subseteq S\}$$

where

$$e_{I_1} \leq e_{I_2} \text{ if and only if } I_1 \subseteq I_2.$$

Moreover $P(e_I) = P_I$ and $J_I = Ge_IG$, then$M(J_I)$ is actually a submonoid of $\mathcal{M}$. $\mathcal{M}$ is called a *canonical monoid* for GL_n. Renner and the author [23] construct canonical monoids for any reductive group or group of Lie type.

Next consider the representation ψ of SL_n given by

$$\psi(a) = \wedge^1(a) \oplus \wedge^2(a) \oplus \cdots \oplus \wedge^n(a).$$

Then ψ is of degree $2^n - 1$. Let $\tilde{\mathcal{M}}$ denote the Zariski closure of $F \cdot \psi(SL_n)$. We call $\tilde{\mathcal{M}}$ a *dual canonical monoid* for GL_n. The $\mathcal{J}$-class and idempotent structure of $\tilde{\mathcal{M}}$ is that of $\mathcal{M}$ turned upside down. The cross-section lattice

$$\Lambda = \{1\} \cup \{e_I | I \subseteq S\}$$

where

$$e_{I_1} \leq e_{I_2} \text{ if and only if } I_2 \subseteq I_1.$$

Moreover the lattice of diagonal idempotents is isomorphic to the Coxeter complex of the symmetric group and the lattice of $\mathcal{R}$-classes is isomorphic to the Tits building of GL_n. For all idempotents $e \neq 1, e\tilde{\mathcal{M}}e$ is an inverse monoid. We refer to [15] for the general theory of reductive monoids. These monoids are being discussed in more detail by Solomon in his article [35] in these proceedings. We will therefore now concentrate on monoids of Lie type.

2. Definition, structure and examples

Monoids of Lie type were introduced by the author [16] under the name 'regular split monoids'. Finite reductive monoids discussed by Renner [29] in these proceedings, provide an important source of examples. Let G be a finite group of Lie type with opposite Borel subgroups B_0, B_0^- and Weyl group W. We refer to [2], [6] for details. A finite regular monoid M with unit group G is a *monoid of Lie type* if every local monoid is of the form

$$M(G, P, P^-, L/K)$$

where P, P^- are opposite parabolic subgroups of G and $K \triangleleft L = P \cap P^-$. Let $E(M)$ denote the idempotent set of M and let the *cross-section lattice*

$$\Lambda = \Lambda(M) = \{e \in E(M) | B_0 \subseteq P(e), B_0^- \subseteq P^-(e)\}$$

where

$$P(e) = \{g \in G | ge = ege\},$$
$$P^-(e) = \{g \in G | eg = ege\}.$$

We also let

$$L(e) = \{g \in G | ge = eg\},$$

$$W(e) = \{x \in W | xex^{-1} = e\}.$$

Then $W(e)$ is the Weyl group of $L(e)$. The lattice of *diagonal idempotents*,

$$\mathcal{D} = \mathcal{D}(M) = \bigcup_{x \in W} x\Lambda x^{-1}.$$

Theorem 2.1 ([16])

(i) *For all* $e, f \in \mathcal{D}, ef = fe$.

(ii) Λ *and* $\mathcal{D}$ *are lattices.*

(iii) $M = \bigsqcup_{e \in \Lambda} GeG$.

(iv) *If* $e, f \in \Lambda$, *then* $f \in MeM$ *if and only if* $f \leq e$.

(v) *If* $x, y \in W, e, f \in \Lambda$, *then* $xex^{-1} \geq yey^{-1}$ *if and only if* $e \geq f$ *and* $x \cdot W(e) \cap y \cdot W(e) \neq \emptyset$.

Now $B_0 e$ is a Borel subgroup of the unit group of eMe. Let

$$W_e = \{x \in W(e) | x \cdot Be \cdot x^{-1} = Be\}.$$

Both $W(e)$ and W_e are standard parabolic subgroups of W and

$$W(e) = W_e \times \widetilde{W(e)}$$

where $\widetilde{W(e)}$ is the Weyl group of the unit group of eMe. The *Renner monoid*

$$R = R(M) = \langle W, \Lambda \rangle = W\mathcal{D} = \mathcal{D}W$$

where for $e \in \Lambda, x \in W(e)$, and we have by definition that

$$xe = e \text{ if and only if } x \in W_e.$$

Theorem 2.2 ([16])

(i) R *is an inverse monoid with unit group* W *and idempotent set* $\mathcal{D}$.

(ii) $R = \bigsqcup_{e \in \Lambda} WeW$.

(iii) *The local monoids of* R *are* $M(W, W(e), W(e), \widetilde{W(e)}), e \in \Lambda$.

(iv) $M = \bigsqcup_{\sigma \in \Lambda} B_\sigma B$.

We will now give several examples of monoids of Lie type.

Example 2.1 $M = M_n(\mathbb{F}_q)$ is a finite reductive monoid and hence a monoid of Lie type with unit group $G = GL_n(\mathbb{F}_q)$. The cross-section lattice

$$\Lambda = \{0 = e_0 < e_1 < \ldots < e_n = 1\}$$

where $e_r = \left[\begin{smallmatrix} I_r & 0 \\ 0 & 0 \end{smallmatrix}\right]$. The local monoids are

$$M(G, P_r, P_r^-, GL_r)$$

where $P_r = P(e_r), P_r^- = P^-(e_r)$. Now

$$G \cong \tilde{G} = \{A \oplus \alpha | A \in G, \alpha \in F, \alpha \cdot \det A = 1\} \subseteq GL_{n+1}(\mathbb{F}_q).$$

For $r = 0, 1, \ldots, n$, let $\tilde{e}_r = e_r \oplus 1$ and let

$$\tilde{\Lambda} = \{\tilde{e}_r | r = 0, \ldots, n\}.$$

Then

$$M^+ = \langle \tilde{G}, \tilde{\Lambda} \rangle$$

is a monoid of Lie type with unit group $\tilde{G} \cong G$ and cross-section lattice $\tilde{\Lambda} \cong \Lambda$. The local monoids of M^+ are

$$M(G, P_r, P_r^-, GL_r \times F^*).$$

Next for singular matrices $A_1, A_2 \in M_n(\mathbb{F}_q)$ define $A_1 \sim A_2$ if $A_2 = \alpha A_1$ for some $\alpha \in \mathbb{F}_q^*$. Then

$$M^- = M/\sim$$

is also a monoid of Lie type with unit group G and cross-section lattice Λ. The local monoids are now

$$M(G, P_r, P_r^-, PGL_r).$$

We have surjective idempotent separating homomorphisms

$$M^+ \to M \to M^-.$$

We call M_+ the universal monoid and M^- the fundamental monoid. M^+ and M^- are not reductive monoids. One advantage that M^+ has over M is that by [21], every irreducible modular representation of G of degree > 1 has a non-trivial extension to M^+. This is not true of M.

Example 2.2 Let $G = GL_n(\mathbb{F}_q), 1 \leq r < n$. Then

$$M = M(G, P_r, P_r^-, GL_r)$$

is a monoid of Lie type. The universal monoid

$$M^+ = M(G, P_r, P_r^-, GL_r \times GL_{n-r})$$

and the fundamental monoid

$$M^- = M(G, P_r, P_r^-, 1).$$

Example 2.3 Let G be a group of Lie type and P, P^- opposite parabolic subgroups, $L = P \cap P^-$. Then we have the universal monoid

$$M^+ = M(G, P, P^-, L)$$

and the fundamental monoid

$$M^- = M(G, P, P^-, 1).$$

Example 2.4 (**[23]**) Let G be a group of Lie type. Then we can construct a canonical reductive monoid $\mathcal{M} = \mathcal{M}(G)$ with cross-section lattice

$$\Lambda = \{0\} \cup \{e_I | I \subseteq S\}$$

where S is the generating set of simple reflections of W. For $I_1, I_2, \subseteq S$,

$$e_{I_1} \leq e_{I_2} \text{ if and only if } I_1 \subseteq I_2.$$

The universal monoid $\mathcal{M}^+$ has local monoids

$$M(G, P_I, P_I^-, L_I), I \subseteq S$$

and each one of them is a submonoid of $\mathcal{M}^+$. The fundamental monoid $\mathcal{M}^-$ has local monoids

$$M(G, P_I, P_I^-, L_I/Z(L_I)), I \subseteq S$$

where $Z(L_I)$ is the center of L_I. We have surjective idempotent separating homomorphisms

$$\mathcal{M}^+ \to \mathcal{M} \to \mathcal{M}^-.$$

Example 2.5 (**[13, 21]**) Let G be a group of Lie type. The dual canonical reductive monoid $\tilde{\mathcal{M}}$ can be constructed with cross-section lattice

$$\Lambda = \{e_I | I \subseteq S\} \cup \{1\}$$

where for $I_1, I_2 \subseteq S$,

$$e_{I_1} \leq e_{I_2} \text{ if and only if } I_2 \subseteq I_1.$$

For any idempotent $e \neq 1, e\tilde{\mathcal{M}}e$ is an inverse monoid. The lattice of $\mathcal{R}$-classes of $\tilde{\mathcal{M}}$ is isomorphic to the Tits building of G and the lattice of diagonal idempotents $\mathcal{D}$ is isomorphic to the Coxeter complex of the Weyl group W. The local monoids of the universal monoid $\tilde{\mathcal{M}}^+$ are

$$M(G, P_I, P_I^-, L_I/L_I'),\ I \subseteq S$$

where L_I' is the subgroup of L_I generated by its unipotent elements. Usually L_I' is the commutator subgroup of L_I. The fundamental monoid $\tilde{\mathcal{M}}^-$ has local monoids

$$M(G, P_I, P_I^-, 1),\ I \subseteq S.$$

We have surjective idempotent separating homomorphisms

$$\tilde{\mathcal{M}}^+ \to \tilde{\mathcal{M}} \to \tilde{\mathcal{M}}^-.$$

$\tilde{\mathcal{M}}^+$ has the property that any irreducible modular representation of G of degree > 1 has a unique non-trivial extension to $\tilde{\mathcal{M}}^+$.

Example 2.6 Let G be a group of Lie type and $\theta : G \to GL(V)$ and irreducible modular representation in the natural characteristic of G. Let E consist of all idempotents e in End V such that

$$P = P(e) = \{a \in G | \theta(a) \cdot e = e \cdot \theta(a) \cdot e\},$$
$$P^- = P^-(e) = \{a \in G | e \cdot \theta(a) = e \cdot \theta(a) \cdot e\}$$

are opposite parabolic subgroups of G with

$$\theta(R_u(P)) \cdot e = \{e\} = e \cdot \theta(R_u(P^-))$$

where $R_u(P)$ and $R_u(P^-)$ are the unipotent radicals of P and P^- respectively. Then the monoid generated by $\theta(G)$ and E is a monoid of Lie type with idempotent set E.

3. Classification

Fix a finite group of Lie type G with Weyl group W and a generating set S of simple reflections. Then S has the structure of a graph, called the Coxeter graph (cf. [7]). If $s, t \in S$, then s and t are adjacent if and only if $st \neq ts$. The parabolic subgroups of G have a rich geometric structure of a Tits building (cf. [36, 31]). In particular two parabolic subgroups P_1, P_2 are *incident* if $P_1 \cap P_2$ is parabolic. If P is a parabolic subgroup and B a Borel subgroup, then by [36]

$$\mathrm{proj}_P(B) = (P \cap B) \cdot R_u(P)$$

where $R_u(P)$ is the unipotent radical of P, is a Borel subgroup contained in P. If P, P^- are opposite parabolic subgroups, then for some $x \in G, x^{-1}Px = P_I, x^{-1}P^-x = P_I^-$. Call I the *type* of the pair P, P^-.

Let M be a monoid of Lie type with unit group G and cross-section lattice Λ. We have a *type map* [22],

$$\lambda : \Lambda \to 2^S$$

where for $e \in \Lambda, P(e) = P_{\lambda(e)}$. So $\lambda(1) = S$. The idempotent set $E(M)$ has a biordered structure [10]:

$$f \leq_r e \text{ if } ef = f, f \leq_\ell e \text{ if } fe = f$$

where if $f \leq_r e$ then $fe \in E(M)$, and if $f \leq_\ell e$ then $ef \in E(M)$.

Now let Λ be a finite lattice and $\lambda : \Lambda \to 2^S$ a map with $\lambda(1) = S$. Let

$$\mathcal{D}_\lambda = \{(a, xW_{\lambda(a)}) | a \in \Lambda, x \in W\}.$$

For $f = (a, xW_{\lambda(a)}), e = (b, yW_{\lambda(b)}) \in \mathcal{D}_\lambda$ define $f \leq e$ if

$$a \leq b \text{ and } xW_{\lambda(a)} \cap yW_{\lambda(b)} \neq \emptyset.$$

Call λ *regular* if $(\mathcal{D}_\lambda, \leq)$ is a lattice. Let

$$E_\lambda = \{(a, P, P^-) | a \in \Lambda, P, P^- \text{ are opposite parabolic subgroups of } G \text{ of type } \lambda(a)\}.$$

For $f = (a, P, P^-), e = (b, Q, Q^-) \in E_\lambda$, we define a potential biordered structure on E_λ as:

$$f \leq_r e \text{ if } a \leq b \text{ and } P, Q \text{ are incident.}$$

In such a case define

$$ef = f, fe = (a, P, P')$$

where if B is a Borel subgroup contained in $P \cap Q$ and B^- an opposite contained in P^-, then P' is the unique opposite of P such that

$$\text{proj}_{Q^-}\, \text{proj}_Q(B^-) \subseteq P'.$$

Similarly

$$f \leq_\ell e \text{ if } a \leq b \text{ and } P^-, Q^- \text{ are incident.}$$

In such a case define

$$fe = e, ef = (a, P_1, P^-)$$

where if B^- is a Borel subgroup contained in $P^- \cap Q^-$ and if B is an opposite of B^- contained in P, then P_1 is the unique opposite of P^- such that

$$\text{proj}_Q\, \text{proj}_{Q^-}(B) \subseteq P_1.$$

Theorem 3.1 ([16]) *λ is the type map of some monoid of Lie type if and only if λ is regular. In such a case $\mathcal{D}(M) \cong \mathcal{D}_\lambda$ and $E(M) \cong E_\lambda$.*

In particular the type map λ completely determines the biordered set $E(M)$ of idempotents. For reductive monoids this was proved by Renner and the author [22]. The following description of regular type maps in often useful.

Theorem 3.2 ([16, 1]) *λ is regular if and only if the following conditions are satisfied:*

(i) *If $a, b, c \in \Lambda$ with $a < b < c$, then any component of $\lambda(b)$ is contained in $\lambda(a)$ or $\lambda(c)$.*

(ii) *For all $a, b \in \Lambda, \lambda(a) \cap \lambda(b) \subseteq \lambda(a \wedge b)$.*

(iii) *If $s \in S, a \in \Lambda$, then $\max\{b \in \Lambda | b \leq a, s \in \lambda(b)\}$ exists.*

(iv) *If $a_1, a_2 \in \Lambda, a_1 < a_2, X$ a two element discrete subset of $\lambda(a_1) \cup \lambda(a_2)$, then $X \subseteq \lambda(b)$ for some $b \in \Lambda$ with $a_1 \leq b \leq a_2$.*

Example 3.1 Let $M = M_n(\mathbb{F}_q)$ be as in Example 2.1. Then

$$\lambda(e_r) = S - \{(r\ r+1)\}.$$

Example 3.2 For the canonical monoid $\mathcal{M}$ or the dual canonical monoid $\tilde{\mathcal{M}}$ (see Examples 2.4, 2.5)

$$\lambda(e_I) = I,\ I \subseteq S.$$

We now wish to classify monoids of Lie type up to isomorphism. Fix a regular type map $\lambda : \Lambda \to 2^S$. For a monoid of Lie type M of type $\lambda, e \in \Lambda$, let

$$K_e = K_e(M) = \{a \in b | ae = ea = e\} \triangleleft L(e).$$

Define the $\mathcal{K}$-*structure* of M,

$$\mathcal{K} = \mathcal{K}(M) = \{(e, K_e) | e \in \Lambda\}.$$

For $e, f \in \Lambda$,

$$f \leq e \text{ implies } K_e \subseteq K_f.$$

Hence $\mathcal{K}$ can be viewed as a semilattice of groups. Given two $\mathcal{K}$-structures $\mathcal{K}_1 = \mathcal{K}_1(M_1), \mathcal{K}_2 = \mathcal{K}_2(M_2)$, define $\mathcal{K}_1 \leq \mathcal{K}_2$ if

$$K_e(M_1) \subseteq K_e(M_2) \text{ for all } e \in \Lambda.$$

Theorem 3.3 ([16])

(i) *A monoid of Lie type is determined up to isomorphism by its $\mathcal{K}$-structure.*

(ii) *If M_1 and M_2 are monoids of Lie type of type λ, then the identity map on $G \cup \Lambda$ extends to a homomorphism from M_1 to M_2 if and only if $\mathcal{K}(M_1) \leq \mathcal{K}(M_2)$.*

Fundamental semigroups exist for abstract regular semigroups. For monoids of Lie type there are also universal monoids.

Theorem 3.4 ([16, 21])

(i) *There is a universal monoid $M^+ = M^+(\lambda)$ and fundamental monoid $M^- = M^-(\lambda)$ such that for all monoids of Lie type M of type $\lambda, \mathcal{K}(M^+) \leq \mathcal{K}(M) \leq \mathcal{K}(M^-)$.*

(ii) *For $e \in \Lambda$, let $K_e^+ = K_e(M^+)$ and $K_e^- = K_e(M^-)$. Then K_e^+ is the normal subgroup of $L(e)$ generated by $U_{\lambda(f)} \cap L(e), f \in \Lambda, f \geq e$. For $e \neq 1$,*

$$K_e^- = \bigcap_{\substack{x \in L(f) \\ e \geq f}} x(L(e) \cap L(f))x^{-1}.$$

(iii) *Choose $K_e \triangleleft L(e), e \in \Lambda, K_1 = \{1\}$. Then $\mathcal{K} = \{(e, K_e) | e \in \Lambda\}$ is the $\mathcal{K}$-structure of a monoid of Lie type if and only if for all $e, f \in \Lambda, e \geq f$ implies $K_e \subseteq K_f$ and $K_e^+ \subseteq K_e$.*

For $e \in \Lambda$, let $\lambda^+(e)$ denote the union of all connected components A of $\lambda(e)$ such that $A \not\subseteq \lambda(f)$ for some $f \in \Lambda$ with $f \geq e$. Let $\lambda^-(e)$ denote the union of all connected components of $\lambda(e)$ such that $A \subseteq \lambda(f)$ for all $f \leq e$. Then $\lambda^+(e) \subseteq \lambda^-(e)$. The following characterization of K_e^+ and K_e^- is quite useful.

Theorem 3.5 ([25]) *Let $e \in \Lambda, e \neq 1$. Then*

(i) *If* $\lambda^+(e) = I$, *then* $K_e^+ = L_I'$, *the subgroup of* L_I *generated by its unipotent elements.*

(ii) *If* $\lambda^-(e) = J$, *then* $K_e^- = L_J$.

We note that usually L_I' is just the commutator subgroup of L_I.

Example 3.3 Let λ denote the type map of $M_n(\mathbb{F}_q)$. Then for $1 \le r < n$,

$$K_{e_r}^+ = \left\{ \begin{bmatrix} I_r & 0 \\ 0 & A \end{bmatrix} | A \in SL_{n-r} \right\},$$

$$K_{e_r}^- = \left\{ \begin{bmatrix} \alpha I_r & 0 \\ 0 & A \end{bmatrix} | \alpha \in F^*, A \in GL_{n-r} \right\}.$$

There is a very nice duality theory for monoids of Lie type that is being developed by Renner and the author [25] that turns monoids upside down and establishes a duality between idempotent separating morphisms and surjective morphisms.

4. Renner monoids

In the same way that the Weyl group W is the core of a reductive group or a group of Lie type, the Renner monoid R is the core of a reductive monoid or a monoid of Lie type. The length function and Bruhat order are defined on W as follows (cf. [7]). If $x \in W$, then $\ell(x)$ is the smallest non-negative integer m such that $x \in S^m$. If $x = s_1 \dots s_m, s_i \in S, m = \ell(x)$, then $y \le x$ if

$$y = s_{i_1} \dots s_{i_r} \text{ with } i_1 < i_2 < \dots < i_r.$$

For $e \in \Lambda$, let

$$\begin{aligned} W(e) &= \{x \in W | xe = ex\}, \\ W_e &= \{x \in W | xe = e\} \lhd W(e), \\ D(e) &= \{x \in W | x \text{ is of minimum length in } xW(e)\}, \\ D_e &= \{x \in W | x \text{ is of minimum length in } xW_e\}. \end{aligned}$$

We recall that the local monoids of R are

$$M(W, W(e), W(e), \widetilde{W(e)}), e \in \Lambda$$

where $W(e) = W_e \times \widetilde{W(e)}$ and that

$$R = \bigsqcup_{e \in \Lambda} WeW.$$

Now each $\sigma \in WeW$ can be uniquely written as:

$$\sigma = xey^{-1}, x \in D_e, y \in D(e).$$

We call this the *standard form* of σ. If w_0, v_0 are respectively the longest elements of W and $W(e)$, then $w_0 v_0$ is the longest element of $D(e)$. Following Solomon [32], [34] and Renner [28], we let

$$\ell(e) = \ell(w_0 v_0)$$

and for $\sigma = xey^{-1}$ in standard form

$$\ell(\sigma) = \ell(x) + \ell(e) - \ell(y).$$

It is shown in [28] that

$$\ell(\sigma) = 0 \text{ if and only if } \sigma B = B\sigma.$$

If $\sigma = xey^{-1}$ and $\theta = sft^{-1}$ in standard form, then following Renner and the author [24], we define $\sigma \leq \theta$ if

$$e \leq f \text{ and } x \leq sw, tw \leq y \text{ for some } w \in W(f)W_e.$$

It is shown in [24] that for reductive monoids

$$\sigma \leq \theta \text{ if and only if } B\sigma B \subseteq \overline{B\theta B}.$$

If $\sigma \in WeW$ and $\theta \in WeW$, then

$$\sigma \leq \theta \text{ implies } \ell(\sigma) \leq \ell(\theta).$$

Example 4.1 For $M = M_n(\mathbf{F}_q)$, W is the symmetric group and R is the symmetric inverse semigroup.

Example 4.2 Let G be a group of Lie type, $I \subseteq S$,

$$\begin{aligned} M^+ &= M(G, P_I, P_I^-, L_I), \\ M^- &= M(G, P_I, P_I^-, 1). \end{aligned}$$

The corresponding Renner monoids are

$$\begin{aligned} R^+ &= M(W, W_I, W_I, W_I), \\ R^- &= M(W, W_I, W_I, 1). \end{aligned}$$

Example 4.3 Let G be a group of Lie type and let $\mathcal{M} = \mathcal{M}(G)$ be as in Example 2.4. The local monoids of the corresponding Renner monoid R are of the form

$$M(W, W_I, W_I, W_I)$$

and these are submonoids of R.

Example 4.4 Let G be a group of Lie type and let $\tilde{\mathcal{M}} = \tilde{\mathcal{M}}(G)$ be as in Example 2.5, R its Renner monoid. Then for all idempotents $e \neq 1$, eRe is a semilattice and local monoids of R are of the form

$$M(W, W_I, W_I, 1).$$

If M, M' are finite monoids then call M a *central extension* of M' if there is a surjective homomorphism $\theta : M \to M'$ such that for all subgroups H of M, the kernel of θ restricted to H is contained in the center of H. For finite reductive monoids, M^+ is a central extension of M^- and all monoids of this type have the same Renner monoid. This is not true in general as Example 4.2 shows. However given a Renner monoid R, among the monoids of Lie type with this Renner monoid, there is a largest monoid of Lie type M_1 and a smallest monoid of Lie type M_2 such that M_1 is a central extension of M_2 and for any monoid of Lie type M with Renner monoid R, we have surjective idempotent separating homomorphisms

$$M_1 \to M \to M_2.$$

See [25] for details.

5. Complex representation theory

The starting point here is the following result of Okniński and the author [11]. It was proved using Harish-Chandra's theory of cuspidal representations, cf. [1], and the universal canonical monoid $\mathcal{M}^+$.

Theorem 5.1 *Let M be any monoid of Lie type. Then $\mathbb{C}[M]$ is semisimple.*

It follows by [4, Chapter 5] that

$$\mathbb{C}[M] \cong \oplus_J \mathbb{C}_0[J^0] \qquad (1)$$

where the summation is over the $G \times G$ orbits ($\mathcal{J}$-classes) $J = GeG, e \in \Lambda$. Finding explicitly the irreducible representations of M is equivalent to finding explicitly the unities of $\mathbb{C}_0[J^0]$. This is a difficult open problem.

Since R is the core of M, it is natural to try to relate the representations of M with those of R. As in the case of groups the correct approach is via Hecke algebras. Let

$$\epsilon = \frac{1}{|B|} \sum_{b \in B} b \in \mathbb{C}[G].$$

The Iwahori-Hecke algebra

$$H_{\mathbb{C}}(G, B) = \epsilon\mathbb{C}[G]\epsilon.$$

A theorem of Tits (cf. [2, 6]) states that

$$H_{\mathbb{C}}(G, B) \cong \mathbb{C}[W].$$

This isomorphism was made explicit by Lusztig [9]. Following Solomon [32, 33, 34] we define the *monoid Hecke algebra*

$$H_{\mathbb{C}}(M, B) = \epsilon\mathbb{C}[M]\epsilon.$$

Using Theorem 5.1, Solomon [32, 34] has shown that

$$H_{\mathbb{C}}(M, B) \cong \mathbb{C}[R].$$

The author [19, 20] has made this isomorphism explicit and without the use of Theorem 5.1. As with (1), we have

$$H_{\mathbb{C}}(M,B) \cong \oplus_J H_{\mathbb{C}}(J,B)$$

where the summation is over the $G \times G$ orbits $J = GeG, e \in \Lambda$ and

$$H_{\mathbb{C}}(J,B) = \epsilon \mathbb{C}_0 [J^0] \epsilon.$$

A description of the irreducible representations of $H_{\mathbb{C}}(M,B)$ can be found by explicitly finding the unities of all the algebras $H_{\mathbb{C}}(J,B)$. This has been accomplished by the author [19, 20]. For $\sigma \in WeW$, let

$$A_\sigma = \epsilon\sigma\epsilon \in H_{\mathbb{C}}(J,B), J = GeG.$$

Then $A_\sigma, \sigma \in WeW$ is a basis of $H_{\mathbb{C}}(J,B)$. For $\sigma = xey^{-1}$ in standard form, let

$$R_\sigma = \sum_{z \in W_e} R_{xz,y} \in \mathbb{Z}[q]$$

where $R_{x,y} = R_{x,y}(q)$ are the R-polynomials of Kazhdan and Lusztig [8]. Let $\overline{R}_\sigma$ denote the integer Laurent polynomial in q obtained by replacing q by $1/q$. The author [19, 20] has proved:

Theorem 5.2 *If $e \in \Lambda$ and $J = GeG$, then the unity $\hat{e}$ of $H_{\mathbb{C}}(J,B)$ is given by:*

$$\hat{e} = \sum_{\sigma \in WeW} \overline{R}_\sigma A_\sigma .$$

Let $e \in \Lambda, J = GeG, H$ the unit group of eMe. Since $\mathbb{C}_0[J^0]$ is semisimple, so is the algebra of G-invariants:

$$\mathbb{C}_0[J^0]^G = \{a \in \mathbb{C}_0[J^0] | ag = ga \text{ for all } g \in G\}.$$

Clearly

$$\dim \mathbb{C}_0[J^0]^G = \#\text{Conjugacy classes of } J.$$

Let $\operatorname{Irr} H$ denote the set of irreducible representations of H. If $\theta \in \operatorname{Irr} H$, let θ_+ denote the lift of θ to $P = P(e)$, and θ_+^G the representation of G induced from P to G (cf. [5]).

Theorem 5.3 ([18]) *The number of conjugacy classes of J is equal to*

$$\sum_{\theta \in \operatorname{Irr} H} (\theta_+^G, \theta_+^G) .$$

Applying this to $M_n(\mathbb{F}_q)$ we get

Corollary 5.4 *The number of rational canonical forms of $M_n(\mathbb{F}_q)$ of rank r is equal to $\sum_{\theta \in \operatorname{Irr} GL_r} (\theta_+^G, \theta_+^G)$ where $G = GL_n(\mathbb{F}_q)$.*

Because of Theorem 5.3, studying the conjugacy classes of M is an important problem. For reductive monoids a description of the conjugacy classes has been found by the author [14], [17]. This has also been accomplished by the author [18] for the universal canonical monoid $\mathcal{M}^+$ for any finite group of Lie type G.

Finally we mention that irreducible modular representations of M (in the natural characteristic) also have extremely pleasant properties. For example Renner and the author [23] have shown that the irreducible modular representations of M restrict to irreducible representation of G. This feature is being discussed in more detail by Renner in his article [30] in these proceedings.

References

1. Augustine, M. K. and Putcha, M. S., *Maps into Dynkin diagrams arising from regular monoids*, J. Austral. Math. Soc. (Series A) 47 (1989), 313–321.
2. Carter, R. W., *Finite Groups of Lie Type: Conjugacy Classes and Complex Characters*, Wiley, 1985.
3. Clifford, A. H., *Semigroups admitting relative inverses*, Annals of Math. 42 (1941), 1037–1049.
4. Clifford, A. H. and G. B. Preston, *Algebraic Theory of Semigroups*, Vol. 1, AMS Surveys No. 7, 1961.
5. Curtis, C. W. and I. Reiner, *Methods of Representation Theory*, Volume 1, Wiley, N.Y., 1981.
6. Curtis, C. W. and I. Reiner, *Methods of Representation Theory*, Volume 2, Wiley, NY, 1987.
7. Humphreys, J. E., *Reflection Groups and Coxeter Groups*, Cambridge University Press, 1990.
8. Kazhdan, D. and G. Lusztig, *Representations of Coxeter groups and Hecke algebras*, Invent. Math. 53 (1979), 165–184.
9. Lusztig, G., *On a theorem of Benson and Curtis*, J. Algebra 71 (1981), 490–498.
10. Nambooripad, K. S. S., *Structure of regular semigroups I*, Mem. Amer. Math. Soc. 224 (1979).
11. Okniński, J. and M. S. Putcha, *Complex representations of matrix semigroups*, Trans. Amer. Math. Soc. 323 (1991), 563–581.
12. Putcha, M. S., *A semigroup approach to linear algebraic groups*, J. Algebra 80 (1983), 164–185.
13. Putcha, M. S., *A Semigroup approach to linear algebraic groups III. Buildings*, Canadian J. Math. 38 (1986), 153–163.
14. Putcha, M. S., *Conjugacy classes in algebraic monoids*, Trans. Amer. Math. Soc. 303 (1987), 529–540.
15. Putcha, M. S., *Linear Algebraic Monoids*, London Math. Soc. Lecture Note Series 133, Cambridge Univ. Press, 1988.
16. Putcha, M. S., *Monoids on groups with BN-pairs*, J. Algebra 120 (1989), 139–169.
17. Putcha, M. S., *Conjugacy classes in algebraic monoids II*, Canadian J. Math, to appear.
18. Putcha, M. S., *Hecke algebras and monoid conjugacy classes*, J. Algebra (to appear).
19. Putcha, M. S., *Sandwich matrices, Solomon algebras and Kazhdan-Lusztig Polynomials*, Trans. Amer. Math. Soc. 340 (1993), 415–428.
20. Putcha, M. S., *Monoid Hecke algebras*, to appear.
21. Putcha, M. S., *Classification of monoids of Lie type*, J. Algebra (to appear).
22. Putcha, M. S. and L. E. Renner, *The system of idempotents and the lattice of $\mathcal{J}$-classes of reductive algebraic monoids*, J. Algebra 116 (1988), 385–399.
23. Putcha, M. S. and L. E. Renner, *The canonical compactification of a finite group of Lie type*, Trans. Amer. Math. Soc. 337 (1993), 305–319.
24. Putcha, M. S. and L. E. Renner, *Analogue of the Bruhat decomposition for algebraic monoids III, Order*, to appear.
25. Putcha, M. S. and L. E. Renner, *Classification of monoids of Lie type II. morphisms and duality*, to appear.
26. Renner, L. E., *Classification of semisimple algebraic monoids*, Trans. Amer. Math. Soc. 292 (1985), 193–223.
27. Renner, L. E., *Analogue of the Bruhat decomposition for algebraic monoids*, J. Algebra 101 (1986), 303–338.

28. Renner, L. E., *Analogue of the Bruhat decomposition for algebraic monoids II. The length function and trichotomy*, to appear.
29. Renner, L. E., *Finite reductive monoids*, these Proceedings.
30. Renner, L. E., *Modular representations of finite monoids of Lie type*, these Proceedings.
31. Ronan, M., *Lectures on Buildings*, Academic Press, 1989.
32. Solomon, L., *The Bruhat decomposition, Tits system and Iwahori ring for the monoid of matrices over a finite field*, Geom. Dedicata 36 (1990), 15–49.
33. Solomon, L., *Reductive monoids*, Notes of a talk given at MSRI Workshop on Representations of Reductive Groups over Finite Fields, Berkeley, November 1990.
34. Solomon, L., *Algebraic monoids*, Notes of a talk given at the Canadian Math. Soc. Meeting, Victoria, December 1991.
35. Solomon, L., *An introduction to reductive monoids*, these Proceedings.
36. Tits, J., *Buildings of Spherical Type and Finite BN-Paris*, Lecture Notes in Math. 384, Springer-Verlag, 1974.

FINITE REDUCTIVE MONOIDS

LEX E. RENNER
The University of Western Ontario,
London,
Ontario N6A 5B7,
Canada

1. Introduction

The theory of reductive algebraic monoids has been developed extensively over the last fifteen years by Putcha and the author [4,6,8]. The program was enormously successful, and this allowed Putcha to devise a set of axioms in terms of semigroup notions and groups with BN pairs. The resulting objects are called *monoids of Lie type* [3,7]. The gain here is to obtain the most general theory which has all the important features of reductive monoids, but which does not require all the technical baggage of algebraic geometry. Furthermore, many new classes of monoids of Lie type can be determined with Putcha's axiomatic approach:

(1) monoids arising from modular representations of finite groups of Lie type,
(2) monoids arising from families of subsets of a Dynkin diagram;
(3) monoids arising from Frobenius type morphisms of algebraic monoids (these are the finite reductive monoids).

In each of the above cases one can use Putcha's theory to either contruct synthetically a monoid of Lie type, or take the monoid that arises and show that it is indeed a monoid of Lie type. Furthermore, Putcha's theory is so adequately refined that one can use it to construct monoids of Lie type with preassigned universal properties (e.g., universal central extensions). The entire theory relies on the notion of a type map $\lambda : \mathcal{U}(M) \to 2^S$, from the set of J-classes to the set of subsets of the Dynkin diagram (which can be canonically identified with the set of conjugacy classes of parabolic subgroups). Furthermore, there is a subset $\Lambda \subseteq E(M)$ such that

(1) $\Lambda \cap J = \{e_J\}$ for each $J \in \mathcal{U}(M)$.
(2) $ef = fe$ for $e, f \in \Lambda$.
(3) If $e, f \in \Lambda$ then $ef = fe = e$ if and only if $J_e \leq J_f$.

Λ is called a *cross section lattice* and $\lambda(J_e) =$ type $\big(P(e)\big) \in 2^S$ where

$$P(e) = \{g \in G | ge = ege\}$$

is always a parabolic subgroup of G. See [5] for a discussion of the fundamental ideas here concerning the type map.

The focus of this paper is the class of finite reductive monoids (defined precisely below). This is the class of finite monoids of the Lie type that is most closely related to the theory of reductive algebraic monoids. The main results (structure,

J. Fountain (ed.), Semigroups, Formal Languages and Groups, 369–380.

classification, etc.) about finite reductive monoids are obtained by a systematic analysis of how results about reductive monoids are "carried over" to submonoids of fixed point sets. For example,

$$\begin{aligned} M_n(\mathbb{F}_q) &= \{(a_{ij}) \in M_n(\overline{\mathbb{F}}_q) | a_{ij}^q = a_{ij}\} \\ &= \{A \in M_n(\overline{\mathbb{F}}_q) | \sigma(A) = A\} \end{aligned}$$

where $\sigma\big((a_{ij})\big) = (a_{ij}^q)$. We can say a lot about $M_n(\mathbb{F}_q)$ because it is obtained from the reductive monoid $M_n(\overline{\mathbb{F}}_q)$ and the endomorphism $\sigma : M_n(\overline{\mathbb{F}}_q) \to M_n(\overline{\mathbb{F}}_q)$. For example, let $M_n^r(\mathbb{F}_q) \subseteq M_n(\mathbb{F}_q)$ be the subset of matrices of rank exactly r, and let $\mathcal{I}_n^r$ be the subset of the symmetric inverse semigroup of elements of rank exactly r. There is a length function $l : \mathcal{I}_n^r \to \mathrm{N}$ such that

$$|M_n^r(\mathbb{F}_q)| = q^r(q-1)^{r(r-1)/2} \sum_{\sigma \in \mathcal{I}_n^r} q^{l(\sigma)}.$$

We shall explore some of the main results and open problems associated with counting the elements of a finite reductive monoid.

2. Background

The main object of study in this paper is a certain class of finite monoids defined as follows:

Let $\underline{M}$ be a reductive algebraic monoid with zero defined over $\overline{\mathbb{F}}_p$, and assume that $\sigma : \underline{M} \to \underline{M}$ is a surjective morphism of algebraic monoids. If M is a finite monoid such that $M \cong \{x \in \underline{M} | \sigma(x) = x\}$ we refer to M as a *finite reductive monoid*. In order to understand this class of monoids it is necessary to classify all endomorphisms $\sigma : \underline{G} \to \underline{G}$ of a simple algebraic group $\underline{G}$ such that G is a finite group. These endomorphisms, originally classified by Chevalley [2], are discussed in much detail by Steinberg [12]. The following description is a summary of Steinberg's discussion in 11.6 of [12].

Let $\sigma : \underline{G} \to \underline{G}$ be a surjective endomorphism of $\underline{G}$ such that $G = \{x \in \underline{G} | \sigma(x) = x\}$ is finite. By [12; Corollary 10.10] there is a Borel subgroup $\underline{B}$ of $\underline{G}$ and a maximal torus $\underline{T}$ of $\underline{B}$ such that $\sigma(\underline{T}) = \underline{T}$ and $\sigma(\underline{B}) = \underline{B}$.

For example, consider $\sigma : Sl_n(\overline{\mathbb{F}}_q) \to Sl_n(\overline{\mathbb{F}}_q)$ given by $\sigma(x) = Fr_q(\omega\, {}^t x^{-1} \omega)$, where $\omega = \left(\begin{smallmatrix} 0 & 1 \\ 1 & 0 \end{smallmatrix}\right)$ and $Fr_q\big((x_{ij})\big) = ({x_{ij}}^q)$. Then $\underline{B} = \{x \in Sl_n | x_{ij} = 0 \text{ if } i > j\}$ and $\underline{T} = \{x \in Sl_n | x_{ij} = 0 \text{ if } i \neq j\}$ satisfy the required property.

Let $\Phi \subseteq X(\underline{T})$ and $\Delta \subseteq \Phi$ be, respectively, the roots and the simple roots relative to $\underline{T}$ and $\underline{B}$. Let $\sigma^* : X(\underline{T}) \to X(\underline{T})$ be the map on characters of $\underline{T}$ induced by $\sigma|\underline{T} : \underline{T} \to \underline{T}$. By [12; Proposition 11.2] there exists a bijection $\rho : \Phi \to \Phi$ with $\rho(\Delta) = \Delta$, and a scalar factor $q : \Phi \to \mathrm{N}$ such that

$$(*) \qquad \sigma^*\big(\rho(\alpha)\big) = q(\alpha)\alpha \text{ for all } \alpha \in \Delta.$$

In the above example,

$$X(\underline{T}) = \{\epsilon_1, \ldots, \epsilon_n \mid \sum \epsilon_i = 0\},$$
$$\Delta = \{\epsilon_1 - \epsilon_2, \ldots, \epsilon_{n-1} - \epsilon_n\} \text{ and}$$
$$\Phi = \{\epsilon_i - \epsilon_j\}_{i \neq j}.$$

One verifies that

$$\rho(\epsilon_i - \epsilon_j) = \epsilon_{n+1-j} - \epsilon_{n+1-i}$$

and

$$q(\epsilon_i - \epsilon_j) = q \cdot (\epsilon_i - \epsilon_j).$$

There are essentially two types of σ.

Type I: q is constant. Then ρ is an automorphism of the Coxeter-Dynkin diagram such that $\sigma = Fr_q \circ \rho$ where $Fr_q(t) = t^q$ for all $t \in \underline{T}$. The root systems with automorphisms are A_n, D_n and E_6. $Aut(A_n) = \mathbb{Z}/2\mathbb{Z}$, $Aut(D_n) = \mathbb{Z}/2\mathbb{Z}(n \neq 4)$, $Aut(D_4) = S_3$ and $Aut(E_6) = \mathbb{Z}/2\mathbb{Z}$.

Type 2: q is not constant. Then ρ exchanges long and short roots in such a way that $\rho^2 = id$. Furthermore, $\underline{G}$ is of type C_2, F_4 or G_2. If $\underline{G}$ is of type C_2 then the characteristic is 2 and $q(\alpha) = 2^a$ for α long, $q(\alpha) = 2^{a+1}$ for α short. If $\underline{G}$ is of type F_4 then the characteristic is 2 and $q(\alpha) = 2^a$ for α long, $q(\alpha) = 2^{a+1}$ for α short. If $\underline{G}$ is of G_2 then the characteristic is 3 and $q(\alpha) = 3^a$ for α long, $q(\alpha) = 3^{a+1}$ for α short.

We can depict all sixteen families by the following augmented Coxeter-Dynkin diagrams. The arrows indicate the effect of ρ on Δ.

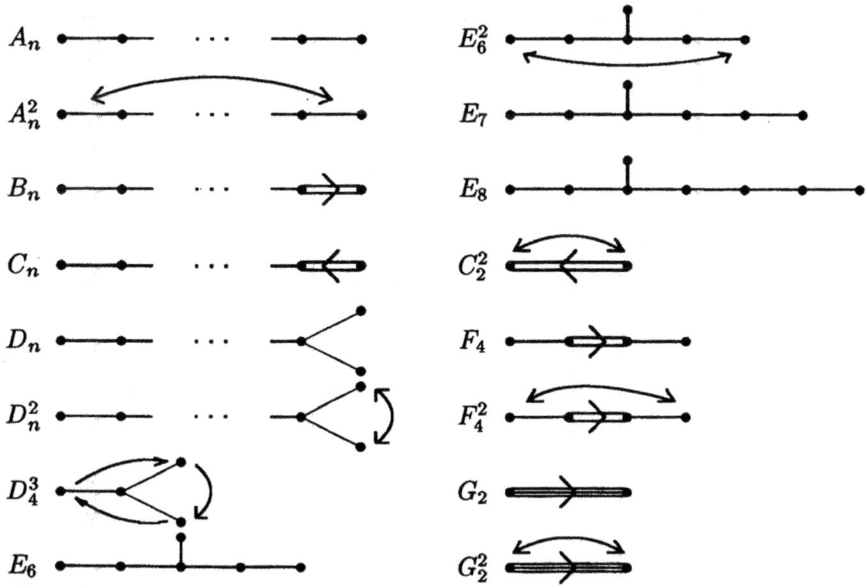

So the above example is of type A^2_{n-1}.

The finite groups $G = \{x \in \underline{G} | \sigma(x) = x\}$ yield sixteen families of finite simple group (parametrized by various finite fields). If we augment the list by adjoining the alternating groups and the cyclic groups of prime order, there are but 26 finite simple groups (the so called sporadic ones) left to account for.

The main tool in the study of finite groups of Lie type, and in the study of finite reductive monoids is *Lang's Theorem* [12; Theorem 10.1].

2.1. Lang's Theorem

Let $\underline{G}$ be a connected algebraic group with endomorphism $\sigma : \underline{G} \to \underline{G}$. Define $1 - \sigma : \underline{G} \to \underline{G}$ by $(1-\sigma)(x) = x\sigma(x)^{-1}$. Then the following are equivalent:

(a) $G = \{x \in \underline{G} | \sigma(x) = x\}$ is finite.

(b) $1 - \sigma : \underline{G} \to \underline{G}$ is surjective.

There are many useful corollaries of 2.1. The reader should consult section 10 of [12], or Chapter E of [1] for all the essential consequences. We consider here only those consequences needed to study finite reductive monoids.

2.2. Homogeneity

Let $\underline{G}$ be a connected algebraic group with endomorphism $\sigma : \underline{G} \to \underline{G}$ such that G is finite. Suppose $\underline{G}$ acts transitively on $\underline{X}$ via $\underline{G} \times \underline{X} \to \underline{X}$. Assume there exists $\gamma : \underline{X} \to \underline{X}$ such that $\gamma(gx) = \sigma(g)\gamma(x)$ if $g \in \underline{G}$ and $x \in \underline{X}$. Then $X = \{x \in \underline{X} | \gamma(x) = x\} \neq \phi$. If further, $\underline{G}_x = \{g \in \underline{G} | gx = x\}$ is connected for $x \in X$, then G acts transitively on X.

2.3. BN Pair Structure on G

Let $\underline{T} \subseteq \underline{B} \subseteq \underline{G}$ be respectively maximal torus and Borel subgroup of $\underline{G}$ such that $\sigma(\underline{T}) = \underline{T}$ and $\sigma(\underline{B}) = \underline{B}$. Observe that $\sigma\left(N_{\underline{G}}(\underline{T})\right) = N_{\underline{G}}(\underline{T})$, and σ induces an isomorphism $\sigma : \underline{W} \to \underline{W}$ $\left(\text{where } \underline{W} = N_{\underline{G}}(\underline{T})/\underline{T}\right)$. Let

$$\begin{aligned} B &= \{x \in \underline{B} | \sigma(x) = x\}, \\ T &= \{x \in \underline{T} | \sigma(x) = x\}, \\ W &= \{\omega \in \underline{W} | \sigma(\omega) = \omega\}, \\ N &= \{x \in N_{\underline{G}}(\underline{T}) | \sigma(x) = x\}. \end{aligned}$$

Then

(a) G is a group with BN pair determined by B, N and W (so every element of W has a representative in N).

(b) Any other pair $\overline{T}' \subseteq \overline{B}'$ with $\sigma(\underline{T}') = \underline{T}'$ and $\sigma(\underline{B}') = \underline{B}'$ is conjugate to (T, B) by some $x \in G$.

Remark. Notice that the Coxeter-Dynkin graph of G may be different than the graph of $\underline{G}$. In the above example we obtain

$\underline{G}$  $n-1$ nodes

G  $\begin{cases} (n-1)/2 \text{ nodes, } n \text{ odd} \\ n/2 \text{ nodes, } n \text{ even} \end{cases}$

The graph for G is obtained by "folding" the graph of $\underline{G}$ according to ρ.

3. Structure and Classification of Finite Reductive Monoids

Let M be the finite reductive monoid associated with $\underline{M}$ and $\sigma : \underline{M} \to \underline{M}$. Since M is finite, so too is G. So the results of section 2 can be applied in this situation.

Theorem 3.1

(a) *M is unit regular.*

(b) *If $e \in E(M)$ and eJf then $e = xfx^{-1}$ for some $x \in G$.*

(c) *If $e \in E(M)$ then $P(e) = \{g \in G | ge = ege\}$ and $P^-(e) = \{g \in G | eg = ege\}$ are opposite parabolic subgroups of G (with the BN pair structure determined as in 2.2).*

(d) *If $e \in E(M)$, and $U \subseteq P(e)$, $U^- \subseteq P^-(e)$ are the unipotent radicals, then $Ue = \{e\} = eU^-$.*

Properties (a), (b), (c) and (d) together imply that M is a *monoid of Lie type* in the sense of Putcha [5]. In particular, there is a type map $\lambda : \mathcal{U}(M) \to 2^S$ which is easily defined by

$$\lambda(J) = \text{type}\left(P(e)\right)$$

where $e \in E(J)$ and type $\left(P(e)\right) \in 2^S$. (The S for G may be different than the S for $\underline{G}$. See example 3.1.)

To indicate how Theorem 3.1 is proved (using 2.2) we focus on the proof of (b).

Let $x \in M$, so that $\sigma(x) = x$. Hence $\sigma(\underline{J}) = \underline{J}$ where $\underline{J} = \underline{G}x\underline{G}$. But from 2.2, we know that $E(J) = \{e \in E(\underline{J}) | \sigma(e) = e\} \neq \phi$. Furthermore, $\underline{G}$ acts transitively on $E(\underline{J})$, and so again by 2.2, G acts transitively on $E(J)$, since $C_{\underline{G}}(e)$ is a connected group for any $e \in E(\underline{M})$.

The existence of a cross section lattice is guaranteed by Putcha's remarkable theory. On the other hand, if we choose $\underline{T} \subseteq \underline{B} \subseteq \underline{G}$ with $\sigma(\underline{T}) = \underline{T}$ and $\sigma(\underline{B}) = \underline{B}$ we define

$$\Lambda = \{e \in E(\overline{\underline{T}}) | \sigma(e) = e \text{ and } \underline{B}e = e\underline{B}e\}.$$

Again using 2.2, one can prove that Λ is a cross section lattice of M. We easily obtain the following result.

Corollary 3.2 *Let $\underline{\Lambda} \subseteq \underline{M}$ be a cross section lattice of $\underline{M}$ such that $\sigma(\underline{\Lambda}) = \underline{\Lambda}$. Then $\Lambda = \{e \in \underline{\Lambda} | \sigma(e) = e\}$ is a cross section lattice of M.*

Example 3.1 The purpose of this example is to illustrate the two significant features while comparing M with $\underline{M}$.

(a) The BN pair structure of $\underline{G}$ and G can be different (ie. $\underline{G}$ and G can be associated with different Coxeter graphs).

(b) It may very well happen that $\Lambda \subsetneqq \underline{\Lambda}$.

Actually for semisimple monoids [8] (a) and (b) are equivalent in the sense that $\Lambda = \underline{\Lambda}$ if and only if G and $\underline{G}$ are associated with the same Coxeter graph.

Our example can be described as follows:

Let $\underline{G} = Sl_4(\overline{\mathbb{F}}_q) \times \overline{\mathbb{F}}_q^*$ and let $\sigma : \underline{G} \to \underline{G}$ be defined by

$$\sigma(x, \alpha) = \Big(Fr_q(\omega\, {}^t x^{-1} \omega^{-1}), \alpha^q\Big)$$

where $Fr_q\big((x_{ij})\big) = (x_{ij}^q)$ and $\omega = \begin{pmatrix} 0 & 0 & 0 & 1 \\ 0 & 0 & 1 & 0 \\ 0 & 1 & 0 & 0 \\ 1 & 0 & 0 & 0 \end{pmatrix}$.

The group G here is a central extension of a unitary group over $\mathbb{F}_q$. See [11] for those details. The maximal torus

$$\underline{T} = \left\{ \left(\begin{pmatrix} s_1 & 0 & 0 & 0 \\ 0 & s_2 & 0 & 0 \\ 0 & 0 & s_3 & 0 \\ 0 & 0 & 0 & s_4 \end{pmatrix}, \alpha \right) \Bigg| \begin{array}{c} \alpha, s_1 s_2 s_3 s_4 \in \overline{\mathbb{F}}_q^* \\ s_1 s_2 s_3 s_4 = 1 \end{array} \right\}$$

has character group $X(\underline{T})$ with the presentation

$$X(\underline{T}) = < \chi_1, \chi_2, \chi_3, \chi_4, \delta | \chi_1 + \chi_2 + \chi_3 + \chi_4 = 0 >$$

where $\chi_i\Big(\big((s_1, s_2, s_3, s_4), \alpha\big)\Big) = s_i$, and $\delta\Big(\big((s_1, s_2, s_3, s_4), \alpha\big)\Big) = \alpha$. One checks that $\sigma|\underline{T}$ is given by

$$\sigma\Big((s_1, s_2, s_3, s_4), \alpha\Big) = \Big((s_4^{-q}, s_3^{-q}, s_2^{-q}, s_1^{-q}), \alpha^q\Big).$$

Thus, $\sigma^* : X(\underline{T}) \to X(\underline{T})$ is given by

$$\begin{aligned} \sigma^*(\chi_1) &= -q\chi_4 \\ \sigma^*(\chi_2) &= -q\chi_3 \\ \sigma^*(\chi_3) &= -q\chi_2 \\ \sigma^*(\chi_4) &= -q\chi_1 \\ \sigma^*(\delta) &= q\delta. \end{aligned}$$

So

$$\sigma^*(\chi_1 - \chi_2) = q(\chi_3 - \chi_4),$$
$$\sigma^*(\chi_2 - \chi_3) = q(\chi_2 - \chi_3) \text{ and}$$
$$\sigma^*(\chi_3 - \chi_4) = q(\chi_1 - \chi_2).$$

In the notation of $(*)$ of section 2, ρ is given on the base $\Delta = \{\chi_1-\chi_2, \chi_2-\chi_3, \chi_3-\chi_4\}$ via

$$(**) \qquad \begin{aligned} \rho(\chi_1 - \chi_2) &= \chi_3 - \chi_4 \\ \rho(\chi_2 - \chi_3) &= \chi_2 - \chi_3 \\ \rho(\chi_3 - \chi_4) &= \chi_3 - \chi_4 \end{aligned}$$

and $q(\alpha) = q$ for any $\alpha \in \Delta$. Thus, σ is of type A_3^2. Notice that the effect of σ on the parabolic subgroups of $\underline{G}$ containing $\underline{B}$ is to permute them according to $(**)$. To construct an interesting reductive monoid $\underline{M}$ here it suffices to find a pointed, normal cone $C \subseteq X(\underline{T})$ such that C generates $X(\underline{T})$ as a group and $\omega(C) = C$ for any $\omega \in \underline{W} = S_4$ (see [8; Theorem 6.5]). Furthermore, in order that $\sigma : \underline{G} \to \underline{G}$ extends to a surjective morphism $\sigma : \underline{M} \to \underline{M}$, it is necessary and sufficient that

(i) $\sigma(C) \subseteq C$.

(ii) $nC \subseteq \sigma^*(C)$ for some $n > 0$.

See [7;Corollary 4.2] and [8;Corollary 4.5]. To this end we let

$$C = \{\chi \in X(\underline{T}) | n\chi \in C_1 \text{ for some } n > 0\}$$

where $C_1 = \langle \chi_i - \chi_j + \delta | i,j = 1,2,3,4, i \neq j \rangle$. Clearly $\sigma^*(C) \subseteq C$ since $\sigma^*(\chi_i - \chi_j + \delta) = q(\chi_{5-j} - \chi_{5-i} + \delta)$. Also $C \subseteq q\sigma^*(C)$ for the same reason.

To summarize, we have constructed a reductive monoid $\underline{M}$ with the following properties:

(i) $\underline{G} = Sl_4(\overline{\mathbb{F}}_q) \times \overline{\mathbb{F}}_q^*$.

(ii) If $\underline{T} \subseteq \underline{G}$ is a maximal torus then $X(\overline{\underline{T}}) = C$.

(iii) The morphism $\delta : \underline{G} \to \underline{G}$ extends (uniquely) to a surjective morphism $\sigma : \underline{M} \to \underline{M}$.

The reader may find the above construction of $\underline{M}$ somewhat abstract or disembodied. A more direct construction of $\underline{M}$ can be obtained as follows. Let $\mathfrak{g}$ be the Lie algebra of $Sl_4(\overline{\mathbb{F}}_q)$, and let $\rho : \underline{G} \longrightarrow Aut(\mathfrak{g})$ be defined by $\rho(g, \alpha) = \alpha Ad(g)$ where Ad is the adjoint representation of $Sl_4(\overline{\mathbb{F}}_q)$. Then $\underline{M}$ is essentially the Zariski closure of the image of ρ. The following properties of $\underline{M}$ are easily verified:

(a) $\underline{M}$ has a unique minimal J-class corresponding to the Weyl group orbit of $\chi_1 - \chi_4 + \delta \in C$ (see 3.1 and 3.2 of [8] for some details relating C to the idempotents of $\overline{\underline{T}}$.).

(b) $\chi_1 - \chi_4$ is in the chamber associated to $\underline{B}$ = the upper triangular group (since it is the highest weight of the adjoint representation). The base associated to $\underline{B}$ is $\Delta = \{\chi_1 - \chi_2, \chi_2 - \chi_3, \chi_3 - \chi_4\} = \{\alpha_{12}, \alpha_{23}, \alpha_{34}\}$.

(c) $\rho(\chi_1 - \chi_4 + \delta) = \chi_1 - \chi_4 + \delta$.

(d) $\{\omega \in W = S_4 | \omega(\chi_1 - \chi_4 + \delta) = \chi_1 - \chi_4 + \delta\} = \{1, \sigma_{23}\}$ where σ_{23} is the transposition interchanging χ_2 and χ_3. So if $e \in \underline{\Lambda}$ is the minimal nonzero idempotent in the cross section lattice of $\underline{B}$ and $\underline{T}$, corresponding to $\chi_1 - \chi_4 + \delta$, then $C_W(e) = \{1, \sigma_{23}\}$.

Thus, $\underline{M}$ is the *J-irreducible* monoid of type $J_0 = \{\alpha_{23}\}$ where the Coxeter graph is α_{12} — α_{23} — α_{34}. See [6] for the entire theory of J-irreducible monoids.

Theorem 4.16 of [6] includes a precise recipe for constructing the lattice of J-classes (and the type map) of a J-irreducible monoid once we know J_0. In this case, $\mathcal{U}(\underline{M})$ is as follows:

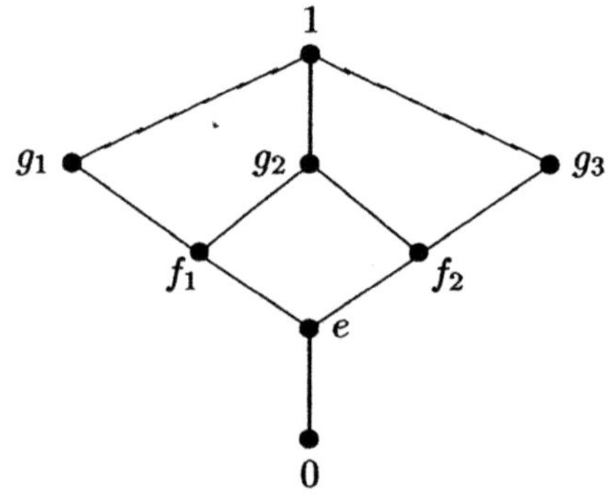

After identifying $\mathcal{U}(\underline{M})$ with $\underline{\Lambda} = \{0, e, f_1, f_2, g_1, g_2, g_3, 1\}$ the type map is qiven as follows:

$$\begin{aligned}
\lambda(0) &= S = \{\alpha_{12}, \alpha_{23}, \alpha_{34}\} \\
\lambda(e) &= \{\alpha_{23}\} \\
\lambda(f_1) &= \{\alpha_{12}\} \\
\lambda(f_2) &= \{\alpha_{34}\} \\
\lambda(g_1) &= \{\alpha_{12}, \alpha_{23}\} \\
\lambda(g_2) &= \{\alpha_{12}, \alpha_{34}\} \\
\lambda(g_3) &= \{\alpha_{23}, \alpha_{34}\} \\
\lambda(1) &= S = \{\alpha_{12}, \alpha_{23}, \alpha_{34}\}.
\end{aligned}$$

Now we have chosen $\underline{\Lambda}$ so that $\sigma(\underline{\Lambda}) = \underline{\Lambda}$. Furthermore, $\sigma(e) = e$, $\sigma(f_1) = f_2$, $\sigma(f_2) = f_1$, $\sigma(g_1) = g_3$, $\sigma(g_3) = g_1$, and $\sigma(g_2) = g_2$. Thus, by Corollary 3.2 we obtain the following description of $\Lambda \subseteq \underline{\Lambda}$:

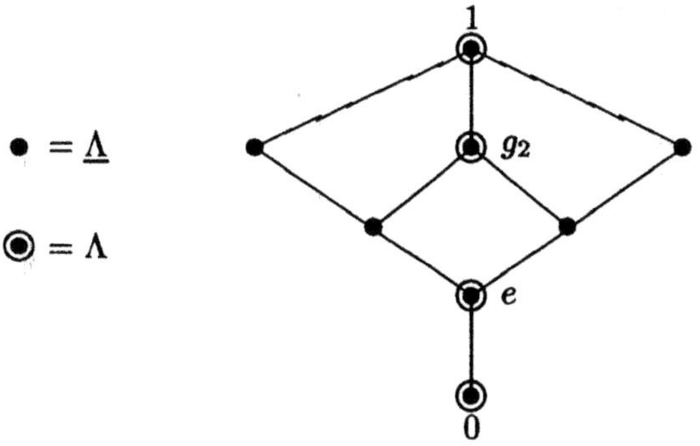

The group G is associated with the diagram $\overset{\bullet}{\alpha}$—$\overset{\bullet}{\beta}$ obtained by "folding" $\overset{\bullet}{\alpha_{12}}$—$\overset{\bullet}{\alpha_{23}}$—$\overset{\bullet}{\alpha_{34}}$ at α_{23} and identifying α_{12} and α_{34}.

The cross section lattice (Λ above) is

and it is easily seen that $\lambda(e) = \alpha$ (corresponding to $\{\alpha_{23}\}$) and $\lambda(g_2) = \beta$ (corresponding to $\{\alpha_{12}, \alpha_{34}\}$).

A basic property of reductive monoids concerns the structure of $E(\overline{\underline{T}})$. It is easily seen that $E(\overline{\underline{T}})$ is isomorphic to the face lattice of the polytope Π which has its vertices at the points $\{\chi_i - \chi_j\}_{i \neq j}$. After a little calculation we can identify $\underline{E} = E(\overline{\underline{T}})$ and $E = \{e \in \underline{E} | \sigma(e) = e\} \subseteq \underline{E}$.

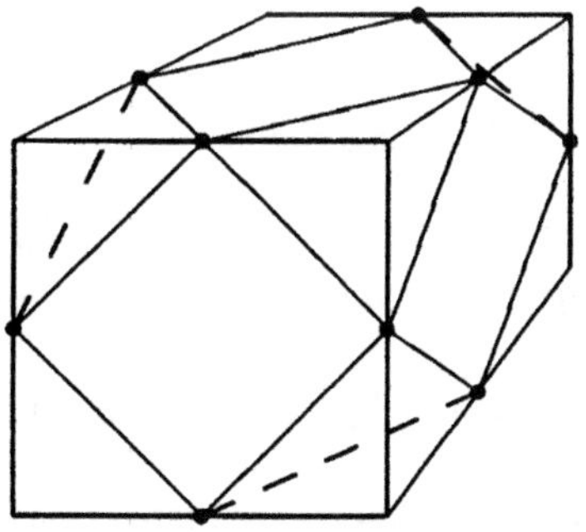

Π

So E can be identified with the subset of the face lattice of Π consisting of the vertices, the square facets, the empty face and the interior of Π.

4. Enumerative Theory of Finite Reductive Monoids

Let $\underline{M}$ be a reductive monoid over $\overline{\mathbb{F}}_q$ with surjective morphism $\sigma : \underline{M} \to \underline{M}$ such that if $\sigma(\underline{T}) = \underline{T}$ and $\sigma(\underline{B}) = \underline{B}$, (with $\underline{T} \subseteq \underline{B}$) then $\sigma(t) = t^q$ for all $t \in \underline{T}$. This restriction is required only for the purposes of exposition since it will allow us to count the elements of M without introducing and developing new techniques. On the other hand, the general counting theory will not differ greatly from the special case considered here.

Let $\underline{T} \subseteq \underline{B}$ be such that $\sigma(\underline{T}) = \underline{T}$ and $\sigma(\underline{B}) = \underline{B}$. Let $\underline{\Lambda} = \{e \in E(\overline{T}) | \underline{B}e = e\underline{B}e\}$, the cross section lattice and let

$$\underline{R} = \overline{N_{\underline{G}}(\underline{T})} \text{ (Zariski closure)}$$
$$\underline{\mathfrak{R}} = \underline{R}/\underline{T}, \text{ the orbit monoid.}$$

Correspondingly, we have for M,

$$\Lambda = \{e \in \underline{\Lambda} | \sigma(e) = e\} \text{ and}$$
$$\mathfrak{R} = \{\gamma \in \underline{\mathfrak{R}} | \sigma(\gamma) = \gamma\}.$$

Our assumption on σ implies that

$$\Lambda = \underline{\Lambda} \text{ and } \mathfrak{R} = \underline{\mathfrak{R}}.$$

On the other hand, by the results of [3] and [9], there is a double coset decomposition

$$(***) \qquad M = \bigsqcup_{\gamma \in \mathfrak{R}} B\gamma B.$$

Furthermore, there is a length function

$$l : \mathfrak{R} \to \mathbf{N}$$

defined as follows:

$$l(\gamma) = \dim(\underline{B}\gamma\underline{B}) - \dim(e\underline{B}e)$$

where $e \in \Lambda$ is J-related to γ.

Proposition 4.1 *For any* $\gamma \in \mathfrak{R}$

$$|B\gamma B| = |eBe| q^{l(\gamma)}.$$

Proof. Using our assumption on σ, and the fact that $\sigma(\underline{B}\gamma\underline{B}) = \underline{B}\gamma\underline{B}$, it is possible to write $\underline{B}\gamma\underline{B} \cong \overline{\mathbb{F}}_q^a \times e\underline{B}e \times \overline{\mathbb{F}}_q^b$ in such a way that σ preserves the factors, and

$$\sigma\Big((\alpha_1, \ldots, \alpha_a), x, (\beta_1, \ldots, \beta_b)\Big) = \Big((\alpha_1^q, \ldots, \alpha_a^q), \sigma(x), (\beta_1^q, \ldots, \beta_b^q)\Big).$$

But then $l(\gamma) = a + b$, and our formula follows by inspection. □

To obtain a formula for $|M|$ it suffices to find one for $|J|$ for each $J \in \mathcal{U}(M)$. Let $\Lambda \cap J = \{e\}$.

Theorem 4.2

$$|J| = |eBe| \sum_{\gamma \in WeW} q^{l(\gamma)}.$$

Proof. From (***) above it follows that $J = \bigsqcup_{\gamma \in \Re \cap J} B\gamma B$. But one can deduce further [9] that $\Re \cap J = WeW$. □

Example 4.1 ([10]) Let $\underline{M} = M_n(\overline{\mathbb{F}}_q)$ and $M = M_n(\mathbb{F}_q)$. Let $M^k = \{x \in M \mid \text{rank}\,(x) = k\}$. Then M^k is the J-class of $e = \begin{pmatrix} 1 & & & & \\ & \ddots & & & \\ & & 1 & & \\ & & & 0 & \\ & & & & \ddots \\ & & & & & 0 \end{pmatrix}$, while eBe is isomorphic to the $k \times k$ upper triangular group over $\mathbb{F}_q$. Thus, $|eBe| = q^k(q-1)^{k(k-1)/2}$ in this case. Hence,

$$|M^k| = q^k(q-1)^{k(k-1)/2} \sum_{\gamma \in \mathcal{I}_{n,k}} q^{l(\gamma)}$$

where $\mathcal{I}_{n,k}$ is the subset of the symmetric inverse semigroup consisting of the elements of rank exactly k.

It follows from Theorem 4.2 that one may write

$$|M| = \sum_{i=0}^{N} a_i q^i$$

for some integers a_i. For example,

$$|M_n(\mathbb{F}_q)| = q^{n^2}.$$

Actually, there is a lot of topological evidence to suggest that the correct form for this enumerative problem is

$$|M \backslash \{0\}| = (q-1) \sum_{i=1}^{n} b_i q^i.$$

For the example $M = M_n(\mathbb{F}_q)$, we obtain

$$|M_n(\mathbb{F}_q) \backslash \{0\}| = (q-1) \sum_{i=0}^{n^2-1} q^i$$

so that $b_i = 1$ if $i = 0, 1, \ldots, n^2 - 1$. It is no accident that these b_i's are the Betti numbers of the projective space $\mathbb{P}^{n^2-1}(\mathbb{C})$.

References

1. A. Borel et al, *Seminar on algebraic groups and related finite groups*, Lecture Notes in Mathematics, vol 131, Springer-Verlag, 1970.
2. C. Chevalley, *Seminaire sur la classification des groupes de Lie algébriques*, Paris, Ecole Norm. Sup., 1956–58.
3. M. Putcha, *Monoids on groups with BN pairs*, Journal of Algebra **120** (1989), 139–169.
4. M. Putcha, *Linear Algebraic Monoids*, London Math. Soc. Lecture Notes **133** (1988), Cambridge University Press.
5. M. Putcha, *Monoids of Lie type*, These Proceedings.
6. M. Putcha and L. Renner, *The system of idempotents and the lattice of J-classes of algebraic monoids*, Journal of Algebra **116** (1988), 385–399.
7. M. Putcha and L. Renner, *The canonical compactification of a finite group of Lie type*, Trans. Amer. Math. Soc. **337** (1993), 305–319.
8. L. Renner, *Classification of semisimple algebraic monoids*, Trans. Amer. Math. Soc. **292** (1985), 193–223.
9. L. Renner, *Analogue of the Bruhat decomposition for algebraic monoids*, Journal of Algebra **101** (1986), 303–338.
10. L. Solomon, *The Bruhat decomposition, Tits system and Iwahori ring for the monoid of matrices over a finite field*, Geometric Dedicata **36** (1990), 15–49.
11. R. Steinberg *Variations on a theme of Chevalley*, Pacific J. Math. **9** (1959), 875–891.
12. R.Steinberg, *Endomorphisms of linear algebraic groups*, Mem. Amer. Math. Soc. **80** (1968).

MODULAR REPRESENTATIONS OF FINITE MONOIDS OF LIE TYPE

LEX E. RENNER
The University of Western Ontario,
London,
Ontario N6A 5B7,
Canada

1. Introduction

The purpose of this paper is to describe the main features of the theory of modular representations of the finite monoid of Lie type M. The unit group G of M has associated with it a naturally defined characteristic. Indeed, if B and B^- are opposite Borel subgroups of G with $H = B \cap B^-$, there is a unique normal subgroup $U \triangleleft B$ such that $UH = B$ and $U \cap H = \{1\}$. It turns out that U is a p-Sylow subgroup of G. This p is referred to as the *characteristic* of G (or M). We consider here only the problem of identifying the irreducible (modular) representations of M over $\overline{\mathbb{F}}_p$. The reader is assured in advance that something truly remarkable happens here that is rarely encountered in semigroup theory.

Let M be a finite monoid of Lie type with zero and unit group G. The main result is the following:

Theorem *Suppose* $\rho : M \to \text{End}(V)$ *is an irreducible representation of* M. *Then* $\rho|G$ *is irreducible.*

Since we know from the work of Richen [3] how to identify the irreducible representations of G (this is described below in section 3) we can identify each irreducible representation of M in terms of the following data:

(i) Apex $(\rho) \in \mathcal{U}(M)$,

(ii) $I(\rho) \in 2^S$,

(iii) $\chi(\rho) : P_{I(\rho)} \to \overline{\mathbb{F}}_q^*$.

Here, $\lambda : \mathcal{U}(M) \to 2^S$ is the type map of M, Apex (ρ) is the smallest J-class of M with $\rho(J) \neq \{0\}$, and $\big(I(\rho),\ \chi(\rho)\big)$ is the invariant associated with $\rho|G$ by Richen's theory.

Furthermore, we can determine exactly which triples $(J, I, \chi) \in \mathcal{U}(M) \times \sum(G)$ can occur for some ρ (where $\sum(G) = \{(I, \chi)| I \in 2^S,\ \chi : L_I \to \overline{\mathbb{F}}_q^*\}$).

J. Fountain (ed.), Semigroups, Formal Languages and Groups, 381–390.

2. Irreducible Representations of Finite Semigroups

Since we are interested only in finite monoids of Lie type, we assume in this section that S is a finite regular semigroup. The reader should consult [2] for the natural generality of the following results.

We consider the well known theory (due to several authors: in particular Clifford, Munn, Ponizovskii and Hewitt-Zuckerman) of irreducible representations of S over a field k. For each irreducible representation $\rho : S \to \mathrm{End}(V)$ there exists a unique smallest $J \in \mathcal{U}(M)$, called the *apex* of ρ such that $\rho(J) \neq \{0\}$. Furthermore, ρ determines an irreducible representation $\overline{\rho} : J^\circ \to \mathrm{End}(V)$ via

$$\overline{\rho}(x) = \begin{cases} \rho(x), & x \in J \\ 0, & x = 0. \end{cases}$$

Conversely, given $J \in \mathcal{U}(S)$ and an irreducible representation $\rho_1 : J^\circ \to \mathrm{End}(V)$, there exists a representation $\rho : S \to \mathrm{End}(V)$, with apex J such that $\overline{\rho} = \rho_1$. Furthermore, ρ is unique up to equivalence.

The second part of this theory identifies the irreducible representations of each J° with those of H, the $\mathcal{H}$-class of (any) $e \in E(J)$. Indeed, if $\rho : J^\circ \to \mathrm{End}(V)$ is an irreducible representation of J° then $\rho|H : H \to \mathrm{End}(e(V))$ is an irreducible representation of H. Conversely, any irreducible representation $\alpha : H \to \mathrm{End}(W)$ of H can be "extended" to an irreducible representation $\rho : J^\circ \to \mathrm{End}(V)$ such that $(\rho|H,\ \rho(e)(V)) \cong (\alpha, W)$. Here again the representation ρ is unique up to equivalence. In practice, the most difficult problem here is relate the dimension of W to the dimension of V. Anyone who has tried to compute the rank of a sandwich matrix knows what this is like.

3. Irreducible Modular Representation of Finite Groups of Lie Type

Let G be a finite group of Lie type. By definition, there exists a reductive group G defined over $\overline{\mathbb{F}}_p$ and an endomorphism $\sigma : \underline{G} \to \underline{G}$ such that $G = \{x \in \underline{G} | \sigma(x) = x\}$. By the results of Richen [3] the irreducible modular representations of G are in one to one correspondence with the set $\{(I, \chi) | \chi : L_I \to \overline{\mathbb{F}}_q^*\}$ where $I \subseteq S$ is a subset of the vertices of the Coxeter graph of G, and L_I is the corresponding Levi subgroup of G. This correspondence has the following properties:

Given $\rho : G \to G\ell(V)$ irreducible, and $B \subseteq G$ a Borel subgroup, there exists a unique line $Y \subseteq V$ such that $\rho(B)Y = Y$. Let $P_I = \{g \in G | \rho(g)Y = Y\}$. Then we obtain $\chi : P_I \to \overline{\mathbb{F}}_q^*$ via $\rho(g)(y) = \chi(g)y$. But χ factors through $P_I \to L_I = P_I/U$ where $U \subseteq P_I$ is the unipotent radical of P_I. The correspondence mentioned above identifies ρ with (I, χ).

There is one more feature that deserves mention. Suppose $\rho : G \to G\ell(V)$ is irreducible. Then there exists an algebraic group $\underline{G}$ defined over $\overline{\mathbb{F}}_p$ with endomorphism $\sigma : \underline{G} \to \underline{G}$ such that

(i) $G \cong \{x \in \underline{G} | \sigma(x) = x\}$

(ii) There is a representation $\underline{\rho} : \underline{G} \to G\ell(V)$ of algebraic groups such that $\underline{\rho}|G = \rho$.

See Theorem 13.3 of [8]. This will be useful in the next section when we construct finite monoids of Lie type from irreducible representations of finite groups of Lie type.

It is impossible in general to specify the "typical" irreducible representation of a finite group G of Lie type. In fact, the (conjectured) dimension formula for the irreducible module of type (I, χ) has yet to be verified in all cases. For $S\ell_2(\mathbb{F}_q)$, one can specify a complete list of modules using Steinberg's tensor product theorem [9] (since the factors are known explicitly). The first step is to obtain the $\otimes$-irreducible modules for $S\ell_2(\mathbb{F}_q)$. These are V_0 (the trivial module), $V_1 = \overline{\mathbb{F}}_q \oplus \overline{\mathbb{F}}_q$, $V_2 = S^2(V_1)$, $V_3 = S^3(V_1), \ldots$ and $V_{p-1} = S^{p-1}(V_1)$, where V_1 has the usual action $\left(\begin{smallmatrix} a & b \\ c & d \end{smallmatrix}\right)\left(\begin{smallmatrix} x \\ y \end{smallmatrix}\right) = \left(\begin{smallmatrix} ax+by \\ cx+dy \end{smallmatrix}\right)$ and S^i denotes the $i-th$ symmetric power. Define

$$\alpha_i : S\ell_2(\overline{\mathbb{F}}_p) \to S\ell_2(\overline{\mathbb{F}}_p) \text{ via } \alpha_i\left(\begin{smallmatrix} a & b \\ c & d \end{smallmatrix}\right) = \left(\begin{smallmatrix} a^\gamma & b^\gamma \\ c^\gamma & d^\gamma \end{smallmatrix}\right)$$

where $\gamma = p^i$. Then any module (ρ, V) for $S\ell_2(\overline{\mathbb{F}}_p)$ yields other modules (ρ^{α_1}, V), (ρ^{α_2}, V),
$\ldots$ via $\rho^{\alpha_i}(x)(v) = \rho(\alpha_i(x))(v)$. We denote (ρ^{α_i}, V) by V^{α_i}.

Steinberg's $\otimes$-product theorem for $S\ell_2(\overline{\mathbb{F}}_q)$ and $S\ell_2(\mathbb{F}_q)$

(a) *Any irreducible module of $S\ell_2(\overline{\mathbb{F}}_q)$ can be written uniquely as*

$$V_{i_0}^{\alpha_0} \otimes V_{i_1}^{\alpha_1} \otimes \cdots \otimes V_{i_S}^{\alpha_S}$$

where $0 \le i_j \le p-1$. Conversely, any such module is irreducible for $S\ell_2(\overline{\mathbb{F}}_q)$.

(b) *Any irreducible module of $S\ell_2(\mathbb{F}_q)$ is the restriction of some irreducible module of $S\ell_2(\overline{\mathbb{F}}_q)$. Any such module can be written uniquely as $V_{i_0}^{\alpha_0} \otimes V_{i_1}^{\alpha_1} \otimes \cdots \otimes V_{i_S}^{\alpha^S}$ where $0 \le i_j \le p-1$ and $\sum_{j=0}^{S} p^j i_j \le q-1$. Conversely, any such module is irreducible for $S\ell_2(\mathbb{F}_q)$.*

Thus, the irreducible modules of $S\ell_2(\mathbb{F}_q)$ can be indexed by the non-negative integers $m \le q-1$. To compute $V(m)$, write $m = \sum_{j=0}^{S} p^j i_j$ where $0 \le i_j \le p-1$ (p-adic expansion). Then $V(m) \cong \bigotimes_{j=0}^{S} V_{i_j}^{\alpha_{i_j}}$ and $\dim(V(m)) = \prod_{j=0}^{S} (i_j + 1)$.

In the terminology of Richen's theory (for $m > 0$), $V(m)$ is the irreducible module of type (ϕ, χ) where $\chi : L_\phi \to \mathbb{F}_q^*$ is given by $\chi(t) = t^m$, and $L_\phi = \left\{\left(\begin{smallmatrix} t & 0 \\ 0 & t^{-1} \end{smallmatrix}\right) | t \in \mathbb{F}_q^*\right\}$ is the Levi factor of the upper triangular group (which is the only proper parabolic subgroup of $S\ell_2(\mathbb{F}_q)$, up to conjugacy).

We can depict the pattern here for $S\ell_2(\mathbb{F}_9)$ using the "3-adic Pascal triangle"

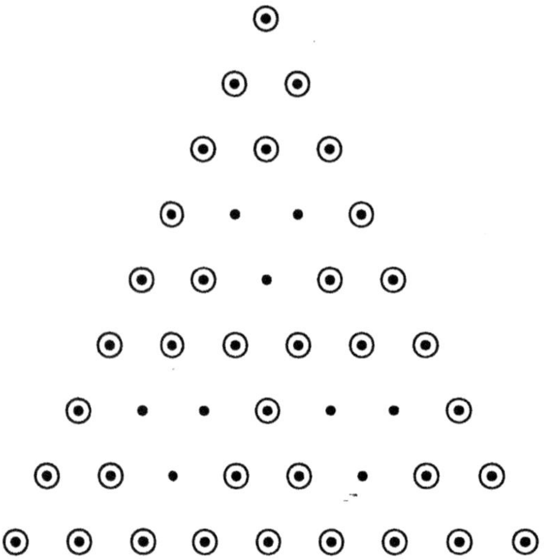

module	dimension
0	1.
1	2
2	3
3	2
4	4
5	6
6	3
7	6
8	9

Each row represents the irreducible module with given label and has the indicated dimension (= the number of $\odot$'s in that row). If we label the positions in the n-th row $0, 1, 2, \ldots n$, then there is a $\odot$ in the i-th position if and only if $\binom{n}{i} \not\equiv 0$ mod (3).

4. Irreducible Representations of Finite Monoids of Lie Type

Let M be a finite monoid of Lie type with unit group G of characteristic p. We wish to explain the following theorem. What transpires is a truly remarkable blend of semigroup theory and group theory.

Theorem 4.1 *Suppose $\rho : M \to \mathrm{End}(V)$ is an irreducible representation of M. Then $\rho|G$ is irreducible.*

Theorem 4.1 is proved in two steps. The first step is to prove the result for the *canonical monoid* $\mathcal{M} = \mathcal{M}(G)$ (see [5]). This monoid is characterized (via Theorem 3.8(ii) of [4]) by the following two properties:

(a) The type map $\lambda : \mathcal{U}(\mathcal{M}) \to 2^S$ satisfies

(i) $\lambda : \mathcal{U}(\mathcal{M})\backslash\{0\} \to 2^S$ is bijective

(ii) $\lambda(J_{ef}) = \lambda(J_e) \cap \lambda(J_f)$, where $\mathcal{U}(\mathcal{M})$ has been identified in the usual way with Λ.

(b) For each $e \in \Lambda$, $\{g \in G | eg = ge = e\} = \{1\}$.

$\mathcal{M}$ also enjoys the following extremely useful property:

(c) $J_eJ_f \subseteq J_{ef} \cup \{0\}$ if $e, f \in \Lambda$. In particular, if $\lambda(J) = I \in 2^S$, then $\mathcal{M}_I = G \cup J \cup \{0\}$ is a subsemigroup of $\mathcal{M}$.

Furthermore, $\mathcal{M}_I$ has the following universal property:

(d) Let M be a monoid of Lie type with three J-classes $M = G \cup J' \cup \{0\}$ and such that $\lambda(J') = I$. Then for any $e \in E(J')$ there exists a unique morphism $\varphi : \mathcal{M}_I \to M$ such that $\varphi(e_J) = e$ and φ is the identity on G.

To prove 4.1 for $\mathcal{M}$ we first "construct" a collection of distinct irreducible representations of $\mathcal{M}$ all of which restrict to irreducible representations of G. We then count these up and find out (via the results of section 2) that *all* irreducible representations of $\mathcal{M}$ have been accounted for.

Let $\rho : G \to G\ell(V)$ be the irreducible representation of G of type (I, χ) (as guaranteed by Richen's theorem; see section 3). So $\chi : L_I \to \overline{\mathbb{F}}_p^*$. By Theorem 13.3 of [8] ρ is the restriction of an irreducible representation φ of some reductive group $\underline{G}$. It is then possible to construct a submonoid M_φ of End(V) such that $M_\varphi = G \cup J' \cup \{0\}$ and $\lambda(J') = I$. Hence, by (*d*) above there exists a morphism $\mathcal{M}_I \to M_\varphi$, and so V becomes an irreducible $\mathcal{M}_I$-module. Write $\mathcal{M}_I = G \cup J_I^\circ$, where $J_I^\circ = J_I \cup \{0\}$. Now let $\Omega = \cup\{J_K \subseteq \mathcal{M} | I \not\subseteq K\}$. Then $\Omega \subseteq \mathcal{M}$ is an ideal, and $J_I^\circ \subseteq \mathcal{M}/\Omega$ is an ideal. But then (by the results of section 2) V is an $\mathcal{M}/\Omega$-module and hence an $\mathcal{M}$-module with apex J_I. Since $J_I V \neq 0$ we obtain that $J_K V \neq 0$ for any $K \supseteq I$. Hence, V is an irreducible $\mathcal{M}_K$-module. But, as above, V becomes an irreducible $\mathcal{M}$-module (in a different way) with apex J_K. In each case, the restriction to G is the same irreducible module (of type (I, χ)).

Hence, for each irreducible module (ρ, V) of G of type (I, χ) we have constructed $2^{|S \setminus I|}$ distinct irreducible modules of $\mathcal{M}$ whose restriction to G is (ρ, V). In this way, we obtain a total of

$$\sum_{I \subseteq S} 2^{|S \setminus I|} \alpha_I \tag{1}$$

distinct irreducible representations of $\mathcal{M}$, where $\alpha_I = |\{\chi : L_I \to \overline{\mathbb{F}}_p^*\}|$. On the other hand the number of irreducible representations of L_I is $\sum_{K \subseteq I} \alpha_K$. So by the results of section 2, there are exactly

$$\sum_{I \subseteq S} \sum_{K \subseteq I} \alpha_K \tag{2}$$

inequivalent irreducible modular representations of $\mathcal{M}$. But the numbers (1) and (2) are equal! So all irreducible representations of $\mathcal{M}$ are accounted for by those that restrict to irreducible representations of G.

Now let M be any finite monoid of Lie type, and let V be an irreducible M-module. Then there is a minimal $J \in \mathcal{U}(M)$ such that $JV \neq \{0\}$. Let $\Omega = \{a \in M | J \not\subseteq MaM\}$. Then Ω is an ideal of M. Since J° is also an ideal of M/Ω, V is an irreducible J°-module. Thus V is an irreducible M'-module, where $M' = G \cup J^\circ \subseteq M/\Omega$. But, again by (*d*) above, V is an irreducible $\mathcal{M}_I$-module (where $I = \lambda(J)$). However, as we saw in step one, any irreducible $\mathcal{M}_I$-module is naturally the restriction of an irreducible $\mathcal{M}$-module. But such modules (as we have seen in step one) remain irreducible when restricted to G.

The special case of Theorem 4.1 where $M = M_n(\mathbb{F}_q)$ was obtained independently by Doty, Harris-Kuhn and Kropp. The methods employed are entirely different.

We now turn to the task of identifying the irreducible representations of a finite monoid of Lie type. Any irreducible representation $\rho : M \to \mathrm{End}(V)$ of M yields an apex $J \in \mathcal{U}(M)$ and (thanks to Theorem 4.1) an irreducible representation of G of type, say (I, χ). On the other hand, by the results of section 2, the representation ρ is uniquely associated with the irreducible representation $\rho|H(e) : H(e) \to \mathrm{End}(\rho(e)V)$ where $H(e)$ is the $\mathcal{H}$-class of $e \in E(J)$. Furthermore, $C_G(e) \to H(e)$, $g \longmapsto ge$, is surjective so that $\gamma = \rho|H(e)$ can be considered as a representation of $C_G(e)$ with the property that $\gamma(g) = \gamma(h)$ if $ge = he$. We thus obtain, as a first step, the following description:

Proposition 4.2 *There is a one to one correspondence between irreducible representations of M, and pairs (e, γ) where $e \in \Lambda \cong \mathcal{U}(M)$ and γ is an irreducible representation of $C_G(e)$ with $\gamma(ge) = \gamma(he)$ if $ge = he$.*

But this does not solve all our basic problems:

(a) To which irreducible representation of G does a given (e, γ) correspond?

(b) What are the conditions on a triple (J, I, χ), with $\chi : L_I \to \overline{\mathbb{F}}_q^*$ and $J \in \mathcal{U}(M)$ so that there is an irreducible representation ρ of M such that

(i) $\rho|G$ is of type (I, χ)

(ii) Apex $(\rho) = J$?

To answer these questions we need some notation. Let M be a finite monoid of Lie type with $e \in \Lambda$. Then $C_G^r(e) = P_{\lambda(J)}$ where $\lambda(J) \subseteq S$ and $\{e\} = \Lambda \cap J$. If $H = \{g \in G | ge = e\}$ and $B \subseteq C_G^r(e)$ is a Borel subgroup, then $BH \subseteq C_G^r(e)$ is a parabolic subgroup containing B. Hence, we define

$$\nu(J) \in 2^S \text{ via } BH = P_{\nu(J)}.$$

We can now identify the irreducible representations of M in a way that clarifies the entire situation (in particular, answering questions (a) and (b) above).

Theorem 4.3 *Let $I \in 2^S$ and $J \in \mathcal{U}(M)$. Assume that $\nu(J) \subseteq I \subseteq \lambda(J)$. Define*

$$\alpha_{I,J} = \{\chi : L_I \to \overline{\mathbb{F}}_q^* | \chi(g) = \chi(h) \text{ if } e_J g = e_J h\}.$$

Then there is a one to one correspondence between the irreducible representations of M and the set

$$\bigsqcup_{\substack{I \in 2^S,\ J \in \mathcal{U}(M) \\ \nu(J) \subseteq I \subseteq \lambda(J)}} \alpha_{I,J}.$$

Under this correspondence $\chi \in \alpha_{I,J}$ corresponds to the unique irreducible representation $\rho : M \to \mathrm{End}(V)$ such that

(i) *Apex* $(\rho) = J$.

(ii) *There is a line* $Y \subseteq V$ *such that* $\{g \in G | \rho(g)Y = Y\} = P_I$.

(iii) *If* $g \in P_I$ *and* $y \in Y$ *then* $\rho(g)(y) = \chi(g)y$.

There are essentially two steps to the proof of Theorem 4.3.
Step One: Identify (in terms of Richen's theory of section 3) the irreducible representations of M with apex J with

$$\bigsqcup_{\nu(J) \subseteq I \subset \lambda(J)} \alpha_{I,J}.$$

Step Two: Show that each irreducible representation $\rho : M \to \mathrm{End}(V)$ with apex J satisfies conditions (ii) and (iii) of Theorem 4.3 for the χ obtained in Step One.

We indicate the proof of Step Two (omitting the relationship with Step One). The proof of Step One is more straightforward.

So let $\rho : M \to \mathrm{End}(V)$ be irreducible with apex $J \in \mathcal{U}(M)$. If $e \in \Lambda \cap J$ then $e(V) \neq 0$ and this is an irreducible eMe-module. Furthermore, $\rho(e)(V) \subseteq V^U = \{v \in V | \rho(u)(v) = v \text{ for all } u \in U\}$ where $U \subseteq P(e)$ is the unipotent radical. It follows from a result of Cabanes [1] that V^U is an irreducibe $P(e)$-module and hence an irreducible $C_G(e)$-module. But $\rho(e)(V)$ is $C_G(e)$-submodule of V^U since $\rho(e) : V \to V$ is a $C_G(e)$-module endomorphism. We conclude that $\rho(e)(V) = V^U$.

Now $\rho(e)(V)$ is an irreducible $C_G(e)$-module. Thus, by the results of Richen (section 2), for any Borel subgroup B_1 of $C_G(e)$ there is a unique line $Y \subseteq \rho(e)(V)$ such that $\rho(B_1)Y = Y$. But $B_1U = B$ is a Borel subgroup of $P(e)$ such that $\rho(B)Y = Y$.

Now let $g \in G$ and assume that $\rho(g)Y = Y$. Let $0 \neq \rho(e)(v) \in Y$. Thus, $0 \neq \rho(g)\rho(e)(v) = \gamma\rho(e)(v)$ for some $\gamma \in \overline{\mathbf{F}}_q^*$, and so $0 \neq \rho(g)\rho(e)(v) = \rho(ege)(v)$. Then $ege \in J$ since otherwise $ege \in \Omega = \{a \in M | J \not\subseteq MaM\}$, while $\rho(\Omega)V = \{0\}$, contradicting the fact that $\rho(ege)V \neq 0$. But then $ege\ \mathcal{H}e$, and from this it follows easily that $g \in P^-(e)P(e)$. Then, by the Bruhat decomposition, $g \in P(e)$ (since $\{g \in G | \rho(g)Y = Y\}$ is a subgroup). To summarize,

> if $P_1 \subseteq H(e)$ is the stabilizer of Y in $H(e)$ then $\varphi^{-1}(P_1) \subseteq P(e)$ is the stabilizer of Y in G, where $\varphi : P(e) \to H(e)$, $\varphi(g) = ge$.

We now recall $\nu(J) \in 2^S$. This was defined above so that $\varphi^{-1}(B_1) = P_{\nu(J)}$ where $B_1 \subseteq H(e)$ is a Borel subgroup. From what we have just proved it follows easily that any irreducible representation $\rho : M \to \mathrm{End}(V)$ with apex $J \in \mathcal{U}(M)$ has the following property.

> If $e \in \Lambda \cap J$ then there is a line $Y \subseteq \rho(e)(V)$ stabilized by B such that if $P_I = \{g \in G | \rho(g)Y = Y\}$, then $\nu(J) \subseteq I \subseteq \lambda(J) = \text{type } (P(e))$. Furthermore, if $y \in Y$ and $\chi : L_I \to \overline{\mathbf{F}}_q^*$ is defined by $\rho(g)(y) = \chi(g)y$ then χ vanishes on $K_J = \{g \in G | ge = eg = e\}$.

This essentially completes our discussion of the proof of Theorem 4.3.

Problem 4.1 There is a lot of evidence to suggest that, at least for finite reductive monoids M, there should be a very nice formula that counts the set

$$IR(M) = \{\rho : M \to \mathrm{End}(V) | \rho \text{ is irreducible}\}/_{\sim}$$

where "$\sim$" denotes equivalence of representations. It should be possible to write

$$|IR(M)| = \sum_{i=0}^{N} a_i(q-1)^i$$

where $\mathbb{F}_q$ is the "field of definition" of M. In particular, one might hope to interpret the a_i's in terms of λ, Λ and other invariants of M that are independent of q. Rather than making a conjecture, we consider a familiar example. Let $M = M_n(\mathbb{F}_q)$. By the theory of Munn and Ponizovskii,

$$|IR(M)| = \sum_{k=0}^{n} |IR(G\ell_k(\mathbb{F}_q))|,$$

while by the formulas of Richen (for $k > 0$),

$$\begin{aligned}
|IR(G\ell_k(\mathbb{F}_q))| &= \sum_{\substack{(k_1,\ldots,k_s) \\ k_1+\cdots+k_s=k}} |\mathrm{Hom}(\prod_{i=1}^{s} G\ell_{k_i}(\mathbb{F}_q), \mathbb{F}_q{}^*)| \\
&= \sum_{\substack{(k_1,\ldots,k_s) \\ k_1+\cdots+k_s=k}} (q-1)^s \\
&= \sum_{i=0}^{k-1} \binom{k-1}{i} (q-1)^{i+1} \\
&= (q-1)q^{k-1}.
\end{aligned}$$

Thus, combining these formulas,

$$|IR(M)| - 1 = (q-1)\sum_{k=1}^{n} q^{k-1} = q^n - 1.$$

But we can also write

$$\begin{aligned}
|IR(M)| - 1 &= \sum_{k=1}^{n} |IR(G\ell_k(\mathbb{F}_q))| \qquad \text{(as above)} \\
&= \sum_{k=1}^{n} \sum_{i=0}^{k-1} \binom{k-1}{i} (q-1)^{i+1} \\
&= \sum_{i=0}^{n} \left(\sum_{k=i=1}^{n} \binom{k-1}{i} \right) (q-1)^{i+1}.
\end{aligned}$$

Notice that the coefficient is independent of q. Expanding the first formula we obtain

$$\begin{aligned} q^n - 1 &= (1 + (q-1))^n - 1 \\ &= \sum_{i=1}^{n} \binom{n}{i} (q-1)^i \\ &= \sum_{i=0}^{n-1} \binom{n}{i+1} (q-1)^{i+1}. \end{aligned}$$

Comparing coefficients we obtain a well known identity

$$(*) \qquad \binom{n}{i+1} = \sum_{k=i+1}^{n} \binom{k-1}{i}.$$

We now interpret this coefficient.

An elementary calculation shows that if $J \subseteq M$ is a J-class (say $J = \{x \in M \mid \operatorname{rank}(x) = k\}$) and $\nu(J) \subseteq I \subseteq \lambda(J)$ then

$$|\alpha_{I,J}| = (q-1)^{|\lambda(J)\backslash I|+1}.$$

We conclude, using Theorem 4.3 and (*) above that

$$|IR(M)| - 1 = \sum_{i=0}^{n} \binom{n}{i+1} (q-1)^{i+1} \qquad \text{where}$$

$$\binom{n}{i+1} = \left| \left\{ (I, J) \in 2^S \times \mathcal{U}(M) \;\middle|\; \begin{array}{c} \nu(J) \subseteq I \subseteq \lambda(J) \\ |\lambda(J)\backslash I| = i \end{array} \right\} \right|.$$

5. The Amazing Monoid $\mathcal{M}^*$

Using Putcha's theory of monoids of Lie type [4] it is possible to construct, for each group G of Lie type a monoid $\mathcal{M}^*$ with the following properties:

(i) $\lambda : \Lambda \cong 2^S \sqcup \{\lambda(1)\}$ where $\lambda(ef) = \lambda(e) \cup \lambda(f)$ if $e, f \in \Lambda\backslash\{1\}$.

(ii) For each $e \in \Lambda\backslash\{1\}$, say $\lambda(e) = I$, $H(e) = eL_I \cong L_I/L_I'$ where L_I' is the subgroup generated by the elements of L_I of order p^n for some $n \geq 0$.

$\mathcal{M}^*$ is called the *dual canonical monoid* of G. By Putcha's theory, $\mathcal{M}^*$ is uniquely determined by properties (i) and (ii). Furthermore, $\mathcal{M}^*\backslash G$ is a locally inverse semigroup. It is easily verified that $\nu(J) = \lambda(J)$ for each $J \in \mathcal{U}(\mathcal{M}^*)$ (i.e., the maximal subgroups of $\mathcal{M}^*$ are as small as possible). Combining this observation with properties (i) and (ii) above, and applying Theorem 4.3 above yields the following remarkable theorem.

Theorem 5.1 *Let $\rho : G \to G\ell(V)$ be a nontrivial irreducible representation. Then there exists a unique representation $\rho^* : \mathcal{M}^* \to \operatorname{End}(V)$ such that*

(i) $\rho^*|G = \rho$

(ii) $\rho^*(\mathcal{M}^*\backslash G) \neq \{0\}$.

Furthermore, if ρ *is of type* (I, χ) *then the apex of* ρ^* *is the unique* J*-class* $J \in \mathcal{U}(M)\backslash\{G\}$ *such that* $\lambda(J) = I$.

This monoid $\mathcal{M}^*$ will very likely be of significance in studying the irreducible modular representations of G.

References

1. M. Cabanes, *Irreducible modular representations and Levi supplements*, J. of Alg. **90** (1984), 84–97.
2. A. H. Clifford and G. B. Preston, *The Algebraic Theory of Semigroups*, AMS Surveys #7 **1**, Providence, 1961.
3. C. W. Curtis, *Modular representations of finite groups with split BN pairs* Springer Lecture Notes **131**, 57–95, 1970.
4. M. S. Putcha, *Monoids on groups with BN-pairs*, J. Algebra **120** (1989), 139–169.
5. M. S. Putcha, *Monoids of Lie type*, These proceedings.
6. M. S. Putcha and L. Renner, *The canonical compactification of a finite group of Lie type*, Trans. Amer. Math. Soc. **337** (1993), 305–319.
7. L. Renner, *Finite reductive monoids*, These proceedings.
8. R. Steinberg, *Endomorphisms of linear algebraic groups*, Mem. Amer. Math. Soc. **80** (1968).
9. R. Steinberg, *Representations of algebraic groups*, Nagoya Math. J. **22** (1963), 33–56.

CASCADE DECOMPOSITION OF ARBITRARY SEMIGROUPS

CHRYSTOPHER L. NEHANIV *
School of Computer Science and Engineering
The University of Aizu
Aizu-Wakamatsu City, 965 Japan

For Echo and Perseverance

1. Introduction

1.1. Krohn-Rhodes Theory

In this paper "semigroup" will always mean "non-empty semigroup". By (X, S) we denote a **(right) transformation semigroup** or **automaton**, that is, a set X called **states** and a semigroup S which acts on the right of X. Formally, there is a function $X \times S \mapsto X$, the **action**, written $(x, s) \rightarrow x \cdot s$, satisfying

$$(x \cdot s) \cdot s' = x \cdot (ss')$$

for all x in X and all s, s' in S.

From a computational point of view, a generating set for S can be considered as **inputs** to the automaton (X, S) and the action describes the state transitions under strings of inputs (which yield products in S).

A transformation semigroup (X, S) is **faithful** if $x \cdot s = x \cdot t$ for all $x \in X$ implies $s = t$. Thus, S may be regarded as a subsemigroup of the semigroup of all functions from X to itself with composition as the associative binary operation. *For us, all transformation semigroups are faithful unless otherwise stated.* Hence whenever we construct an automaton, we shall be obliged to prove that the action is faithful in case this is not obvious. If (X, S) is not faithful, we may introduce a congruence on S

$$s \equiv s' \iff \forall x \in X : x \cdot s = x \cdot s'.$$

Then the induced action of $S/\equiv$ is well-defined and faithful. Furthermore, $\equiv$ is the unique congruence with this property: any other congruence on S which makes the action faithful clearly contains $\equiv$ and cannot be larger lest well-definedness be lost.

For an arbitrary semigroup S, one canonically obtains a faithful transformation semigroup (S^1, S). Here, S^1 denotes the underlying set of the monoid (also denoted S^1) obtained by adjoining to S an identity element if and only if S is not itself a monoid. S acts on the right of S^1 by multiplication in S^1: $s_1 \cdot s = s_1 s$. Notice

* This is work is based on a portion of the author's Ph.D. thesis at the Mathematics Department of the University of California, Berkeley under the inspiring supervision of Professor John L. Rhodes. It was supported in part by a U.S. Department of Education National Need Fellowship.

J. Fountain (ed.), Semigroups, Formal Languages and Groups, 391–425.

that adjoining a new identity if necessary guarantees that this action is faithful: if $s, t \in S$ with $x \cdot s = x \cdot t$ for all $x \in S^1$, then $s = t$ since $t = 1 \cdot t = 1 \cdot s = s$. Also let S^I be the semigroup obtained from S by adjoining a new identity *regardless of whether or not S already has one;* that is, $Is = sI = s$ for all $s \in S$.

For an arbitrary transformation semigroup (X, S), let $(X, \overline{S})$ be obtained by including all constant maps on X not already in (the action of) S.

In the case of finite state automata and finite semigroups, the theory of Krohn-Rhodes developed in the 1960's shows that all such finite structures can be obtained in a feedback-free cascade (wreath product) of fundamental building blocks: (1) finite groups (or the corresponding permutation automata), which can be built by cascading simple groups, and (2) semigroups whose subgroups are all trivial, which can be built by cascading "flip-flops". A flip-flop is the automaton with two states $\{a, b\}$ and three transformations: the identity mapping and the two constant maps.[1]

The starting point for this paper is the Krohn-Rhodes Theorem (cf. [9, 4]) for finite transformation semigroups.

Theorem 1.1 (Krohn-Rhodes) *Let (X, S) be a finite transformation semigroup. Then there exist finite transformation semigroups (X_i, S_i), where each S_i is a group or has only trivial subgroups, such that*

$$(X, S) \longleftarrow (Z, T) \leq (X_n, S_n) \circ \cdots \circ (X_1, S_1).$$

In the Krohn-Rhodes decomposition, we say the (possibly non-faithful) transformation semigroup (Z, T) *covers* or *emulates the computation of* (X, S). Computation can be *lifted* to the larger automaton (Z, T), which has a simple global coordinatization in terms of cascaded pieces. See §1.3.

The *wreath product* $(X_n, S_n) \circ \cdots \circ (X_1, S_1)$ has states $(x_n, \ldots, x_1)$ where $x_i \in X_i$ and semigroup consisting of all maps f such that $(x'_n, \ldots, x'_1) = (x_n, \ldots, x_1)f$, where $x'_i = x_i \cdot s_i$ for an $s_i \in S_i$ depending only on f and the initial segment $(x_{i-1}, \cdots, x_1)$ of coordinates with index strictly less than i. Such a coordinatization has that remarkable property that it is *sequential*, that is, projection onto any initial segment of coordinates in the wreath product is a morphism.

In view of what follows it is crucial to observe that to specify a wreath product, it is enough to specify an indexing of a family of transformation semigroups by elements of finite ordinal.

The Krohn-Rhodes theorem may be proved by expressing a finite semigroup S as the union of two proper subsemigroups V, a left ideal ($SV \subseteq V$), and T. One shows how to cover S by a wreath product of V and T (with constants and new identities adjoined). Since V and T have smaller cardinality than S, by iterating the process on each of V and T one finally obtains a wreath product of subsemigroups of S which are either cyclic or have no proper left ideals. The latter is finite and

[1] Since the inputs to an automaton or finite state machine $\mathcal{A}$ map its set of states Σ to Σ, closing these mappings under composition generates a semigroup, *the semigroup of the automaton* $\mathcal{A}$, whose properties ones study algebraically to obtain information about $\mathcal{A}$. Conversely, taking the right regular representation of a semigroup S yields a transformation semigroup (S^1, S). Due to the consequences of this connection between semigroups and automata, it is common for researchers loosely to speak of semigroups as automata.

left-simple[2], hence necessarily of the form $X^\ell \times G$, where G is a group and X^ℓ is left-zero semigroup ($\forall x, y \in X : xy = x$). For more details of the classical '$V \cup T$' proof, see Krohn, Rhodes, Tilson [9] or Eilenberg [4].

1.2. The Infinite Case

In the 1970's Rhodes and Goodwin attempted to generalize this "V union T" method to infinite semigroups. Goodwin obtained a result (Lemma 4.3 below) implying that except for cyclic and left-simple semigroups, a semigroup with a maximal L-class could be written $V \cup T$. Of course, an infinite semigroup need not have a maximal L-class. But even worse, they saw that for infinite S, as the following example shows, there may be no such V and T.

Example 1.1 (Rhodes) Let $\mathbb{Q}(0,x)$ denote the rational numbers in the open interval $(0,x)$ under multiplication where x is a positive real number in the unit interval; similarly define $\mathbb{Q}(0,x]$ for a half-open interval. Then $\mathbb{Q}(0,1)$ cannot be written as the union of a proper left ideal V and a proper subsemigroup T.

Proof: $\mathbb{Q}(0,1)$ is commutative and the ideals of $\mathbb{Q}(0,1)$ are $\mathbb{Q}(0,y)$ or $\mathbb{Q}(0,y]$ for $y \in (0,1)$, since $x \cdot \mathbb{Q}(0,1) = \mathbb{Q}(0,x)$ for all $x \in \mathbb{Q}(0,1)$. Further $\forall z \in (0,1)$, $\mathbb{Q}(z,1)$ generates $\mathbb{Q}(0,1)$ since $(z_1,1) \cdot (z_2,1) = (z_1 z_2, 1)$ and $\lim_{n\to\infty} z^n = 0$. So $(0,1) = \bigcup_n (z^n,1) = \bigcup_n (z,1)^n$. Hence if $\mathbb{Q}(0,1) = V \cup T$ and $V \neq \mathbb{Q}(0,1)$ then $V = \mathbb{Q}(0,x)$ with $x < 1$, so choosing $z \in \mathbb{Q}(0,1)$ so $x < z < 1$, we have $\mathbb{Q}(z,1) \subseteq T$, so

$$\mathbb{Q}(0,1) = \langle \mathbb{Q}(z,1) \rangle \subseteq T \subseteq \mathbb{Q}(0,1),$$

and so $T = \mathbb{Q}(0,1)$. □

When confronted with S not writable as $V \cup T$, Goodwin and Rhodes attempted to expand S using the Rhodes expansion $\widehat{S}$, which must have a maximal L-class. Expanding removed the obstruction at a particular stage, but created instead apparently insurmountable set-theoretic obstructions. Their attempt at proving a theorem like our Theorem 4.3 could not be completed.

To prove a nice decomposition theorem for arbitrary semigroups, there are several new ingredients we need.

(§2) Cascade Integral and Cascade Transform. We replace the notion of wreath product of transformation semigroups by the more general notion of cascade integral. Lemmas about the associativity of the wreath product and transitivity ("the cover of a cover is a cover") *et cetera* carry over to the cascade integral, as shown in the Appendix. The cascade transform is a notational tool for writing down theorems in this setting.[3]

---

[2] Throughout this paper, we make use of **Green's relations** L,R,H, and J when discussing semigroups. Recall that that $s \geq_L t$ if there is an $\ell \in S^1$ such that $t = \ell s$. Thus the principal left ideal $S^1 s$ generated by s contains the principal left ideal $S^1 t$ generated by t. Dually define $s \geq_R t$. Write $s \equiv_H t$ if s and t are both L-equivalent and R-equivalent. Finally we say $s \geq_J t$ if $S^1 t S^1 \subseteq S^1 s S^1$; that is, the principal ideal generated by s contains the principal ideal generated by t. Define L_s to be the L-equivalence class of $s \in S$. Similarly define R_s, H_s, and J_s. For a detailed development of these important relations see any of [9, 3, 10, 8].

[3] The cascade transform also as a deeper significance in the tree interpretation of wreath product and cascade integration [14, 13] that we cannot go into here. Briefly, there is a natural tree order

(§3) Cascadable Systems. Instead of trying to decompose a semigroup as $V \cup T$, we shall also allow decompositions into subsemigroups, possibly in other ways. Cascadable systems correspond to nice decompositions into cascades of subsemigroups.

(§3.2) Restrictions on Covering. In the 1960's Dennis Allen, Jr. showed how every countable semigroup divides a wreath product of two copies of $(\mathbf{N}, \mathbf{N})$. His construction is Example 3.5. If we are to have a non-trivial structure theory in the infinite case, this decomposition is must be considered pathological. Therefore we motivate the aperiodic, faithful, and group-lifting restrictions – unnecessary in the finite theory – and show them to be satisfied for decompositions arising from cascadable systems.

(§4.1) Refinement. We introduce an ordering on cascadable systems, hence on decompositions. Instead of step-by-step iteration of decomposition, we are concerned with refinement: better decompositions are higher in the refinement order. Instead of iterating a decomposition technique, we shall demand (and get) the existence of refinements for cascadable systems whenever the building block subsemigroups can be further reduced. Maximal cascadable systems are shown to exist (§4.2), and these correspond to the best decompositions via subsemigroups.

We discuss in §4 how the notions of cascadable system and refinement allow us to avoid any set-theoretic obstructions. The paper culminates in §4.5 with the Cascade Decomposition Theorem for Arbitrary Semigroups: An arbitrary semigroup S can be nicely covered by a totally ordered cascade of very simple subsemigroups of S (cyclic or left-zero $\times$ group). Each subgroup of S lifts to an isomorphic subgroup of the covering semigroup. The division is aperiodic, faithful, and a lift of any state can be described by writing *finitely* many coordinates (although infinitely many transformation semigroups may have been cascaded). In particular, the Krohn-Rhodes Theorem for finite semigroups is a corollary. Thus our main theorem is a direct generalization from finite to infinite automata, yielding *global sequential coordinates* on *arbitrary* semigroups and automata.

Applications to arbitrary union-of-groups semigroups, the bicyclic monoid and free semigroups are in §5.

1.3. The Interpretation of Covering

We now say formally what covering means for finite or infinite semigroups and automata. This has a natural interpretation in terms of emulation of one automaton by a covering automaton.

Let (X, S) and (Y, T) be transformation semigroups. A **division** or **covering morphism of transformation semigroups** $w : (Y, T) < (X, S)$ is a surjective partial function $w : X \twoheadrightarrow Y$ satisfying

$$\text{(1)} \qquad \forall t \in T \text{ there exists } \tilde{t} \in S, (x \cdot \tilde{t})w = xw \cdot t \text{ whenever } xw \text{ is defined.}$$

Note: If xw is defined, this implies $(x \cdot \tilde{t})w$ is also defined, for all $\tilde{t}$ as in (1). Write $\widetilde{Y} = (Y)w^{-1}$.

---

on the segments of states of a cascade integral. The cascade transform of a semigroup element in a cascade integral (or wreath product) is a labelling of such a tree. Such labelled trees are multiplied by the Convolution Rule in §2.

The **algebraic cover of T by S induced by w**, also denoted w, is the relation

$$w = \{(s,t) \in S \times T : \forall x \in \widetilde{Y} : (x \cdot s)w = xw \cdot t\}.$$

Note that $(t)w^{-1}$ is the set of all $\tilde{t}$ that work in (1) for t. If $\tilde{a} \in (a)w^{-1}$, we say $\tilde{a}$ is a **lift** of a and that $\tilde{a}$ **covers** a.

Also if $(s,t_1),(s,t_2) \in w$, then since $w : X \twoheadrightarrow Y$ is surjective for all $y \in Y$, there exists $x \in X$ with

$$y \cdot t_i = xw \cdot t_i = (x \cdot s)w \text{ for } i = 1,2,$$

so by faithfulness of (Y,T) we conclude that $t_1 = t_2$. Thus we may interpret w as a partial function on S. Therefore we shall also write

$$sw = t \text{ and } (t)w^{-1} = \{s \in S : \forall x \in \widetilde{Y},\ (x \cdot s)w = xw \cdot t\}.$$

The relation w is a subsemigroup of $S \times T$ projecting fully onto T: for if $(s,t),(s',t') \in w$ choose any x with xw defined; we have

$$\begin{aligned} xw \cdot tt' &= (x \cdot s)w \cdot t' \\ &= (x \cdot ss')w, \end{aligned} \tag{2}$$

whence $(ss',tt') \in w$.

In functional notation, $sw = t$ and $s'w = t' \Rightarrow (ss')w = tt'$, i.e. $(ss')w = (sw)(s'w)$.

It follows that $(T)w^{-1}$ is a subsemigroup $\widetilde{T}$ of S and that $w : S \to T$ is a surjective homomorphism when restricted to $\widetilde{T}$. In standard terminology [4, 22], we say $w^{-1} : T \to S$ (or $w : T < S$) is a **division of semigroups** and that T **divides** S, meaning that w is a homomorphism from a subsemigroup of S onto T. We shall especially be interested in the case when $w : \widetilde{T} \twoheadrightarrow T$ is a 'nice' homomorphism. For instance, we may require that for each idempotent $e^2 = e$ in T, that the subsemigroup $w^{-1}(e)$ is torsion and contains no non-trivial groups. A surmorphism having this property is called an **aperiodic** map. A covering w satisfying this condition is called an **aperiodic** division.

The idea behind covering morphisms is that the computation of $y \cdot t$ in the transformation semigroup (Y,T) can done using (X,S). By lifting $y \in Y$ and $t \in T$ to $x \in (y)w^{-1}$ and $s \in (t)w^{-1}$, respectively, we can calculate $x \cdot s$. By the covering property (1), $x \cdot s$ is a lift of $y \cdot t$. Since w is a partial function on X, $x \cdot s$ is not a lift of any other $y' \in Y$. Thus, the covering morphism w gives a way for (X,S) to "emulate" the computation of (Y,T). In other words, (Y,T) is "blown-up" via w to $(\widetilde{Y},\widetilde{T})$, which may be more desirable to compute in than (Y,T) if (X,S) has nice properties. It may happen that $(\widetilde{Y},\widetilde{T})$ fails to be faithful.

If w is fully defined on both X and S, this also has a computational interpretation: we can "simulate" or "approximate" the computation of (X,S) by using (Y,T), since covering morphism condition (1) becomes

$$xw \cdot sw = (x \cdot s)w, \text{ for all } x \in X, s \in S.$$

Note that a fully defined injective covering is necessarily an isomorphism. For a development of covering morphisms of finite *partial* transformation semigroups see Eilenberg [4]. For a detailed study of the relationship between coverings in the categories of monoids and small categories, see Tilson [22].

It is easy to verify that the composition of coverings is a covering and that the identity map on a transformation semigroup is a covering. Since composition of coverings is also associative, we have a category $\mathcal{TS}$ of transformation semigroups and covering morphisms.

When considering coverings of (Y,T), one would like to find a sub-transformation semigroup $(\widetilde{Y},\widetilde{T})$ of (X,S) such that

$$(Y,T) \overset{w}{\longleftarrow\!\!\!\!+} (\widetilde{Y},\widetilde{T}) \leq (X,S).$$

In the finite case, faithfulness of $(\widetilde{Y},\widetilde{T})$ is not usually required; however we shall see it be guaranteed for our decompositions.

Here we do not require that $w^{-1}(t)$ contain all possible lifts of t in S, but only lifts $\tilde{t}$ lying in $\tilde{T}$. In addition to w being aperiodic or 'nice', one desires that (X,S) be 'easily globally computed' and that the embedding of $(\tilde{Y},\tilde{T})$ into (X,S) not be too 'wild'. The generalization of the Krohn-Rhodes Theorem proved in §4.5 accomplishes this and more for arbitrary semigroups.

2. Cascade Integration and Wreath Products

The first new concept introduced here is the *cascade integral* of any family of transformation semigroups indexed by elements of a strict partial order. This generalizes the wreath product of a finite totally ordered collection of transformation semigroups. Possibly infinitely many transformation semigroups with ***arbitrary*** loop-free dependency relations among them are combined in such a way that the partial order codes their dependencies. We say that the new transformation semigroup produced in this way is the cascade integral of the family of transformation semigroups over the order. The classical wreath product then is a cascade integral over a finite ordinal, and the finite series-parallel combination of automata is also a special case, namely a cascade integral over a finite partial order.[4] Additional elementary lemmas on cascade integration are collected in the appendix.

2.1. Partial Orders and Cascade Integration

A **(strict) partial order** $\mu = (V, \prec)$ is pair consisting of a set V with an associated transitive, irreflexive relation $\prec$ on V. It is an easy exercise to see that $\prec$ is necessarily anti-symmetric. If in addition for all $\alpha, \beta, \gamma \in V$,

$$(\alpha \prec \gamma \text{ and } \beta \prec \gamma) \Rightarrow (\alpha \prec \beta \text{ or } \beta \prec \alpha \text{ or } \alpha = \beta),$$

then μ is a **tree order**. If μ is a partial order such that for all $\alpha, \beta \in V$,

$$\alpha \prec \beta \quad \text{or} \quad \beta \prec \alpha \quad \text{or} \quad \alpha = \beta,$$

[4] An interesting related concept is the *diagram of ultimate effects* in the work of W. Ross Ashby [1, 2] from the early days of cybernetics.

then μ is a **linear** or **total** order. For $\prec$ any partial order, we define α is **$\prec$-less** than β to mean $\alpha \prec \beta$. We say α is **$\prec$-minimal** if the set of $\beta \in V$ with $\beta \prec \alpha$ is empty.

The cascade integral is the most generic way to combine arbitrary transformation semigroups in such a way that their dependencies on one another are coded by the order μ.

In our main result, we shall only need cascade integrals over *total orders* and shall restrict to a small faithful sub-transformation semigroup of the cascade integral. Partial orders may also be used to obtain a cascade-parallel version of the theorem. The theory for integrals over arbitrary partial orders presented here includes the totally ordered case, while following exactly the same development.

In the sequel, we use the following notational conventions: Suppose $\{X_\alpha\}_{\alpha \in V}$ is a family of sets indexed by elements of a partial order $(V, \prec)$. If V' is a subset of V closed under less than, *i.e.*

$$(\beta \prec \alpha \text{ and } \alpha \in V') \Rightarrow \beta \in V',$$

then V' is called a **segment** of μ. For any $\alpha \in V$, the set of elements strictly less than α is called the **α-segment** of μ.

Similarly, a **segment** of $x \in X = \prod_{\beta \in V} X_\alpha$ is the tuple of coordinates of x whose indices lie in a segment of μ. Also for each $\gamma \in V$, let x_γ denote the image of x under the coordinate projection $\prod_{\alpha \in V} X_\alpha \to X_\gamma$.

Thus, the **α-segment** $x_{\beta \prec \alpha}$ of x is the image of x under the projection $\prod_{\beta \in V} X_\beta \to \prod_{\beta \in V, \beta \prec \alpha} X_\beta$. A tuple z is also called an **α-segment** for X if z occurs as the α-segment $x_{\beta \prec \alpha}$ of some $x \in X$, $\alpha \in V$. Note that for any $\alpha \in V$ which is $\prec$-minimal, the α-segment of all $x \in X$ is the same, namely, the empty segment.

Key Definition 1 Let a partial order $\mu = (V, \prec)$ be given along with a family F of transformation semigroups $F(\alpha) = (X_\alpha, S_\alpha)$ indexed by elements α of V. We show how to integrate the family F over V using μ to obtain a new transformation semigroup (X, S): Define the **cascade integral of the (X_α, S_α)'s over V with respect to μ** to be the transformation semigroup

$$(X, S) = \int_V F(\alpha)\, d\mu = \int_V (X_\alpha, S_\alpha)\, d\mu$$

with state space

$$X = \prod_{\alpha \in V} X_\alpha$$

and semigroup S equal to a certain subset of maps f from X to itself under function composition. S acts on the right of X by evaluation $x \cdot f = xf$. The elements of S are characterized by: $f \in X^X$ is an element of S if and only if

for all $\alpha \in V$ and all α-segments $x_{\beta \prec \alpha}$,
there exists $s_\alpha \in S_\alpha$,
such that whenever $y \in X$ with $y_{\beta \prec \alpha} = x_{\beta \prec \alpha}$,
then $(yf)_\alpha = y_\alpha \cdot s_\alpha$.

This restriction (3) says exactly that what f does in the α-coordinate $x_\alpha \in X_\alpha$ of a given $x \in X$ is given by an $s_\alpha \in S_\alpha$ which depends only on f and the α-segment of x (coordinates of x with indices $\prec$-less than α). *In particular, since $\prec$ is irreflexive, s_α does not depend on x_α.*

Let us re-state the crucial properties of elements of S:

[**sequential form**] to apply $f \in S$, the operation performed to obtain the new α-coordinate $(xf)_\alpha$ depends only on f and coordinates of x which are $\prec$-less than α.

[**α-component action lies in S_α**] the action on the α-coordinate is computed by applying an element s_α of the α-th transformation semigroup to x_α.

Examples: 1. The cascade integral is a generalization of the classical wreath product of transformation semigroups, which is the special case of integrating over a finite ordinal. In particular, let μ be the ordering $0 < 1$, and let $F(0) = (X, S)$ and $F(1) = (Y, T)$, then $\int_{\{0,1\}} F(\alpha)\, d\mu$ is $(Y, T) \circ (X, S)$, the **wreath product**. More generally, this term refers to any cascade over a finite ordinal.

2. In the case of non-linear partial orderings, we get interesting behaviour. For instance if ϕ is the empty partial order $<$ on $\{0, 1\}$ (neither $0 < 1$ nor $1 < 0$ hold under ϕ), we have $\int_{\{0,1\}} F(\alpha)\, d\phi = (Y, T) \times (X, S)$, the **parallel connection** of (Y, T) and (X, S). Similarly, we get a parallel connection of any collection of transformation semigroups by integrating over the empty partial order.

3. Let $\mu' = (V', <')$ be a suborder of $\mu = (V, \mu)$ such that $\alpha \in V'$ and $\beta < \alpha$ implies that also $\beta \in V'$. Then there is an obvious **canonical projection** of transformation semigroups

$$\int_V (X_\alpha, S_\alpha)\, d\mu \twoheadrightarrow \int_{V'} (X_\alpha, S_\alpha) d\mu',$$

induced by projecting each state x onto its V'-segment (*i.e.* forgetting all coordinates x_α with $\alpha \in V \setminus V'$).

The integral notation has been chosen to suggest formal similarities with the direct integral of abstract harmonic analysis [11, 5] and to suggest the integration of parts into a transcending whole with respect to dependency relations encoded by μ.

Lemma 2.1 *In any cascade integral $(X, S) = \int_V (X_\alpha, S_\alpha)\, d\mu$ over a partial order $\mu = (V, \prec)$, corresponding to a fixed $f \in S$ and an α-segment $x_{\beta \prec \alpha}$ there exists a unique s_α as in (3).*

Proof: Since $f \in S$, some such s_α exists. To show uniqueness, suppose $s_\alpha, t_\alpha \in S_\alpha$ both work. Then

$$y_\alpha \cdot s_\alpha = (yf)_\alpha = y_\alpha \cdot t_\alpha \text{ for all } y \in X \text{ with } y_{\beta \prec \alpha} = x_{\beta \prec \alpha}.$$

Since μ is irreflexive, the α-coordinate of x is not a component of $x_{\beta\prec\alpha}$; hence the α-coordinate of y's in X with $y_{\beta\prec\alpha} = x_{\beta\prec\alpha}$ can take any value in X_α without affecting s_α. Thus s_α and t_α have the same action on all $y_\alpha \in X_\alpha$. By the hypothesis that (X_α, S_α) is faithful, it follows that the $s_\alpha = t_\alpha$. □

The verification that S is closed under composition is analogous to the finite case, but since the construction is new, we shall carry it out explicitly.

Theorem 2.1 *The cascade integral of [faithful] transformation semigroups over any subset of an partial order $(V, \prec)$ is itself a [faithful] transformation semigroup.*

Proof: By the above remarks, we have only left to verify closure of S and faithfulness of the action. Let f, g be elements of S. Given α and $x_{\beta\prec\alpha}$, we must find s_α as required by (3) corresponding to the composition f followed by g. For any $y \in X$ with $y_{\beta\prec\alpha} = x_{\beta\prec\alpha}$, we see

$$(3)\qquad \begin{aligned} (y \cdot fg)_\alpha &= (yf \cdot g)_\alpha \\ &= (yf)_\alpha \cdot s''_\alpha && \text{for some } s''_\alpha \in S_\alpha \text{ by (3) and } g \in S \\ &= (y_\alpha \cdot s'_\alpha) \cdot s''_\alpha && \text{for some } s'_\alpha \in S_\alpha \text{ by (3) and } f \in S \\ &= y_\alpha \cdot (s'_\alpha s''_\alpha) \end{aligned}$$

Using f and g and an α-segment, we have constructed $s_\alpha = s'_\alpha s''_\alpha \in S_\alpha$. Now by (3), s''_α depends only on $(yf)_{\beta\prec\alpha}$ and g, and s'_α depends only on $y_{\beta\prec\alpha}$ and f. The β-coordinates (where $\beta \prec \alpha$) of $(yf)_{\beta\prec\alpha}$ depend just on y_β, $y_{\gamma\prec\beta}$ and f. But $\gamma \prec \beta \prec \alpha$ implies $\gamma \prec \alpha$, so s_α depends only on f, g, and $y_{\beta\prec\alpha} = x_{\beta\prec\alpha}$,

Furthermore s_α is independent of the the factorization of fg: Suppose we had factored fg differently and had obtained t'_α and t''_α (instead of s'_α and s''_α, respectively), then for all y_α in X_α, there is a $y \in X$ with α-coordinate y_α, and we have $y_\alpha \cdot (s'_\alpha s''_\alpha) = (y \cdot fg)_\alpha = y_\alpha \cdot (t'_\alpha t''_\alpha)$. So by faithfulness of (X_α, S_α), we conclude $s'_\alpha s''_\alpha = t'_\alpha t''_\alpha$, so s_α is independent of the the factorization of fg and depends only on $x_{\beta\prec\alpha}$ and fg. This proves fg is an element of S. Thus S is closed under composition.

The action of S is faithful since S was defined to be a subset of X^X, the semigroup of all maps from X to itself. □

2.2. Cascade Transform

This section introduces the notion of *cascade transform*, which describes the "infinitesimal behavior" of a transformation at a given state. We use this concept as a language tool to develop the elementary theory of cascade integration and to understand, in a *component-wise* manner, the covering of one automaton by another and the limit of a chain of coverings describing a automaton in successively better coordinates.

The cascade transform is strongly connected with monoids acting on trees. Indeed the states of a wreath product or cascade integral may be regarded as maximal paths in a rooted tree, and each transformation s of the wreath product or cascade integral may be regarded as a labelled tree (the cascade transform of s) encoding in a global sequential manner a depth preserving endomorphism of the rooted tree. Full details will appear in an expanded version of this paper, but they are also in [14, 13]. This viewpoint of semigroups elements in the wreath product or cascade integral as labelled trees is due to the author and first appears in [14]. In the case of finite and depth $(\mathbf{N}, <)$ trees, the viewpont of states in the wreath product as paths in (graph-theoretic) trees is due to Rhodes [18], whose work is streamlined and extended to order-theoretic trees in Nehaniv [14, 16, 13].

Now define a function $\overline{f}$, the **cascade transform** of f, from the segments of X to the disjoint union of the S_α's by letting $(x_{\beta\prec\alpha})\overline{f}$ be the unique s_α of (3) corresponding to f and $x_{\beta\prec\alpha}$.

Lemma 2.2 (Convolution Rule) *In any cascade integral* $(X, S) = \int_V F(\alpha)\, d\mu$, *for all* $f, g \in S$,

$$(x_{\beta\prec\alpha})\overline{fg} = (x_{\beta\prec\alpha})\overline{f}\,((x_\beta \cdot (x_{\gamma\prec\beta})\overline{f})_{\beta\prec\alpha})\overline{g},$$

where $(x_\beta \cdot (x_{\gamma\prec\beta})\overline{f})_{\beta\prec\alpha}$ *denotes the α-segment whose β-coordinate has the value* $x_\beta \cdot (x_{\gamma\prec\beta})\overline{f}$ *in* X_β *for each* $\beta \prec \alpha$.

Define $(x_{\beta\prec\alpha})(\overline{f} * \overline{g})$ *to be the right hand side. Then*

$$\overline{fg} = \overline{f} * \overline{g}.$$

Proof: Note that in (3) above $s_\alpha = (x_{\beta\prec\alpha})\overline{fg}$, $s'_\alpha = (x_{\beta\prec\alpha})\overline{f}$, and $s''_\alpha = ((xf)_{\beta\prec\alpha})\overline{g}$, then since $s_\alpha = s'_\alpha s''_\alpha$,

$$(x_{\beta\prec\alpha})\overline{fg} = (x_{\beta\prec\alpha})\overline{f}\,((xf)_{\beta\prec\alpha})\overline{g} \qquad [\text{ multiplication in } S_\alpha\ ]. \tag{4}$$

By definition of cascade transform, $(xf)_\beta = x_\beta \cdot (x_{\gamma\prec\beta})\overline{f}$. Therefore,

$$(xf)_{\beta\prec\alpha} = (x_\beta \cdot (x_{\gamma\prec\beta})\overline{f})_{\beta\prec\alpha}.$$

Substituting this into (4) on the right hand side (for the argument of $\overline{g}$) yields the result. □

Theorem 2.2 (Cascade Transform Inversion) *For a cascade integral,*

$$(X, S) = \int_V (X_\alpha, S_\alpha)\, d\mu,$$

any $f \in S$ *is uniquely determined by its cascade transform* $\overline{f}$. *Conversely, any* h *mapping segments of* X *to the* S_α*'s satisfying* $(x_{\beta\prec\alpha})h \in S_\alpha$ *determines a unique* $\hat{h}$ *in* S. *This correspondence is one-to-one.*

Proof: By (3), $(xf)_\alpha = x_\alpha \cdot (x_{\beta\prec\alpha})\overline{f}$ so f is completely determined by $\overline{f}$. Conversely, specifying a mapping h from segments of X to the S_α's such that $(x_{\beta\prec\alpha})\overline{h} \in S_\alpha$, determines a unique $\hat{h} \in S$ defined by $(x\hat{h})_\alpha = x_\alpha \cdot (x_{\beta\prec\alpha})h$. Since $\hat{\overline{f}} = f$ and $\overline{\hat{h}} = h$, the correspondence is bijective. □

Let $\overline{S}$ denote the collection of mappings from segments of X to the S_α's. That is, $h \in \overline{S}$ if and only if for all α-segments $x_{\beta\prec\alpha}$ of X, $(x_{\beta\prec\alpha})h \in S_\alpha$. Use the multiplication $*$ on $\overline{S}$ defined by the rule in Lemma 2.2.

Theorem 2.3 *There is an isomorphism of semigroups,* $(S, \circ) \cong (\overline{S}, *)$.

Proof: Lemma 2.2 says that the cascade transform is a homomorphism from S under composition $\circ$ to $\overline{S}$ under convolution $*$. The theorem implies that this map is a bijective. In particular, since $(\overline{S}, *)$ is a homomorphic image of $(S, \circ)$, it is a associative. □

The appendix further develops the elementary theory of cascade integration, making use of the cascade transform notation.

3. Cascadable Systems

We now introduce the *cascadable systems* and then show how they encode cascade decompositions by subsemigroups.

3.1. Definition and Examples

Key Definition 2 Let S be any semigroup, Σ a partition of S into equivalence classes, and suppose $<$ is a linear ordering on Σ. Denote by $[s]_\Sigma$ the equivalence class of $s \in S$ under Σ. If no confusion can result, we suppress the subscript and write $[s]$.

A **cascadable system** $F : \Sigma \to SUB(S)$ for a semigroup S is a function F from the equivalence classes of Σ into subsemigroups of S, if it satisfies:

(i) $G_s \subseteq F[s]$,
where G_s is the (unique) maximal subgroup of S containing s, unless s is not contained in any subgroup of S, in which case $G_s = \{s\}$.

(ii) $[s] > [t] \Rightarrow \forall t' \in F[t],\ [t's] > [t]$.

(iii) $[s] > [t] \Rightarrow t \notin F[s]$.

Condition (i) guarantees that the largest subgroup containing s (or at least s itself) is contained in $F[s]$. Condition (ii) is the most important: Intuitively, if s is 'heavier' than t (i.e. $[s] > [t]$) and s hits anything (t') in the semigroup $F[t]$ that covers the class of t, then s drags that thing to a level $[t's]$ also 'heavier' than t. This corresponds to the sequential property of the wreath product or cascade integral. Condition (iii) says $[t]$ is the last ($>$-greatest) class of Σ whose F-image contains t. Also related to the concept of cascadable system is an earlier concept developed for finite semigroups, "system of semigroups for S" which was used giving an algebraic proof of the Krohn-Rhodes Theorem in [9], Ch. 5.

Example 3.1 For any semigroup S, there is a cascadable system for S defined by $\Sigma = \{S\}$ and $F(S) = S$.

Given a cascadable system for S we shall decompose S into a cascade of $F[s]$'s over the ordering $(\Sigma, <)$. That a cascadable system for S yields such a decomposition is shown in the Cascade Covering Theorem below. For such a decomposition to have any content, we shall require non-triviality of the cascadable system:

Definition 4 A cascadable system F for S is called **non-trivial** if Σ has a least two partition blocks ($|\Sigma| \geq 2$) and for no Σ-block $[s]$ does $F[s]$ equal S.

Example 3.2 (Krohn-Rhodes) [This reformulates the famous Krohn-Rhodes "V-union-T' technique [9, 4] in terms of cascadable systems.]

If $S = V \cup T$ with V a proper left ideal and T proper subsemigroup, then S has a non-trivial cascadable system defined by

$$\Sigma = \{V, T \setminus V\} \text{ with } V > T \setminus V$$

$$F[V] = V,\ F[T \setminus V] = T.$$

Proof: To see that this is a cascadable system: (i) $s \in V$, implies L_s hence G_s is in $V = F[s]$ since V is a left ideal. For $t \in T \setminus V$, observe that $G_t \subseteq T = F[t]$; for if $L_t \cap V \neq \emptyset$ then $t \in L_t \subseteq V$. But then $S = V \cup T$, implies L_t hence G_t is in T. (ii) If $[s] > [t]$, then necessarily $s \in V$ and $t \in T$, whence $F[t]s \subseteq V$ since V is a left ideal, so $t's \in V$ for for all $t' \in F[t]$; *i.e.* $[t's] = V > T \setminus V = [t]$ as required. (iii) If $[s] > [t]$ and $t \in F[s] = V$ then $t \in V$, contradicting $t \in T \setminus V$. So $t \notin F[s]$. □

It is true that a *finite* semigroup S has cascadable system of the form in Example 3.2 if and only if it is not cyclic or left-simple. Simply let L be a maximal L-class and define $T = \langle L \rangle$ and $V = S \setminus L$. Then [the Rees-Sushkevych Theorem or] Goodwin's Lemma (see section 4.3) implies $T \neq S$. The converse is trivial.

In fact, the same argument applies to infinite semigroups *with a maximal L-class.* However, in the infinite case, there may be no such "V-union-T" cascadable system, even when S is neither cyclic nor left simple. We saw this in Example 1.1. To decompose a semigroup with no maximal L-class, the following technique can be used to obtain a cascadable system:

Example 3.3 (Nehaniv [14]) A non-empty semigroup S without maximal L-class has a non-trivial cascadable system.

Proof: Let C be a maximal chain $C = \{I_\alpha\}$ of proper left ideals under strict inclusion $\subsetneq$. Also, for $s \in S$, let

$$I_s = \bigcap \{I_\alpha \in C : s \in I_\alpha\}.$$

We claim I_s is a proper left ideal. Since all the I_α's are, this follows once we show s lies in at least one I_α. Suppose it does not; then clearly $I = \bigcup I_\alpha \neq S$, and so $I \cup S^1 s$ is a left ideal – which is proper, lest s lie in maximal L-class. Thus extending the chain C by $I \cup S^1 s$ yields a longer chain, contradicting maximality.

By maximality of the chain, I_s occurs in C. Also I_s contains s and hence L_s.

Let Σ denote the L-equivalence relation on S; that is, $[s] = L_s$. Totally order Σ by

$$[s] > [t] \text{ if } I_s \text{ is properly contained in } I_t.$$

[Note this is equivalent to $t \notin I_s$.] Define $F[s] = I_s$.

We verify that this is a non-trivial cascadable system: (i) $G_s \subseteq L_s \subseteq I_s = F[s]$. (ii) If $[s] > [t]$ and $t' \in I_t$, then $t's \leq_L s$. Hence, $t's \in I_s$ and so $I_{t's} \subseteq I_s \subsetneq I_t$ Thus $[t's] > [t]$.

(iii) If $[s] > [t]$ then $I_s \subsetneq I_t$. Hence by definition of I_t, $t \notin I_s$.

Since $F[s] = I_s$ is a proper left ideal, $F[s] \neq S$ for all $[s]$. Fix any $s, t \in S$ and $t >_L s$ [this is possible since S has no maximal left ideal], then clearly $I_s \subsetneq I_t$ and $[s] > [t]$, showing that the cascadable system is non-trivial. □

A modification of this technique (Lemma 4.4 below) will allow us to decompose all but cyclic and left-simple semigroups.

Our definition of cascadable system is justified by the following theorem which says that a cascadable system may be integrated over μ to obtain a nice cover of S.

Theorem 3.1 (Cascade Covering) *Let $F : \Sigma \to SUB(S)$ be a cascadable system for S. Then*

$$(S^1, S) \longleftarrow (Q, \widetilde{S}) \leq \int_{\alpha \in \Sigma} (F(\alpha)^I, \overline{F(\alpha)^I})\, d\mu.$$

where Q consists of all states with only finitely many coordinates not equal to I and $(Q, \widetilde{S})$ is faithful.

(Note that trivially $|\Sigma| \leq |S|$.) *Proof:* Given $s \in S$, lift s to $\tilde{s}$ by specifying the cascade transform of $\tilde{s}$:

$$(5)\ \ (x_{\beta<\alpha})\bar{\tilde{s}} = \begin{cases} x_{\beta_n} \ldots x_{\beta_1} s & \text{there exist only the finitely many indices} \\ & \beta_1 < \ldots < \beta_n \text{ less than } \alpha \text{ for which } x_\beta \neq I; \\ & \text{and } [x_{\beta_n} \ldots x_{\beta_1} s] = \alpha; \\ & \text{but for no } \alpha' < \alpha \text{ does this case apply} \\ \\ \textit{constant } I & \text{for no } \alpha' \leq \alpha \text{ does the first case hold;} \\ & \text{and there exist only finitely many indices } \beta < \alpha \\ & \text{for which } x_\beta \neq I \\ \\ I & \text{for some } \alpha' < \alpha \text{ the first case holds;} \\ & \text{or there exist infinitely many indices } \beta < \alpha \\ & \text{for which } x_\beta \neq I \end{cases}$$

In the first case, we understand $x_{\beta_n} \ldots x_{\beta_1} s$ as s if $x_\beta = I$ for all $\beta < \alpha$.

Suppose z – a state of the cascade integral – has finitely many, say n, coordinates not equal to I. Define a covering map down to (S^1, S) by $w(z) = z_{\alpha_n} \ldots z_{\alpha_1}$ if the z_{α_i} denote the coordinates not equal to I and $\alpha_1 < \ldots < \alpha_n$. (If $n = 0$, then $w(z) = 1$.) This is surjective since each state $z \in S$ has a lift to the state $\tilde{z}$ with z in the $[z]$ coordinate and I's elsewhere. Otherwise, infinitely many coordinates are non-trivial and $w(z)$ is not defined.

To lift a semigroup element: If $n = 0$, case one of the definition of $\tilde{s}$ applies exactly when $\alpha = [s]$. Otherwise, $n \geq 1$, and for $1 \leq i \leq n$ we define $\gamma_0 = [s]$ and $\gamma_i = [z_{\alpha_i} \ldots z_{\alpha_1} s]$. Let i_* be the least number i, $0 \leq i \leq n-1$, for which $\gamma_i \leq \alpha_{i+1}$, if any such i exists. Then in applying $\tilde{s}$ to z, case one of the definition of $\tilde{s}$ applies if and only if $\alpha = \gamma_{i_*}$. If no such i exists, case one applies if and only if $\alpha = \gamma_n$. In any event, there is a unique α for which case one applies when calculating $z \cdot \tilde{s}$.

Notation: if $w(z)$ is defined, we write $z = (z_m, \ldots, z_1)$, where all but finitely many of the coordinates equal to I have been suppressed and all of the finitely many coordinates not equal to I occur.

Let $\alpha \in |\mu|$ be the unique coordinate at which case one applies. We may assume z_α is among the z_i.[5]

$$(z_m, \ldots, z_1) \cdot \tilde{s} = (z_m, \ldots, z_{i+1}, z_i \cdot z_{i-1} \ldots z_1 s, I, \ldots, I), \tag{6}$$

where $z_i \cdot z_{i-1} \ldots z_1 s$ occurs in the αth position. Any positions with lesser coordinate indices have been reset to I under the second case of the definition of $\tilde{s}$, and any positions with greater indices have been multiplied by I according to the third case. So clearly if $w(z)$ is defined, so is $w(z \cdot \tilde{s})$ since acting on z by $\tilde{s}$ changes the number of coordinates not equal to I by at most a finite number.

The semigroup of the cover is the subsemigroup $\widetilde{S}$ of the cascade integral generated by the lifts $\tilde{s}$ for $s \in S$. So we have $Q \cdot \widetilde{S} \subseteq Q$. Also it is obvious that $w(z \cdot \tilde{s}) = z_m \ldots z_1 s = w(z) \cdot s$, so this is a covering morphism.

To see that $(Q, \widetilde{S})$ is faithful, we use Lemma 2.1 which guarantees that $(X, \widetilde{S})$ is faithful: given $\psi \neq \phi \in \widetilde{S}$, we know there is an $x \in X$ with

$$x \cdot \psi \neq x \cdot \phi.$$

Hence these two states disagree in some coordinate, say α:

$$(x_{\beta<\alpha})\overline{\psi} \neq (x_{\beta<\alpha})\overline{\phi}.$$

If $x_\beta \neq I$ for infinitely many $\beta < \alpha$, then it follows that from the definition of the $\tilde{s}$ that

$$(x_{\beta<\alpha})\overline{\phi} = (x_{\beta<\alpha})\overline{\psi} = I,$$

and so

$$(x \cdot \phi)_\alpha = x_\alpha \cdot (x_{\beta<\alpha})\overline{\phi} = x_\alpha \cdot I = x_\alpha.$$

And similarly $(x \cdot \psi)_\alpha = x_\alpha$, contradicting the choice of α. Hence $x_\beta \neq I$ for only finitely many $\beta < \alpha$, and so replacing x by $x' \in Q$ with $x'_\beta = x_\beta$ for $\beta \leq \alpha$ and $x'_\beta = I$ for $\beta > \alpha$, we have (by sequential form of the action)

$$(x' \cdot \phi)_\alpha = (x \cdot \phi)_\alpha \neq (x' \cdot \psi)_\alpha = (x \cdot \psi)_\alpha.$$

This proves $\widetilde{S}$ acts faithfully on Q. □

Note that only property (ii) of a cascadable system is used in this proof. Properties (i) and (iii) will be used to guarantee that the morphism $\widetilde{S} \twoheadrightarrow S^1$ is aperiodic and that subgroups of S lift to isomorphic subgroups of $\widetilde{S}$.

[5] If necessary we may insert $z_\alpha = I$ into the tuple representing z without changing the element denoted.

3.2. Properties of the Covering Map

There is a danger in a decomposition theory that one may be too rough with that which one intends to understand. This proves to be a real danger in the realm of infinite semigroup theory; for if conditions are not imposed on the decompositions the delicate gems one hopes to see via a decomposition may be crushed to sand and dust.

In this section we motivate the aperiodicity restriction on covering. The desirable faithful and group-lifting properties of coverings are also defined. Without some such restrictions, we will see that the cascade decomposition theory even for countable semigroups would be trivial. We then show that for integrals of cascadable systems the restrictions are satisfied.

Recall that a pseudo-variety is a family of semigroups closed under taking finite direct products, subsemigroups, and homomorphic images. (This differs from a common definition of pseudo-variety which allows only *finite* semigroups.)

Fact 1 *The collection $\mathcal{A}$ of torsion semigroups containing no non-trivial subgroups is the largest pseudo-variety of semigroups not containing any non-trivial groups.*

Proof: It is easy to check that $\mathcal{A}$ is a pseudo-variety. Suppose we add a new semigroup S to this pseudo-variety. Since it was not there originally, it either contains a non-trivial group or it contains a non-torsion element s. In the second case, $\langle s \rangle$ is infinite and so has each finite cyclic group $\mathbb{Z}_n$ as a homomorphic image. In either case every pseudo-variety containing S also contains non-trivial groups. The conclusion follows. □

Recall from §1.3 that a morphism is aperiodic if the the inverse image of each idempotent is an aperiodic semigroup. For finite semigroups, $\widetilde{S} \twoheadrightarrow S$ aperiodic implies that $\widetilde{S}$ and S have the same complexity in the sense of Krohn-Rhodes [23].

Example 3.4 (The Shift Map) Let $2^{\mathbf{Z}}$ be set of all maps f from the integers to the set $2 = \{0, 1\}$. $2^{\mathbf{Z}}$ can be viewed as all bi-infinite sequences of zeroes and ones, and also as the state space of the cascade integral $\int_{\mathbf{Z}}(\{0,1\}, U)\, d{<}$, where U consists of a reset to 0 and a reset to 1.

Consider the *shift map* $\sigma : 2^{\mathbf{Z}} \to 2^{\mathbf{Z}}$ given by

$$(f \cdot \sigma)_n = f_{n-1}.$$

Observe

$$(f_{m<n})\overline{\sigma} = \begin{cases} \text{reset } 0 & \text{if } f_{n-1} = 0 \\ \text{reset } 1 & \text{if } f_{n-1} = 1 \end{cases}$$

Thus, the shift map σ lies in the above cascade integral.

Any finite cyclic group $\mathbb{Z}_n$ divides $(2^{\mathbf{Z}}, \langle \sigma \rangle)$: Simply let Y_n denote all *n-periodic sequences* in $2^{\mathbf{Z}}$

$$Y_n = \{f : f_{k+n} = f_k \text{ for all k } \}.$$

We obtain

$$(Y_n, \mathbb{Z}_n) < (2^{\mathbb{Z}}, \langle\sigma\rangle) \le \int (\{0,1\}, U)\, d\mathbb{Z},$$

Now $\langle\sigma\rangle$ is an infinite cyclic semigroup, hence idempotent-free. $(Y_n, \langle\sigma\rangle)$ is not faithful, but making it faithful yields a covering morphism to $(Y_n, \mathbb{Z}_n)$ taking σ to a generator of $\mathbb{Z}_n$. The lifts of identity of $\mathbb{Z}_n$ all have trivial action on Y_n and form an infinite cyclic semigroup, σ^n, $\sigma^{2n}, \ldots$. Hence the map is not aperiodic. Nor is there any subgroup of the covering semigroup that maps onto $\mathbb{Z}_n$.

In the finite case, a group can always be lifted to a group in the cover; but in contrast this covering creates a group "out of nothing".

Example 3.5 (D. Allen, Jr.) Any countable semigroup S divides $(\mathbf{N}, \mathbf{N}) \circ (\mathbf{N}, \mathbf{N})$, although the division is not aperiodic unless S is idempotent-free.

Proof: It suffices to show that the free semigroup on two letters 2^+ divides this wreath product, since the free semigroup on countably many letters ω^+ is a subsemigroup of 2^+ mapping homomorphically onto each countable semigroup.

Elements of 2^+ may be represented as binary strings of 0's and 1's. So a word w in 2^+ maps to a unique positive integer $|w|$, where we interpret letters farther right as multiplied by higher powers of 2. For example,

$$|01100| = 0 \times 2^0 + 1 \times 2^1 + 1 \times 2^2 + 0 \times 2^3 + 0 \times 2^4 = 6.$$

But distinct words may represent the integer because of right trailing zeroes.

Let $len(w)$ denote the number of letters in w. Lift a state w to $(|w|, len(w))$, this lift uniquely determines w. Define $()\overline{\tilde{w}} = len(w)$ and $(n)\overline{\tilde{w}} = 2^n|w|$. Then

$$(|w'|, len(w')) \cdot \tilde{w} = (|w'| + 2^{len(w')}|w|, len(w') + len(w)) = (|w'w|, len(w'w)),$$

which is the lift of $w'w$.

Hence any countable S divides $\omega^+ \le 2^+ \le \mathbf{N} \circ \mathbf{N}$.

Since 2^+ has no idempotents, the lifts of an idempotent of S cannot be aperiodic. Hence this division is aperiodic only for idempotent-free S. □

These examples may lead one to doubt (as many many researchers did) that any reasonable decomposition theory was possible in the case of infinite semigroups. They certainly establish the necessity of restrictions such as aperiodic division in such a theory. Without such restrictions, even in the countable case, the decomposition theory has little content; for every countable semigroup divides a wreath product of two infinite cyclic monoids.

In addition to non-aperiodicity, these "bad" divisions share an additional property: a group does not lift to a subgroup of the cover. In the finite case, this is impossible. Also, in the shift example, the action of the lift semigroup is not faithful on the lifted states.

If all groups do lift for a division, we call it ***group-lifting***. If the lifted transformations act faithfully on the lifted states, we call it a ***faithful division***. These two properties along with aperiodicity (while failing for such pathological decompositions) will be attained for all integrals of cascadable systems for arbitrary semigroups and hence in our main theorem. We have already shown a division arising from the Cascade Covering Theorem is faithful; now let us establish the

Aperiodicity Theorem Under the conditions of the Cascade Covering Theorem, if $e^2 = e \in S$, then each lift $\tilde{e}$ of e generates a finite semigroup containing no non-trivial groups. Therefore the lifts of e form an aperiodic semigroup.

In the Cascade Covering Theorem, we did not use the property that $[s]$ is the greatest $\alpha \in \Sigma$ for which $s \in F(\alpha)$. This will now be used in showing that global sequential coordinates in the covering just described are good ones.

Claim 1 *An element $\widetilde{s_1} \ldots \widetilde{s_n}$ of $\widetilde{S}$ acts on an element $z = (z_m, \cdots, z_1)$ [notation as above] via*

$$(7) \quad (z_m, \cdots, z_1) \cdot \widetilde{s_1} \ldots \widetilde{s_n} = (z_m, \cdots, z_{j+1}, z_j \cdot z_{j-1} \cdots z_1 s_{i_0} \cdots s_{i_1}, \\ I \cdot constant\, s_{i_1+1} \cdots s_{i_2}, \cdots, I \cdot constant\, s_{i_{k-1}+1} \cdots s_{i_k}),$$

where, $1 \leq j \leq m+1$, and $1 = i_0 \leq i_1 \leq \ldots \leq i_{k-1} \leq i_k = n$.

(Here when $j-1=0$ we understand $z_{j-1} \ldots z_1$ as I and that z_j may be I. Also if $j-1=m$ we understand all z's to the left of $z_{j-1} \ldots z_1$ to be I. If $k=1$ then the constant actions do not occur.)

Observe that the new state has coordinates that are products of the pieces of sequence $z_m, \cdots, z_1, s_1, \cdots, s_n$ after it has been appropriately chopped up.

Proof: The validity of this claim is clear from (6) and induction. □ (Claim1)

Suppose $w(\tilde{s_1} \ldots \tilde{s_n}) = e = e^2$, then $e = s_1 \ldots s_n$. To show aperiodicity of the inverse image of e, we must show that for each such factorization there is a positive exponent p such that $(\widetilde{s_1} \ldots \widetilde{s_n})^p = (\widetilde{s_1} \ldots \widetilde{s_n})^{p+1}$.

The set of results of products of contiguous substrings w of the infinite sequence

$$s_1, \cdots, s_n, s_1, \cdots, s_n, s_1, \cdots, s_n \cdots$$

is finite. In fact, its cardinality is clearly not more than $N = 3n-1$. Since a word w of length $3n$ or more contains a substring $s_1, \ldots, s_n$ and either to its right or left must contain an adjacent copy of $s_1, \cdots, s_n$; so $e^2 = e$ implies that w can be shortened by n letters without changing the result. Therefore the set $\mathcal{N}$ of indices $[s]$, where s is such a product, is no more than $3n-1$.

Therefore it follows that $(\widetilde{s_1} \ldots \widetilde{s_n})^r$ for large enough r does not change any of the coordinates involving the z_k's, except possibly the one involving z_1, and in the latter case the new coordinate also involves z_1. Furthermore the only coordinates with indices less than that of z_1 which are not equal to I must be of the form $[s_i \ldots s_j]$ where $s_i, \cdots, s_j$ is a substring of $(s_1, \cdots, s_n)^r$. Hence it follows that r may be chosen independent of the state z. (This is also true even if z is allowed to be a state of the cascade integral with infinitely many non-trivial coordinates.)

Claim 2 *For any $s_1, \ldots, s_n$ with $s_1 \ldots s_n$ idempotent, we have*

$$z \cdot (\widetilde{s_1} \ldots \widetilde{s_n})^{r+N} = z \cdot (\widetilde{s_1} \ldots \widetilde{s_n})^{r+N+1},$$

Proof of Claim 2: By (7) the least non-trivial coordinate of $(z \cdot (\widetilde{s_1} \ldots \widetilde{s_n})^r) \cdot \widetilde{s_1} \ldots \widetilde{s_n}$ is given by multiplying by $s_i \ldots s_n$ or is a given by a constant $s_i \ldots s_n$ depending only the fact that the lesser coordinates are all ones. In the constant case, after applying $\widetilde{s_1} \ldots \widetilde{s_n}$, they are still all ones, this coordinate is unchanged under iterating further applications of $\widetilde{s_1} \ldots \widetilde{s_n}$. In the non-constant case, observe that $i = 1$ and the next iteration of $\widetilde{s_1} \ldots \widetilde{s_n}$ just multiplies the least non-trivial coordinate by $s_1 \ldots s_n = e$ again, so we have $(\widetilde{s_1} \ldots \widetilde{s_n})^r = (\widetilde{s_1} \ldots \widetilde{s_n})^{r+1}$.

Therefore we may assume the constant case; applying $\widetilde{s_1} \ldots \widetilde{s_n}$ again does not change the first coordinate so the the second coordinate is always acted on the same way under further iterations of $\widetilde{s_1} \ldots \widetilde{s_n}$. Inductively, continuing with further applications of $\widetilde{s_1} \ldots \widetilde{s_n}$ shows that after $r + m$ iterations, the first m non-trivial coordinates will be unchanged by more iterations, and the trivial coordinates with indices less than any of these remain trivial. The action must necessarily change only a coordinate whose index lies in the set $\mathcal{N}$, so at most N of the non-trivial coordinates do not involve the z_i's; and thus after at most $r + N$ iterations we enter the case where the initial segment not involving the z_i's remains fixed. The action of $\widetilde{s_1} \ldots \widetilde{s_n}$ on this initial segment is the same for all further iterations since it is always given by constant maps depending on coordinates which cannot change under further iterations.

After this initial segment, there is action by $s_i \ldots s_n s_1 \ldots s_{i-1}$ [where we understand $s_0 = s_n$ in the case $i = 1$], and the action in this coordinate is the same for all further iterations since it depends only on $\widetilde{s_1} \ldots \widetilde{s_n}$ and the unchanging coordinates of lower index. All coordinates with greater indices are therefore multiplied by I. The action on this coordinate of the next iteration is again given in the same way since the initial coordinates have not changed and the action is sequential. But now

$$s_i \ldots s_n s_1 \ldots s_{i-1} s_i \ldots s_n s_1 \ldots s_{i-1} = s_i \ldots s_n e s_1 \ldots s_{i-1}$$

and so

$$s_i \ldots s_n s_1 \ldots s_{i-1} (s_i \ldots s_n s_1 \ldots s_{i-1})^k = s_i \ldots s_n e s_1 \ldots s_{i-1}.$$

So this coordinate is also fixed by all further iterations. □ (Claim 2)

Now since how N and r were chosen depends only on the factorization $s_1, \ldots, s_n$ of $s_1 \ldots s_n$ and not on z, this establishes the claim that the inverse image of an idempotent is aperiodic. □

Note that in the case of finite semigroups and finite μ, the last assertion of the Theorem implies that the covering is complexity preserving in the sense of Krohn-Rhodes (*cf.* [19, 9, 4].)

Corollary 3.2 *The covering morphism of the Cascade Covering Theorem is injective on subgroups of $\tilde{S}$.*

Proof: If G is a subgroup of the covering semigroup $\tilde{S}$, then and g lies in the kernel of w restricted to G. Then there is an M such that $g^M = g^{M+1}$, since g is in the inverse image of an idempotent. It follows that g is the identity element of G. □

Corollary 3.3 *For every subgroup G of S, there is an isomorphic subgroup $\tilde{G}$ in the cover $\tilde{S}$ mapping onto S via the covering morphism w.*

Proof: A cascadable system is such that $[g] = [g']$ whenever g and g' lie in a single subgroup of S. To see this, observe that $g, g' \in G_g = G_{g'} \subseteq F[g], F[g']$. If, say, $[g] > [g']$, then $g' \notin F[g]$, a contradiction. Hence $[g] = [g']$.

Now let e denote the identity element of G. Choose the idempotent $\tilde{e}^p$ in the inverse image of e. Then any element of the form $\tilde{e}^p\tilde{g}\tilde{e}^p$ with $g \in G$ maps down to $ege = g$. Furthermore $\tilde{e}^p\tilde{g}\tilde{e}^p\tilde{g'} = \tilde{e}^p\widetilde{gg'}$, since $\tilde{g}$ acts on $z \cdot \tilde{e}^p$ by multiplying the $[g]$-coordinate by g; and then $\tilde{e}^p$ acts by multiplying that coordinate by e (p times) which leaves it fixed; whereupon g' a multiplies this same coordinate by g'. It follows that $\tilde{e}^p\tilde{g}\tilde{e}^p\tilde{g'} = \tilde{e}^p\widetilde{gg'}$, and thus

$$\tilde{e}^p\tilde{g}\tilde{e}^p\tilde{e}^p\tilde{g'}\tilde{e}^p = \tilde{e}^p\tilde{g}\tilde{e}^p\tilde{g'}\tilde{e}^p = \tilde{e}^p\widetilde{gg'}\tilde{e}^p.$$

Therefore $g \mapsto \tilde{e}^p\tilde{g}\tilde{e}^p$ is an embedding of G into $\tilde{S}$, which is a section of the covering map. □

In the language of algebraic connectivity (see Nehaniv [12]), we record the following consequence,

Corollary 3.4 *Let $\tilde{S}$ denote semigroup of lifts of elements of S in Cascade Covering Theorem. The surmorphism $\tilde{S} \twoheadrightarrow S$ is an aperiodic-homotopy equivalence.*

Proof: This follows from the Aperiodicity Theorem and Theorem 6.2 of [12]. □

4. Refining the Decomposition Forever

Let Each One Go To His Limit.
- The Mahābhārata

Speaking informally, we would like to obtain a decomposition in terms of irreducible pieces by iterating forever the Cascade Covering Theorem on the covering subsemigroups $F[s]$. In carrying this out, we would like to preserve the nice properties of the map down and also be sure that the procedure converges.

The notion of a cascadable system evolved from successively refined concepts designed to fulfill these requirements. We show in this section that it works.

If one naively iterates a cascade decomposition on factors of a decomposition already obtained and then cascades the covering semigroups, one is lead into set-theoretic difficulties. Conceivably, a subsemigroup in the decomposition could be isomorphic to the semigroup we started with. Such a process, even with transfinite iteratation, may be non-terminating.

Another problem is excessive redundancy: since factors of a decomposition may intersect non-trivially, a subsemigroup in the intersection must be multiply decomposed. This makes it difficult (impossible?) to bound the final cardinality of the order we cascade over.

Even with termination, in the limit (whatever that might mean), some factors may become smaller and smaller and, in fact, empty.

Two intuitions to overcome these difficulties are captured in the notions of cascadable systems and their refinements that we introduce now. The first intuition is that to improve a decomposition, it is not necessary to cover each semigroup used in the decomposition at the current stage; rather, it is only necessary cover with smaller subsemigroups than at the current stage. Thus it need not be true that $S' < S''$, if $S < S'$ in our decomposition at one stage and $S < S''$ in the decomposition at a later stage. But it is true that the later decomposition is better. Given a decomposition $S < S'$, where S' is given by a cascadable system, we shall merely use cascadable systems on the $F[s]$ to get a more refined cascadable system on S.

The second intuition is that while we want to decompose each $F[s]$, we should in some sense "restrict" this decomposition to the elements of S that we are concerned with emulating; that is, to those in $[s]$. We need not be concerned about other the elements of $F[s]$. Our main concern is that in the limit each one, each $s \in [s]$, must be emulated by some element in a subsemigroup of $F[s]$. As we iterate decompositions, the subset $[s]$ becomes smaller and smaller, while its emulating semigroup $F[s]$ must become a smaller and smaller subsemigroup of S (but never empty since it contains s).

The notion of refinement on cascadable systems is a partial ordering. It is crucial (in avoiding the above-mentioned difficulties) to note that the cascadable systems on S comprise a *set*. We order this set by demanding that higher elements give more refined decompositions. We show maximal elements exist in this ordering, and then, via a single application of the Cascade Covering Theorem, that these yield the desired decompositions.

4.1. Refinements

We introduce the following ordering on the collection of cascadable systems for S. Take $F : \Sigma \to SUB(S)$, $F' : \Xi \to SUB(S)$. Define $F \leq F'$ (F' **refines** F)

1. (Partition Refinement) $s\Xi t \Longrightarrow s\Sigma t$ (*i.e.* Ξ refines Σ)
2. (Isotone Covering)
   a. $[s]_\Sigma > [t]_\Sigma \Rightarrow [s]_\Xi > [t]_\Xi$.
   b. $[s]_\Xi > [t]_\Xi \Rightarrow [s]_\Sigma \geq [t]_\Sigma$.
3. (Semigroup Refinement)
   a. $\bigcup_{t \in [s]_\Sigma} F'([t]_\Xi) \subseteq F([s]_\Sigma)$.
   b. $F([s]_\Sigma) = F'([s]_\Xi) \Rightarrow [s]_\Sigma = [s]_\Xi$.

This is easily seen to be transitive, antisymmetric and reflexive.

NB: $F < F'$ does not imply that $\int_\Sigma F d\mu$ divides $\int_\Xi F' d\mu'$. But it does imply that S divides each of these cascade integrals and that each for each s, $F'[s]$ divides $F[s]$ (in fact $F'[s]$ is a subsemigroup).

Refinement captures – or, rather, improves upon – the idea of iterated decomposition: The following lemma shows how to obtain a refinement F', from F given a non-trivial cascadable system on $F[s]$.

Lemma 4.1 *Let $F : \Sigma \to SUB(S)$ be a cascadable system for S, $s \in S$, and $H : \Xi \to SUB(F[s])$ a non-trivial cascadable system for $F[s]$. Then we obtain a new cascadable system $F' : \Sigma' \to SUB(S)$ on S with $F' > F$ as follows:*

Refine the partition Σ to obtain a new partition Σ':

$$[t]_{\Sigma'} = \begin{cases} [t]_\Sigma & t \notin [s]_\Sigma \\ [t]_\Xi \cap [s]_\Sigma & t \in [s]_\Sigma. \end{cases}$$

(Henceforth write $[t]'$ for $[t]_{\Sigma'}$).

Define $[t_1]' > [t_2]'$ if

$$[t_1]_\Sigma > [t_2]_\Sigma \text{ or } ([t_1]_\Sigma = [t_2]_\Sigma = [s]_\Sigma \,\&\, [t_1]_\Xi > [t_2]_\Xi).$$

Define

$$F'[t]' = \begin{cases} F[t]_\Sigma & t \notin [s]_\Sigma \\ H[t]_\Xi & t \in [s]_\Sigma \end{cases}$$

Proof: First we show F' is a cascadable system for S:

(i) For $t \notin [s]_\Sigma$, $G_t \subseteq [t]_\Sigma \subseteq F[t] = F'[t]'$

For $t \in [s]_\Sigma$, $G_t \subseteq [s]_\Sigma \subseteq F[s]_\Sigma$, so $[t]_\Xi \supseteq G_t$; whence $[t]' = [t]_\Xi \cap [s]_\Sigma$ contains G_t.

(ii) We verify $[t_1]' > [t_2]' \Rightarrow \forall t_3 \in F'[t_2]', [t_3 t_1]' > [t_2]'$:
case 1: $t_i \notin [s]_\Sigma$ for at least one $i \in \{1,2\}$

From the definition of the ordering on Σ', we know $[t_1]_\Sigma > [t_2]_\Sigma$. Now since F is a cascadable system, for all $t_3 \in F[t_2]$ we have $[t_3 t_1]_\Sigma > [t_2]_\Sigma$; whence by definition of the ordering, $[t_3 t_1]' > [t_2]'$.

case 2: $t_1, t_2 \in [s]_\Sigma$

By definition of the ordering, we must have $[t_1]_\Xi > [t_2]_\Xi$. For each $t_3 \in F'[t_2]' = H[t_2]_\Xi$, we have $[t_3 t_1]_\Xi > [t_2]_\Xi$. In particular, $[t_3 t_1]_\Sigma = [t_2]_\Sigma = [s]_\Sigma$. So by definition, $[t_3 t_1]' > [t_2]'$.

(iii) If $[t_1]' > [t_2]'$, we must show $t_2 \notin F'[t_1]'$.

If $t_1, t_2 \in [s]_\Sigma$, then by definition $[t]_\Xi > [t_2]_\Xi$, so by (3) for H we now have $t_2 \notin H[t_1]_\Xi$.

Otherwise, we know $[t_1]_\Sigma > [t_2]_\Sigma$, so $t_2 \notin F[t_1]_\Sigma$; but $F'[t_1]'$ is always a subset of $F[t_1]_\Sigma$, so we are done.

This completes the verification that F' is a cascadable system for S.

Next we verify that $F' > F$:

Partition Refinement: This is clear. Moreover, since Ξ has a least two classes, Σ' properly refines Σ.

Isotone Covering:

a. If $[s]_\Sigma > [t]_\Sigma$, then by definition $[s]_{\Sigma'} > [t]_{\Sigma'}$.

b. Given $[s]_{\Sigma'} > [t]_{\Sigma'}$, then : if $[s]_\Sigma = [t]_\Sigma$, then of course $[s]_\Sigma \geq [t]_\Sigma$. Otherwise the definition of the ordering on Σ' implies $[s]_\Sigma > [t]_\Sigma$.

Semigroup Refinement:

a. If $[t]_\Sigma = [s]_\Sigma$,

$$F[t]_\Sigma = F[s]_\Sigma = \bigcup_{t_1 \in F[s]_\Sigma} G[t_1]_\Xi \supseteq \bigcup_{t_1 \in [s]_\Sigma} F'[t_1]'.$$

b. If $[t]_\Sigma \neq [s]_\Sigma$, $F[t]_\Sigma = F'[t_1]'$ for all $t_1 \in [t]_\Sigma$.

Whence,

$$F[t]_\Sigma \supseteq \bigcup_{t_1 \in [t]_\Sigma} F'[t_1]'.$$

Finally if $F'[t]' = F[t]_\Sigma$, then : if $t \notin [s]_\Sigma$, then by definition $[t]' = [t]$; if on the other hand $t \in [s]_\Sigma$, then $G[t]_\Xi = F[t]_\Sigma = F[s]_\Sigma$, contradicting non-triviality of G.

□

4.2. Existence of Maximal Cascadable Systems

Lemma 4.2 *A $<$-chain of cascadable systems $F_\alpha : \Sigma_\alpha \to SUB(S)$ has an upper bound: namely,*

$$F : \Sigma \to SUB(S),$$

$$\textit{where } F[s] = \bigcap F_i[s]$$

$$\textit{and } \Sigma = \bigcap \Sigma_\alpha \textit{ i.e. } [s] = [t] \textit{ if and only if for all } \alpha,\ [s]_\alpha = [t]_\alpha$$

with Σ ordered by

$$[s] < [t] \textit{ if and only if there exists } \alpha,\ [s]_\alpha < [t]_\alpha.$$

Therefore, since the cascadable systems on S comprise a set, Zorn's Lemma implies the existence of maximal cascadable systems.

Proof: [Note that the ordering on Σ is well-defined since if $s' \in [s]$ and $t' \in [t]$, then $s' \in [s]_\alpha \subseteq [s]$ and $t' \in [t]_\alpha \subseteq [t]$ by definition of Σ. Thus, $[s']_\alpha = [s]_\alpha < [t]_\alpha = [t']_\alpha$.]

We must show that F is a cascadable system with $F_\alpha \leq F$ for all α.

Clearly Σ is a partition on S refining the Σ_α's. Also $<$ is a total order on Σ: antisymmetric, irreflexive, transitive are easily verified. If $[s], [t] \in \Sigma$, then either $[s]_\alpha = [t]_\alpha$ for all α and thus $[s] = [t]$; or without loss of generality $[t]_\alpha < [s]_\alpha$ holds for some α, and thus $[t] < [s]$. This verifies trichotomy.

(i) Since G_s is contained in each of the $F_\alpha[s]$, it is contained in their intersection $F[s]$. (ii) Given $[s] > [t]$ and $t' \in F[t]$, we must verify $[t's] > [t]$. Since $[s] > [t]$, we know there exists α_0 with $[s]_{\alpha_0} > [t]_{\alpha_0}$. Now $t' \in F[t]$ is contained in $F_{\alpha_0}[t]$, whence $[t's]_{\alpha_0} > [t]_{\alpha_0}$ so also $[t's] > [t]$. (iii) $[s] > [t] \Rightarrow t \notin F_{\alpha_0}[s]$ whence $t \notin F[s]$. Hence F is a cascadable system.

Next we verify $F \geq F_{\alpha_0}$. If $[s]_{\alpha_0} > [t]_{\alpha_0}$, we have $[s] > [t]$ by definition. If $[s]_{\alpha_0} = [t]_{\alpha_0}$, but $[s] \not> [t]$, then for no α does $[s]_\alpha > [t]_\alpha$ hold. Hence for all $F_\alpha > F_{\alpha_0}$, we have $[s]_\alpha = [t]_\alpha$. Also for all $F_\alpha \leq F_{\alpha_0}$, we know $s, t \in [s]_{\alpha_0} \subseteq [s]_\alpha$. Thus $[s]_\alpha = [t]_\alpha$ for all α, and $[s] = [t]$.

Given $t \in [s]_{\alpha_0}$ and $t' \in F[t]$, we show $t' \in F_{\alpha_0}[s]_{\alpha_0}$: we have $t' \in F[t]_\alpha$ for all α, therefore $t' \in F[t]_{\alpha_0} = F[s]_{\alpha_0}$ as $[t]_{\alpha_0} = [s]_{\alpha_0}$.

Finally, if $F[s] = F_{\alpha_0}[s]_{\alpha_0}$, then $F_{\alpha_0} = \bigcap F_\alpha[s]_\alpha$. Moreover, for all $F_{\alpha_1} \geq F_{\alpha_0}$, $F_{\alpha_1}[s]_{\alpha_1} \subseteq F_{\alpha_0}[s]_{\alpha_0}$, so $F_{\alpha_1}[s]_{\alpha_1} = F_{\alpha_0}[s]_{\alpha_0}$. Whence, $[s]_{\alpha_1} = [s]_{\alpha_0}$. Since for $F_\alpha < F_{\alpha_0}$, we have $[s]_\alpha \supseteq [s]_{\alpha_0}$, and thus

$$[s] = \bigcap [s]_\alpha = \bigcap_{F_{\alpha_1} \geq F_{\alpha_0}} [s]_{\alpha_1} = [s]_{\alpha_0}.$$

Thus $F[s] = F_{\alpha_0}[s]_{\alpha_0}$ implies $[s] = [s]_{\alpha_0}$ as required.

In applying Zorn's Lemma, note that the set of cascadable systems on S is never empty: for letting Σ have a single partition block and defining $F(S) = S$ yields a cascadable system. □

4.3. Goodwin's Lemma

Lemma 4.3 (Goodwin [6]) [6] *Suppose S is not cyclic. If S is generated by a subset of an L-class then S is left-simple (and trivially conversely).*

Caution: If W is a semigroup and L is an L-class of W, we are not asserting $S = \langle L \rangle$ is left-simple or cyclic. Members of L need *not* stay L-equivalent in $\langle L \rangle$. The same remarks apply if L is replaced by any subset of L.

Proof: Let X be a subset of an L-class of S and suppose $\langle X \rangle = S$. We can assume X has two or more elements and the distinct members of X are $x_1, x_2, x_3, \ldots$, since otherwise S is cyclic. By hypothesis on X

$$\forall i, j \geq 1,\ i \neq j,\ \exists w_{ij} \in X^+ \text{ (the non-empty finite strings over } X)$$

[6] This beautiful unpublished lemma of Goodwin for arbitrary semigroups is stronger than any corresponding result I know of in the literature. The closest contender occurs in Lallement [10], where the result is established in the case of *stable* S.

such that

$$\overline{w}_{ij}x_j = x_i \text{ for } i \neq j \tag{8}$$

where $\overline{w}_{ij}$ is w_{ij} considered in S (i.e. the result of multiplying the elements of w_{ij} in S). By (8) all right segments of $\overline{w}_{ij}x_j$ containing x_j are L-equivalent to x_i, so all right segments of $\overline{w}_{ij}x_j$ containing x_j are L-equivalent since all members of X are L-equivalent.

Now consider x_jx_i (perhaps with $i = j$). We want to show x_jx_i is L-equivalent to (any member of) X. Let $k \neq i$ and let $x_{j'}x_i$ be the segment of $w_{ki}x_i$ of length two. By the above, $x_{j'}x_i$ is L-equivalent to X. Thus if $\geq$ denotes the L-order and L denotes L-equivalence.

$$x_i \geq x_jx_i \geq \overline{w_{j'j}}x_jx_i = x_{j'}x_i \, L \, x_i,$$

so $x_jx_i \, L \, x_i$. Hence, since L is a right congruence, for arbitrary $s = t_m \ldots t_1$ with $t_i = x_{k_i}$ we have

$$s = (t_mt_{m-1})t_{m-2}\ldots t_1 \, L \, t_{m-1}\ldots t_1 \, L \cdots L \, t_1 \in X.$$

This proves S is left-simple. □

4.4. Properties of Maximal Cascadable Systems

The goal of this section is to establish the following result.

Proposition 4.1 *For a maximal cascadable system on S, each $F[s]$ is either*
(1) a cyclic semigroup $\langle s \rangle$, or
(2) a direct product of a left-zero semigroup and a group.

We shall prove now several lemmas that imply this.

Lemma 4.4 ($\mathcal{L}$-Decomposition) *If S is neither cyclic nor left-simple, then S has a non-trivial cascadable system.*

Proof: Define Σ to be the set of L-classes of S ordered totally in a way that respects L-ordering (Zorn). For example, use an ordering as obtained in example 3.3 but by taking *a maximal chain of non-empty left ideals.*

Thus $[s] = L_s$, and $[s] > [t]$ implies $t \not\leq_L s$.

Define $F[s] = \langle L_s \rangle$.

Verify that this is a cascadable system: (1) Since all elements of G_s are L-equivalent to s, we have $G_s \subseteq F[s]$.

(2) Suppose $[s] > [t]$ and $t' \in F[t]$ Then $t' \in \langle L_s \rangle$, hence $t' \leq_L t$ (for $t' = t_1 \ldots t_n$ with t_iLt, there is a $u \in S^1$ with $t_n = ut$.) Therefore by right multiplicativity of $\leq_L$:

$$t's \leq_L ts \leq_L s$$

So $[t's] \geq [s] > [t]$. And $[t's] > [t]$ as required.

(3) If $[s] > [t]$ but $t \in F[s] = \langle L_s \rangle$, then $t = us$ for some u, and $t \leq_L s$ contradicting $[s] > [t]$. Therefore, $t \notin F[s]$. Thus we have a cascadable system.

Finally, by Goodwin's Lemma and the hypotheses, $F[s] \neq S$. Since S is not left-simple, Σ has at least two elements. This shows F is non-trivial. □

Left simple semigroups are of the form $S = X^\ell \times G$ if they have an idempotent [20, 3]. Otherwise the global structure of S has been described by Teissier [21] and in a more complete modern form by the author in [15] as embedding into – using *left* transformation semigroups – a Baer-Levi automaton reverse wreath a left-zero automaton. We may decompose idempotent-free left-simple semigroups by the following.

Lemma 4.5 *An idempotent-free left-simple semigroup has a non-trivial cascadable system.*

Proof: Let S be left-simple and idempotent-free . Clearly S has infinitely many elements. As Teissier observed, S has no right stabilizers: for if $xy = x$, then by left-simplicity there is z with $zx = y$, whence

$$y = zx = z(xy) = (zx)y = yy,$$

contradicting idempotent-freeness. Also observe aRb implies $a = b$, lest $au = b$ and $bv = a$ imply uv is right stabilizer. Thus S is also R-trivial: each R-class R_s is a singleton $\{s\}$.

Let Σ denote the collection of R-classes of S. Let $F[s] = \langle R_s \rangle$. By R-triviality, $F[s]$ is the infinite cyclic semigroup $\langle s \rangle$. Analogously with the what we did above, order Σ so that

$$s <_R t \Rightarrow [s] > [t].$$

If $[s] > [t]$ and $t' \in \langle R_t \rangle$, we may write $t' = t^k, k \geq 1$. Now $t's = t^k s \leq_R t$, but equality cannot hold lest t have a right stabilizer. Therefore $[t's] > [t]$. □

Lemma 4.6 *If S is cyclic or left-simple and contains an idempotent, then S has no non-trivial cascadable system.*

Proof: Clearly any non-trivial cascadable system refines the trivial cascadable system with $|\Sigma| = 1$. If $S = \langle s \rangle$ is cyclic, then the trivial system $F(S) = S$, cannot be properly refined, for $F'[s]$ must contain $\langle s \rangle = S$, hence by condition 3b of refinement $[s] = S$. If S is left-simple and contains an idempotent, then $S \cong X^\ell \times G$; so in a proper refinement F' of the trivial system, there exist $[(x, g)] > [(x', g')]$. We have $[(x', g'g)] = [(x', g')(x, g)] > [(x', g')]$ by property (ii) of cascadable systems. Hence property (iii) implies $(x', g') \notin F[(x', g'g)]$. This is a contradiction, since $\{x'\} \times G \subset F'[(x', g'g)]$ by property (i). □

It follows that if $F[s]$ is cyclic or $X^\ell \times G$ that $F'[s]' = F[s]$ for any F' refining F.

Corollary 4.2 *A cascadable system $F : \Sigma \to SUB(S)$ is maximal if and only if every $F[s]$ is cyclic or is a left-simple semigroup containing an idempotent.*

Proof: If some $F[s]$ is neither cyclic nor left-simple, $F[s]$ has a non-trivial cascadable system by Lemma 4.4. If $F[s]$ is left-simple and idempotent-free, the same is true by Lemma 4.5. Applying Lemma 4.1 produces a $F' > F$; so F was not maximal.

Conversely if each $F[s]$ is $\langle s \rangle$ or $X^\ell \times G$, and $F' \geq F$, then the arguments of Lemma 4.6 show that $F'[s]' = F[s]$ and $[s] = [s]'$, whence $F' = F$. □

Proposition 4.1 follows from this corollary.

4.5. Main Result

We have shown that every semigroup S has a maximal cascadable system whose components $F[s]$ are as described in Proposition 4.1. Now applying the Cascade Covering Theorem and the Aperiodicity Theorem to such a maximal system yields our main theorem:[7]

Theorem 4.3 (Cascade Decomposition of Arbitrary Semigroups) *An arbitrary semigroup S aperiodically divides a cascade integral*

$$(S^1, S) \leftarrow\!\!\leftarrow (Q, \widetilde{S}) \leq \int_{|\mu|} (F(\alpha)^I, \overline{F(\alpha)^I})\, d\mu,$$

where each $F(\alpha)$ is a cyclic or left-simple (with idempotent) subsemigroup of S and where $|\mu|$ has no more elements than S, $(Q, \widetilde{S})$ is faithful, and all subgroups of S lift to $\widetilde{S}$. Moreover, each state of the cover can be finitely denoted (by its finitely many non-trivial coordinates and their indices).

Corollary 4.4 (Krohn-Rhodes Theorem) *A finite S aperiodically divides a finite cascade of finite cyclic semigroups and finite left-simple semigroups $X^\ell \times G$ which occur as subsemigroups of S, where constants and new identities are adjoined to the subsemigroups. The length of this decomposition does not exceed $|S|$.*

The usual form of the Krohn-Rhodes Theorem says the a finite semigroup divides a wreath product of finite aperiodics and simple group divisors of S. This form can be obtained from ours by standard techniques: a finite cyclic semigroup divides a nilpotent cyclic semigroup wreath a cyclic group, a product divides the wreath product of its factors, $\overline{S^I}$ divides S wreath an aperiodic (product of flip-flops). These methods may be found in Krohn, Rhodes, and Tilson [9], Ch. 5. It should be noted that taking the constant maps off $\overline{G}$ *cannot* be done with an aperiodic division; so the usual Krohn-Rhodes division is *not* aperiodic. Finally, to decompose a group into a wreath product of simple groups, we may take a maximal subnormal series of G and introduce Lagrange coordinates as in [9] for finite groups, or Nehaniv [14] or [13] for arbitrary groups.

Now to extend the Cascade Decomposition Theorem from an arbitrary semigroup S to an arbitrary faithful transformation semigroup (X, S), simply consider the

[7] In the original version of this theorem in [14], the left-simple idempotent-free factors were not eliminated completely, but each such factor was decomposed into a left-simple Baer-Levi semigroup wreath large products of any nontrivial monoid. See [15] for an even stronger decomposition of any idempotent-free left-simple semigroup (as described above).

transformation semigroup (not necessarily faithful) associated to each - possibly non-faithful – orbit (xS^I, S), for $x \in X$. (S^I, S) maps equivariantly onto this orbit via gluing: namely, lifting a state x to I in the copy of (S^I, S) corresponding to x and lifting each transformation s to multiplication by s in each (S^I, S) uniquely determines the covering map. So by taking a direct product of $|X|$-many copies of the decomposition for (S^I, S) we obtain one for (X, S). Obviously, if $Y \subseteq X$ with $YS = X$, it suffices to glue only $|Y|$-many copies, one for each element of the generating set Y.

5. Some Decompositions

We illustrate the application of Theorem 4.3 to several semigroups. In each case we explicitly give a maximal cascadable cover.

5.1. Unions-of-Groups

A ***union-of-groups*** is a semigroup such that each element is contained in a subgroup. Assume S is a union-of-groups (also called *completely regular*) semigroup. Let $[s]$ be the L-class of s, order these in a way compatible with a maximal chain of left ideals, so that $[s] > [t]$ implies $t \notin I_s$ as in example 3.3 above. Let $F[s] = L_s$, which is isomorphic to $A^\ell \times G$, since S is a union-of-groups. Each $F[s]$ is left-simple, so this describes a maximal cascadable system for S. Hence S aperiodically divides the cascade integral over an L-compatible total order $\mathcal{L}$ of its L-classes which are subsemigroups given by a left-zero semigroup direct product a group.

$$(S^1, S) \longleftarrow \int_{L-classes} ((A_i^\ell \times G_i)^I, \overline{(A_i^\ell \times G_i)^I})\, d\mathcal{L}$$

5.2. Rationals and the Bicyclic Monoid

$\mathbb{Q}(0,1)$. For our ordering μ we use the reverse of $\mathbb{Q}(0,1)$ ordered in the usual way by $<$. Define $F(q) = \langle q \rangle$ and $[q] = q$. Obviously $F([q])$ contains q and for no $<^{rev}$-greater index i does $F(i)$ contain q, and $[q] >^{rev} [p]$ implies $[p^n q] >^{rev} [p]$ as required. Since each $F(q)$ is cyclic this cascadable system is maximal. So this semigroup aperiodically divides

$$\int_{q \in \mathbb{Q}(0,1)} (q^*, \overline{q^*})\, d<^{rev}.$$

(Note that $F(q) = \langle q \rangle$ is the infinite free cyclic semigroup, so that $\langle q \rangle^I$ is the free monoid q^* on one generator.) The action is described by

$$(\dots, q_n^{k_n}, \dots, q_i^{k_i}, \dots, q_1^{k_1}, \dots) \cdot \tilde{q} = (\dots, q_n^{k_n}, \dots, p^k, \dots),$$

where $p^k = q_i^{k_i} \dots q_1^{k_1} q$.

Bicyclic Monoid. The bicyclic monoid, presented by

$$B = \langle a, b \,|\, ba = 1 \rangle,$$

has "eggbox picture" (rows are R-classes, columns are L-classes, intersections of these are H-classes) as in Table I.

1	b	b^2	b^3	b^4	$\dots$
a	ab	ab^2	ab^3	ab^4	$\dots$
a^2	a^2b	a^2b^2	a^2b^3	a^2b^4	$\dots$
$\vdots$	$\vdots$	$\vdots$	$\vdots$	$\vdots$	$\ddots$

TABLE I
Eggbox Picture for the Bicyclic Monoid

The idempotents are the diagonal elements of this table. The L-classes of the bicyclic are the columns, and the L-relation totally orders them

$$L_1 >_L L_b >_L L_{b^2} >_L \dots >_L L_{b^n} >_L L_{b^{n+1}} >_L \cdots$$

Let $[a^k b^n] = L_{b^n}$ and $F[a^k b^n] = \langle L_{b^n} \rangle$. Reversing the ordering, yields a cascadable system by Lemma 4.4; but this far from maximal.

We refine F as follows: further partition each L-class according to

$$L_{b^n} = \{a^n b^n\} < \{a^k b^n : k > n\} < \{b^n\} < \{ab^n\} < \cdots \{a^{n-1} b^n\}.$$

Note that $\{a^k b^n : k > n\}$ is an infinite cyclic semigroup generated by $a^{n+1} b^n$. With the partition, let

$$F'[a^m b^n] = \langle a^{min(n+1,m)} b^n \rangle,$$

which is finite if and only if $m = n$. Suppressing set-brackets around singletons, the ordered partition is:

$$1 < \langle a \rangle < ab < \langle a^2 b \rangle < b < a^2 b^2 < \langle a^3 b^2 \rangle < b^2 < ab^2 < \dots$$

$$\dots < a^{n-2} b^{n-1} < a^n b^n < \langle a^{n+1} b^n \rangle < b^n < ab^n \dots < a^{n-1} b^n < \dots$$

The partition is shown in Table II. The least partition block is numbered $P0$, second least $P1$, etc.

1^{P0}	b^{P4}	${b^2}^{P7}$	${b^3}^{P11}$	${b^4}^{P16}$	$\dots$
a^{P1}	ab^{P2}	${ab^2}^{P8}$	${ab^3}^{P12}$	${ab^4}^{P17}$	$\dots$
a^2	${a^2b}^{P3}$	${a^2b^2}^{P5}$	${a^2b^3}^{P13}$	${a^2b^4}^{P18}$	$\dots$
a^3	a^3b	${a^3b^2}^{P6}$	${a^3b^3}^{P9}$	${a^3b^4}^{P19}$	$\dots$
$\vdots$	$\vdots$	$\vdots$	${a^4b^3}^{P10}$	${a^4b^4}^{P14}$	$\ddots$
$\vdots$	$\vdots$	$\vdots$	$\vdots$	${a^5b^4}^{P15}$	$\ddots$

TABLE II
A Cascadable System for the Bicyclic Monoid

This is a cascadable system: (i) Trivially $G_s \subseteq F'[s]$. (ii) If $[a^m b^n] > [a^p b^q]$, we must show $a^p b^q \notin F[a^m b^n]$. In case $n > q$, this follows since all elements of $F'[a^m b^n]$

have b-exponent at least n. If $n = q$, then necessarily $m > p$, and so $p \leq n$. For $p = n$, a^pb^q is an idempotent a^nb^n and clearly not in $F[a^mb^n]$. For $p < n$, then either $(m < n)$ the powers of a^mb^n have higher b-exponent than n or, $(m > n)$ the elements of $F'[a^mb^n]$ are powers of $a^{n+1}b^n$ and so have higher a-exponents that a^pb^q, or $(m = n)$ so $F'[a^mb^n] = \{a^nb^n\}$. In all cases, the conclusion holds. (iii) Finally, we must show $[s] > [t]$ implies $[t's] > [t]$ for all $t' \in F'[t]$. If s and t are in different L-classes, this is clear since F was a cascadable system. So we may assume sLt. Write $s = a^kb^n$, $t = a^\ell b^n$. Case 1: $k < n$. Then $ts = a^\ell b^n a^k b^n$ has b-exponent greater than n, so $[ts] > [t]$. If $\ell \geq n$, then t' is a a^nb^n or a power of $a^{n+1}b^n$, so the same argument show $[t's] > [t]$. If $\ell < n$, then $t' = t^{r+1}$ has b-exponent exceeding n, which multiplying by s on the right cannot decrease so $[t's]$ lies a lower L-class than whence $[t's] > [t]$. Case 2: $k = n + r \geq n$. We have $s = a^{n+r}b^n$; and since $[s] > [t]$ we conclude $t = a^nb^n$, whence $t' \in F'[t]$ is just t. We have $[t's] = [ts] = [s] > [t]$, as required.

5.3. Free Semigroups

We now study decompositions for 2^+ arising from a Theorem 4.3. We know 2^+ has only singleton L-classes and so the same is true of any subsemigroup. Therefore in subsemigroups in the decomposition are all necessarily cyclic and idempotent-free, and the are generated by single words. Now these cyclic semigroups are integrated over a linear ordering to give a decomposition of 2^+, and each $w \in 2^+$ must occur at some unique greatest depth. It follows the if $u = v^n$ then the subsemigroup generated by v must occur at a coordinate of lesser index than u. Therefore the integral is taken over an ordering of a subset Σ of 2^+ linearly ordered in this way, such each w occurs in a greatest $F(\alpha) = \langle v \rangle$, and every word occurs in some $F(\alpha)$. A natural way to do this is to let $\Sigma = 2^+$ be ordered lexicographically. Thus lexicographical ordering on the free semigroup arises in the cascade decomposition theory.

$$2^+ < \int_{2^+} (\langle w \rangle^I, \overline{\langle w \rangle^I}\,)\, d(lexicographic).$$

Let us look more closely at the action. A state of the cover has the form (only non-trivial coordinates are shown):

$$(..., w_n^{k_n}, ..., w_i^{k_i}, .., w_1^{k_1}, ..)$$

and where only finitely many entries are not I and w_i precedes w_j in the ordering whenever $i < j$. If lift $\tilde{w}$ of a word w acts on this state, it acts at the last position for which $\langle v \rangle$ contains w (namely $v = w$), unless some w_i before the w-coordinate intervenes. Generally for a lift of w we have a new state of the form

$$(..., w_n^{k_n}, ..., w_{i+1}^{k_{i+1}} \cdot w_i^{k_i} \ldots w_1^{k_1} u_1^{j_1}, u_2^{j_2}, \cdots, u_t^{j_t}),$$

where $w = u_1^{j_1} u^{j_2} \ldots u_t^{j_t}$ and $u_1 > u_2 > \ldots > u_t$ (possibly $w_{i+1}^{k_{i+1}} = I$)

We see that $\widetilde{2^+}$ remembers the order and counts powers of words which come in as long as the words coming in successively do not increase in the lexicographic order. Once a lexicographically greater word comes in, all information about the entry order for smaller words is lost.

This cascade decomposition reveals interesting properties of global sequential coordinatizations of the free semigroup 2^+. A decomposition is obtained for A^+ in the same way. These new coordinatizations may prove useful in the study of word problems as one often encounters questions of which word is a power of another in considering such problems. In particular such arguments are crucial in the studies by Novikov, Ol'shanskii, Ivanov, and McCammond of the Burnside groups and semigroups.

6. Concluding Remarks

We have applied cascade integration to give a cascade decomposition of arbitrary transformation semigroups covered by a cascade integral of very simple pieces. In particular, this yields a direct generalization to arbitrary semigroups of 1962 Krohn-Rhodes Theorem. If finite semigroup theory is finite group theory plus the flip-flop (plus finite ordinals), then the present paper implies that semigroup theory is group theory plus the flip-flop plus the infinite cyclic semigroup (and total orders).

We have proceeded in four steps: (i) replace wreath products with cascade integration, (ii) replace 'V-union-T' decompositions by cascadable systems, and show they lead to nice coverings, (iii) define an order on cascadable systems, which captures the notion of successive refinement of decomposition; (iv) show existence of maximal systems, yielding the main theorem.

A method of proof related to the above-mentioned "V union T" proof of the finite Krohn-Rhodes theorem appears in in Chapter 5, sec. 4 of [9]. This proof proceeds using a primitive fore-runner of our cascadable systems. In his Ph.D. thesis [14] the author introduced a version of cascadable system and took chains of left ideals (as in example 3.3) to circumvent the difficulties of the "V union T" approach, obtaining the first proof of the main result. This paper significantly streamlines the result by modifying the definition of cascadable system and using the new method of taking a cascadable system based on the L-order and subsemigroups generated by the L-classes (Lemma 4.4 $\mathcal{L}$-Decomposition).

A semigroup is *finite J-above* if for each element s there are only a finite number of elements t such that $s \in S^1 t S^1$. If S contains an infinite group or an element without torsion then S is not finite J-above. In the special case of finite J-above semigroups S, nice "holonomy" decompositions have been obtained by Rhodes [18] and collaborators [7]. In these decompositions groups also lift, but the theorems had not extended as nicely to arbitrary semigroups, which were first expanded to obtain finite J-above semigroups. Such expansion must necessarily destroy all infinite subgroups of S. Forthcoming work of the author and G.Z. Elston [17] using cascade integration shows how to extend the holonomy decomposition to an arbitrary stable semigroup S by embedding $\widehat{S}$ into a permutation-reset cascade integral with the Zeiger property.

Also, using partial instead of total orders leads to cascade-parallel versions of the results of this paper.

Appendix

A. Properties of Cascade Integration

In this appendix we establish some useful lemmas relating covers and integration. These generalize to arbitrary partial orders several well-known folklore results in the case of covering by finitely many finite transformation semigroups. Notice how the language of the cascade transform makes it easy to write concise formal proofs. This technical section may be omitted on first reading.

First we show that *the integral of covers is a cover*; that is, we generalize the fact that if (X_i, S_i) covers (Y_i, T_i) for $i = 1,2$ then $(X_2, S_2) \circ (X_1, S_1)$ covers $(Y_2, T_2) \circ (Y_1, T_1)$.

The next result, "transitivity of covers", of proposition A.1, generalizes the fact that if $(Y_2, S_2) \circ (Y_1, S_1)$ covers (Z, T) and each (Y_i, S_i) is covered by $(X_{i2}, R_{i2}) \circ (X_{i1}, R_{i1})$, for $i = 1,2$ respectively, then $((X_{22}, R_{22}) \circ (X_{21}, R_{21})) \circ ((X_{12}, R_{12}) \circ (X_{11}, R_{11}))$ covers (Z, T). We also generalize the fact that parentheses in the last sentence are actually superfluous since the wreath product operation is associative on the class of transformation semigroups. Generalized associativity is the essential content of the second part of proposition A.1 below.

Lemma A.1 (Integration of Covers) *Let $\mu = (V, \prec)$ be any partial order and let F and G be two families of transformation semigroups indexed by V such that for all $\alpha \in V$ there is a covering morphism*

$$F(\alpha) \overset{w_\alpha}{\dashrightarrow} G(\alpha)$$

then there exists a covering morphism $w = \int_V w_\alpha \, d\mu$ with

$$\int_V F(\alpha)\, d\mu \overset{w}{\dashrightarrow} \int_V G(\alpha)\, d\mu.$$

Proof: Write $F(\alpha) = (X_\alpha, S_\alpha)$ and $G(\alpha) = (Y_\alpha, T_\alpha)$. Then the state-space of $\int F$ is $X = \prod X_\alpha$ and that of $\int G$ is $Y = \prod Y_\alpha$. Define $w : \prod X_\alpha \rightarrow \prod Y_\alpha$ component-wise by

(9) $$(xw)_\alpha = (x_\alpha)w_\alpha \text{ if for all } \alpha \in V, (x_\alpha)w_\alpha \text{ is defined.}$$

This is a partial surjective function since by hypothesis each w_α is.

The semigroup of $\int F$ is a subsemigroup S of X^X and that for $\int G$ is a subsemigroup T of Y^Y. We must show how to lift each $t \in T$ to an $f \in S$ with

$$xw \cdot t = (x \cdot f)w.$$

We use the cascade transform $\overline{t}$ of t to lift t "component-wise" as well.

Suppose $x_{\beta \prec \alpha}$ is an segment for X with $(x_\gamma)w_\gamma$ defined for all $\gamma \prec \alpha$. Let $y_{\beta \prec \alpha} = (xw)_{\beta \prec \alpha}$. Then $(y_{\beta \prec \alpha})\overline{t}$ lies in T_α. Let s_α be any lift with respect to w_α of this element to S_α. Now to construct (the cascade transform of) a lift f of t, define

$$(x_{\beta \prec \alpha})\overline{f} = s_\alpha.$$

(Since s_α depends in part on a choice, in general many $\overline{f}$'s will exist.)

By cascade transform inversion, this $\overline{f}$ determines a unique $f = \hat{\overline{f}} \in S$. We verify f is a lift of t: Let $x \in X$ with xw defined, then $\forall \alpha \in V$, $(x_\alpha)w_\alpha = (xw)_\alpha$ is defined. So by definition of the cascade transform

$$
\begin{aligned}
(xw \cdot t)_\alpha &= (xw)_\alpha \cdot ((xw)_{\beta \prec \alpha})\overline{t} \\
&= (x_\alpha)w_\alpha \cdot (y_{\beta \prec \alpha})\overline{t} \\
&= (x_\alpha \cdot s_\alpha)w_\alpha \\
&= (x_\alpha \cdot (x_{\beta \prec \alpha})\overline{f})w_\alpha \\
&= ((x \cdot f)_\alpha)w_\alpha \\
&= ((x \cdot f)w)_\alpha
\end{aligned} \tag{10}
$$

We have proved component-wise that

$$xw \cdot t = (x \cdot f)w, \text{ whenever } xw \text{ is defined.}$$

Therefore f is a lift of t. Hence w is satisfies the covering morphism property (1). □

In iterated covering, *i.e.* covering each of the transformation semigroups in a cover of a given transformation semigroup and integrating this collection of coverings to obtain a cover of the original cover, we need an appropriate notion to express what covers what.

Definition 5 An **isotone covering** of partial orders $\varphi : (V, \prec) \to (W, \prec')$ is a surjective partial function from V to W with the property

$$
\begin{aligned}
\alpha \prec \alpha' &\Rightarrow \alpha\varphi \prec' \alpha'\varphi \text{ or } \alpha\varphi = \alpha'\varphi \\
\alpha'\varphi \prec' \alpha\varphi &\Rightarrow \alpha' \prec \alpha
\end{aligned} \tag{11}
$$

for all α, α' in V if both $\alpha\varphi$ and $\alpha'\varphi$ are defined.

That φ is an isotone covering means that φ neither creates nor destroys dependencies, although points may be collapsed. Observe that the composition of isotone coverings is an isotone covering.

For example, that map $(\mathbb{R}, <) \to (\mathbb{Z}, <)$ given by taking the integer part $r \mapsto \lfloor r \rfloor$, is an isotone cover.

Proposition A.1 *Let F and G be a families of transformation semigroups $F(\alpha) = (X_\alpha, S_\alpha)$ and $G(\beta) = (Y_\beta, T_\beta)$ indexed respectively by partial orders $\mu = (V, \ll)$, and $\nu = (W, \prec)$. Suppose $\varphi : \mu \to \nu$ is an isotone covering map of partial orders and there are coverings*

$$w_\beta : \int_{\beta\varphi^{-1}} F(\alpha)\, d\mu_{|\beta\varphi^{-1}} \to G(\beta).$$

Then 1. (Transitivity of Covers) The map

$$w = \int_W w_\beta \, d\nu : \int_W \Big(\int_{\beta\varphi^{-1}} F(\alpha) \, d\mu_{|\beta\varphi^{-1}} \Big) \, d\nu \to \int_W G(\beta) \, d\nu$$

is a covering map.

2. (Generalized Associativity of Cascade Integral) Integrating over the domain $W\varphi^{-1}$ *of* φ, *we have*

$$\int_{W\varphi^{-1}} F(\alpha) \, d\mu_{|W\varphi^{-1}} \cong \int_W \Big(\int_{\beta\varphi^{-1}} F(\alpha) \, d\mu_{|\beta\varphi^{-1}} \Big) \, d\nu,$$

Hence by 1., we have a covering

$$\int_{W\varphi^{-1}} F(\alpha) \, d\mu_{|W\varphi-1} \to \int_W G(\beta) \, d\nu.$$

Proof: The first assertion is just a straight-forward application of the above lemma. To obtain the second we construct an isomorphism. Let (X, S) denote the transformation semigroup on the left hand side. Write $(Y_\beta, T_\beta) = \int_{\alpha \in \beta\varphi^{-1}} (X_\alpha, S_\alpha) d\mu_{|\beta\varphi^{-1}}$. Then $Y_\beta = \prod_{\alpha \in \beta\varphi^{-1}} X_\alpha$, and write $(Y, T) = \int_W (Y_\beta, T_\beta) d\nu$, the transformation semigroup on the right hand side. The state-space $X = \prod_{\beta \in W} \prod_{\alpha \in \beta\varphi^{-1}} X_\alpha$ is canonically isomorphic to $Y = \prod_{\alpha \in W\varphi^{-1}} X_\alpha$ via the component-wise mapping

$$(y_\beta)_\alpha \mapsto x_\alpha \text{ for } \alpha \text{ with } \alpha\varphi = \beta.$$

To see the isomorphism of the semigroups, take any $f \in T$ and $\beta = \alpha\varphi$, then

$$(y \cdot f)_\beta = y_\beta \cdot \overline{(y_{\beta' \prec \beta}) f}$$

Now y_β lies in Y_β and $\overline{(y_{\beta' \prec \beta}) f}$ lies in T_β. Moreover $\overline{(y_{\beta' \prec \beta}) f}$ depends only on f and the β'-coordinates of y which are $\prec$-less than β. Observe,

$$\gamma\varphi = \beta' \prec \beta = \alpha\varphi \iff \gamma \ll \alpha \text{ and } \gamma\varphi \neq \alpha\varphi \tag{12}$$

Since (Y_β, T_β) is itself a cascade integral, we may for any α in $\beta\varphi^{-1}$ express the α-coordinate of $y_\beta \cdot \overline{(y_{\beta' \prec \beta}) f}$ using the cascade transform as

$$(y_\beta \cdot \overline{(y_{\beta' \prec \beta}) f})_\alpha = (y_\beta)_\alpha \cdot \overline{((y_\beta)_{\alpha' \ll_\beta \alpha}) \overline{(y_{\beta' \prec \beta}) f}},$$

where $(y_\beta)_{\alpha' \ll_\beta \alpha}$ is the α-segment of y_β, so that $\alpha' \in \beta\varphi^{-1}$.

Now note that the semigroup element $\overline{((y_\beta)_{\alpha' \ll_\beta \alpha}) \overline{(y_{\beta' \prec \beta}) f}}$ lies in S_α and only on depends f, $y_{\beta' \prec \beta}$, and the μ-coordinates of y_β which are $\ll$-less than α; that is, on f and the set of $(y_{\gamma\varphi})_\gamma$ with

$$\begin{aligned} & (\gamma\varphi \prec \alpha\varphi) \text{ or } (\gamma\varphi = \alpha\varphi \text{ and } \gamma \ll \alpha) \\ \iff & (\gamma \ll \alpha \text{ and } \gamma\varphi \neq \alpha\varphi) \text{ or } (\gamma\varphi = \alpha\varphi \text{ and } \gamma \ll \alpha) \text{ (by (12))} \\ \iff & (\gamma \ll \alpha) \end{aligned} \tag{13}$$

Thus this element of S_α depends only on f and $x_{\gamma\ll\alpha}$, where x is the image of y under isomorphism of state-spaces. So corresponding to any f in T, we have g in S whose cascade transform is given by

$$(x_{\gamma\ll\alpha})\overline{g} = ((y_\beta)_{\alpha'\ll_\beta\alpha})\overline{(y_{\beta'\prec\beta})\overline{f}} \tag{14}$$

We have

$$\begin{aligned}(x\cdot g)_\alpha &= x_\alpha\cdot(x_{\gamma\ll\alpha})\overline{g}\\ &= x_\alpha\cdot((y_\beta)_{\alpha'\ll_\beta\alpha})\overline{(y_{\beta'\prec\beta})\overline{f}}\\ &= (y_\beta\cdot(y_{\beta'\prec\beta})\overline{f})_\alpha\\ &= ((y\cdot f)_\beta)_\alpha\end{aligned} \tag{15}$$

This gives to each element of T an element of S whose action on X respects the isomorphism of state-spaces. Conversely, given $g\in S$ define an f in T by

$$(y_\beta\cdot(y_{\beta'\prec\beta})\overline{f})_\alpha = x_\alpha\cdot(x_{\gamma\ll\alpha})\overline{g},$$

where $\alpha\in\beta\varphi^{-1}$. Since for each such α, the element $(x_{\gamma\ll\alpha})\overline{g}\in S_\alpha$, we have that $(y_{\beta'\prec\beta})\overline{f}$ lies in T_β and depends only on g and $y_{\beta'\prec\beta}$. One sees that the correspondence between S and T is one-one and respects the isomorphism on state-spaces. □

References

1. W. Ross Ashby. *Design for a Brain: the origin of adaptive behaviour.* Chapman & Hall Ltd., London, 1952.
2. W. Ross Ashby. *An Introduction to Cybernetics.* University Paperbacks, London, 1968.
3. A. H. Clifford and G. B. Preston. *The Algebraic Theory of Semigroups - Volume I.* Number 7 in Mathematical Surveys. American Mathematical Society, 1961.
4. S. Eilenberg. *Automata, Languages and Machines,* volume B. Academic Press, New York, 1976.
5. J.M.G. Fell and R.S. Doran. *Representations of $*$-Algebras, Locally Compact Groups, and Banach $*$-Algebraic Bundles,* volume 1. Academic Press, 1988.
6. Jerrold R. Goodwin. [unpublished results]. *circa* 1970.
7. K. Henckell, S. Lazurus, and J. Rhodes. Prime decomposition theorem for arbitrary semigroups: General holonomy and synthesis theorem. *Journal of Pure and Applied Algebra,* 55:127–172, 1988.
8. J.M. Howie. *An Introduction to Semigroup Theory.* Academic Press, 1976.
9. K. Krohn, J. Rhodes, and B. Tilson. Lectures on finite semigroups. In M. Arbib, editor, *Algebraic Theory of Machines, Languages, and Semigroups.* Academic Press, 1968.
10. G. Lallement. *Semigroups and Combinatorial Applications.* Wiley, New York, 1979.
11. G.W. Mackey. Borel structure in groups and their duals. *Trans. Amer. Math. Soc.,* pages 134–165, 1957.
12. C.L. Nehaniv. Algebraic connectivity. *International Journal of Algebra and Computation,* 1(4):445–471, 1991.
13. C.L. Nehaniv. Cascade decomposition and synthesis of groups. *preprint.*
14. C.L. Nehaniv. *Global Sequential Coordinates on Semigroups, Automata and Infinite Groups.* PhD thesis, University of California at Berkeley, University Microfilms International (Ann Arbor, Michigan), 1992.

15. C.L. Nehaniv. Global structure of left simple semigroups. *Semigroup Forum*, submitted for publication.
16. C.L. Nehaniv. Monoid and group acting on trees: Characterizations, gluing and applications of the depth preserving actions. *International Journal of Algebra and Computation*, to appear.
17. C.L. Nehaniv and G.Z. Elston. Holonomy decomposition and embedding of arbitrary stable semigroups. *preprint*, 1993.
18. J. Rhodes. Monoids acting on trees: Elliptic and wreath products and the holonomy theorem for arbitrary monoids. *International Journal of Algebra and Computation*, 1(2), 1991.
19. J. Rhodes and C.L. Nehaniv. Complexity theories in mathematics. *in preparation.*
20. A. Sushkevych. *Theory of Generalized Groups.* Hos. Nauk. Tekh. Izd. Ukranii, Kharkiv, 1937.
21. M. Teissier. Sur les demi-groupes admettant l'existence du quotient d'un côté. *Comptes Rendus Hebdomadaires des Séances de l'Academie des Sciences*, 236(11):1120–1122, Mars 1953.
22. B. Tilson. Categories as algebras: an essential ingredient in the theory of monoids. *Journal of Pure and Applied Algebra*, 48:83–198, 1987.
23. B. Tilson. Chapter XII. In Samuel Eilenberg, editor, *Automata, Languages and Machines*, volume B, New York, 1976. Academic Press.

Index

GPSR Compliance

The European Union's (EU) General Product Safety Regulation (GPSR) is a set of rules that requires consumer products to be safe and our obligations to ensure this.

If you have any concerns about our products, you can contact us on ProductSafety@springernature.com

In case Publisher is established outside the EU, the EU authorized representative is:

Springer Nature Customer Service Center GmbH
Europaplatz 3
69115 Heidelberg, Germany

Batch number: 10755284

Printed by Printforce, the Netherlands